Vertebrate Endocrinology

THIRD EDITION

Vertebrate Endocrinology

THIRD EDITION

David O. Norris

Department of Environmental, Population, and Organismic Biology
University of Colorado
Boulder, Colorado

ACADEMIC PRESS

An Imprint of Elsevier

San Diego   London   Boston   New York   Sydney   Tokyo   Toronto

Front Cover Photographs: Top, left, courtesy of Image Club Graphics. Bottom left
And right, © 1996 by Digital Stock, Inc.

This book is printed on acid-free paper. ⊚

Academic Press
An imprint of Elsevier Science
525 B Street, Suite 1900, San Diego, California 92101-4495, USA
http://www.academicpress.com

Academic Press
84 Theobald's Road, London WC1X 8RR, UK
http://www.academicpress.com

Library of Congress Cataloging-in-Publication Data

Norris, David O.
        Vertebrate Endocrinology / by David O. Norris. – 3rd.
                p.   cm.
        Includes bibliographical references and index.
    International Standard Book Number: 0-12-521670-X (alk. paper)
    1. Endocrinology.   2. Vertebrates—Physiology.   I. Title.
    [DNLM: 1. Hormones—physiology.   2. Vertebrates—physiology.   WK
102 N854v   1996]
QP187.N67   1996
596'.0142—dc20
DNLM/DLC
For Library of Congress   96-33146
                                                           CIP

PRINTED IN THE UNITED STATES OF AMERICA
04 05 06 07 08 09   9   8   7   6   5

CONTENTS

4. Organization of the Mammalian Hypothalamo–Hypophysial Axis

5. Comparative Aspects of the Hypothalamo–Hypophysial System in Nonmammalian Vertebrates

6. Neurohormones of the Pars Nervosa and the Epithalamus

7. The Hypothalamo–Hypophysial–Thyroid Axis of Mammals

8. Comparative Aspects of Vertebrate Thyroids

9. The Mammalian Adrenal Glands: Cortical and Chromaffin Cells

10. Comparative Aspects of Vertebrate Adrenals

11. The Endocrinology of Mammalian Reproduction

12. Comparative Aspects of Vertebrate Reproduction

13. Regulators of the Gastrointestinal Tract

14. Chemical Regulators of Metabolism

15. Regulation of Calcium and Phosphate Homeostasis

PREFACE

THIS TEXT is a biased attempt to present an unbiased synthesis of modern vertebrate endocrinology for the advanced undergraduate or beginning graduate student who needs a general understanding of chemical regulation or who plans to pursue a research career in endocrinology. A basic background in cellular and organismic physiology and a basic knowledge of chemistry is assumed. The subjects covered reflect as much as possible the expanding scope of what traditionally has represented the field of endocrinology to include the interfaces of classical endocrinology with neurobiology, immunology, cellular biology, and molecular biology. All areas are not treated equally and certain areas have been favored deliberately. Furthermore, I have chosen to emphasize generalizations on the one hand and unique exceptions on the other in order to show both unity and diversity in the patterns of vertebrate chemical regulation.

Each year there are approximately 60,000 scientific journals published in 65 languages with about 1,000,000 new papers. Furthermore, about 300,000 monographs and conference proceedings also appear each year. Within this deluge of scientific reports, there is a vast and exponentially expanding endocrine database that constitutes the primary literature of endocrinology. This is represented by published reports and reviews appearing in only a few hundred journals. Furthermore, the traditional disciplines of endocrinology, neurobiology, and immunology have become hopelessly and forever intertwined, expanding this database still further. It is barely possible even with modern abstracting systems and computerized library searches to keep oneself abreast of new developments. New data appear rapidly and cause each of us to constantly revise our thoughts and conclusions about chemical regulation and its evolution. Textbooks such as this one represent only meager introductions to the basic concepts of chemical regulation as a key to entering the greater kingdom, for it is mastery of the primary literature that holds the keys to future understanding. It is only while delving intelligently into the primary literature that the student will develop a deep and lasting fascination for the intricate mechanisms and importance of endocrinology.

Only by becoming self-learners can students begin to deal with the masses of information being generated today. To wade into the primary literature as soon as possible is of the utmost importance. Conventional textbooks are outdated in many respects before they leave the printing presses. A college course no longer can cover all aspects of endocrinology fairly, just as no single book can include everything that users may desire. All we can hope for is to stimulate the student's innate curiosity through courses and textbooks and provide an avenue for her/him to pursue knowledge and to interpret it. In our teaching, we must focus more on how to find the necessary information and how to analyze and interpret it. I have deliberately chosen not to emphasize the primary literature in this edition, but rather (1) to provide a general synthesis of many aspects of endocrinology and (2) to cite mostly review articles that will give the curious student a place to begin her/his study of the primary literature. In this day of computerized data searches, the student easily can locate sources of detailed information relating to material presented in this text and evaluate those data for him/herself. This exercise will be all the more valuable to her/his development.

Finally, I must emphasize that this is not just another version of my earlier text, but actually represents a completely different approach with greater emphasis at the cellular and molecular levels. Every topic has been reexamined and new material added. The lapse of more than 10 years since the second edition necessitated this complete overhaul to reflect current knowledge and approaches to endocrinology. Unfortunately, advances are occurring so rapidly, especially on the molecular front, that it is impossible to provide the latest data or latest interpretations. Both mammalian and nonmammalian systems are included here because I believe that the comparative study of nonmammalian systems can only enhance our understanding of mammalian systems and clinical endocrinology, as well as provide information to better understand the physiology, behavior, and ecology of wild populations of animals. The future will be for those who can bridge the gaps between molecular and cellular events and the complex events of physiology and behavior at the organismic level and apply this knowledge to animal functions in natural and altered environments.

This book is dedicated to my mentors Aubrey Gorbman, Howard Bern, and the late Donald S. Dean, as well as my former and present students; all have taught me much and continue to do so every day. I am grateful to many colleagues around the world who prompted me to prepare this edition and to the many people at Academic Press, especially Charles Crumly and Kathy Nida. All of my colleagues who were involved in the earlier editions still are acknowledged for their contributions to the foundation of the third edition. The final version of this edition has profited greatly from the comments of many anonymous reviewers on earlier drafts of these final chapters, and I thank them all. Most of all, I owe a great debt to my close friend and associate, Richard Evan Jones, for his comments and encouragements, alternately harassing and stimulating me, and for his infectious enthusiasm for comparative endocrinology that rubs off on everyone around him.

David O. Norris

1

An Overview of Chemical Regulation

THE STUDY of endocrinology is exciting because it deals with chemical regulation of virtually all biological phenomena and hence is on the "cutting edge" of molecular, cellular, and organismic biology of animals. Further, many endocrine principles have their counterparts in microbial and botanical systems as well. Endocrinology explains the bases for regulation of physiology and behavior at all levels of organization. Instead of a traditional field of specialization that originally developed within organismic and clinical physiology, endocrinology today can be defined to encompass all levels of biological organization including molecular–cellular biology, developmental biology, organismic physiology and behavior, and ecology. Endocrinology no longer follows trends developing in other biological subdisciplines as it did for many decades but is now determining the directions of basic research in many areas.

Endocrinology began as the study of certain glands, called **endocrine glands** or glands of internal secretion, and the actions of their secretions, known as **hormones.** They were called glands of internal secretion because their hormones were released into the blood, through which they were transported to specific **target cells** around the body. A hormone binds to a specific **receptor** molecule located in the target cell and together the hormone–receptor complex produces a measureable effect. In those early days, physiologists viewed the nervous system and the endocrine system as if they were two distinctly separate regulatory mechanisms that helped animals adapt to their environment. The intimate relationship now evident between these two systems, as well as the discovery of chemical regulatory mechanisms in other physiological subdisciplines, such as immunology and developmental biology and their overlap with traditional aspects of endocrinology and neurobiology, has prompted us to bring together a diverse and vast literature on chemical regulation under the umbrella of what we continue to call "endocrinology."

Learning about the intricacies of how the activities of animals are regulated and coordinated by chemicals is one of the most fascinating and complicated endeavors in biology. Everything an animal does is either initiated, modulated, or blocked by chemical regulators. Understanding the endocrine systems of invertebrates and vertebrate animals is essential if we are ever to understand how chemical regulatory systems evolved and how they operate to maintain the vast array of living animal species. Furthermore, the continued health of each ecosystem depends on the continued reproductive success of its component animal species.

The study of chemical regulation in nonmammalian vertebrates also provides useful information applicable to mammalian problems (including those of humans). First, the knowledge of the evolution of a regulatory mechanism from simpler to more complex forms of expressions often helps to unravel events in the mammal and improves our understanding of such endocrine events. Some procedures are more readily performed on nonmammals, whose processes may be spread out stepwise over time, in contrast to the mammal, in which all of these steps may occur at the same time. Furthermore, we have discovered many useful tools for studying physiology and behavior in the form of model systems through examination of nonmammalian vertebrates, including analyses of fishes, amphibians, reptiles, and birds. For example, genetically controlled platyfish strains provide a system for studying the chemical regulation of aging and the reproductive system in relationship to genetic factors. Many other species of fishes are used extensively as models for stress, growth, carcinogenesis, and behavioral studies. The toad urinary bladder is an excellent *in vitro* model for studying the mechanism of action for the mineralocorticoid aldosterone. Similarly, the amphibian ovarian follicle provides an *in vitro* system for studying the chemical regulation of oocyte maturation and the process of ovulation. Studies of amphibians are crucial for understanding mechanisms of tissue induction and chemical regulation of gene activity during embryonic development and differentiation. The lizard *Anolis carolinensis* is an excellent model for studying hormonal and neural control of the alternating pattern of ovulation by the paired ovaries. Avian systems are used extensively for studies of development, neurobiology, immunology, cancer, and molecular genetics. Each of these systems has wide applicability to other vertebrates as well as to the mechanisms of action of other steroid hormones.

A special concern of comparative endocrinologists is the possible disruption of endocrine functions in a natural ecosystem that might affect populations adversely and damage the ecosystem. These **endocrine disrupters** generally are chemicals produced by human activities (anthropogenetic) that mimic natural chemical regulators or prevent their actions. They might induce events at the wrong time or prevent a critical event from occurring on schedule. For example, the pesticide dichlorodiphenyltrichloroethane (DDT) has estrogenic effects in vertebrates whereas one of its metabolites, dichlorodiphenyldichloroethylene (DDE), acts as an antiandrogen. Phytoestrogens are estrogenic compounds synthesized by certain plants. Consumption of these plants has been known to disrupt reproduction and other processes in herbivores. Wastes from wood pulp mills may be especially disruptive of reproductive processes in fishes. One of the actions of polychlorinated biphenyls (PCBs) is to block synthesis of the neurotransmitter serotonin and alter gonadal function through effects at the brain or pituitary level. Polychlorinated biphenyls also may compete with thyroid hormones for thyroid hormone receptors, induce hypothyroidism, and prevent normal development of the nervous system. The cadmium ion is a common aquatic pollutant that enhances release of reproductive hormones and is a good example of a potential inorganic disrupter. Claims for large reductions in human sperm counts worldwide over the last three generations and a marked increase in testicular cancer have been suggested to be consequences of endocrine disrupters. Some researchers propose a link that may exist between endocrine disrupters and the worldwide decline we are observing among amphibian populations.

I. Categories of Chemical Regulators

In addition to traditional endocrine regulation, we now recognize several other patterns of chemical regulation (Fig. 1-1). The first of these involves the nervous system. Special neural cells called **neurons** produce chemical regulators defined as **neurotransmitters.** Each neuron consists of a cell body and one or more elongate processes called **dendrites** and **axons.** The cell bodies of neurons in the brain occur in groups organized for related functions. A group of neuronal cell bodies is called a **nucleus.**

At a specialized connection between cells called a **synapse,** each molecule of neurotransmitter is secreted by a neuron into a specialized extracellular synaptic space or cleft. The neurotransmitter diffuses across that synaptic cleft to a **postsynaptic cell** where it is bound to a specific receptor molecule embedded in the postsynaptic cell membrane. Once bound, the neurotransmitter–receptor complex brings about distinct changes in the postsynaptic cell. The parallelism of endocrine cell/neuron, hormone/neurotransmitter, and target cell/ postsynaptic cell should be obvious.

Some neurobiologists like to separate the original neurotransmitters from what they now call **neuromodulators.** Although the definition of a neuromodular varies somewhat according to different authorities, we will define them simply as regulators secreted by neurons into synapses that produce no effect by themselves but alter the responsiveness of the postsynaptic cell to a typical neurotransmitter. Neuromodulators may produce effects on presynaptic cells, too.

In the middle of this century, additional research into regulatory mechanisms of physiology and behavior led to the conclusion by Ernst Scharrer that some neural regulators were released, like hormones, into the blood. These neural hormones were named **neurosecretions** or **neurohormones** to distinguish them from neurotransmitters, neuromodulators, and the traditional hormones. It seems that it was easier to formulate more definitions than to acknowledge that certain brain regions were also endocrine glands. Neurons that secrete neurohormones are sometimes called **neurosecretory neurons.** The term **neurocrine** has been suggested as a general category to include all of these neural regulators (that is, neurotransmitters, neuromodulators, and neurohormones). Although all neurocrines are actually neurosecretions, that term unaccountably has been reserved for the neurohormones.

Later, the separation of neural and endocrine systems became even more blurred when some established hormones were discovered to be present in the nervous system, where they functioned as neurohormones, neurotransmitters, or neuromodulators (see Table 1-1). It soon became common knowledge that the neural system had control through direct innervation or neurohormones over certain portions of the traditional endocrine system. Likewise, hormones were seen to influence markedly not only the development of neural systems but also their activity.

Discovery of specific chemicals produced by diverse cellular types and released into

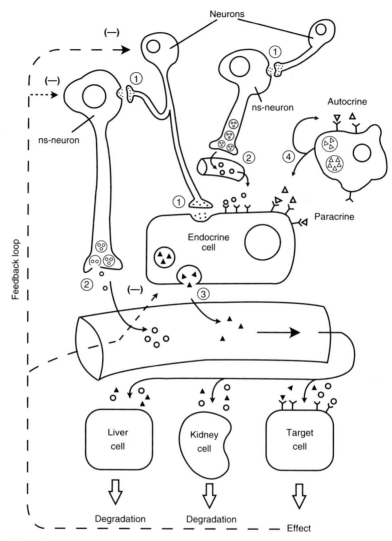

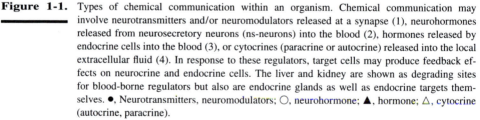

Figure 1-1. Types of chemical communication within an organism. Chemical communication may involve neurotransmitters and/or neuromodulators released at a synapse (1), neurohormones released from neurosecretory neurons (ns-neurons) into the blood (2), hormones released by endocrine cells into the blood (3), or cytocrines (paracrine or autocrine) released into the local extracellular fluid (4). In response to these regulators, target cells may produce feedback effects on neurocrine and endocrine cells. The liver and kidney are shown as degrading sites for blood-borne regulators but also are endocrine glands as well as endocrine targets themselves. ●, Neurotransmitters, neuromodulators; ○, neurohormone; ▲, hormone; △, cytocrine (autocrine, paracrine).

extracellular fluids has broadened the concept of chemical regulation still further to include cell-to-cell chemical communication that is not mediated via blood or at a synapse. These regulators may be lumped together as "local hormones" or **cytocrines.** This category might include locally acting growth factors, mitogenic regulators, embryonic inducing substances, secretion-enhancing secretogogues, inhibitors, and immune regulators. If cyto-

Table 1-1. Some Mammalian Neurocrine Regulators[a]

Class of regulator	Example
Nonpeptides	Acetylcholine (ACh)
	Carbon monoxide (CO)
	Dopamine (DA)
	Epinephrine (E)
	γ-Aminobutyric acid (GABA)
	Glutamate
	Nitric oxide (NO)
	Norepinephrine (NE)
	Serotonin (5-HT)
Hypothalamic releasing peptides	Corticotropin-releasing hormone (CRH)
	Gonadotropin-releasing hormone (GnRH)
	Melanotropin release-inhibiting hormone (MRIH)
	Prolactin-releasing hormone (PRH)
	Prolactin release-inhibiting hormone (PRIH)
	Somatostatin (SS or GHRIH)
	Thyrotropin-releasing hormone (TRH)
Other neuropeptides	Angiotensin II (ANG-II)
	Arginine vasopressin (AVP)
	Atrial natriuretic peptide (ANP)
	Brain natriuretic peptide (BNP)
	Cholecystokinin (CCK_8)
	Insulin
	Neuropeptide Y (NPY)
	Neuropeptide YY (PYY)
	Peptide histidine isoleucine (PHI)
	Substance P (SP)
	Vasoactive intestinal peptide (VIP)

[a] Some of these molecules may function only as a neurotransmitter, neuromodulator, neuro-hormone, or paracrine regulator whereas others may perform multiple roles.

crines also affect the emitting cell, they are sometimes termed **autocrines.** When they affect other cell types, they can be called **paracrines.** However, what we are calling the cytocrine mode of local secretion into the general extracellular fluids is sometimes loosely termed paracrine secretion.

Chemicals released from neurons into the cerebrospinal fluid (CSF) do not quite fit the definition of neurohormone since they are released into a filtrate of blood or a special extracellular fluid. However, the CSF is continuous throughout the ventricular system within the brain and the central canal of the spinal cord, forming a sort of circulatory system within the central nervous system. Therefore, these molecules do not quite fit the definition of paracrine secretion, either. For our purposes, we will arbitrarily refer to them as neurohormones.

For completeness, we might include the intracellular chemical regulators that govern intracellular events as a subclass of acceptable chemical regulators, especially since many of these are linked to the actions of intercellular chemical regulators. It has been suggested that these intracellular messengers be called **intracrines.** Some of these intracrine regulators (e.g., the second messengers) are discussed in Chapter 2.

In its broadest sense, the study of the chemical regulation of physiology and behavior includes specific regulators released by one organism into its environment that may affect the physiology or behavior of other individuals of that species or even of another species.

A good name for these substances might have been "exocrines," but that term had already been assigned to the products of **exocrine glands** that secrete their products into ducts through which they are conveyed to their sites of action in such places as the digestive tract or the surface of the skin. Examples of exocrine glands include salivary glands, sweat glands, mammary glands, and portions of the liver and pancreas. We can group these external chemical regulators under the category of "ectohormones" or **semiochemicals** (Gr. *semio,* signal). Two subclasses of semiochemicals have been identified on functional bases. **Pheromones** are semiochemicals that act only on other members of the same species. **Primer pheromones** usually initiate a series of physiological events such as gonadal maturation. Signal or **releaser pheromones** trigger immediate behavioral responses such as sexual attraction or copulation. **Allelomones** are interspecific semiochemicals and are further separated into two types. If only the emitter of the chemical benefits from the effect on the other species, the allelomone is called an **allomone.** The well-known odor released by skunks is a dramatic example of an allomone that protects the skunk from many would-be predators. When only the recipient species benefits, the allelomone may be termed a **kairomone.** The release of the simple metabolite L-lactate in sweat of humans attracts female mosquitos, who obtain a blood meal necessary for their reproduction. L-Lactate, then, could be classified as a kairomone, for there is no obvious benefit to the emitter.

In summary, intraorganismal chemical regulation can be classified as endocrine (hormones), neurocrine (neurotransmitters, neuromodulators, neurohormones), paracrine (cytocrines, autocrines), or intracrine (intracellular regulatory messengers). Semiochemicals (pheromones and allelomones) are specialized for interorganismal communication. A listing of the types of chemical regulators and their definitions are provided in Table 1-2.

Most chemical regulators are peptides, proteins, or derivatives of amino acids. Some are lipids (e.g., steroids) and still others are nucleotides or nucleotide derivatives. A discussion of the chemical nature of regulators, how they are synthesized, how they produce their effects on targets, and how they are metabolized is the subject of Chapter 2. However, before examining these regulators more closely, we must consider some more general features of chemical regulatory systems.

Table 1-2. Types of Regulators

Agent	Description	Examples
Neurotransmitter	Secreted by neurons into synaptic space	Acetylcholine, dopamine, substance P, GABA
Neuromodulator	Secreted by neurons into synaptic space; modulates sensitivity of postsynaptic cell to other neurotransmitters	Endorphins and various other neuropeptides
Neurohormone	Secreted by neurons into the blood or CSF; may be stored in neurohemal organ prior to release	TRH, CRH, oxytocin, dopamine
Hormone	Secreted by specialized nonneural cells into the blood	Thyroxine, GH, insulin
Cytocrine	Secreted by cells into the surrounding extracellular fluid; these local regulators typically travel short distances to nearby target cells	Somatostatin, norepinephrine
Paracrine	Secreted by cells that affect other cell types	Embryonic inducers, somatostatin, interleukins
Autocrine	Secreted by cells that affect emitting cells	Mitogenic agents, interleukins
Intracrine	Intracellular messengers; typically mediators of other regulators that bind to membrane receptors	cAMP, DAG, IP_3, cGMP, calmodulin, calcium ions
Semiochemical	Secreted into environment	Pheromones, allelomones

II. Organization of the Endocrine System

Traditionally, we have separated the endocrine system arbitrarily into two major categories, based on whether the brain and pituitary gland are integral parts. Chapters 4–12 of this textbook deal with the brain–pituitary systems, including the thyroid, adrenal, and gonadal organs. Chapters 13–15 deal with independent chemical regulators, those not directly involved with the brain and pituitary. Our approach will be to focus on the mammalian systems and then discuss similarities and differences found in nonmammalian vertebrates.

The portion of the brain we focus on is the **hypothalamus,** that portion of the diencephalic region of the mammalian brain directly above the pituitary gland. In addition, the **preoptic area (POA),** located in the posterior part of the telencephalon, is also involved. In fact, it is so closely associated both anatomically and functionally that it is often treated as if it were part of the hypothalamus. The organization of this system is described in Chapter 4. Special nuclei of neurosecretory neurons in the hypothalamus produce neurohormones. Some of these neurohormones control the secretion of peptide and protein hormones called **tropic hormones** in the anterior pituitary. These tropic hormones regulate the activities of the thyroid gland, adrenal cortex, and the gonads as well as general aspects of growth, metabolism, and reproduction. The neurohormones vasopressin and oxytocin are stored in the posterior pituitary until they are required. Vasopressin influences kidney function and oxytocin plays many reproductive roles. Among the independent endocrine glands are the parathyroids, controlling calcium balance; the endocrine pancreas, controlling glucose and lipid metabolism; and the gastrointestinal hormones, regulating digestion and some postabsorptive events. Other systems involved with chemical regulation include the pineal gland, the thymus, and the cellular elements of the immune system, as well as substances of more general origins such as growth factors, developmental inducers, kinins, and eicosanoids. A brief listing of chemical regulators and some of their characteristics is provided in Table 1-3.

III. Cytological and Histological Organization of Endocrine Cells

Certain endocrine cells can be identified easily on the basis of their cytological features as specialized secretory cells. Peptide-secreting cells have well-developed rough endoplasmic reticula and typically contain many protein-filled storage granules or vesicles (see Appendix B for a brief description of cellular structures). Mitochondria of peptide-secreting cells have flat, platelike cristae. The morphology and content of the protein storage granules may be used to differentiate specific types of endocrine cells. For example, the various tropic hormone-secreting cells of the anterior pituitary can be partially identified by the differential sizes of their storage granules (Fig. 1-2) as well as by special chemical methods (see Chapter 3). In contrast, steroid-secreting cells have well-developed smooth endoplasmic reticula, and their mitochondria have tubular cristae (compare steroid- and peptide-secreting cells shown in Figs. 1-2 and 1-3, respectively). Steroids are usually not stored but lipid droplets containing cholesterol, the precursor steroid for their synthesis, are commonly observed.

Neurosecretory neurons and regular neurons are not only specialized, elongated cells that are readily identifiable but also contain discrete **synaptic vesicles** containing neurocrine products in their axonal tips that characterize them as secretory cells. Neurosecretory neurons tend to be much larger than ordinary neurons. Neurons secreting peptides contain larger, denser granules than those secreting nonpeptides such as catecholamines

Table 1-3. The Mammalian Endocrine System and Major Secretions[a]

Source[b]	Target	Action
Anterior pituitary: Produces tropic hormones		
Glycoprotein tropic hormones		
Thyrotropin (thyroid-stimulating hormone; TSH)	Thyroid gland	Synthesis and release of thyroid hormones
Luteinizing hormone (lutropin, LH)	Gonads	Androgen synthesis; progesterone synthesis; gamete release
Follicle-stimulating hormone (follitropin, FSH)	Gonads	Gamete formation; estrogen synthesis
Nonglycoprotein tropic hormones		
Growth hormone (somatotropin, GH)	Liver, connective tissues, muscle	Synthesis of IGF, proteins
Prolactin (mammotropin, PRL)	Mammary glands, epididymus	Synthesis of proteins
Corticotropin (adrenocorticostimulating hormone, ACTH)	Adrenal cortex	Synthesis of corticosteroids
Melanotropin (melanocyte- or melanophore-stimulating hormone, MSH)	Melanin-producing cells	Synthesis of melanin
Hypothalamus: Neurosecretory nuclei produce neurohormones		
Hypothalamic releasing hormones		
Thyrotropin-releasing hormone (TRH)	Anterior pituitary	Releases TSH
Gonadotropin-releasing hormone (GnRH)	Anterior pituitary	Releases LH/FSH
Corticotropin-releasing hormone (CRH)	Anterior pituitary	Releases ACTH
Somatostatin (GH-RIH or SS)	Anterior pituitary	Inhibits GH release
Somatocrinin (GHRH)[c]	Anterior pituitary	Releases GH
Prolactin release-inhibiting hormone (PRIH)	Anterior pituitary	Inhibits PRL release
Prolactin-releasing hormone (PRH)[c]	Anterior pituitary	Releases PRL
Melanotropin release-inhibiting hormone (MRIH)[c]	Anterior pituitary	Inhibits MSH release
Melanotropin-releasing hormone (MRH)[c]	Anterior pituitary	Releases MSH
Other neurohormones		
Arginine vasopressin[d] (antidiuretic hormone, AVP)	Kidney / Brain	Water reabsorption / Drinking behavior
Oxytocin (OXY)	Uterus, vas deferens	Smooth muscle contraction
Endorphins/enkephalins	Pain neurons	Desensitizes neurons
Thyroid gland		
Thyroid hormones		
Thyroxine (T_4) and triiodothyronine (T_3)	Most tissues	Increases metabolism; controls development and differentiation
Calcitonin (thyrocalcitonin, CT)[e]	Bone	Prevents resorption caused by parathyroid hormone
Gonads		
Ovary		
Estrogens (e.g., estradiol)	Primary and secondary sexual structures / Brain	Stimulates development / Reproductive behavior

continues

Table 1-3. —*Continued*

Source[b]	Target	Action
Progesterone	Uterus	Stimulates secretion by uterine glands
Inhibin	Anterior pituitary	Blocks FSH release
Testis		
Androgens (e.g., testosterone)	Primary and secondary sexual structures	Stimulates development and secretion
	Brain	Reproductive behavior[f]
Inhibin	Anterior pituitary	Blocks FSH release
Adrenal gland		
Adrenal cortex		
Aldosterone (A)	Kidney	Sodium reabsorption; potassium secretion into urine
Corticosterone (B)/Cortisol (F)	Liver, muscle	Conversion of protein into carbohydrates
Adrenal medulla		
Epinephrine/norepinephrine	Liver, muscle, heart	Glycogen breakdown to glucose
Parathyroid gland		
Parathyroid hormone (parathormone, PTH)	Bone	Bone resorption
	Kidney	Calcium reabsorption and phosphate secretion into urine
Endocrine pancreas		
Insulin	Liver	Glycogen storage
	Muscle	Glucose uptake
	Adipose tissue	Inhibits fat hydrolysis
Glucagon	Liver, adipose tissue	Antiinsulin actions
Pancreatic polypeptide	Liver, muscle?	??
Somatostatin (paracrine substance)	Endocrine pancreas	Blocks release of pancreatic hormones
Gastrointestinal system		
Stomach		
Gastrin	Gastric glands of stomach	Stimulates acid secretion into lumen
Small intestine		
Secretin	Exocrine pancreas	Release of basic juice into duodenum
Cholecystokinin (CCK; same as pancreozymin-cholecystokinin, PZCCK)	Exocrine pancreas	Release of enzymes into duodenum
	Gall bladder	Contraction to eject bile into duodenum
Gastrin-releasing peptide	Stomach gastrin cells	Release of gastrin

continues

Table 1-3. —*Continued*

Source[b]	Target	Action
Glucose-dependent insulinotropic peptide (Gastric inhibitory peptide, GIP)	Endocrine pancreas	Release of insulin
Motilin	Stomach	Stimulates pepsinogen secretion and gastric motility
Somatostatin (paracrine action)	Small intestine	Inhibits release of other regulators
Vasoactive intestinal peptide (VIP)	Visceral blood vessels	Increases blood flow to intestines
Liver		
Insulin-like growth factors (IGF-I, IGF-II)	Many tissues	Mitogenic effects
Synlactin	Mammary gland	Participates in action of PRL on gland
Kidney		
Erythropoietin	Bone marrow	Stimulates RBC formation
Renin	Renin substrate in blood	Produces angiotensin
1,25-dihydroxycholecalciferol (1,25-DHC)[g]	Small intestine	Stimulates calcium absorption
Pineal gland		
Melatonin (neurohormone)	Brain	Controls puberty, thyroid, adrenal, and reproductive rhythms
Immune system		
Thymus		
Thymosins	Lymphocyte-producing tissue	Production of lymphocytes
Macrophages/lymphocytes		
Interleukin 1 (autocrine/cytocrine)	Helper T cell	Activation
Interleukin 2 (autocrine/cytocrine)	Cytotoxic T cell	Activation

[a] In some cases, closely related molecular forms may be present and will be discussed at the appropriate time. One or more alternative names may be used for a regulator and some of these alternates appear within parentheses.

[b] Alternate names are given in parentheses, along with the most common abbreviation.

[c] Exact chemical nature not clear (see Chapter 4 for more details).

[d] Some mammals may rely on a different nonapeptide (e.g., lysine vasopressin or phenypressin; see Chapter 6).

[e] Secretory cells derived from ultimobranchial gland in mammals become incorporated into the thyroid (see Chapter 11).

[f] May require conversion into estrogens within certain brain target cells before effect is observed (see Chapter 11).

[g] Cholecalciferol is made in skin and converted in liver to precursor kidney uses to make 1,25-DHC (see Chapter 15).

or acetylcholine. For example, the synaptic vesicles for acetylcholine are 30–45 nm in diameter, those for norepinephrine are about 70 nm, and peptide-containing vesicles are 100–300 nm.

In the nervous system, the cell bodies of neurons are localized in nuclei, and their axons often form specific tracts connecting to other nuclei, blood vessels, or the cerebrospinal fluid, or they may exit from the central nervous system as nerves. In fact, neurosecretory neurons were first identified in part because their cell bodies in **neurosecretory nuclei** and

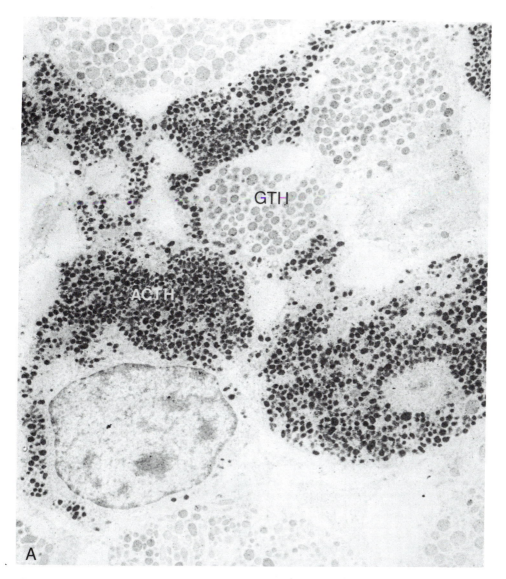

Figure 1-2. Cytology of polypeptide-secreting cells. (A) These cells from the pituitary gland of the frog *Rana temporaria*, contain dense granules that represent stored polypeptide secretions. The electron density of the storage granules in the ACTH-secreting cell has been enhanced further by using an antibody specific for ACTH, which results in deposition of an electron-dense reaction product (see discussion of immunocytochemistry in Chapter 3). The storage granules of gonadotropic hormones in the GTH cell have not been enhanced. Courtesy of Jeannine Doerr-Schott, Institut de Zoologie et de Biologie Generale, Strasburg, France. (B) A growth hormone-secreting cell from coho salmon (*Oncorhynchus kisutch*) showing several features characteristic of polypeptide-secreting cells: electron-dense secretory granules, well-developed Golgi bodies, and mitochondria with platelike cristae. Courtesy of Howard A. Bern and Richard Nishioka, University of California, Berkeley.

axons in **neurosecretory tracts** contained materials that stain with certain dyes, distinguishing these cells from ordinary neurons. However, these general methods usually did not distinguish between different kinds of neurosecretory neurons. Modern immunological

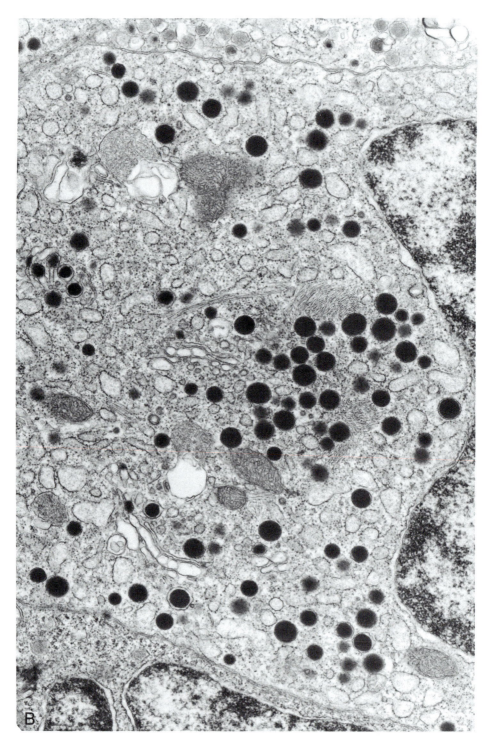

Figure 1-2   (*continued*).

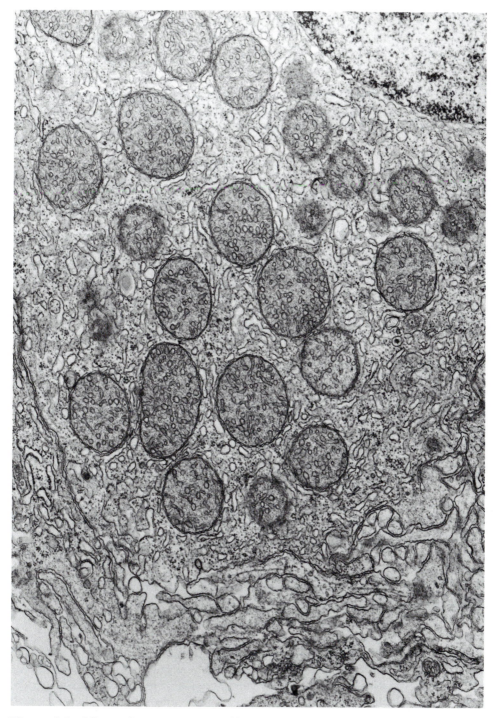

Figure 1-3. Microscopic appearance of a steroid-secreting cell. Adrenocortical cells (interrenal) from juvenile coho salmon showing mitochondria with tubular cristae and an abundance of smooth endoplasmic reticulum. Courtesy of Howard A. Bern and Richard Nishioka, University of California, Berkeley.

techniques now allow us to identify each type of neurosecretory or ordinary neuron with respect to its particular secretions (see Chapter 3).

Another factor that helped in the early identification of endocrine cells was their anatomical relationship to one another. Many endocrine cells are specialized epithelial cells that tend to be clumped in groups that are organized in one of the following ways (see Appendix B for descriptions of epithelia and other tissue types). The most common orientation of secretory cells is to form folded sheets or **cords** of cells as seen in the pituitary gland or the adrenal cortex (Fig. 1-4). In a few cases, the cells may form a spherical mass of one cell layer surrounding a fluid-filled space or lumen. This arrangement is termed a **follicle** and occurs in the thyroid gland and in the anterior pituitaries of some vertebrates (Fig. 1-4). The lumen provides a unique storage site for secretions of the follicular cells. Sometimes, the endocrine cells will be separated into scattered clumps or **islets** of a few cells (Fig. 1-4). Mixed islets containing several secretory cell types are best known in the pancreas, where they are called the **islets of Langerhans.**

However, many cells that secrete chemical regulators are not so cytologically distinct and were not identified until precise biochemical techniques were developed. One of the reasons it took so long to identify the sources of gastrointestinal hormones was due to the tendency for these secretory types to occur as **isolated cells** mixed in with many other cell types in the stomach and intestinal walls.

Another critical feature of organization is the presence of an extensive vascular supply for endocrine cells and neurosecretory neurons. Endocrine glands typically are highly vascularized such that no secretory cell is far from a blood vessel. The axonal endings of many neurosecretory neurons end collectively in masses of capillaries to form what is called a **neurohemal organ.** These neurosecretory neurons release their neurohormones into the blood that flows through the neurohemal organ. Many other neurons may release their products individually into the cerebrospinal fluid, however, and do not form aggregates at common release sites.

Most regulatory cells employ **merocrine secretion,** whereby secretory products are released by exocytosis with no damage to the cell. In **apocrine secretion,** the apical portion or tip of the cell is sloughed along with stored secretions whereas **holocrine secretion** involves lysis and death of the secretory cell. These latter two patterns are more characteristic of exocrine glands such as the mammary gland (apocrine) or the sebaceous glands of the skin (holocrine). **Cytogenous secretion** is the release of entire cells such as sperm release from testes or ovum release from ovaries. These secretory patterns are summarized in Table 1-4.

IV. The Origins of Chemical Regulation and Endocrine Cells

Regulatory chemicals were probably essential for the survival of the first living cells, both for coordination of internal events and for cell-to-cell interactions. Secretions that favored survival of the secreting cell no doubt led to further evolution of new chemical regulators. Thus, chemical regulation probably had its origin in paracrine- and intracrine-type secretions. The evolution of multicellular aggregates and eventually of multicellular organisms allowed further paracrine and intracrine regulation but more importantly allowed evolution of endocrine and neurocrine regulation. The earliest appearance of neurons is noted in the primitive cnidarian invertebrates (Phylum Cnidaria), and these neurons secrete both peptide and nonpeptide regulators. There is also evidence of paracrine regulation in cnidarians.

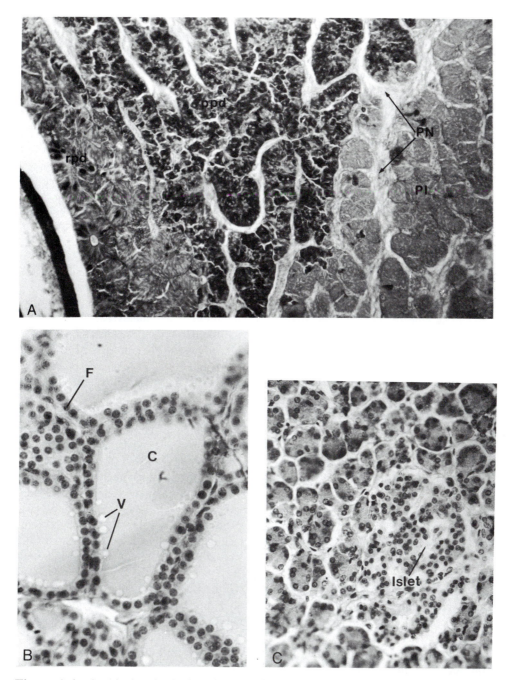

Figure 1-4. Organization of endocrine cells. Typically, endocrine cells are organized into (1) folded cords of cells, often associated with sinusoidal capillaries, (2) follicles consisting of a single layer of cells surrounding a lumen, or (3) small clumps of cells, called *islets*, within a tissue. In addition, isolated cells may constitute diffuse endocrine glands, functioning in a coordinated manner even though they do not form a continuous mass. This latter condition is commonly observed for several types of endocrine cells in the small intestine of vertebrates and for the C cells of the mammalian thyroid. (A) Pituitary of rainbow trout, *Oncorhynchus mykiss*; (B) thyroid gland of dog; (C) pancreatic islet of rat.

Table 1-4. Cellular Patterns of Secretion

Secretory pattern	Description	Example
Endocrine	Product secreted into the blood for transport internally to target tissues	Hormones
Exocrine	Product secreted into a duct that opens onto an external or internal surface	Sweat
Exocytosis	Product released from secretory cell via a process essentially the reverse of pinocytosis	Peptide hormone release
Merocrine	Product secreted without visible damage to the secretory cell (involves exocytosis)	Thyroxine secretion
Apocrine	Product released by sloughing of "outer" or apical portion of secretory cell	Mammary gland milk
Holocrine	Product released through cell death and lysis	Sebaceous gland secretion
Cytogenous	Release of whole, viable cells	Spermatozoa

By definition, true endocrine glands that secrete hormones did not appear until the type of internal transport mechanisms we call *blood vascular systems* evolved, although the gastro-vascular systems of cnidarians and flatworms (Phylum Platyhelminthes) can be considered primitive circulatory systems since they perform similar functions. Distinct endocrine glands are present in annelids (Phylum Annelida), mollusks (Phylum Mollusca), insects, arachnids, and crustaceans (Phylum Arthropoda), in addition to complicated neurosecretory systems integrated with the endocrine components into a coordinated regulatory scheme similar to what we find in vertebrates.

Among the vertebrates, we have made extensive analyses of the embryonic origins of endocrine glands that have provided a better understanding of the evolution of vertebrate endocrine systems. It was once thought that neurosecretory cells and the posterior pituitary were derived from neural ectoderm; the anterior pituitary was derived from nonneural ectoderm; the liver, thyroid, parathyroids, thymus, and gastrointestinal endocrine cells were endodermal in origin; and the gonads and kidneys developed from mesoderm (see Appendix B, if necessary, for a brief description of these embryonic tissues). Endocrinologists believed that the fundamental differences among these groups of glands reflected their embryological origins until modern biochemical techniques demonstrated that many of the ectodermal and endodermal glands produced similar and sometimes identical secretory products.

Considerable revision in thinking followed the discovery that many of these ectodermally and endodermally derived cells possessed a property previously ascribed only to amine-secreting neurons. This property, the ability to accumulate amino acid precursors and convert these precursors into biologically active amine neurotransmitters by removing their carboxyl acidic group, was termed **amine content and amine precursor uptake and decarboxylation;** shortened to **APUD.** This property also was shared by melanin-producing skin cells that, similar to the amine-secreting neurons, were derived from special embryonic nerve cells called **neural crest** cells. During early development, neural crest cells develop as paired masses of cells at intervals along the embryonic nerve cord, migrate to other locations, and give rise to the sympathetic ganglia, melanin-producing skin cells, the adrenal medulla, and components of the skull and branchial skeletal elements. The discovery of this APUD property in many peptide-secreting endocrine and neuroendocrine cells prompted A. G. E. Pearse to propose that all cells with APUD characteristics have the same neural origins including parathyroid cells, calcitonin-secreting cells of the ultimobranchial body, gastrointestinal endocrine cells, and even some anterior pituitary cells. Later, elegant experiments by N. LeDouarin and C. LeLievre verified that the calcitonin-secreting cells of the avian ultimobranchial glands were of neural crest origin although the matrix of the gland itself developed as an outpocketing of gut endoderm. They transplanted quail neural crest cells with a cytological marker into chicken embryos and observed them migrating to and finally residing in the ultimobranchial glands. Similar evidence was obtained for the parathyroid gland secretory cells in the frog *Rana temporaria*. Earlier efforts that showed a neural origin for the endocrine cells of the anterior pituitary as well as the neurosecretory neurons of the hypothalamus apparently have been confirmed in the laboratory of Sakae Kikuyama by studies of transplants of pigmented neural ridge tissue into albino Japanese toads, *Bufo japonicus*.

V. Homeostasis

Chemical regulatory mechanisms are the bases for controlling all physiology and behavior. It is through these mechanisms that homeostatic balance and hence survival in a harsh and dangerous environment is possible. Although Claude Bernard formulated the concept of homeostasis in the nineteenth century, it was the American physiologist Walter B. Cannon who in 1929 coined the term **homeostasis** to describe balanced physiological systems operating in the organism to maintain a dynamic equilibrium; that is, a relatively constant steady state maintained within certain tolerable limits. In Cannon's words:

> When we consider the extreme instability of our bodily structure, its readiness for disturbance by the slightest application of external forces and the rapid onset of its decomposition as soon as favoring circumstances are withdrawn, its persistence through many decades seems almost miraculous. The wonder increases when we realize that the system is open, engaging in free exchange with the outer world, and that the structure itself is not permanent but is being continuously broken down by the wear and tear of action, and as continuously built up again by processes of repair....
>
> The constant conditions which are maintained in the body might be termed equilibria. That word, however, has come to have fairly exact meaning as applied to relatively simple physico-chemical states, in closed systems, where known forces are balanced. The coordinated physiological processes which maintain most of the steady states in the organism are so complex and so peculiar to living beings—involving, as they may, the brain and nerves, the heart, lungs, kidneys and spleen, all working cooperatively—that I have suggested a special designation for these states, homeostasis. The word does not imply something immobile, a stagnation. It means a condition—a condition which may vary, but which is relatively constant. (Walter B. Cannon, 1929, "The Wisdom of the Body")

Cannon's original formulation of the homeostatic mechanism emphasized the maintenance of blood parameters such as osmotic pressure, volume, hydrostatic pressure, and levels of various chemicals such as calcium, sodium, and glucose. Today, we have expanded his viewpoint to include all manner of physiological regulation at the organismal level as well as at the molecular–cellular level.

When attempting to comprehend physiological systems, it is helpful to employ simplified models that simulate the various components of the system in a way that is easy to grasp and at the same time provide insights into how the system works as well as predictions on how it will respond to disturbances. In the following section, we consider a very simple model of a basic regulatory mechanism operating for all physiological systems and provide some insight on how to use this model to understand more complicated, integrated systems such as those discussed in later chapters.

A. A Homeostatic Reflex Model

In this model, we define **information (I)** as any stimulus that can provide quantitative or qualitative cues detectable is some way by the system we are examining. The model is depicted in Fig. 1-5. The information is detected by a **receptor (R)** or transducer of some sort that translates (transduces) this information into the biological language of the regulatory system. For example, pressure may cause sodium ions to enter a neuron, depolarizing the cell membrane and inducing an action potential that in turn causes a nerve impulse to be generated. The transduced information is now called **input (I′)** and is translocated to an integrating center called the **controller (C).** The controller uses a preprogrammed set of cellular instructions to compare the input with a **set point** and determines whether any adjustments are warranted. If the controller ascertains adjustments are needed to maintain or reattain homeostatic balance, it will direct a message called **output** to one or more **effectors (E)** that will perform some specific action **(effect)** that, in turn, will bring about corrective changes in the system. The responsiveness of the controller may be influenced by input received from other homeostatic regulators (Fig. 1-5). These additional inputs may enhance or reduce the output of the controller through altering its sensitivity to other input or by adjusting the set point.

Corrective changes signaled through output from the controller will alter the nature of the information perceived by the receptors, and the controller will be notified immediately if the response has been sufficient and appropriate. This pathway whereby the controller is apprised of its effectiveness is called **feedback.** If the response was insufficient, the controller may increase its output to elevate the effector response. If an overcorrection or **overshoot** has

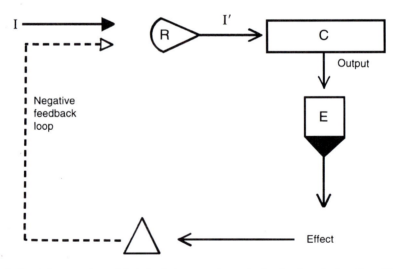

Figure 1-5. A homeostatic model. A simple homeostatic mechanism involves the detection of information (I) by a receptor (R), which converts the information into a biologically recognizable cue or input (I′) that is transmitted to the controller (C). The role of the controller is to compare input with a programmed set point and make physiological adjustments if needed. The controller can send messages (Output) to effectors (E) that respond by producing an effect, which in turn causes a change in the system (Δ) that feeds back through the same or a different receptor as a signal to tell the controller if it made an appropriate response (i.e., has the system moved closer to the set point). This type of feedback is termed *negative feedback*.

occurred, the controller will alter its output accordingly and may even generate new output to other effectors to bring the system into line with the preprogrammed set point. Therefore, the set point represents the optimal physiological condition. The feedback loop that drives a physiological system toward the preprogrammed set point is called **negative feedback.** Thus, any disturbance causes a homeostatic reflex to be activated to maintain homeostasis.

We might consider a controlled temperature room as a physical model of a simple control system similar to what we have described for a biological system. The programmable thermostat represents both the receptor and controller components; an air conditioner and a heater are the effectors. Mechanical deformations produced in a bimetal strip (receptor) exposed to the air temperature (information) of the room are transduced into electrical current (input). The controller compares this input with the preprogrammed temperature (set point) and electrically turns on or off (output) the appropriate effector to maintain a constant temperature (homeostasis). The new air temperature will be detected by the same receptor and new input will be sent to the controller, which will continue to make adjustments to drive the system to the set point (negative feedback).

Under some conditions, feedback may drive a system away from the preprogrammed condition. Such a feedback loop is called **positive feedback,** and it drives a system to a different level of activity. Positive feedback is invoked where a rapid change is required, may be associated with an emergency-type response, or might be responsible for short-term adaptations to complete a series of changes. The rapid influx of sodium during generation of an action potential, the physiological stress response, certain events during ovulation, and the induction of sexual maturation are examples of events that employ positive feedback. In general, positive feedback is important for short-term events but is detrimental over longer time periods and can lead to the death of the animal if it persists. In contrast, long-term negative feedback generally is advantageous to survival as it helps maintain homeostatic balance in the face of environmental or internal changes. Negative feedback is the most common type of feedback.

In certain instances, changes in a regulated variable are anticipated through **feedforward regulation,** which accelerates homeostatic responses and minimizes fluctuations in the regulated variable. For example, the regulation of internal body temperature involves a classical negative-feedback loop based on the temperature of the blood flowing to the brain. However, changes in body surface temperature can send input to the brain that is interpreted to begin making appropriate adjustments in temperature production and conservation or dissipation of heat to ward off anticipated changes in internal body temperature. This would be an example of feedforward regulation.

To apply this homeostatic reflex model to any regulatory mechanism, one must ask a series of simple questions. First of all, on what information does the system cue? What are the receptors and where are they located? What sort of input is generated by this stimulus and how is it conducted to the controller? What is the controller and where is it located? What set point is involved? What sort of output is generated? What are the effectors and what effects are produced? What change in the system results and how does feedback occur? Is there negative, positive, or feedforward regulation?

When examining physiological systems, you will soon discover that complicated responses involve the integration of many different simple homeostatic reflexes. For example, the "controller" for one system may actually be the "effector" of another homeostatic reflex, and there may be pathways that modulate the responsiveness of a controller or effector to other stimuli. An excellent example of the integration of individual reflexes is the **neuroendocrine reflex.**

A typical vertebrate neuroendocrine reflex is represented by the adrenal endocrine axis in Fig. 1-6. The kidney cell can be considered an effector whose activity is regulated by a steroid hormone from the adrenal cortex. This would make the adrenal cortex the control-

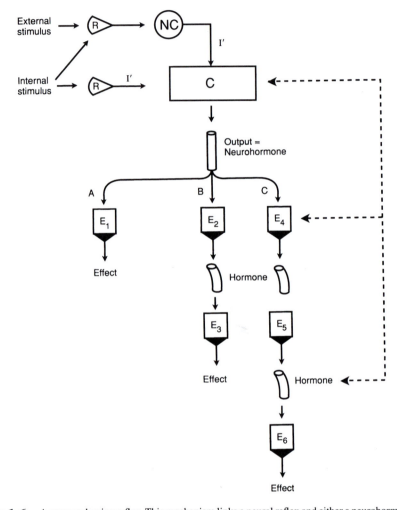

Figure 1-6. A neuroendocrine reflex. This mechanism links a neural reflex and either a neurohormonal reflex or an endocrine reflex so that various external or internal stimuli can cause or inhibit release of a hormone through activation of a neural circuit. Feedback is not illustrated (see Chapter 4 for details) but may occur through blood-borne messages or via nervous stimulation involving various components of the system. Suckling of an infant at its mother's breast is an example involving two neuroendocrine reflexes. Mechanical stimulation of receptors (R) in the nipple sends neural impulses to a neural center (NC) in the brain, which relays neural impulses to a neurosecretory nucleus in the hypothalamus (the normal controller, C). The first reflex (pathway A) releases the neurohormone oxytocin, which travels through the blood and stimulates milk ejection from the breast (E_1). The second reflex (pathway B) involves a neurohormone (see Chapter 4 for details) that evokes release of the hormone prolactin from the pituitary gland (E_2), which in turn stimulates milk synthesis in the breast (E_3) prior to the next feeding. Pathway C illustrates a more complex reflex involving the pituitary (E_4), which in turn stimulates another endocrine gland (E_5) that produces a hormone that induces an effect on a nonendocrine target (E_6). This reflex also is described in Chapter 4.

ler, but it is also an effector for the anterior pituitary, which controls its activity through output of a tropic hormone, corticotropin or ACTH. Yet, the anterior pituitary is also an effector for a neurohormone (corticotropin-releasing hormone, CRH) from the hypothalamus (which arbitrarily is labeled the controller in this diagram). Only the major or primary feedback loop is shown, but other feedback loops are actually operating (see Chapters 4, 5, and 9). In this neuroendocrine reflex, we have indicated that the hypothalamic controller receives information via the blood (Internal stimulus) but it also can be affected by information accumulated through a variety of receptors associated with other reflexes (External stimulus). Two neural reflex loops are shown, each with its own receptor and controller, that might send modulating input to the hypothalamic controller, altering the set point or telling the hypothalamus to ignore the set point altogether. In spite of the apparent complexity of this neuroendocrine reflex as compared to our basic model, one simply asks the same series of questions given above for each level in the system until an understanding is achieved of the integrated whole.

Suggested Reading

Books

General/Mammalian

Baulieu, E.-E., and Kelly, P. A. (1990). "Hormones." Chapman & Hall, New York.

Bolander, F. F. (1994). "Molecular Endocrinology," 2nd Ed. Academic Press, San Diego.

Brown, R. E. (1994). "An Introduction to Neuroendocrinology." Cambridge Univ. Press, New York.

DeGroot, L. J., Besser, G. M., Cahill, G. F., Jr., Marshall, J. C., Nelson, D. H., Odell, W. D., Potts, J. T., Jr., Rubenstein, A. H., and Steinberger, E. (1989). "Endocrinology," 2nd Ed., Vols. 1–3. Saunders, Philadelphia, PA.

Gass, G. H., and Kaplan, H. M. (1996). "Handbook of Endocrinology," 2nd Ed. CRC Press, Boca Raton, FL.

Griffin, J. E., and Ojeda, S. R. (1996). "Textbook of Endocrine Physiology," 3rd Ed. Oxford Univ. Press, New York.

Hadley, M. E. (1996). "Endocrinology," 4th Ed. Prentice Hall, Upper Saddle Park, NJ.

Malvern, P. V. (1993). "Mammalian Neuroendocrinology." CRC Press, Boca Raton, FL.

Martin, C. (1995). "The Dictionary of Endocrinology and Related Biomedical Sciences." Oxford Univ. Press, New York.

Nemeroff, C. B. (1992). "Neuroendocrinology." CRC Press, Boca Raton, FL.

Norman, A. W., and Litwack, G. (1987). "Hormones." Academic Press, San Diego.

Timiras, P. S., Quay, W. B., and Vernadakis, A. (1995). "Hormones and Aging." CRC Press, Boca Raton, FL.

Comparative

Barrington, E. J. W. (1979). "Hormones and Evolution," Vols. 1 and 2. Academic Press, New York.

Becker, J. B., Breedlove, S. M., and Crews, D. (1992). "Behavioral Endocrinology." MIT Press, Cambridge, MA.

Chester-Jones, I., Ingleton, P. M., and Phillips, J. G. (1987). "Fundamentals of Comparative Vertebrate Endocrinology." Plenum, New York.

Davey, K. G., Peter, R. E., and Tobe, S. S. (1994). "Perspectives in Comparative Endocrinology." National Research Council of Canada, Ottawa.

Epple, A., and Stetson, M. H. (1980). "Avian Endocrinology." Academic Press, New York.

Gorbman, A., and Bern, H. A. (1962). "A Textbook of Comparative Endocrinology." Wiley, New York.

Gorbman, A., Dickhoff, W. W., Vigna, S. R., Clark, N. B., and Ralph, C. L. (1983). "Comparative Endocrinology." Wiley, New York.

Matsumoto, A., and Ishii, S. (1992). "Atlas of Endocrine Organs: Vertebrates and Invertebrates." Japan Society for Comparative Endocrinology and Springer-Verlag, Berlin.

Nelson, R. J. (1995). "An Introduction to Behavioral Endocrinology." Sinauer, Sunderland, MA.

Norris, D. O. (1985). "Vertebrate Endocrinology," 2nd Ed. Lea & Febiger, Philadelphia, PA.

Pang, P. K. T., and Epple, A. (1980). "Evolution of Vertebrate Endocrine Systems." Texas Tech Press, Lubbock, TX.

Pang, P. K. T., and Schreibman, M. P. (1986–). "Vertebrate Endocrinology: Fundamentals and Biomedical Implications," Vols. 1–. Academic Press, San Diego.

Ralph, C. L. (1986). "Comparative Endocrinology: Developments and Directions." Alan R. Liss, New York.

Scanes, C. G., Ottinger, M. A., Kenny, A. D., Balthazart, J., Cronshaw, J., and Chester-Jones, I. (1982). "Aspects of Avian Endocrinology: Practical and Theoretical Implications." Texas Tech Press, Lubbock, TX.

Schreibman, M. P., and Scanes, C. G. (1989). "Development, Maturation, and Senescence of Neuroendocrine Systems: A Comparative Approach." Academic Press, San Diego.

Schreibman, M. P., Scanes, C. G., and Pang, P. K. T. (1993). "The Endocrinology of Growth, Development, and Metabolism in Vertebrates." Academic Press, San Diego.

Sharp, P. J. (1993). "Avian Endocrinology." The Society for Endocrinology, Bristol, UK.

Clinical

Ambrecht, H. J., Coe, R. M., and Wongsurawat, N. (1990). "Endocrine Function and Aging." Springer-Verlag, New York.

Becker, K. L. (1995). "Principles and Practice of Endocrinology and Metabolism," 2nd Ed. Lippincott-Raven, Philadelphia, PA.

Brook, C. G. D. (1995). "Clinical Paediatric Endocrinology," 3rd Ed. Blackwell, Oxford.

Collu, R., Brown, G. M., and Van Loon, G. R. (1988). "Clinical Neuroendocrinology." Blackwell, Oxford.

Grossman, A. (1992). "Clinical Endocrinology." Blackwell, Oxford.

Mazzaferri, E. L., and Samann, N. A. (1993). "Endocrine Tumors." Blackwell, Oxford.

Serials

Annual Review of Medicine
Annual Review of Physiology
Cell Signalling
Endocrine Reviews
Frontiers in Neuroendocrinology
Pharmacological Reviews
Recent Progress in Hormone Research
Trends in Endocrinology and Metabolism
Trends in Neurosciences
Vitamins and Hormones

Articles

Blaustein, A. R., and Wake, D. B. (1995). The puzzle of declining amphibian populations. *Sci. Am.* **April,** 52–57.

Brabant, G., Prank, K., and Schofl, C. (1992). Pulsatile patterns in hormone secretion. *Trends Endocrinol. Metab.* **3,** 183–190.

Bern, H. A. (1990). The "new" endocrinology: Its scope and its impact. *Am. Zool.* **30,** 877–885.

Grimmelikhuijzen, C. J. P., Carstensen, K., Darmer, D., Moosler, A., Nothacker, H.-P., Reinscheid, R. K., Schmutzler, C., Vollert, H., McFarlene, I., and Rinehart, K. L. (1992). Coelenterate neuropeptides: Structure, action and biosynthesis. *Am. Zool.* **32,** 1–12.

Guillette, L. J., Jr. (1995). Endocrine disrupting environmental contaminants and developmental abnormalities in embryos. *Hum. Ecol. Risk Assess.* **1,** 25–36.

Hose, J. E., and Guillette, L. J., Jr. (1995). Defining the role of pollutants in the disruption of reproduction in wildlife. *Environ. Health Perspect.* **103**(Suppl. 4), 87–91.

Matt, K. S. (1993). Neuroendocrine mechanisms of environmental integration. *Am. Zool.* **33,** 266–274.

Robash, M., and Hall, J. C. (1989). The molecular biology of circadian rhythms. *Neuron* **3,** 387–398.

Sharpe, R. M., and Shakkebaek, N. E. (1993). Are oestrogens involved in falling sperm counts and disorders of the male reproductive tract? *Lancet* **341,** 1392–1395.

Schofl, C., Prank, K., Wiersinga, W., and Brabant, G. (1995). Pulsatile hormone secretion: Analysis and biological significance. *Trends Endocrinol. Metab.* **6,** 113–114.

2

The Molecular Bases for Chemical Regulation

THE CHEMICAL properties of each regulatory molecule are keys to understanding much of its physiology. They not only prescribe how a regulator is synthesized but also how it is secreted and transported to its target site, how it produces its effects on a specific target, and how it is metabolized or inactivated. Most chemical regulators, such as monoamines, small peptides, polypeptides, or proteins, are at home in aqueous media. In sharp contrast, some chemical regulators such as steroids, thyroid hormones, and eicosanoids have low solubility in aqueous media, and, unlike peptides, readily pass through cell membranes. These hydrophobic regulators have markedly different synthetic pathways and processes of secretion, transport, and action on target cells. Because the features of each group of chemical regulators are uniquely tied to their chemical composition, each major type is discussed separately in this chapter. We will examine the following groupings: (1) amino acids, amines, peptides, and proteins, (2) steroids, (3) thyroid hormones, (4) eicosanoids, and (5) other kinds of molecules.

I. Amino Acids, Amines, Peptides, and Proteins

Most neurotransmitters, neuromodulators, neurohormones, and classical hormones as well as many chemical regulators common to interstitial fluids are composed of linear sequences of amino acids linked together by peptide bonds (see Table 2-1). These peptides vary in length from the tripeptide known as thyrotropin-releasing hormone (TRH) to large molecules of 200 or more amino acids (e.g., growth hormone, GH). Some common chemical regulators are single amino acids, or modified amino acids such as the catecholamines and indoleamines.

A. Catecholamines

A **catechol** is a six-carbon unsaturated ring (phenolic group) with two hydroxyl groups attached to adjacent carbons (dihydroxyphenol; see Fig. 2-1). Attachment of the catechol ring to a side chain with an amine group characterizes a **catecholamine.** Catecholamines are synthesized in the following manner (see Fig. 2-2). Addition of a phenolic group to the three-carbon amino acid alanine results in formation of another amino acid, phenylalanine. Combining one hydroxyl to the opposite end of the phenolic group converts phenylalanine into the amino acid tyrosine. Addition of one more hydroxyl to the phenolic ring followed by removal of the carboxyl group from the alanine portion yields a catecholamine.

Three important catecholamines are synthesized from tyrosine by neurons and cells of

Table 2-1. Amino Acid Composition of Some Peptide Regulators

Name	Primary role	Structure
Thyrotropin-releasing hormone	Neurohormone	Pyro-Glu-His-Pro-CONH$_2$
Substance P	Neuromodulator	NH$_2$-Arg-Pro-Lys-Pro-Gln-Gln-Phe-Phe-Gly-Leu-Met-CONH$_2$
Glucagon	Hormone	NH$_2$-His-Ser-Gln-Gly-Thr-Phe-Thr-Ser-Asp-Tyr-Ser-Lys-Tyr-Leu-Asp-Ser-Arg-Arg-Ala-Gln-Asp-Phe-Val-Gln-Trp-Leu-Met-Asn-Thr-COOH

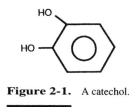

Figure 2-1. A catechol.

the adrenal medulla (Fig. 2-3). The addition of an OH⁻ group to tyrosine by the enzyme **tyrosine hydroxylase** yields **dihydroxyphenylalanine** or **Dopa.** Next, the carboxyl group is removed by the enzyme **Dopa decarboxylase** to form the catecholamine **dopamine (DA),** a common neurotransmitter in the central nervous system. In some neurons, an additional enzyme, **dopamine β-hydroxylase,** converts DA to the catecholamine neurotransmitter **norepinephrine (NE)** by addition of an OH⁻ to the former alanine side chain. NE also is called phenylethanolamine. In still other neurons, NE is further altered by addition of a methyl group to the amine group to make the catecholamine neurotransmitter **epinephrine (E).** This last conversion is catalyzed by the enzyme **phenylethanolamine N-methyltransferase (PNMT).** Tyrosine hydroxylase immunoreactivity is often used as a

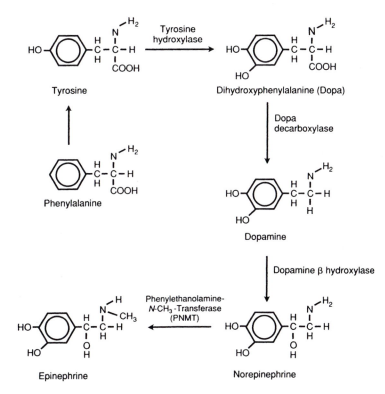

Figure 2-2. Synthesis of catecholamines. Catecholamines may be synthesized from either of the amino acids phenylalanine or tyrosine. The rate-limiting enzyme for this pathway is tyrosine hydroxylase. Depending on which enzymes are active in a cell, the final secretory product may be dopamine, norepinephrine, or epinephrine.

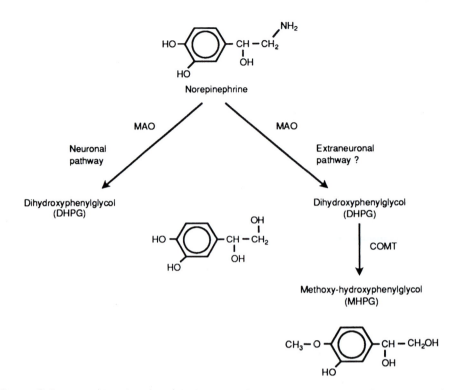

Figure 2-3. Metabolism of norepinephrine in the central nervous system. The end product of norepinephrine metabolism in neurons is thought to be DHPG, whereas in glial cells (extraneuronal pathway) the major product is MHPG. MAO, Monoamine oxidase; COMT, catechol O-methyltransferase.

histochemical marker to locate catecholamine-secreting cells although it does not indicate whether the final secretory product of that cell is DA, NE, or E.

Catecholamines are more than just neurotransmitters. DA, NE, and E can all be released into the circulation where they function as hormones. The hypothalamus releases DA that acts as a neurohormone to inhibit release of prolactin from the pituitary gland (see Chapter 4). The adrenal medulla, a modified sympathetic ganglion, secretes NE and E into the blood in response to neural signals directed via sympathetic nerve pathways from the hypothalamus. Although traditionally recognized as hormones, these adrenal medullary secretions also could be called neurohormones. Finally, in the central nervous system, there is evidence that NE is acting as a paracrine regulator, too.

Release of catecholamine neurotransmitters is followed by their partial reuptake and recycling by the secreting neuron to free the postsynaptic receptors and turn off the postsynaptic cell. The intraneuronal enzyme complex called **monoamine oxidase (MAO)** is responsible for degrading catecholamines. A second enzyme, **catechol O-methyltransferase (COMT),** associated with glial cells in the nervous system, is important in degrading catecholamines, too. A summary of brain catecholamine metabolism and structures of their metabolites are provided in Fig. 2-3. NE and E in the peripheral circulation are metabolized by liver MAO and aldehyde oxidase to produce slightly different metabolites than those found in the brain (Fig. 2-4).

Numerous agonists and antagonists for catecholamines have been developed as well as inhibitors of MAO and COMT. Other drugs have been developed that block release of the

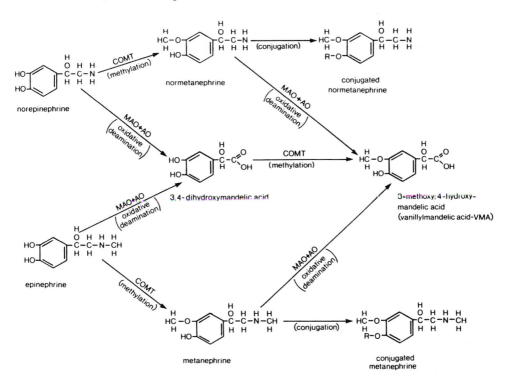

Figure 2-4. Peripheral metabolism of norepinephrine and epinephrine. AO, Amine oxidase; COMT, catechol O-methyltransferase; MAO, monoamine oxidase.

catecholamines. A partial listing of these compounds is provided in Table 2-2. Use of such drugs has allowed us to investigate the roles of catecholamine neurotransmitters in the regulation of endocrine function by the brain.

Table 2-2. Some Pharmacological Compounds That Alter Catecholamine Activity

Compound	Action
Propranolol	Antagonist that binds competitively to β_1- and β_2-adrenergic receptors and prevents actions of epinephrine or norepinephrine; consequently, propranolol is called a β-blocker
Phentolamine	An α-antagonist that competively blocks norepinephrine action
Haloperidol	Antagonist that binds competitively to dopaminergic receptors and blocks effects of dopamine on its target cells
Isoproterenol	General β-agonist that mimicks both epinephrine and norepinephrine
Dobutamine	A relative selective β_1-agonist that is an effective heart stimulant
Ephedrine	An agonist that binds to α- and β-adrenergic receptors
Amphetamine	Potent agonist of α- and β-adrenergic receptors that binds especially well to central nervous system receptors but poorly to peripheral receptors
Reserpine	Blocks reuptake of catecholamines by presynaptic neurons
Pyrogallol	Blocks the enzyme catechol O-methyltransferase (COMT), an enzyme necessary for metabolism of catecholamines
Harmaline	Inhibitor of monoamine oxidase (MAO), an important enzyme for degradation of catecholamines

B. Indoleamines

Synthesis of the indoleamines is outlined in Fig. 2-5. The amino acid tryptophan is hydroxylated by the enzyme tryptophan hydroxylase to yield 5-hydroxytryptophan. This metabolite is converted by L-aromatic amino acid decarboxylase to the neurotransmitter 5-hydroxytryptamine or **serotonin (5-HT)**. A form of MAO degrades serotonin to inactive **5-hydroxyindoleacetic acid (5-HIAA).** In the pineal gland, the enzyme **N-acetyltransferase (NAT)** alters serotonin to an N-acetylated form, which in turn is altered by **hydroxyindole-O-methyltransferase (HIOMT)** to the pineal neurohormone, **melatonin** or N-acetyl-5-methoxytryptamine (see Chapter 6). NAT is considered to be the rate-limiting enzyme for melatonin synthesis.

Melatonin is secreted primarily during the dark **(scotophase)** and appears to be important in regulating cyclical functions as well as having negative influences on thyroid and reproductive functions. Daytime levels **(photophase)** are very low. For example, cessation of melatonin secretion in children can lead to precocial sexual development (see Chapter 11). Some drugs that alter or mimic serotonin and melatonin functions are provided in Table 2-3.

C. Peptides

Similar to most peptides and proteins destined for export from the cell, peptide and protein regulators are synthesized at the ribosome-studded rough endoplasmic reticulum (RER) in the cytoplasm according to directions encoded in nuclear genes. Each nuclear gene is represented by a unique linear portion of a double-stranded molecule of **deoxyribonucleic acid** or **DNA** consisting of alternating sequences of nucleotides called **exons**

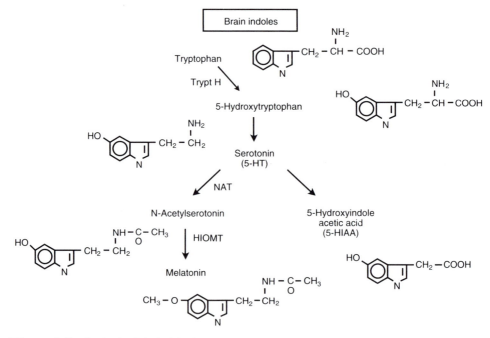

Figure 2-5. Synthesis of the indolamines: serotonin and melatonin. 5-HIAA is a principal metabolite of serotonin in brain tissue. The rate-limiting enzyme for melatonin synthesis is N-acetyltransferase (NAT). TryptH, Tryptophan hydroxylase; HIOMT, hydroxyindole-O-methyltransferase.

Table 2-3. Some Pharmacological Compounds That Alter
Indoleamine Activity

Compound	Action
Methysergide	Antagonist of serotonin at receptor sites
LSD	Antagonist of serotonin at receptor sites
Fenfluramine	Blocks reuptake of serotonin by presynaptic neurons
Harmaline	Inhibitor of monoamine oxidase (MAO), an important enzyme for degradation of indoleamines

and **introns.** During the process of nuclear **transcription,** the sequence of nucleotides in one strand of DNA is rewritten as a sequence of nucleotides in a **ribonucleic acid** or **RNA** molecule.

1. The Transcription Process

Transcription is the process whereby the sequence of nucleotides in one strand of DNA that represents a gene is rewritten into a single strand of complementary RNA nucleotides. DNA consists of two strings of nucleotides that are **complementary;** that is, one strand is a mirror image of the other strand in a special way (Fig. 2-6). The two strands are said to be antiparallel since they run in opposite directions. Each DNA strand consists of nucleotides that are each constructed of a phosphate group, a five-carbon sugar called deoxyribose and one of four bases: adenine, guanine, cytosine, or thymine. When a DNA polymer of nucleotides is formed, the individual nucleotides are joined through the phosphate–sugar backbone. At one end of a DNA strand, there is a free hydroxyl group on the sugar, attached to the number 3′ carbon (3′-carbon), and at the other end of the strand is a free hydroxyl group attached to the 5′-carbon. Two nucleotides are joined by replacing the hydrogen of the hydroxyl on the 5′-carbon with a bond joining a phosphate that links to the 3′-hydroxyl of an adjacent nucleotide.

Adenine of one strand of nucleotides can form hydrogen bonds only with thymine in the other strand and guanine with cytosine. Similarly, the presence of cytosine or thymine in the first strand dictates the nucleotide that must be opposite in the other strand of DNA, hence complementarity. Two complementary strands line up with the 3′ end of one opposite the 5′ end of the other in antiparallel fashion.

New RNA is constructed during transcription as a single strand of nucleotides. Each nucleotide consists of a phosphate group, a five-carbon sugar called ribose and one of four bases: adenine, guanine, cytosine, or uracil. The last base, uracil, replaces thymine in RNAs and is complementary to adenine as was thymine in DNA.

To synthesize RNA, the portion of two strands of DNA representing the gene must separate or unwind. Transcription begins at a special sequence of DNA nucleotides, called a **promoter,** on one of the separated strands. This site is the only place where special RNA-synthesizing enzymes called **RNA polymerases** can attach and assemble a complementary strand of RNA determined by the sequence of nucleotides in that DNA strand. That sequence of RNA nucleotides eventually determines the sequence of amino acids in a polypeptide. Following transcription, certain segments (intron related) of the transcribed RNA are clipped out by special enzymes and the remaining segments (exon related) are joined together to produce a smaller molecule known as **messenger RNA** or **mRNA.**

RNA polymerase does not always have ready access to promoter sites and one or more **transcription factors** may be necessary to expose the promoter to RNA polymerase. Al-

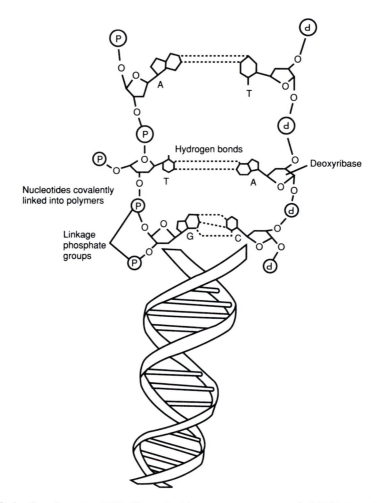

Figure 2-6. Complementary DNA. The nucleotide sequence on one strand of DNA determines the base sequences on the opposite strand.

though a variety of transcription factors have been identified, the predominant type consists of special peptides complexed to zinc ions (Zn^{2+}). The interaction of certain amino acid residues with Zn^{2+} causes the peptide to develop special loops that have been called **zinc fingers** (Fig. 2-7). Interaction of the zinc fingers with the major grooves of DNA unmasks the promoter site and allows transcription to occur. Steroid and thyroid receptors that directly affect gene transcription (see Sections II,D and III,D) have zinc fingers that are highly specific for binding only to certain gene promoters. For example, the estrogen receptor has a sequence of 80 amino acids that form two zinc fingers that bind to a particular DNA sequence, called a **hormone response element (HRE),** allowing only a specific promoter site to be exposed.

2. Translation and Posttranslational Events

Once the clipping and splicing are completed, the mRNA leaves the nucleus and travels to the RER, where the sequence of nucleotides is translated into a linear sequence of amino

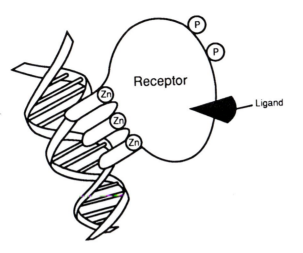

Figure 2-7. Zinc fingers. Certain sequences of amino acids can fold around zinc ions to form projections called "zinc fingers." These zinc fingers are associated with the DNA-binding domains of steroid receptors and facilitate binding to HREs on the DNA. Only a monomer is depicted here.

acids. For products destined for export from the cell, this sequence of amino acids is called a **prepropeptide.** This process is called **translation** and takes place on ribosomes. Prepropeptides have a special sequence of amino acids located at the amino terminal end, which is termed the **signal peptide.** The newly synthesized signal peptide sequence is synthesized first and recognized quickly by a special particle called a **signal recognition particle.** Attachment of the signal recognition particle to the translation complex halts further translation. The signal recognition particle also recognizes a specific **docking protein,** a special receptor embedded in the RER membrane. Thus, the prepropeptide is attached to the RER membrane, the signal recognition particle detaches, and translation is resumed on membrane-bound ribosomes of the RER (Fig. 2-8). While the hydrophobic signal peptide is firmly attached to the membrane, the remainder of the preproptide, called the **propeptide,** is synthesized and intruded through the membrane into the cisterna of the RER. Once in the cisterna, the propeptide can migrate to vesicle-forming regions of the RER and be packaged into vesicles for translocation to the Golgi apparatus.

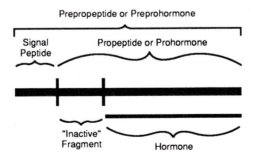

Figure 2-8. Synthesis of export proteins. The product of mRNA at the ribosome is the preprohormone. The signal peptide is necessary to transport the prohormone into the cisterna of the endoplasmic reticulum. The prohormone is later cleaved to produce an "inactive fragment" and the definitive hormone. Typically, both the inactive fragment and the hormone will be released from the cell.

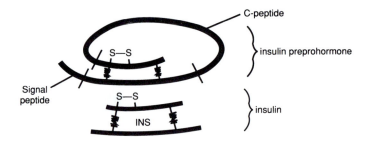

Figure 2-9.   Insulin synthesis. The hormone insulin is synthesized by folding a single peptide chain and
cleaving it in three places (|) to yield a signal fragment, a connecting C-peptide fragment, and
the hormone insulin that now appears to be made of two separate polypeptide chains.

Additional posttranslational processing of the propeptide may occur within the RER, in
the Golgi apparatus, or possibly in the storage granules prior to release from the cell. Typically, enzymes will remove a portion or portions of the propeptide to produce the final
peptide or peptides destined for export. Insulin is produced in this way following removal
of a large segment of the propeptide (Fig. 2-9). Another type of posttranslational event is
the joining of two or more polypeptides to form the definitive product such as occurs in the
preparation of gonadotropins [follicle-stimulating hormone (FSH), luteinizing hormone
(LH)] and thyrotropin (TSH). For certain products, other compounds such as carbohydrates
or lipids may be added as well. Posttranslational processing may involve additions to the
basic prohormone as well as deletions. In the synthesis of glycoprotein hormones (gonadotropins and TSH) in the anterior pituitary, carbohydrates are complexed to two separate
propeptides, which are then joined together to yield the biologically active hormones. In
other cases, there is no posttranslational processing of the translated peptide and the definitive secretory product enters the RER ready for export.

Endocrinologists refer to the precursor forms of a peptide hormone as **preprohormone**
(prepropeptide) and **prohormone** (propeptide). For example, the prohormone for the pancreatic hormone insulin consists of a long polypeptide folded through the formation of
disulfide bonds between cysteine residues located in different parts of the chain. A special
enzyme cleaves off a connecting sequence known as the **C-peptide,** leaving what appears
to be two separate peptides (A-peptide and B-peptide) joined together by disulfide bonds.
This remaining molecule is known as insulin (see Fig. 2-9). Both the C-peptide and insulin
are released from the cell, although no peripheral function is known for the C-peptide.

In another example, enzymatic cleavage occurs at several sites along the prohormone

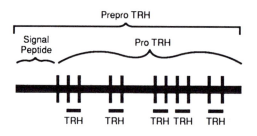

Figure 2-10.   Synthesis of thyrotropin-releasing hormone, TRH. Five copies of the TRH tripeptide are produced from each prohormone by hydrolyzing it at multiple sites.

for the tripeptide, thyrotropin-releasing hormone (TRH) (Fig. 2-10). This results in production of five identical TRH tripeptides from each prohormone molecule, thus amplifying the amount of neurohormone synthesized.

Another variation in posttranslational processing can produce more than one biologically active species from the same preprohormone depending on what processing enzymes are involved in different cell types. For example, a prohormone called **proopiomelanocortin (POMC)** is cleaved to produce the pituitary tropic hormone corticotropin (ACTH) in a corticotropic cell of the anterior pituitary, whereas different processing enzymes in pars intermedia cells of the pituitary release the melanocyte-stimulating hormone (MSH) from proopiomelanocortin (see Chapter 4 for details about proopiomelanocortin and other enzymatic products).

3. Homologies in Peptide and Protein Structure

Analysis of regulatory peptides and proteins shows that it is possible to group molecules with common amino acid sequences into families of related peptides. In many cases, the genes responsible for directing the synthesis of structurally related or homologous peptides are also similar structurally with respect to their nucleotide sequences that ultimately code for these peptides and proteins. Some examples of homologous families are provided in Tables 2-4 and 2-5 and Fig. 2-11. Many argue that the peptides or proteins that represent a

Table 2-4. Families of Peptide Chemical Regulators[a]

Family	Members	Number of AA Residues	Number of AAs common to molecule 1, or other homologous feature
Neurohypophysial octapeptides	1. Arginine vasopressin	9	9
	2. Lysine vasopressin	9	8
	3. Arginine vasotocin	9	8
	4. Mesotocin	9	7
	5. Oxytocin	9	7
	6. Phenypressin	9	8
Glucagon–secretin	1. PACAP-27	27	27
	2. VIP	28	19
	3. Glucagon	29	10
	4. Secretin	27	11
	5. PHI	27	14
	6. GHRH	44	8
	7. GIP	42	6
Insulin-like peptides	1. Insulin ($\alpha + \beta$ chain)	51	51
	2. Relaxin ($\alpha + \beta$ chain)	54	Insulin-like
	3. IGF-I (single chain)	70	24, like insulin
	4. IGF-II (single chain)	67	29, like insulin
Endothelins	1. Endothelin 1	21	21
	2. Endothelin 2	21	18
	3. Endothelin 3	21	15
	4. Sarafotoxin S6b	21	14
CRH-like peptides	1. hCRH	41	41
	2. oCRH	41	34
	3. Sauvagine	39	18
	4. Urotensin I (carp)	41	22

continues

Table 2-4. — *Continued*

Family	Members	Number of AA Residues	Number of AAs common to molecule 1, or other homologous feature
POMC-related peptides			
hACTH-like	1. ACTH	39	
	2. α-MSH		1–13 of ACTH
	3. CLIP		18–39 of ACTH
hLPH-like	1. β-LPH	89	
	2. γ-LPH		1–56 of γ-LPH
	3. β-Endorphin		59–89 of γ-LPH
	4. Dynorphin A		16 AA with 1–4 like β-endorphin
	5. Met-enkephalin		59–63 of β-endorphin and 1–5 of dynorphin
Gastrin–CCK	1. Gastrin-34 (big gastrin)	34	Common amino-terminal pentapeptide sequence
	2. Gastrin-17 (little gastrin)	17	Common amino-terminal pentapeptide sequence
	3. Cholecystokinin (CCK)	33	Common amino-terminal pentapeptide sequence
	4. CCK-8	8	Common amino-terminal pentapeptide sequence
Tachykinins	1. Substance P	11	Common amino-terminal pentapeptide sequence
	2. Neurokinin A	10	Common amino-terminal pentapeptide sequence
	3. Neurokinin B	10	Common amino-terminal pentapeptide sequence
	4. Physalaemin	11	Common amino-terminal pentapeptide sequence
	5. Eledoisin	11	Common amino-terminal pentapeptide sequence
	6. Kassinin	12	Common amino-terminal pentapeptide sequence
Bombesin	1. Bombesin	14	Common amino-terminal hexapeptide sequence
	2. Gastrin-releasing peptide	27	Common amino-terminal hexapeptide sequence
	3. Ranatensin	11	Common amino-terminal hexapeptide sequence
	4. Neuromedin B	32	Common amino-terminal hexapeptide sequence

[a] See text for explanation of abbreviations.

single family had, in an evolutionary sense, a common ancestral gene that, through duplications and subsequent independent, random mutations, evolved via natural selection into a family of closely related genes. In the case of the glycoprotein pituitary hormones that are composed of two peptide subunits each, there has been little change in the gene coding for the α subunit, but the β subunit shows considerable variation related to the different functions of these hormones (see Chapter 4 for more details).

Smaller variations may exist between particular peptides or proteins isolated from different species. For example, the primary structure of pituitary corticotropin (ACTH) iso-

Table 2-5. Families of Human Protein Chemical Regulators

Family	Members	Molecular weight	Common features
Pituitary/placental glycoproteins	1. hTSH	32,000	α Subunit of 89 AA, β subunit, 112 AA
	2. hFSH	32,000	α Subunit of 89 AA, β subunit, 115 AA
	3. hLH	32,000	α Subunit of 89 AA, β subunit, 115 AA
	4. hCG	38,000	Most like LH in structure and action
Nonglycoprotein pituitary/placental hormones	1. hGH	22,650	191 AA
	2. hProlactin	23,510	198 AA
	3. hPL (hCS)	22,000	191 AA
TGF-β superfamily	1. TGF-β1		Composed of α and β subunits
	2. Bone morphogenic proteins (BMPs)		Composed of α and β subunits
	3. Müllerian-inhibiting substance (MIS)		Composed of α and β subunits
	4. Inhibins		Composed of α and β subunits
	5. Activins		Composed of α and β subunits

lated from humans differs by only one amino acid from those produced in sheep and pigs (Fig. 2-12).

The sequence of amino acids gives rise to the primary structure of a peptide or protein.

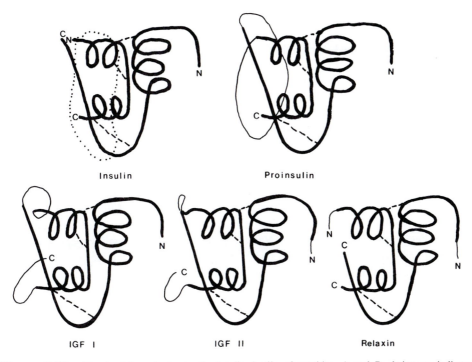

Figure 2-11. Structural homologies in the insulin family of peptides. A and B chains are indicated by heavy lines. Dashed lines represent the disulfide bonds between A and B chains. The C-peptide or extensions of the molecule homologous to parts of the insulin C-peptide are represented by the thin lines. [From Bolander, F. F. (1989). "Molecular Endocrinology," p. 265. Academic Press, San Diego.]

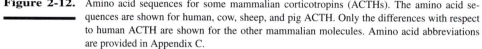

		5	10	15	20	25	30	35	39
Human				S-Y-S-M-E-H-F-R-W-G-K-P-V-G-K-K-R-R-P-V-L-V-Y-P-N-G-A-E-R-E-S-A-E-A-F-P-L-E-F					
Bovine							E	S		
Ovine							Q	S		
Porcine							E	L		

Figure 2-12. Amino acid sequences for some mammalian corticotropins (ACTHs). The amino acid sequences are shown for human, cow, sheep, and pig ACTH. Only the differences with respect to human ACTH are shown for the other mammalian molecules. Amino acid abbreviations are provided in Appendix C.

The secondary and tertiary structures, which are based largely on the primary structure, also may be similar, resulting in considerable overlap in their three-dimensional structures. This may cause overlap with respect to the receptors that bind these similar molecules. Such overlap in binding may yield confusing experimental results, especially when large doses of a regulator are applied. In such cases, the applied regulator normally may not compete effectively for binding to receptors of a chemically related regulator but, by adding excessive amounts, sufficient binding to a related receptor may produce actions that are not physiologically associated with that particular regulator.

4. Transport of Peptide Regulators to Target Cells

Completed peptide and protein regulators along with peptide fragments separated during posttranslational processing are packaged into secretory vesicles or granules at the Golgi apparatus and are retained or stored in the cytoplasm until released from the cell through exocytosis into the interstitial fluids or blood. Regulators released into interstitial fluids (i.e., cytocrines, autocrines, neurotransmitters, neuromodulators) rely on diffusion and any movement of the interstitial fluid that might occur to allow them to reach their target sites more rapidly. In the blood, peptides may complex to specific plasma proteins that enhance their transportability in the blood, slow their rate of metabolism by blood peptidases, and may even facilitate interactions with target cells.

The time for removal or rate of disappearance of any chemical regulator from the blood through degradation, excretion, or removal by target cells is termed its **biological half-life.** This usually is determined by following the rate of removal of radioactively labeled molecules from the blood. The time required to remove or clear half of the labeled dose from the blood is the half-life. In general terms, larger peptides have longer biological half-lives than smaller peptides (see Table 2-6). Notable exceptions are the smallest peptides such as TRH that are modified during posttranslational processing in a manner that slows their enzymatic degradation while in the blood. A number of hormones possess special modifications or activities discussed below that prolong their presence in the blood.

D. Mechanisms of Action for Amine and Peptide Regulators

Although amino acids, amines, peptides, and proteins are soluble in blood plasma and interstitial fluids, they cannot readily pass through cell membranes, which are composed largely of phospholipids and cholesterol. Instead, they bind to **receptor proteins** embedded in the target cell membrane. A target cell for a peptide regulator, by definition, is a cell possessing receptor proteins in its external membrane that specifically bind the regulator.

Table 2-6. Biological Half-Life of Some Mammalian Hormones

Hormone	Size[a]	Species	Approximate half-life (min)
TRH	3	Mice	2
		Human	5
OXY	8	Rat	2
		Human	3
AVP	8	Rat	3–4
α-MSH	13	Dog	2
GnRH	10	Pig	12
		Dog	5
		Human	2–5
Gastrin	17	Human	7–8
ACTH	29	Rat	1–4
		Pig	5–7
		Human	5–29
Glucagon	29	Human	5–10
β-Endorphin	31	Human	15
Calcitonin	32	Rat	2
		Human	3
		Pig	2–5
CCK	33	Human	2
Insulin	51	Pig	6
		Human	3–4
Proinsulin	82	Pig	20
		Human	18–25
GH	200	Human	20–30
PRL	198	Rabbit	16
		Cow	29
LH	32,000[b]	Human	136
FSH	32,000[b]	Human	220
Cortisol	300[b]	Human	90
Aldosterone	300[b]	Human	35
T$_4$	777[b]	Human	7 days
T$_3$	651[b]	Human	24 hr

[a]Number of amino acids or [b]molecular weight; except for glycoproteins, the number of amino acids $\times$ 100 gives an approximate estimate of molecular weight of any peptide (120 $\times$ is more accurate but not as easy to calculate).

A receptor has a three-dimensional **binding site** for the regulator that recognizes the shape of this regulator and will not bind any other molecule unless it closely or exactly conforms to that shape (Fig. 2-13). This relationship is like the lock-and-key fit so often described for an enzyme (the receptor, in this case) and its substrate (the regulator, here). **Ligand** is a general term for any molecule that binds to a receptor-binding site. Once a ligand is bound to a receptor, the receptor changes from an unoccupied to an **occupied receptor.** A responsive target cell typically has about 10,000 receptors for a specific ligand. Many nontarget cells may express a low number of receptors for the same ligand, but are unresponsive for reasons discussed below.

There are several classes of receptor, each of which is specific for different ligands or groups of closely related ligands. For example, insulin, insulin-like growth factor I (IGF-

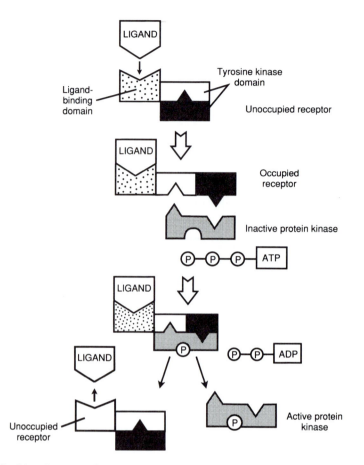

Figure 2-13. Ligand–receptor fit. As occurs for enzymes and substrates, ligands bind to particular domains on receptor molecules and typically cause conformational changes in the receptor that are important for initiating a response in a target cell. This hypothetical illustration imagines a ligand that binds to a receptor with tyrosine kinase activity. Once occupied, the receptor changes shape and now interacts with ATP and a protein kinase. The protein kinase is activated by phosphorylation and now can produce other effects in the cell. The receptor releases its ligand and returns to its unoccupied state.

I), and several other growth factors have similar receptors possessing **tyrosine kinase** enzymatic activity.

Tyrosine kinases are a general class of enzymes that phosphorylates tyrosine residues in proteins. A number of receptors have been characterized as having innate protein kinase activity with the ability to phosphorylate tyrosine residues in other proteins. Among these are the receptors for insulin and a number of growth factors such as insulin-like growth factor I (IGF-I), epidermal growth factor (EGF), colony-stimulating factor (CSF), and platelet-derived growth factor (PDGF). These receptors are called **receptor tyrosine kinases** and are composed of four major types of molecular domains (Fig. 2-14). A **domain** is a region of the molecule to which a particular function may be ascribed. The **extracellular domain** is responsible for binding a specific ligand and is represented by the

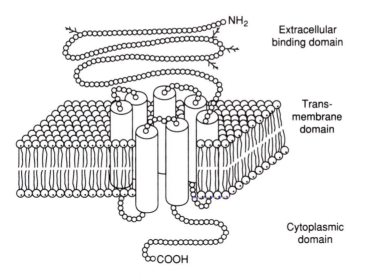

Figure 2-14. Representation of a serpentine transmembrane receptor. The extracellular domain is responsible for binding the specific ligand. The transmembrane domain traverses the membrane seven times before ending in the cytoplasmic domain, which can interact with a G-protein. Occupied receptors often form dimers as part of the mechanism of forming second messengers.

α subunit. This domain is unique for each type of receptor and the ligand it binds preferentially. A second domain, the **transmembrane domain,** crosses the plasma membrane once to connect the extracellular domain to the β subunit that possesses the **tyrosine kinase domain.** The receptor also contains one or more regulatory domains. There is also a juxtamembrane region that separates the **intracellular domain** from the plasma membrane and the transmembrane domain.

Receptor tyrosine kinases for growth factors form dimers after binding their ligand, and experimental studies show that formation of dimers is necessary to activate fully the tyrosine kinase. The receptor tyrosine kinases for insulin and insulin-like growth factors apparently are effectively dimerized prior to addition of ligand.

Studies employing mutant forms of insulin and EGF receptors that fail to undergo internalization when occupied show that internalization is not a requisite for transduction of the external signal into the cell. Once occupied, these mutant receptors exhibit tyrosine kinase activity followed by the typical intracellular events normally associated with that ligand. The receptor tyrosine kinases undergo rapid autophosphorylation, which makes them capable of interacting with other proteins. Phosphorylation of tyrosine residues facilitates binding of other proteins to the receptor tyrosine kinase. There is evidence for interactions of receptor tyrosine kinase with intracellular serine and threonine protein kinases, too, which may carry the message to other intracellular sites.

We said that internalization was not necessarily a requirement for action of receptor tyrosine kinases. However, the activity of these receptors on the membranes of endosomes may be important in maintaining sustained effects of a ligand following internalization of the ligand–receptor complex.

Molecules in the secretin–glucagon family of peptides, which includes vasoactive intes-

Table 2-7. Classification of Receptor Types

I.   Receptors coupled to G proteins[a]
A. G_s protein that produces cAMP
Activated for example by ACTH, GH-RH, TSH, LH, FSH, glucagon, β-adrenergic
agonists
B. G_i protein inhibits cAMP production
Activated for example by SS (GH-RIH), IGF-II, α_2-adrenergic agonists, m_2-muscarinic
agonists
C. G_q protein causes increase in IP_3
Activated for example by V_1-type vasopressin receptors
II. Receptors with enzymatic activity
A. Tyrosine kinases (phosphorylate tyrosine residues in self or in other proteins)
Activated, for example, by insulin, many different growth factors
Many of these are linked to protooncogenes
B. Serine/threonine kinases (phosphorylate serine or threonine residues)
Activated for example by activin, inhibin, TGF-β
C. Guanylyl cyclase (forms cGMP from GTP)
Activated by ANP
III. Receptors coupled to ion channels
A. Directly connected to ion channel:
Glutamate receptors
B. Indirectly connected through second messenger

[a] Some type II and III receptors are also G-protein coupled.

tinal peptide (VIP), pituitary adenylate cyclase-activating polypeptides (PACAP), glucagon, and secretin, all have similar sequences of amino acids and often bind to the same receptor proteins. Growth hormone and prolactin (PRL) from the pituitary have similar molecular structures as do their receptors. Similarly, there may be more than one kind of receptor for a given ligand, of which each differs slightly in its composition and binding characteristics. Thus, different tissues may contain different receptor forms for the same ligand (Table 2-7).

1. Multiple Receptor Types

Each chemical regulator must have a specific receptor associated with its target cell to produce a localized and specific effect. However, there may be more than one kind of receptor for a given regulator resulting in different tissue-specific responses to the same regulator. Cell membrane receptors are often complexes of several proteins. As we learn more about regulator actions, we note that there are small differences in cell membrane receptors for a given regulator in different tissues that influence binding and may even be connected to different second messenger systems. Typically, because of variances in the molecular composition of receptors, there is differential sensitivity of receptors to agonists or antagonists. These features often become incorporated into the name of the receptor type.

For example, there are two types of **cholinergic receptors** for the neurotransmitter **acetylcholine** in heart muscle and skeletal muscle. The cholinergic receptors of heart muscle cells are stimulated by a substance obtained from certain mushrooms called **muscarinic acid,** but those of skeletal muscles are not. In turn, skeletal muscle cholinergic receptors are activated by **nicotine,** whereas heart muscle receptors are not. Consequently, we refer to the heart receptors as **muscarinic cholinergic receptors** and those of skeletal muscle as **nicotinic cholinergic receptors. V_2 receptors** for arginine vasopressin are linked to a different cellular mechanism of action than are V_1 **receptors,** and arginine vasopressin produces markedly different intracellular events in different tissues, depending on the type of receptor present.

In the case of receptors for the catecholamine neurotransmitters norepinephrine and epinephrine, we designate the receptors as **α-receptors** or **β-receptors.** These receptors involve several proteins each and exist in a variety of subforms (for example, β_1 and β_2). Norepinephrine has a higher affinity for the α-receptors, whereas epinephrine binds well to either α- or β-receptors. Biochemists have synthesized specific agonists (mimics) and antagonists (blockers) for each adrenergic receptor and its subtypes (see Table 2-2). Dopamine **D_1-receptors** are distinct from **D_2-receptors** and also are characterized by binding different synthetic agonists and antagonists.

The mechanisms of action for these multiple receptor types may be very different with one type operating through one mechanism and another type through a different mechanism of action. However, the biological meaning of variance in receptor types is not always so obvious. In some cases, we may be looking at evolving receptors moving toward greater selectivity and tissue specificity for ligands. Consequently, we may not be able to assign an "adaptive value" at this time, leading to confusion about what their roles might be. Nevertheless, from the pharmacological point of view, this diversity in receptor types has provided mechanisms through which we can experimentally investigate how regulators interact with cells and produce their dramatic effects and has provided avenues for specific drug therapies of clinical disorders.

2. The Second Messenger Concept

Following binding of the regulator to its receptor, a new set of events is initiated in the target cell, which are specific to that regulator and its receptor. How is this orchestrated? Generally speaking, the chemical regulator can be considered the **first messenger** carrying a signal from the secreting cell to a target cell. The first messenger binds to its receptor and may initiate the synthesis of internal **second-messenger** molecules that carry the signal into the interior of the cell. The message of relatively few molecules of first messenger is thereby amplified as many second-messenger molecules are produced for each first messenger bound to a receptor and for as long as the receptor is occupied by the ligand (Fig. 2-15). The need for this amplification, at least in part, relates to the short half-life of second messengers.

In the 1960s, Earl Sutherland and coworkers discovered the first of several second messengers while attempting to elucidate the mechanism of epinephrine's action on glycogen breakdown in skeletal muscle and liver cells. Epinephrine has its actions on muscle and liver cells by first binding to a cell membrane-bound receptor, just like peptides do. The importance of this discovery was highlighted by the award of a Nobel Prize to Sutherland in 1972. Much later, the discovery of a membrane-bound intermediate **G protein** operating between the first and second messengers led to a second Nobel Prize in 1994 for Alfred Gilman and Martin Robdell. Although the G protein logically could be called a second messenger, this latter term has been retained for the type of cytosolic messenger originally described by Sutherland's group.

Several types of second-messenger systems have been discovered, all of which involve the interaction of an occupied receptor with a specific molecule of a class of membrane-bound proteins called **G proteins,** so named because when activated they can bind and hydrolyze the nucleotide **guanosine triphosphate** or **GTP.** The G proteins are intermedi-

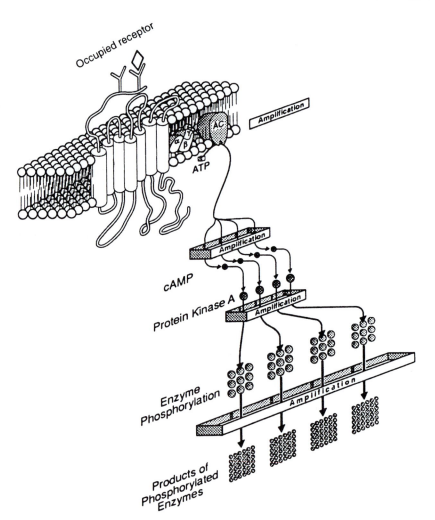

Figure 2-15. Amplification of signal by second messengers. In this example, the occupied receptor interacts with a G_s protein (consisting of α, β, and γ subunits) that activates a critical enzyme, adenylate cyclase (AC), which in turn stimulates production of many cAMP molecules. At each following step there is an amplification of the signal through the increased production of more activated molecules until finally there is a substantial change in the cell initiated in response to the single chemical regulator that originally occupied the receptor. [Modified from R. E. Brown, "An Introduction to Neuroendocrinology." © Cambridge University Press, 1994. Reprinted with permission of Cambridge University Press.]

aries in the production of second messengers. However, these intermediaries were not discovered until after the term "second messenger" had been coined. It is critical that there be enough membrane receptors and G proteins available to amplify the message sufficiently to bring about specific internal changes in the target cell.

Activated G proteins may interact directly with **ion channels** in the cell membrane or with a **signal-generating enzyme** adjacent to or located in the membrane but on the cytosolic side. In the case of an ion channel, its opening or closing can influence entrance or exit of ions, which, if a sufficient number of channels are affected, may in turn alter intracellular or membrane events. Hence, an ion such as calcium (see Section I,D,2,d) that enters the cell and causes specific changes can be called a second messenger. When a signal-generating enzyme is activated, it catalyzes the synthesis of second messengers from a specific substrate, which then initiate intracellular events normally associated with the actions of the first messenger on that cell. Thus, through the second-messenger system, a small amount of chemical regulator can be amplified to hundreds of active molecules that bring about specific changes in a target cell. Because the ligand–receptor complex is not located at a fixed site in the membrane, it can migrate within the membrane and contact more than one G protein, thus possibly amplifying its effect even further.

a. The Cyclic AMP Second-Messenger System

Seconds after epinephrine binds to the target cell, a unique compound is synthesized. The new molecule is a form of adenosine monophosphate (AMP), the end product of the hydrolysis of the energy rich molecule, adenosine triphosphate or ATP. However, it is found to occur in a novel form where the phosphate forms two bonds instead of one to the adenine base (Fig. 2-16). This novel form is called **cyclic AMP (cAMP,** commonly pro-

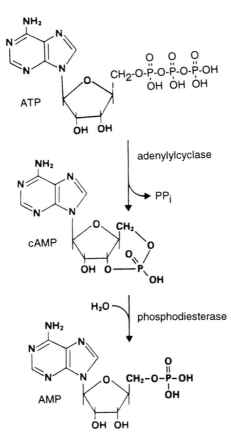

Figure 2-16. Formation and degradation of cAMP. ATP is converted by adenylylcyclase to cAMP. One of several phosphodiesterases (see Table 2-8) inactivates cAMP by converting it to ordinary AMP.

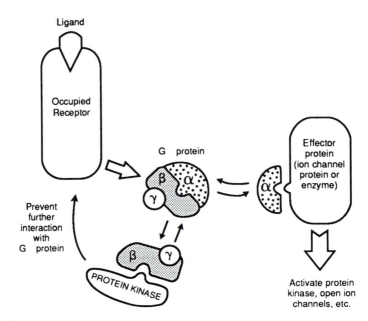

Figure 2-17. G proteins consist of three subunits. The α subunit, which has innate GTP-binding and hydrolyzing capacity, can separate from the other subunits following interaction with an appropriate, occupied receptor. The free α subunit interacts with a membrane channel protein or an enzyme that generates a second messenger. Once the GTP has been hydrolyzed, the subunits recombine. The complex of free β and γ subunits may interact with the occupied receptor and prevent its interaction with intact G proteins as reported for the β-adrenergic receptor (see text).

nounced as the word "camp") and is formed by a cell membrane-associated, signal-generating enzyme called **adenylate cyclase (AC).** Different forms of adenylylcyclase are known that may be stimulated, inhibited, or unaffected by calcium ions. Epinephrine-occupied receptor interacts with a particular G protein called **G$_s$ protein,** which in turn interacts with adenylate cyclase, which then generates many molecules of cAMP (amplification). Additional studies show that administration of cAMP can mimic all the actions of epinephrine, supporting its role as a true intracellular messenger.

When the receptor is unoccupied, the adenylate cyclase is inactive and GDP is bound to the G$_s$ protein. The G$_s$ protein consists of three peptide subunits: α, β, and γ subunits (Fig. 2-17). The occupied receptor interacts with the β and γ subunits of the G$_s$ protein, allowing the dissociation of GDP and the binding of GTP to the α subunit of G$_s$ protein. The α subunit then dissociates from the β and γ units to interact with the catalytic portion of the adenylate cyclase and generate cAMP. Intrinsic GTPase degrading activity of the α subunit converts the GTP back to GDP, reducing its stimulatory effect on the catalytic domain of adenylate cyclase. The α subunit then returns to its association with the other G$_s$ subunits.

The specific cellular response to cAMP as a second messenger depends on the cell type involved. Once generated in a liver or skeletal muscle cell, cAMP can repeatedly combine with **protein kinase A,** a cytosolic enzyme that phosphorylates the inactive enzyme, **phosphorylase kinase,** which in turn converts **phosphorylase b** into its active form, **phosphorylase a.** Active phosphorylase a triggers the enzymatic breakdown of glycogen to provide

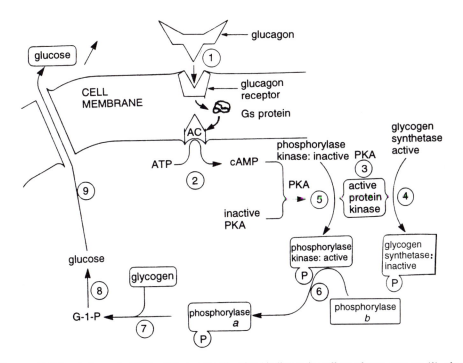

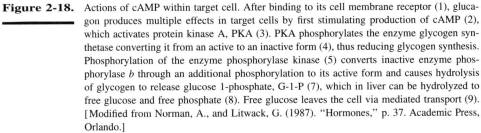

Figure 2-18. Actions of cAMP within target cell. After binding to its cell membrane receptor (1), glucagon produces multiple effects in target cells by first stimulating production of cAMP (2),
which activates protein kinase A, PKA (3). PKA phosphorylates the enzyme glycogen synthetase converting it from an active to an inactive form (4), thus reducing glycogen synthesis.
Phosphorylation of the enzyme phosphorylase kinase (5) converts inactive enzyme phosphorylase *b* through an additional phosphorylation to its active form and causes hydrolysis
of glycogen to release glucose 1-phosphate, G-1-P (7), which in liver can be hydrolyzed to
free glucose and free phosphate (8). Free glucose leaves the cell via mediated transport (9).
[Modified from Norman, A., and Litwack, G. (1987). "Hormones," p. 37. Academic Press,
Orlando.]

glucose for energy production in the target cell (see Fig. 2-18). In an adipose or fat cell, the
cAMP-activated protein kinase activates a **hormone-dependent lipase** that results in hydrolysis of fats. cAMP also may produce effects via other cellular mechanisms associated
with the same or different first messengers. For example, in cardiac muscle cells, cAMP
not only activates protein kinase A but also binds to calcium channels, allowing an influx
of Ca^{2+} that prolongs depolarization of the cell membrane and influences the degree of
cardiac muscle contraction.

Cell membranes often contain another G protein called a **G_i protein** that can inhibit
the cAMP mechanism in the following manner. Apparently, certain regulators can bind
to a receptor that employs the G_i protein which, like the G_s protein, consists of three subunits. The G_i protein has a unique α subunit but its β and γ subunits are very similar to
those of the G_s protein. One hypothesis suggests that occupied receptor allows GTP to
bind the G_i protein, which frees the β and γ subunits to bind up to the G_s protein's α subunit. This prevents the G_s protein's α subunit from combining with the catalytic unit of
adenylate cyclase and prevents cAMP generation. Alternatively, the G_i protein could interact directly with the adenylate cyclase catalytic domain. Several regulators that antagonize

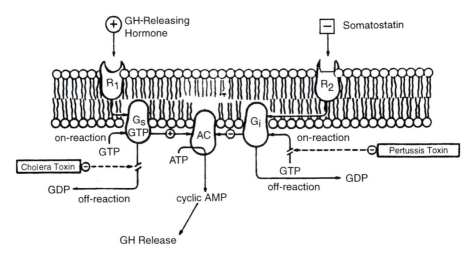

Figure 2-19. G-protein interactions and inhibition of cellular reactions. Growth hormone (GH)-releasing hormone binds to its receptor (R_1) and activates the G_s protein that turns on adenylate cyclase (AC) to synthesize cAMP from ATP. cAMP acts as a second messenger to mediate release of GH. Somatostatin (SS), after binding to the R_2 receptor, works through an inhibitory G_i protein to prevent the activation of AC. Thus, in the presence of SS, it is difficult to stimulate GH release except through addition of exogenous cAMP. A similar mechanism operates in the antagonism of norepinephrine (NE) by acetylcholine (ACh) in cardiac muscle. [Modified from Frohman, L. A., and Jansson, J.-O. (1986). Growth-hormone releasing hormone. *Endocr. Rev.* **7**, 223–253. © The Endocrine Society.]

the actions of cAMP-dependent peptide regulators are known to employ this mechanism (see Fig. 2-19).

Several drugs have proven very useful for studying cAMP second-messenger mechanisms. Adenylylcyclase can be activated directly by a diterpene called **forskolin,** obtained from an Indian medicinal herb, *Coleus forskohli.* This plant had been used with some success for centuries in India to treat heart disease, respiratory ailments, convulsions, and insomnia. **Cholera toxin,** obtained from pathogenic bacteria, causes ADP ribosylation (binding of ADP) of an arginine residue on the α subunit of the G_s protein, blocking its GTPase activity. When GTP is bound to the α subunit, it activates adenylylcyclase, but after conversion of GTP to GDP, the α subunit no longer activates the enzyme. Cholera toxin prevents conversion of α-subunit GTP to GDP, which would prevent dissociation from adenylylcyclase and allows persistent stimulation of cAMP production. It is this action in the digestive tract that leads to continuous diarrhea resulting in excessive water losses and eventually the death of the infected person.

Another bacterial toxin, **pertussis toxin,** causes binding of ADP ribosylation of a cysteine on the α subunit of the inhibitory G_i protein, reducing its affinity for GTP (a different effect than described above for cholera toxin on G_s protein). This prevents the G_i α subunit from interacting normally, causing a slight rise in cAMP levels and preventing the action of regulators that normally work through the G_i protein.

Certain mutations in the α-subunit gene cause a loss of GTPase activity in the resultant G-protein α subunit. This produces a G protein that is always stimulated even in the absence of a first messenger. Such a cell expressing this defective gene would be constantly turned on.

In addition to the cytoplasmic regulatory elements associated with cAMP, a **cAMP regulatory element binding protein (CREB)** has been isolated from the nucleus and identified as a transcription factor that binds to DNA and can regulate gene transcription. This could explain some of the delayed effects of peptide hormones on protein synthesis that have been observed following the initial enzymatic activation through cAMP. Regulatory elements like CREB as well as other transcription factors may be considered "third messengers" (the first contacts the cell, the second mediates cytoplasmic events, and the third mediates nuclear events).

b. Inositol Trisphosphate and Diacylglycerol as Second Messengers

Some receptors such as those for acetylcholine, certain receptors for vasopressin, and for certain eicosanoids interact with yet another G protein, **G_q protein,** that activates an additional signal-generating enzyme complex called **protein kinase C (PKC).** The substrate for phospholipase C is **phosphatidylinositol biphosphate (PIP_2),** which is cleaved to produce *two* second messengers, **inositol trisphosphate (IP_3)** and **diacylglycerol (DAG),** each of which has independent actions.

IP_3 interacts with IP_3 receptors on the external membrane of the endoplasmic reticulum to release stored calcium into the cytosolic compartment of the cell. In muscle cells, these calcium ions bind to special proteins that initiate contraction. In nonmuscle cells, the resulting increase in intracellular calcium usually involves a cytoplasmic calcium-binding peptide called **calmodulin** (Fig. 2-20). Calmodulin can produce a variety of cellular effects depending on the cell type involved. It can activate certain enzymes and also can increase the rate of degradation of cAMP, thus antagonizing the actions of any first messenger, such as epinephrine or norepinephrine, that operates through cAMP as a second messenger. Calmodulin actually combines with and activates the major enzyme necessary for cAMP degradation (see Section I,E,2).

DAG and calcium ions apparently activate the enzyme PKC, which in turn may activate other cytoplasmic enzymes, but all the details of this action are not clear. Furthermore, DAG may serve as a substrate for production of arachidonic acid, a precursor for the synthesis of eicosanoids, chemical regulators themselves (see Section IV).

c. cGMP as a Second Messenger

The role of **cyclic guanosine monophosphate (cGMP)** as a second messenger has not been studied to the same extent as the cAMP system. The signal-generating enzyme **guanylyl cyclase** is activated by **atrial natriuretic peptide (ANP)** and produces cGMP from GTP. The enzyme appears to be an integral part of the receptor itself. Furthermore, ANP lowers cAMP production. Initially, it was observed that several regulators with antagonistic actions to other regulators working through a cAMP mediator caused an elevation in cGMP (e.g., inhibition by acetylcholine of norepinephrine's action in cardiac muscle). However, other work suggests the observed increase in cGMP may be a consequence of other secondary events and cGMP may not directly antagonize cAMP-dependent events. More work is needed to clarify this important issue.

d. Calcium Flux as an Intracellular Messenger

Studies of hormone action have discovered a common theme involving calcium ion cycling across the cell membrane as a potential mechanism for sustaining a cellular response, even hours after the initial binding of the hormone. This has led to development of a model system by Howard Rasmussen and colleagues using mammalian adrenal cortical cells to

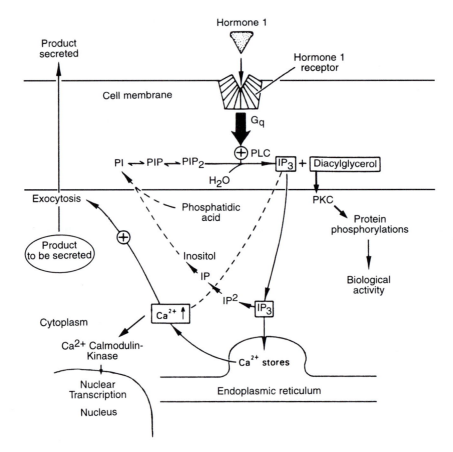

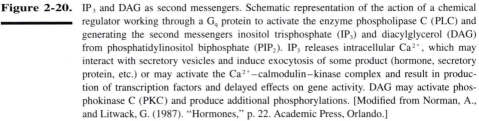

Figure 2-20. IP$_3$ and DAG as second messengers. Schematic representation of the action of a chemical regulator working through a G$_q$ protein to activate the enzyme phospholipase C (PLC) and generating the second messengers inositol trisphosphate (IP$_3$) and diacylglycerol (DAG) from phosphatidylinositol biphosphate (PIP$_2$). IP$_3$ releases intracellular Ca^{2+}, which may interact with secretory vesicles and induce exocytosis of some product (hormone, secretory protein, etc.) or may activate the Ca^{2+}–calmodulin–kinase complex and result in production of transcription factors and delayed effects on gene activity. DAG may activate phosphokinase C (PKC) and produce additional phosphorylations. [Modified from Norman, A., and Litwack, G. (1987). "Hormones," p. 22. Academic Press, Orlando.]

study the involvement of calcium ions in the action of certain regulators. Initial observations of the action of the octapeptide angiotensin II (ANG-II) on the sustained release of the mineralocorticoid aldosterone found only a transient rise in cytosolic calcium following binding of ANG-II to cells of the adrenal cortex, yet a prolonged release of aldosterone was observed. However, it was noted that there was an increase in calcium influx as well as efflux. They discovered that this calcium cycling was caused by a calcium–calmodulin-dependent protein kinase. Furthermore, not only does this enzyme phosphorylate and activate the calcium pump, it also behaves as a plasma membrane-associated transducer that converts the calcium cycling message into a sustained cellular event: release of aldosterone.

When a ligand–receptor complex interacts with the enzyme PKC, the substrate PIP$_2$ is converted to IP$_3$ and DAG (see Section I,C,2). The IP$_3$ interacts with the endoplasmic reticulum causing a transient increase in intracellular calcium. These calcium ions bind to calmodulin, resulting in the activation of PKC, which in the presence of DAG becomes associated with the plasma membrane and stimulates its calcium pump. As long as DAG remains in the plasma membrane, the PKC remains there as well. All of these events lead to a submembrane increase in calcium ions that is believed to facilitate the phosphorylating activities of PKC and to bring about a prolonged response.

Aldosterone is responsible for directing reabsorption of Na^+ from the urine by kidney cells and returning Na^+ to the blood. Appropriate levels of Na^+ and K^+ in blood and extracellular fluids are essential for maintaining normal plasma membrane potentials on cells. Inappropriate release of aldosterone could lead to death caused by neural and muscular dysfunction. This is prevented by the sensitivity of the calcium channels in adrenocortical cells to circulating levels of K^+. High K^+ hyperpolarizes the plasma membrane of the adrenal cell and its voltage-sensitive calcium pumps, so that ANG-II cannot produce sustained calcium cycling. Likewise, lowered K^+ partially depolarizes the calcium channels and makes them more responsive to ANG-II.

A similar mechanism involving calcium cycling across the plasma membrane has been described for the regulation of insulin release from the pancreatic β cell. Opening of the calcium channels is facilitated by high circulating levels of glucose so that more insulin is released to bring about a decrease in blood glucose. An identical system involving calcium cycling, DAG, and PKC has been described for the stimulation of arterial and trachial smooth muscle contraction although the connection between PKC and the final phosphorylated proteins responsible for the contraction is unproven.

E. Turning Off the Response

Once a regulator binds to its receptors, the target cell is stimulated to produce second messengers until the occupied receptor becomes unoccupied again. To terminate the action on the target cell, the regulator must be separated from its receptor and the second messengers also must be inactivated. Several mechanisms have evolved in cells to turn off the response of the target cell once it has been stimulated.

1. Fate of Membrane-Bound Ligands

The earliest clues to the fate of first messenger ligands came from studies of the action of nonpeptide neurotransmitters (acetylcholine and norepinephrine) on their target cells. In the case of acetylcholine, an enzyme called **acetylcholinesterase (AChE),** associated with the postsynaptic cell membrane, was found to degrade acetylcholine to acetic acid and choline, freeing the receptors from ligand until more acetylcholine was released from the presynaptic cell. Other studies showed that the neurotransmitter norepinephrine was either retaken up by the presynaptic neuron where it was recycled or it was taken up by nearby glial cells, which degraded it intracellularly. However, the fate of peptide ligands soon was shown to follow a unique pathway.

The first reports of finding peptide regulators intracellularly in target cells were assumed to be artifacts of the methods employed to demonstrate their presence, since it was well known that they did not enter cells but bound only to the cell membrane. We now know that subsequent to occupation and activation of receptors, occupied receptors with their ligands migrate along the surface of the cell membrane to regions called **clathrin-coated pits.** These pits are membrane regions where the protein clathrin has accumulated and are sites for internalization of occupied receptors through endocytosis or phagocytosis

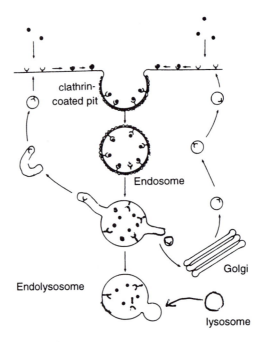

Figure 2-21. Down-regulation of occupied receptors. Occupied receptors migrate along the cell membrane
to clathrin-coated pits, which form endosomes. Fusion of endosomes with lysosomes to form
endolysosomes results in degradation of both ligand and most or all of the receptors. Formation of small vesicles containing some receptors allows for possible recycling of receptors directly to the cell surface or indirectly via the Golgi apparatus. [Modified from
Bolander, F. F. (1989). "Molecular Endocrinology." Academic Press, San Diego.]

(Fig. 2-21). Internalized vesicles of cell membrane or **endosomes** (also called phagosomes)
result from endocytosis, and consist of a small sphere of cell membrane with occupied
receptors extending into its lumen. Lysosomes containing hydrolytic enzymes fuse with
the endosomes to form **endolysosomes** (or phagolysosomes). The hydrolytic enzymes may
degrade both the receptors and the ligands. In some cases, the ligand is destroyed, but some
of the receptors may be recycled back to the cell surface (Fig. 2-21).

   The discovery of this internalization of occupied receptors explains the phenomenon of
down-regulation of receptors observed in most chemical regulatory systems following
arrival of the ligand at its target. Down-regulation refers to the observed reduction in receptor number that typically occurs as evidenced by reduced sensitivity of the target cell
following the initial stimulation by the peptide. A later increase in receptors or an **up-
regulation** of receptors may follow ligand-stimulated protein synthesis. In many developmental systems, the first contact of a ligand with the target cell may result in up-regulation
of receptors and accelerated binding and subsequent actions of the ligand on the target cell.
Once the target cell has responded, down-regulation occurs frequently, so that the cell cannot be restimulated. Up- and down-regulation can be considered mechanisms for altering
the dynamic range of a hormone-sensitive system.

2. Receptor and G-protein Interaction

Phosphorylation of the β-adrenergic receptor prevents it from interacting with the G_s protein. Once activated, the β-subunit and γ-subunit complex of the G_s protein can activate a

Table 2-8. Classification of Mammalian Phosphodiesterases (PDEs)[a]

Class	Unique features of the PDE	Factors that increase PDE activity
1. Calcium-calmodulin-activated PDE (5 forms described)	Higher affinity for cGMP	GnRH, muscarinic cholinergic agonists
2. cGMP-activated PDE	Have allosteric cGMP-binding site	ANP
3. cGMP-inhibited PDE (3 forms described)	cGMP increases can inhibit cAMP action	Insulin, glucagon, dexamethasone[b]
4. cAMP-specific PDE	cAMP is the intracellular moderator	FSH, PGE, TSH, β-adrenergic agonists
5. cGMP-specific PDE (3 forms described)	Found in retina of eye; employs special G protein called transducin	Light

[a] See text for abbreviations.
[b] Dexamethasone is a very potent synthetic glucocorticoid.

protein kinase called **β-adrenergic receptor kinase (βark),** which then allows a cytosolic protein, **β-arrestin,** to bind the phosphorylated receptor. In this state, the occupied receptor is unable to continue its interaction with the G_s protein. We might expect that similar mechanisms for preventing other kinds of occupied receptors from interacting with their G proteins will be discovered.

3. Fate of Second Messengers

Just as it is important to remove the first messengers to allow a cell to stop responding to a regulator, so second messengers also must be degraded. Specific enzymes are present in the cytosol to degrade second messengers. The best known of these enzymes are the **phosphodiesterases** (PDEs), which degrade cyclic nucleotides (see Table 2-8). One such phosphodiesterase has a high affinity for cAMP and rapidly destroys it. Binding of cAMP to protein kinase A and activation of a specific metabolic pathway in a target cell are thus dynamic events governed by the generation of cAMP and its rate of degradation by phosphodiesterase. Certain drugs called **methylxanthines** are potent inhibitors of phosphodiesterase and can be used to potentiate the actions of a regulator by prolonging the half-life of cAMP. One of these drugs, **caffeine,** is used commonly in experimental studies. Another drug, **theophylline,** is administered to augment the natural action of epinephrine in order to alleviate asthmatic symptoms in humans.

IP$_3$ is reduced progressively to inositol by **phosphomonoesterases.** These esterases are inhibited by lithium ions. The successful treatment of certain forms of mental depression with lithium may be related to this ion's actions on IP$_3$ metabolism.

4. Inactivation in the Blood and/or Excretion of Regulators

In addition to the inactivation of peptide regulators by their target cells, many of the smaller peptides are degraded rapidly by peptidase enzymes that are present in the blood. These peptidases attack one end of the peptide and remove amino acids one at a time until the ability of the peptide to be recognized by its receptor is diminished and then lost. Eventually, the entire peptide is demolished. Even larger peptides may be partially degraded while in the blood. Consequently, many peptides of differing numbers of amino acids derived from a secreted regulator may be present in the circulation at the same time, all or only some of which may be fully active. For example, fragments of parathyroid hormone as small as 34 amino acids have the same biological activity as the parent regulator of 84 amino acids, whereas smaller fragments have little or no activity.

F. Effects of Membrane-Bound Regulators on Nuclear Transcription

In addition to the early actions on target cells by regulators that employ membrane-bound receptors, delayed effects on protein synthesis were well known long before we could explain how this effect might be mediated. The physical distance separating the cell membrane from the nucleus is very small by our standards, only about 20 to 30 μm, but this is a huge separation on a molecular scale. Discoveries of specific **transcription factors** in prokaryotic and simple eukaryotic cellular systems paved the way for understanding that similar mechanisms are employed in vertebrates. Transcription factors are cytoplasmic proteins that migrate into the nucleus, bind to the regulatory regions of a gene, and enhance transcription of that gene. Studies of protein kinase actions have led to the development of one hypothesis to link the cell membrane to nuclear events. It is called the **kinase cascade hypothesis** and has evolved from studies of cell division regulation. Cell biologists in many laboratories working on the **mitogenic actions** (i.e., stimulation of cell division) of growth factors such as **epidermal growth factor (EGF)** provided the first connections.

Mitogens, such as EGF and many hormones, can cause hyperplasia of target cells (e.g., the effect of TSH on goiter formation in the thyroid; see Chapter 7). New evidence suggests that mitogens induce a cascade of protein kinase-dependent intracellular events not unlike those described for second messenger systems (Fig. 2-22). The EGF receptor is one of a

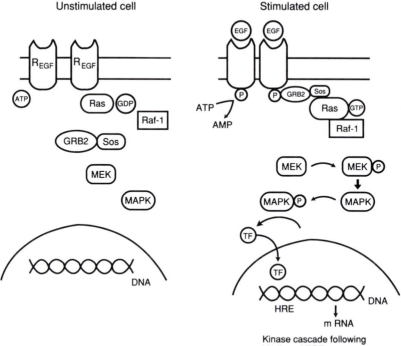

Figure 2-22. The kinase cascade induced by epidermal growth factor (EGF). Occupied receptors interact with a series of protein kinases resulting in production of transcription factors (TF) that as "third messengers" enter the nucleus and alter transcription. See text for abbreviations.

family of membrane receptors that when occupied forms a dimer with tyrosine kinase activity. The occupied receptor dimer phosphorylates itself, allowing it to interact with an intracellular heterodimer consisting of two component proteins: **growth factor receptor-binding protein (GRB2)** and the product of a gene first named the **son of sevenless (SOS)** for its role in regulating eye development in fruit flies. Once the heterodimer is activated by occupied EGF receptor, the SOS component interacts with a specific GTP-binding protein (G protein) called **Ras,** previously identified as a common relay protein for cell growth factors. Ras then exchanges its GDP for a GTP and now binds to a cytoplasmic protein kinase called **Raf-1 protein,** which in turn phosphorylates another kinase termed **MEK.** Once activated, MEK phosphorylates a cytoplasmic complex of enzymes known as **mitogen-activated protein kinases (MAPKs).** Among the many proteins phosphorylated by these MAPKs are certain transcription factors (TF), which are translocated to the nucleus where they bind to DNA and stimulate cell division.

The kinase cascade hypothesis as outlined here is oversimplified. Evidence suggests there may be additional branches that allow alternative routes for the final activation of the MAPKs. For example, occupied insulin receptors can activate this mitogenic pathway, but apparently employ two unique intracellular proteins that interface with the kinase cascade. Nevertheless, the kinase cascade hypothesis provides the first comprehensive explanation of how interactions with a cell membrane receptor can result in nuclear stimulation.

Additional evidence suggests roles for regulators that activate PKC and stimulate transcription by employing members of the Jun–Fos family of transcription factors. Protein kinase A activated by cAMP apparently employs a different family of transcription factors. These observations imply a general role for protein kinases in all the actions of membrane-bound occupied receptors.

II. Steroids

The chemical term "**steroid**" refers to a variety of lipoidal compounds, all of which possess the basic structure of four carbon rings known as the cyclopentanoperhydrophenanthrene ring or **steroid nucleus** (Fig. 2-23). There are many naturally occurring steroids including cholesterol, 1,25-dihydroxycholecalciferol, the adrenocortical steroids or corticosteroids, and the gonadal sex steroids: androgens, estrogens, and progestogens. The most commonly occurring steroids of these categories are listed in Table 2-9.

1,25-Dihydroxycholecalciferol (1,25-DHC) regulates calcium absorption by cells of the intestinal epithelium. It is synthesized through the sequential cooperation of skin, liver, and kidney. 1,25-DHC is especially important in growing children and pregnant women to increase uptake of adequate calcium for normal bone development and growth. The initial step in its synthesis depends on the actions of sunlight on skin (see Section II,B,1). An intermediate of 1,25-DHC called **vitamin D** (the "sunshine vitamin") is commonly added to milk in many north temperate countries to ensure an adequate supply of 1,25-DHC for absorption of the calcium provided in milk. Cod liver oil is another rich source of vitamin D.

Androgens are defined as compounds that stimulate development of male characteristics, that is, they are masculinizing agents. The primary source for circulating androgens in males is the testis where LH from the adenohypophysis stimulates their synthesis and release into the blood. In females, androgens are synthesized by the adrenal cortex and the ovaries. During pregnancy, the fetal adrenal becomes an important source of androgens for the placenta (see Chapters 9 and 11).

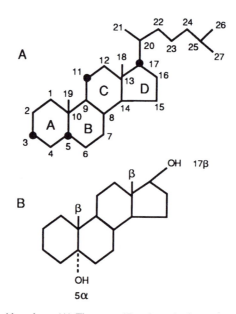

Figure 2-23. The steroid nucleus. (A) There are 17 carbons in the nucleus with two additional carbons (carbons 18 and 19) attached to carbons 13 and 10, respectively. The four rings of the nucleus are labeled A, B, C, and D. The side chain of carbons 20 to 27 is attached to the steroid nucleus at carbon 17 in the β configuration and is indicated as a solid line. Some of the asymmetric carbons of the nucleus are designated as enlarged dots where the lines representing the covalent bonds intersect. (B) Atoms attached to an asymmetric carbon in the α configuration are designated with a dashed line. Those attached in the β configuration (including carbons 18 and 19) are indicated with a solid line.

Estrogens are compounds capable of stimulating cornification in epidermal cells lining the vagina of castrate female rats. Cornification involves deposition of the structural fibrous protein keratin, which is the major protein found in hair. Estrogens also stimulate prolifera-

Table 2-9. Some Common Steroids

Class	Example	Number of Carbons	Site(s) of synthesis
Progestogens	Progesterone	21	Adrenal cortex
			Male: testis
			Female: ovary, adipose tissue
Glucocorticoids	Cortisol	21	Adrenal cortex: zona fasciculata
	Corticosterone	21	Adrenal cortex: zona fasciculata
Mineralocorticoids	Aldosterone	21	Adrenal cortex: zona glomerulosa
Androgens	Testosterone	19	Male: testis
			Female: ovary, adrenal cortex
	Dihydroepiandrosterone (DHEA)	19	Adrenal cortex: zona reticularis
Estrogens	Estradiol	18	Ovary
Vitamin D	Cholecalciferol	25	Skin (in presence of UV light)
	25-Hydroxycholecalciferol	25	Liver (made from cholecalciferol)
	1,25-Dihydroxycholecalciferol	25	Kidney (made from 25-hydroxycholecalciferol)
Cholesterol	Cholesterol	27	Liver, gonads, adrenal cortex

tion and vascularization of the uterine mucosa or endometrium. LH and FSH together stimulate certain ovarian cells to synthesize androgens, convert them into estrogens, and release these into the blood.

A **progestogen** is a compound that maintains pregnancy or the secretory condition of the uterine endometrium during the luteal phase of the ovarian cycle. Progestogens such as pregnenolone and progesterone are produced by all steroidogenic tissues as an intermediate in the synthesis of most of the other steroid hormones. The role of progesterone as a mammalian reproductive hormone, however, is also well established (see Chapter 11).

There are two subcategories of **corticosteroids:** glucocorticoids and mineralocorticoids. **Glucocorticoids** can influence carbohydrate metabolism under certain conditions, and their major physiological role appears to be an influence on peripheral utilization of glucose. **Mineralocorticoids** affect sodium and potassium transport mechanisms in nephrons of the kidney. It must be emphasized that these two terms become rather confusing when applied to nonmammalian vertebrates, because a given molecule that possesses glucocorticoid activity in mammals may have mineralocorticoid activity in nonmammals. Cortisol, for example, is a glucocorticoid in man, but it also is a mineralocorticoid in teleostean fishes. Corticoids are synthesized by the cortical (outer) portions of the adrenal gland or its homologue in nonmammals, the so-called **interrenal tissue,** usually following stimulation by pituitary ACTH (see Chapters 9 and 10).

It should be obvious that these definitions for the various categories of steroid hormones are functionally derived. Although naturally occurring steroids that exhibit a particular hormonal activity in vertebrates (estrogenic, androgenic, etc.) are structurally similar to one another (Fig. 2-24), many structurally unrelated compounds may have similar hormonal activity when tested and could be classified accordingly (see Fig. 2-25). **Diethylstilbestrol** is a powerful estrogen by functional definition and is more potent than any of the naturally occurring estrogens. However, it is not a steroid chemically. The inadvertent, ecological

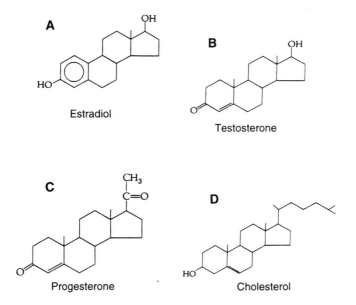

Figure 2-24. Some naturally occurring steroids. (A) Estradiol, an estrogen; (B) testosterone, an androgen; (C) progesterone, a progestogen; (D) cholesterol.

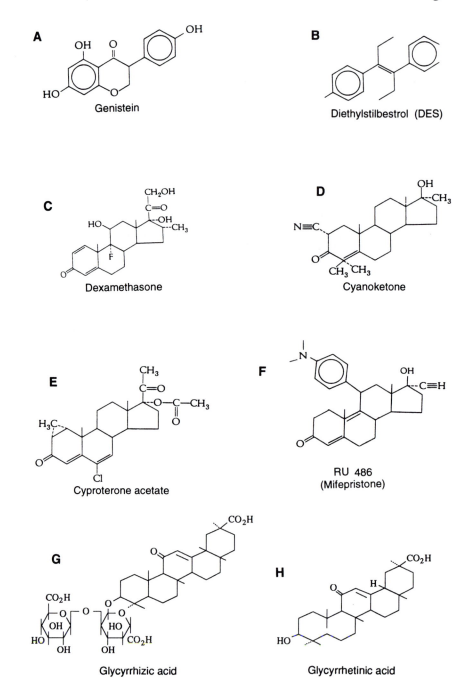

Figure 2-25. Some synthetic steroids and nonsteroids with steroidlike activity. (A) Genistein is a phytoestrogen found in clover and other plants. (B) Diethylstilbestrol is a potent estrogen. (C) Dexamethasone is a synthetic glucocorticoid that is more potent than any of the naturally occurring ones. (D) Cyanoketone is a steroid that inhibits the enzyme that normally converts the steroid pregnenolone to progesterone. (E) Cyproterone acetate is an antiandrogen and blocks androgen binding to receptors. (F) Mifepristine or RU 486 is an antiprogesterone and an antiglucocorticoid. (G) Glycyrrhizic acid and (H) glycyrrhetinic acid are found in licorice and have weak corticosteroid activity.

effects of certain pesticides and other chemicals on vertebrate reproduction may be traceable to estrogen-like effects of these compounds.

A. Steroid Structure and Nomenclature

The literature on steroid hormones is very confusing to the uninitiated, in part because of the multiplicity of trivial and chemical names for a single molecule. Cortisol, for example, in addition to being designated by two different chemical names ($11\beta,17\alpha,21$-trihydroxypregn-4-ene-3,20-dione and $11\beta,17\alpha,21$-trihydroxy-Δ^4-pregnene-3,20-dione) has three trivial names: cortisol, 17-hydroxycorticosterone, and hydrocortisone. Furthermore, cortisol was known as Reichstein's compound M and Kendall's compound F before its chemical structure was elucidated. The common abbreviation for cortisol has become "F" for Kendall's unknown fraction.

To use the established literature, it is necessary to become familiar with the various trivial names as well as to understand the bases for the chemical names. A working knowledge of the chemical nomenclature for steroids is not really necessary for the treatments provided in this textbook, although those students interested in details of synthesis and metabolism of steroids will find knowledge of the chemical nomenclature invaluable. Table 2-10 summarizes special designations used in nomenclature of steroids.

Table 2-10. Summary of Special Designations in Steroid Nomenclature

Designation	Explanation
Δ	Location of double bond
-ene	One double bond in steroid nucleus
-diene	Two double bonds in steroid nucleus
-triene	Three double bonds in steroid nucleus
hydroxy-	Hydroxyl (OH) substituted for hydrogen on nucleus
-ol	Hydroxyl (OH) substituted for hydrogen on nucleus
oxo-	Ketone ($=$O) substituted for hydrogen on nucleus
keto-	Ketone ($=$O) substituted for hydrogen on nucleus
-one	Ketone ($=$O) substituted for hydrogen on nucleus
α	Atom or atoms attached to a given carbon of the steroid nucleus projects away from viewer
β	Atom or atoms attached to a given carbon of the steroid nucleus projects toward viewer
Arabic number	Indicates location of substitution or double bond

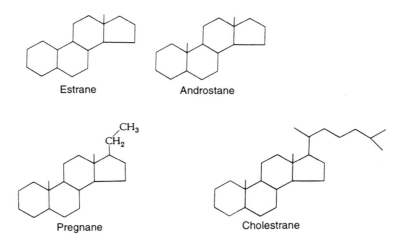

Figure 2-26. Hypothetical steroids employed in steroid nomenclature. These compounds do not exist and are used only for the purposes of constructing the chemical names for the four major groups of steroid hormones.

The steroid nucleus contains 17 carbons, each designated by a number and arranged into four rings (Fig. 2-23). One scheme for naming steroids chemically relates each steroid to a saturated hypothetical parent compound (i.e., one that has no double bonds present) and modifies this name with one or more prefixes and no more than one suffix to designate the specific compound (Fig. 2-26). These hypothetical parent compounds are **estrane** (18 carbons or C_{18}), **androstane** (C_{19}), **pregnane** (C_{21}), and **cholestrane** (C_{27}). Naturally occurring estrogens (all C_{18} compounds) and androgens (C_{19} compounds) possess the basic carbon skeleton of estrane and androstane, respectively, whereas the names for corticoids and progestogens (both C_{21} compounds) are related to pregnane for naming purposes. Cholestrane is employed for cholesterol and the vitamin D compounds.

1. Presence of Double Bonds

The position of double bonds between carbon atoms within the steroid nucleus was formerly designated by the greek letter delta (symbol, Δ) followed by the superscripted number of the lowest numbered steroid carbon with which the bond is involved. Thus, Δ^4 would indicate a double bond between carbons 4 and 5 in the steroid nucleus. Currently, a separate scheme employs the term **ene** to refer to one double bond, **diene** for two double bonds, and **triene** for three double bonds in the nucleus, and the location of each bond is indicated by the number of the carbon preceding the location of the double bond (4-pregnene). Nevertheless, much of our steroid-related terminology still employs the older Δ terminology and this carries over into the naming of certain synthetic pathways (see Section II,B,2,a).

2. Common Substitutions

Substituted groups applied to the steroid skeleton are designated according to the number of the carbon atom in the steroid skeleton to which they are attached. For example, 17-hydroxy refers to the hydroxyl group (–OH) attached to carbon number 17. Hydroxyl groups are usually indicated as prefixes on the chemical name unless they are the only substitution on the parent molecule, in which case the hydroxyl group is designated by the suffix **-ol.** Ketone groups (=O) substituted on the steroid nucleus are indicated by the prefix **oxo-** or **keto-** or by the suffix **-one;** thus, -3,20-dione would refer to two ketone groups attached at positions 3 and 20, respectively. In naming steroids with both hydroxyl and keto groups, the latter takes priority as the suffix (-one) over the hydroxyl (-ol).

3. Stereoisomerism

The carbon atom is capable of forming four covalent bonds. Only two of these bonds are used when a given carbon is incorporated into the steroid nucleus. Because of the necessary bonding angles dictated by the tetrahedral shape of the carbon atom, a steroid does not exist in a flat plane, as most diagrams of their chemical structures would suggest. The remaining two sites for each carbon atom incorporated into the steroid skeleton can form covalent bonds with hydrogens, oxygens, or other carbon atoms. Depending on how and where the carbon atom appears in the nucleus, it may be able to bind hydrogen or oxygen to either side. One of these bonds will project toward the viewer when examining the steroid, as depicted in Fig. 2-23, and the other will project away from the viewer, that is, into the page. Such carbon atoms are referred to as **asymmetric carbons.** There are important asymmetric carbons at positions 3, 5, 11, and 17 in the steroid nucleus.

Two three-dimensional isomers (**stereoisomers**) can be formed by substituting one hydrogen on a given asymmetric carbon with a hydroxyl or other substitution. The chemical formulas would be the same (i.e., the same number of C, H, and O atoms), but the three-dimensional structure of each stereoisomer would differ according to whether the added –OH group projected out from or into the page. A spatial designation of α is used if the substituted group projects away from the viewer, and β is used if the group projects toward the viewer. In estrogens, which have an aromatic A-ring in the steroid nucleus, the carbon at position 3 can be neither α or β. In C_{21} steroids, carbon 20 is connected in the β configuration to carbon 17. Since any other attachment to 17 must be in the α position, it is not necessary to indicate this in the chemical name.

In two-dimensional diagrams the α position is indicated by a slashed or dotted line, whereas the β position is designated with a solid line connecting the substituted group to its carbon (Fig. 2-23). Hydrogen atoms usually are not designated, but their presence is implied. Generally speaking, only one of the two possible stereoisomers associated with substituted groups on the steroid nucleus will exhibit biological activity. Estradiol-17β, for example, is the most potent naturally occurring estrogen in vertebrates, whereas estradiol-17α has little or no estrogenic activity. Henceforth, we shall use "estradiol" to designate estradiol-17β. Notice how the trivial name "estradiol" has been embellished with details derived from its chemical name (diol refers to two OH groups, one of which is on carbon 17, an asymmetric carbon, and in the β conformation). The official chemical name for estradiol is 1,3,5-estratriene-3,17β-diol.

The chemical structures of several steroids are diagrammed in Figs. 2-27 and 2-28, with both chemical and trivial names for some of these steroids indicated in Table 2-11. Can you determine the bases for the chemical name assigned to each steroid?

B. Steroid Synthesis

All vertebrate steroid regulators are synthesized from **cholesterol,** a C_{27} steroid (Fig. 2-27). Cholesterol is synthesized from acetate (acetyl-coenzyme A) produced via glycolysis or via

A Corticosteroids

B Progestogens

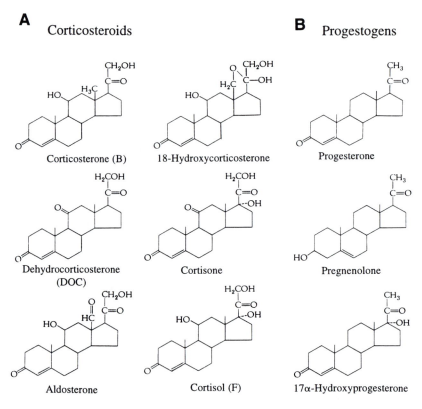

Figure 2-27. Some corticosteroids (A) and progestogens (B).

fatty acid oxidation. The synthesis of the steroid nucleus from acetate units is termed **steroidogenesis** and involves a series of enzymatically catalyzed reactions. Following synthesis of a relatively long hydrocarbon chain, a complex cyclization step results in closure of the carbon skeleton into the steroid nucleus. Cholesterol synthesized in this manner may be used directly in the biosynthesis of the various steroid hormones but cannot be degraded back toward acetate.

Most cholesterol is synthesized in the liver and is released into the blood as lipid droplets coated with protein (for details, see Chapter 14). Adrenal cortex, ovaries, and testes also can synthesize cholesterol but more commonly utilize lipoprotein complexes absorbed from the gut or synthesized in the liver as sources of cholesterol. The pathway for forming 1,25-DHC differs markedly from that for the gonadal and adrenal steroids, and will be described separately.

Cholesterol and the steroid hormones are synthesized primarily in the liver (cholesterol), the gonads (estrogens, androgens, progesterone), the placenta (estrogens, progesterone), and the adrenal cortex (corticosteroids, androgens). Vitamin D compounds are made by the cooperative activities of skin, liver, and kidney. Although we have long known that the brain is a target for steroids and that some of these steroids may be converted by neural enzymes to more active forms for binding to receptors, we must now include the brain as a steroid-

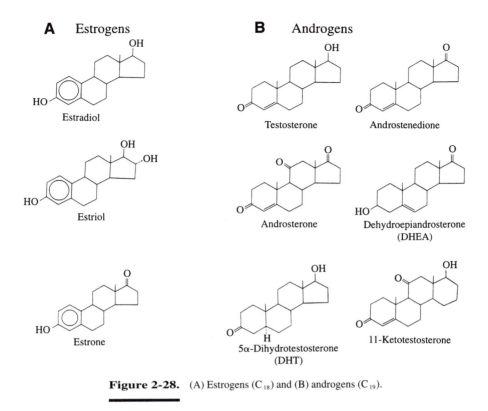

A Estrogens

Estradiol

Estriol

Estrone

B Androgens

Testosterone

Androstenedione

Androsterone

Dehydroepiandrosterone
(DHEA)

5α-Dihydrotestosterone
(DHT)

11-Ketotestosterone

Figure 2-28.   (A) Estrogens (C_{18}) and (B) androgens (C_{19}).

synthesizing tissue. In recent years, it has been demonstrated that steroids, especially progesterone, are synthesized within glial cells of the central and peripheral nervous systems and play important roles in the myelination process. These endogenous steroids have been dubbed **neurosteroids** to emphasize their site of origin and distinguish them from steroids accumulated from the blood or cerebrospinal fluid.

Table 2-11.   Some Vertebrate Steroid Hormones

Category	Trivial name	Chemical name
Androgens	Testosterone	17β-Hydroxy-4-androsten-3-one
	Androstenedione	4-Androstene-3,17-dione
	Dehydroepiandrosterone	3β-Hydroxy-5-androsten-17-one
Corticoids	Aldosterone	11β,21-Dihydroxy-3,20-dioxo-4-pregnen-18-al
	Cortisol	11β,17,21-Trihydroxy-4-pregnene-3,20-dione
	Corticosterone	11β,21-Dihydroxy-4-pregnene-3,20-dione
	11-Deoxycorticosterone	21-Hydroxy-4-pregnene-3,20-dione
Estrogens	Estradiol-17β	1,3,5(10)-Estratriene-3,17β-diol
	Estrone	3-Hydroxy-1,3,5(10)-estratrien-17-one
	Estriol	1,3,5(10)-Estratriene-3,16α,17β-triol
Progestogens	Pregnenolone	3β-Hydroxy-5-pregnen-20-one
	Progesterone	4-Pregnene-3,20-dione

1. Biosynthesis of 1,25-DHC

The common name "vitamin D" often is applied loosely to all of the intermediates between cholesterol and 1,25-DHC. In the presence of ultraviolet light, cholesterol is modified in the skin to **cholecalciferol,** which is also a C_{27} steroid (see Fig. 2-29), hence the designation of vitamin D as the "sunshine vitamin." Cholecalciferol is released into the blood, from which it is removed by the liver, converted into **25-hydroxycholecalciferol** by the addition of one –OH group, and returned to the blood. Next, the kidney removes 25-hydroxycholecalciferol from the blood and adds an additional –OH to produce 1,25-DHC and releases it into the blood. 1,25-DHC enters certain cells of the intestinal mucosa where

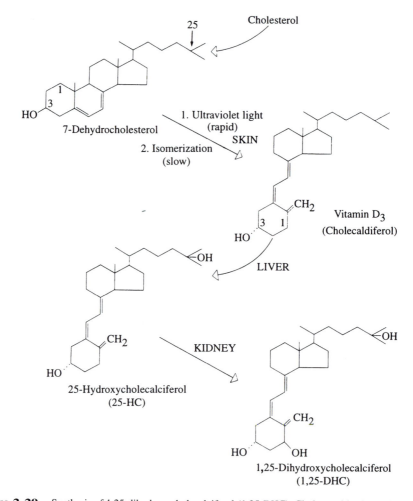

Figure 2-29. Synthesis of 1,25-dihydroxycholecalciferol (1,25-DHC). Cholesterol is changed sequentially by the skin, liver, and kidney. Note the rearrangement of the A-ring as a consequence of opening up the B-ring. The kidney also converts a small amount of 25-HC to 24,25-DHC, which has much weaker biological activity than 1,25-DHC.

it stimulates synthesis of special proteins responsible for calcium uptake from the intestinal contents and transport into the blood. Its mechanism of action is similar to that described for other steroids (see Section II,E).

2. Biosynthesis of Gonadal and Adrenal Steroids

The first series of steps in the biosynthesis of these steroid hormones from cholesterol involves **side-chain hydrolysis** to yield a C_{21} intermediate, **pregnenolone.** In the gonads, pregnenolone may be converted to progesterone (also C_{21}) or altered further to C_{19} androgens that in turn may be modified to C_{18} estrogens. In the adrenal, pregnenolone also may be converted to progesterone (C_{21}) or used directly to form corticosteroids (C_{21}) and some androgens.

A summary of basic pathways for C_{18}-, C_{19}-, and C_{21}-steroid hormone biosynthesis is provided in Figs. 2-30–2-32. After these schemes are examined in some detail, it should not be surprising to learn that many steroidogenic tissues produce more than one class of steroid molecule. The mere presence of a given steroid in a tissue or the ability of a tissue to synthesize a particular steroid does not distinguish between a precursor role of that steroid for synthesis of another steroid and the possibility that the given steroid is secreted into the blood as a hormone. Furthermore, the administration of certain steroids to an animal might lead to their use as substrates and conversion to other steroids. Thus, the observed effect might be due to this conversion product and not to innate activity of the applied steroids. For example, androgens may be converted to estrogens by a number of tissues.

a. Key Enzymes in Gonadal and Adrenal Steroid Biosynthesis

Identification of certain key synthetic enzymes and quantitation of their activity levels are often used as indicators of the biosynthesis of steroid hormones. One key step following side-chain hydrolysis is the conversion of pregnenolone to progesterone, which involves moving the double bond from the B-ring to the A-ring (change from Δ^5 to Δ^4) and converting the β-OH group on carbon 3 to a ketone (=O; see Fig. 2-31). In steroidogenic tissue, catalytic activity of the converting enzyme, **3β-hydroxy-Δ^5-steroid dehydrogenase (3β-HSD),** can be approximated histochemically in frozen sections, thus providing a crude measure of steroid hormone synthesis. Estimation of the rate of steroid hormone synthesis also infers a measure of the rate of release, because steroid hormones are not stored to any appreciable degree in steroidogenic cells and apparently are released into the circulation as they are synthesized.

The enzyme 3β-HSD can work on a variety of steroid substrates. If 3β-HSD acts early in the sequence by converting Δ^5-pregnenolone to Δ^4-progesterone, the subsequent enzymatic transformations of progesterone are referred to as the Δ^4-**pathway.** When pregnenolone is not first converted to Δ^4-progesterone, it is enzymatically transformed along the Δ^5-**pathway.** The significance of these pathways in adrenal, testicular, and ovarian steroid syntheses is discussed in Chapters 9 and 11.

The drug **cyanoketone** is known to competitively inhibit the activity of 3β-HSD because of the structural similarity of cyanoketone (Fig. 2-25) to pregnenolone (Fig. 2-30) and its ability to block access of pregnenolone to the enzyme's binding site. Cyanoketone effectively blocks gonadal steroid biosynthesis and reduces circulating levels of all gonadal steroids. Reduction in circulating steroids reduces negative feedback and causes

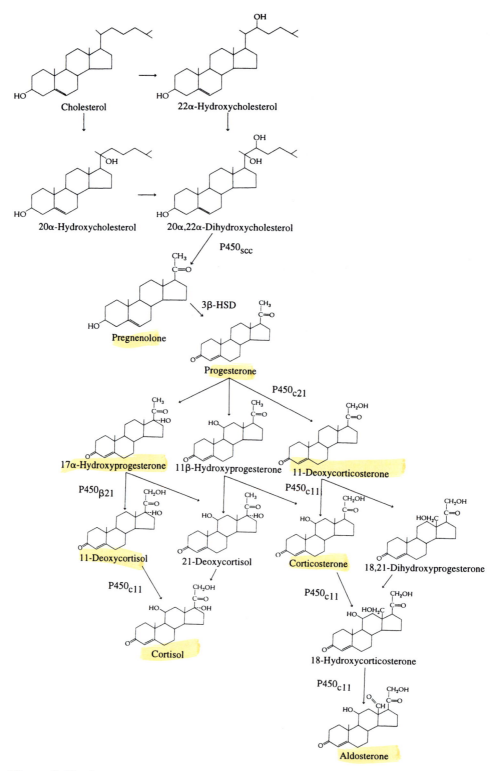

Figure 2-30. Progesterone and corticosteroid synthesis. Cortisol, corticosterone, and aldosterone are the major secretory products with hormonal activity. Identification of the key *P*-450 enzymes can be found in Table 2-12.

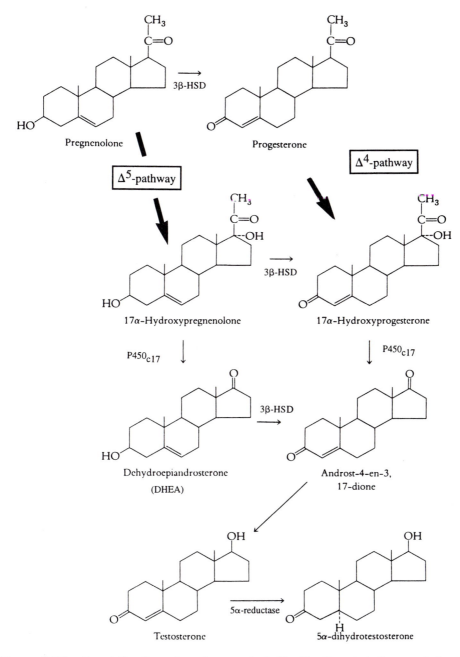

Figure 2-31. Δ^4- and Δ^5-pathways for androgen synthesis. The Δ^5-pathway typically occurs in the adrenal cortex and usually stops with the production of DHEA or DHEAS. In the ovaries of some species, this pathway may lead to testosterone and eventually to estrogen synthesis. Testes employ the Δ^4-pathway.

stimulation of gonadotropes (cells that synthesize the gonadotropins, FSH and LH) in the pituitary, and cyanoketone has proved useful in cytological identification of gonadotropic cells in many species (see Chapter 4).

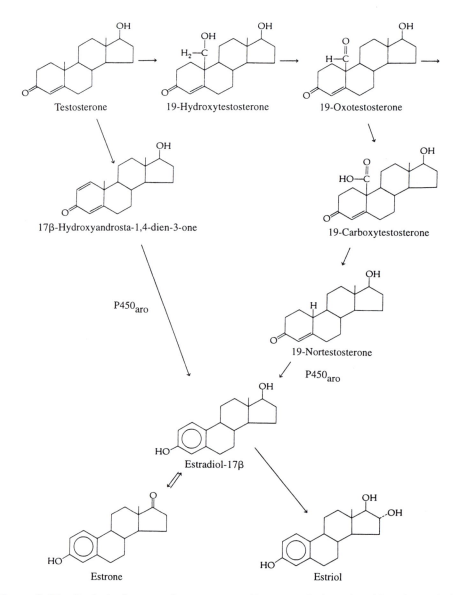

Figure 2-32. Synthesis of estrogens from testosterone. Estrogen synthesis requires either prior synthesis of an androgen or an external source of androgen.

Corticosteroidogenic tissue contains a mitochondrial enzyme, **11β-hydroxylase,** that applies a third hydroxyl group to form cortisol and corticosterone. The activity of this enzyme can be estimated with procedures similar to those described for 3β-HSD. The drug **metyrapone** selectively inhibits 11β-hydroxylase, thus blocking steroidogenesis in adrenocortical cells. Like the case described for cyanoketone, metyrapone has been used successfully to identify the corticotropic cells in the pituitary of many vertebrates.

The basic androgen of most vertebrates is testosterone. The enzyme **5α-reductase** converts testosterone into **5α-dihydrotestosterone (DHT),** an important androgen for development of the penis and scrotum in males. Furthermore, testosterone is converted by the enzyme complex known as **aromatase,** which results in loss of one carbon to yield **estradiol,** a C_{18} estrogen with an aromatic A-ring (Fig. 2-28). The presence of aromatase activity is an indicator of estrogen synthesis.

b. Cytological Aspects of Gonadal and Adrenal Steroid Biosynthesis

Biosynthesis of steroid hormones is, cytologically speaking, a very complex affair involving mitochondrial events as well as the smooth endoplasmic reticulum (SER). The importance of the latter is demonstrated by observations that a major cytological feature of steroidogenic cells is the abundance of SER.

The synthesis of the corticosteroid cortisol illustrates one sequence of cellular events in steroid hormone biosynthesis. Pregnenolone is synthesized in the mitochondria from cholesterol by the action of a **side chain-cleaving enzyme (SCC).** The alteration of pregnenolone to progesterone, however, necessitates that pregnenolone travel for this conversion step to the SER where 3βHSD is located. The addition of hydroxyl groups at the 17 and 20 positions also occurs in the SER, resulting in 11-deoxycortisol. This compound returns to the mitochondria where the 11β-hydrolylase is located that converts 11-deoxycortisol to cortisol. Progesterone conversion to androgens and estrogens requires reentry of progesterone into the mitochondria.

The presence of steroidogenic enzymes in mitochondria led to their recognition as being members of the *P*-450 cytochrome family of proteins (Table 2-12), and this fact is reflected in the abbreviations used for some of the enzymes. For example, the side-chain cleaving enzyme and 11β-hydroxylase are often designated as P-450_{SCC} and P-450_{c11}, respectively.

For many years, it was assumed that cholesterol entered the mitochondrion by simple diffusion. Recent discoveries have identified a **steroidogenic acute regulating protein (StAR)** that facilitates the transfer of cholesterol from the outer mitochondrial membrane to the inner membrane where the SCC enzyme is located. Patients who are unable to con-

Table 2-12. Some Cytochrome *P*-450 Enzymes Involved with Steroid Hormone Synthesis[a]

Enzyme	Cell location	Organ location	Reactions catalyzed
P-450_{scc}	Mitochondria	Adrenal, gonad, liver	Cleaves cholesterol side chain to yield C_{21} steroid, pregnenolone
P-450_{c11}	Mitochondria	Adrenal	Performs 11-hydroxylase, 18-hydroxylase, and 18-methyloxidase reactions; necessary for glucocorticoid and mineralocorticoid synthesis
P-450_{c17}	SER	Adrenal, gonad	Performs 17α-hydroxylase and 17,20-lyase reactions necessary for androgen and cortisol synthesis
P-450_{c21}	SER	Adrenal	Performs 21-hydroxylation necessary for synthesis of cortisol
P-450_{aro}	SER	Gonad	Performs aromatization of A-ring necessary for conversion of testosterone to estradiol

[a] These are oxidative, heme-containing enzymes that show a characteristic shift in absorbance peak from 420 to 450 nm following exposure to carbon monoxide. Most *P*-450 enzymes occur in the smooth endoplasmic reticulum (SER) and often metabolize multiple substrates. The dehydrogenase enzymes that also are important in steroidogenesis are not *P*-450 enzymes.

vert cholesterol to corticosteroids but who have demonstrable SCC enzymes provided the clue to the existence of this protein. These observations suggest that other movements of steroids in and out of cellular organelles also may depend on specific protein interactions.

C. Transport of Steroid Hormones in Blood

Steroids are nonpolar compounds and consequently are not very soluble in aqueous solutions such as blood. Furthermore, free steroids readily pass through cellular membranes and rapidly disappear from the blood and may be destroyed through the activities of the liver and kidneys. The association of circulating steroids with plasma proteins results in their being retained at higher concentrations for longer times in the circulation. These **plasma-binding proteins** reduce removal of active steroid hormones by the liver or kidney and their excretion via the urine. Thus, relatively high titers of steroid hormones can be maintained in the circulation, providing maximal local titers of dissociated free steroid and increasing the probability of their entering appropriate target tissues. As mentioned earlier, plasma-binding proteins may facilitate the entrance of steroids into their target cells.

D. Mechanisms of Steroid Action

Steroids can readily pass through the phospholipid–cholesterol cell membranes to interact with intracellular receptors. These intracellular receptors generally are cytosolic in target cells for corticosteroids and nuclear for the estrogens and androgens. The location of unoccupied progesterone receptors is not confirmed. The original model for steroid hormone action was proposed for estrogens by Elwood Jensen in the early 1970s. Although the model has undergone considerable revision since that time, the basic features are still valid.

Estrogens and androgens are thought to produce their cellular effects through similar mechanisms (Fig. 2-33). Once these steroids enter their target cells, they diffuse into the nucleus where they bind to specific protein receptors. The occupied nuclear receptors behave like transcription factors, bind to specific **hormone-responsive elements (HREs)** or sites on DNA, and initiate new mRNA synthesis. Nuclear adapter proteins may aid binding of occupied receptors to HREs. It is generally agreed that once a receptor is occupied, it becomes phosphorylated and forms a dimer prior to binding to an HRE associated with the promoter region of the gene. A second phosphorylation occurs after binding to the DNA. The resultant synthesis of new proteins by the target cells brings about the events classically associated with the actions of these hormones. Typically, there is considerable delay between contact of the steroid regulator with its target cell and the manifestation of its effects rather than almost immediate changes as described earlier for second messenger systems.

In some androgen-specific target cells, testosterone is first altered chemically in the cytoplasm before it migrates to the nucleus and binds to a receptor. Testosterone may be converted by 5α-reductase to DHT in certain target cells (e.g., in brain, prostate) or by aromatase to estradiol (in brain). Thus, unoccupied nuclear receptors in these cells are not specific for testosterone but for one of its metabolites.

In contrast to estrogen and androgen steroid receptors, the receptors for corticosteroids are definitely cytoplasmic proteins. Whether progesterone receptors are cytoplasmic or nuclear is still being debated, but they have many properties in common with the corticosteroid receptors.

Studies show the glucocorticoid (type 2 corticosteroid receptor), like progesterone, receptors are a complex of several proteins. One protein, the receptor proper, binds the glu-

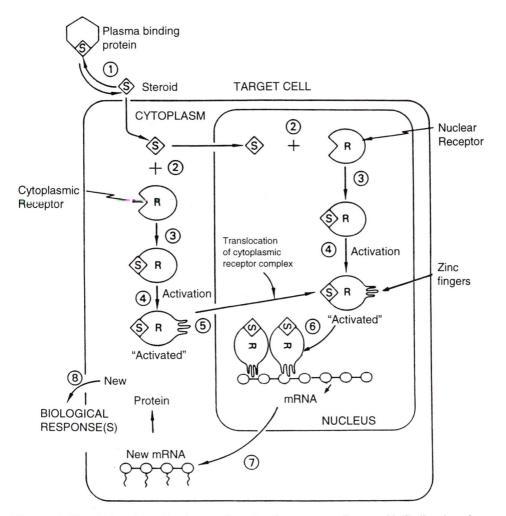

Figure 2-33. Action of steroids via cytosolic and nuclear receptors. Free steroid (S) dissociates from plasma-binding proteins (1), enters the cell, and either binds to a cytosolic or a nuclear receptor, R (2,3). The occupied receptors typically are activated (4) and dimerize (not shown). Cytosolic receptors are translocated to the nucleus (5) and occupied receptor dimers bind via zinc fingers to hormone response elements (HREs) on the DNA (6), facilitating transcription. mRNA leaves the nucleus (7) and directs new protein synthesis associated with the steroid's specific action on that target cell (8). [Modified from Norman, A., and Litwack, G. (1987). "Hormones," p. 37. Academic Press, Orlando.]

cocorticoid. In addition, the unoccupied receptor protein is associated with several other proteins including **heatshock proteins** (hsp56, hsp70, and hsp90; Fig. 2-34). Heatshock proteins were first described in the fruit fly *Drosophila* about 20 years ago and named because they were synthesized in greater amounts following a sudden increase in temperature. Some of the other auxiliary proteins are known as **molecular chaperones** because they are involved in the folding process following translation of the receptor protein.

Various molecular chaperones are associated with each of the steroid receptors (Table 2-13). Hsp90 is found weakly associated with estrogen receptors and is not associated with

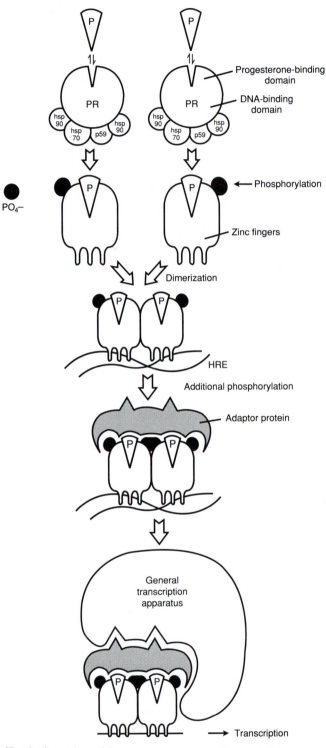

Figure 2-34. Heatshock proteins and the progesterone receptor. Unoccupied progesterone receptor is associated with several molecular chaperones including several heatshock proteins (see also

Table 2-13.   Molecular Chaperones and Steroid Receptors[a]

Proteins	Androgen receptor	Estrogen receptor	Mineralo-corticoid receptor (type 1)	Glucocorticoid receptor (type 2)	Progesterone receptor
hsp90	+	+	+	+	+
hsp70		+		+	+
p60		+		+	+
p48					+
p23		+		+	+
Immunophilins					
FKBP 52 (hsp 56)	+	+		+	+
FKBP 54		+			+
CyP40		+			+

[a]These chaperone proteins facilitate protein folding during synthesis and are retained in association with the receptor protein and are essential for certain aspects of their function. Heatshock proteins (hsp) are proteins whose synthesis is induced by a mild heat shock. Immunophilins are proteins that bind certain immunosuppressant drugs. One group of immunophilins, the cyclosporins (CyP), binds the drug cyclosporin A. Another group binds the immunosuppressant drug FK506 (FK-binding proteins, FKBP). One of the FKBP proteins is also an hsp. The other proteins have not been given special names. In all cases, the number refers to the molecular mass of the protein in kilodaltons.

1,25-DHC receptors. Both of these receptors are considered to reside only in the nucleus. In all cases, association of heatshock proteins with the receptor prevents the steroid–receptor complex from binding with DNA, but binding of the specific ligand causes dissociation of hsp90 from the receptor and allows the ligand–receptor complex to bind to DNA. The roles for the various auxiliary proteins are not understood, but may be involved in translocation of occupied corticosteroid receptor into the nucleus. In the case of glucocorticoid receptor, this complex has been termed a **transportosome** because certain components actually are involved in migration of the ligand–receptor complex into the nucleus.

Numerous steroid actions are very rapid such as the effects of androgens on behavior or of progesterone on oocyte maturation. Researchers began to suspect that not all steroid–target interactions involve intracellular receptors and that there might be immediate membrane-mediated effects. Evidence for membrane receptors for corticosteroids in the mammalian brain and for progesterone in the amphibian oocyte has been with us for some time, but only recently have the first cell membrane-bound steroid receptors been isolated and characterized. In 1991, Miles Orchinik and Frank Moore of Oregon State University isolated what appears to be a cell membrane-bound corticosterone receptor from the newt brain. Since then, David Crews and associates at the University of Texas have reported that large proportions of the estrogen receptors located in the rat brain are surface membrane receptors, too. Presumably, these membranal steroid receptors operate through second messenger systems to produce rapid effects in their target cells similar to what we have already described for peptides and other regulators.

Table 2-13). Once occupied, the receptors are phosphorylated, lose their heatshock proteins, and form dimers. Following binding of the receptor dimer to the HRE site on nuclear DNA, a second phosphorylation occurs and an adapter protein is recruited, which facilitates interaction with the general transcription apparatus (GTA). RNA polymerase activity and hence transcription is thereby modulated. [Based on information in McDonnell, D. P. (1995). Unraveling the human progesterone receptor signal transduction pathway: Insights into antiprogestin action. *Trends Endocrinol. Metab.* **6**, 133–138.]

Steroid class	Starting steroid	Inactivation steps	Steroid structure representations of excreted product	Principal conjugate present
Progestogens	Progesterone	1. Reduction of C-20 2a. Reduction of 4-ene-3-one or 2b. 3β-steroid dehydrogenase	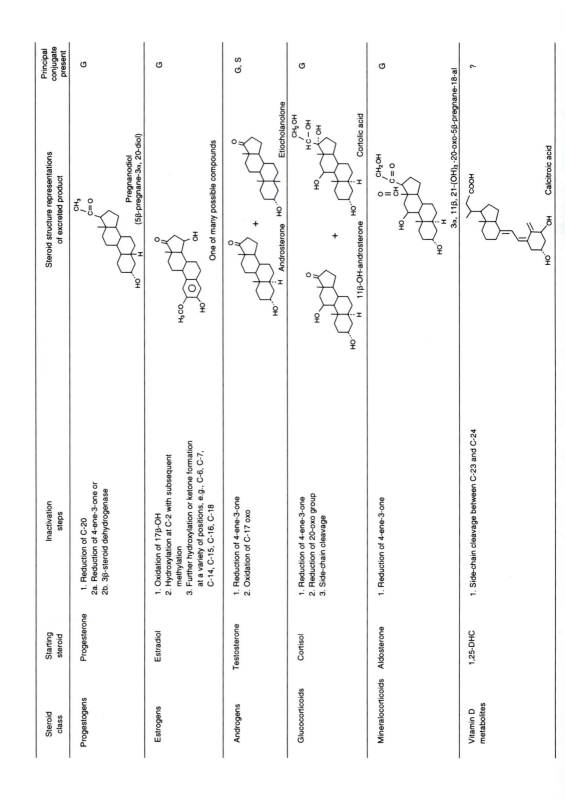 Pregnanediol (5β-pregnane-3α, 20-diol)	G
Estrogens	Estradiol	1. Oxidation of 17β-OH 2. Hydroxylation at C-2 with subsequent methylation 3. Further hydroxylation or ketone formation at a variety of positions, e.g., C-6, C-7, C-14, C-15, C-16, C-18	One of many possible compounds	G
Androgens	Testosterone	1. Reduction of 4-ene-3-one 2. Oxidation of C-17 oxo	Etiocholanolone + Androsterone	G, S
Glucocorticoids	Cortisol	1. Reduction of 4-ene-3-one 2. Reduction of 20-oxo group 3. Side-chain cleavage	11β-OH-androsterone + Cortolic acid	G
Mineralocorticoids	Aldosterone	1. Reduction of 4-ene-3-one	3α, 11β, 21-(OH)₃ -20-oxo-5β-pregnane-18-al	G
Vitamin D metabolites	1,25-DHC	1. Side-chain cleavage between C-23 and C-24	Calcitroic acid	?

E. Metabolism and Excretion of Steroid Hormones

Steroid hormones are metabolized primarily by the liver, which possesses a series of enzymes capable of altering the specific steroids and rendering them biologically inactive and water-soluble. Target cells also may alter steroids metabolically, but it is the liver that performs the major task. Metabolism of steroids typically involves reduction or removal of side chains or attached groups or both, as well as conjugation to glucose as a **glucuronide** or conjugated with **sulfate** (Fig. 2-35). The conjugates are water soluble and when released into the blood will no longer bind effectively to serum proteins and readily appear in the urine. Some of the metabolized steroids are added to the bile and are excreted via the intestinal route.

Steroids also are metabolized via oxidative pathways as well as by reductions. These oxidized metabolites were not identified previously because they occur in a fraction of urine usually discarded when researchers originally isolated urinary steroids for analysis. Oxidized steroid metabolites may represent a substantial portion of the total metabolites for a given steroid. For example, it is estimated that from 12 to 36% of circulating cortisol in humans is converted to oxidized **corotic acids** (cortolic and cortolonic acids, Fig. 2-35).

Androgens of both adrenal and gonadal origin are found in the urine primarily as sulfates and can be measured chemically as **17-ketosteroids.** In addition, some androgens are found as conjugates with glucuronic acid. Estrogens may be excreted as either glucuronides or sulfates such as estriol glucosiduronate or estrone sulfate. **Catechol estrogens** (Fig. 2-35) are also common excretory products (see Section V,A for a chemical explanation of catechols). Progestogens are excreted mainly as glucuronides of pregnanediol or pregnanetriol. However, pregnanediol may be formed from corticosteroid metabolism (that is, from deoxycorticosterone) as well as from progesterone. Pregnanetriol is formed principally from 17α-hydroxyprogesterone, but small amounts also may be produced from 11-deoxycortisol and 17α-hydroxypregnenolone. The corticosteroids are primarily excreted as C_{21} **17-hydroxycorticosteroids** conjugated with glucuronic acid.

Under some conditions, membrane-bound sulfatase enzymes can remove the sulfate and liberate free steroids, which then can enter the cell. Sulfatase activity is very high in the placenta where the sulfated androgen, **dehydroepiandrosterone sulfate (DHEAS),** produced in the fetal adrenal, is desulfated and converted to estrogen. A female fetus is thus protected from the potentially masculinizing actions of fetal androgens by sulfation.

There also are cases known where sulfated steroids can bind to cellular membrane proteins. For example, pregnenolone sulfate produced by certain brain cells is known to bind to and suppress the GABA receptor on nearby neurons.

A number of chemical methods have been employed to identify and quantify various steroid metabolites in urine, but these approaches have given way to radioimmunoassay and sophisticated chromatographic separation and identification techniques (see Chapter 3).

III. Thyroid Hormones

The original molecule isolated from the thyroid gland by Edward C. Kendall in 1915 was believed to be the active thyroid hormone and was named **thyroxine.** It was known to contain four iodide atoms per molecule of a unique compound called **thyronine.** Once

Figure 2-35. Some steroid metabolites. G, Glucuronide; S, sulfate. [Modified from Norman, A. and Litwack, G. (1987). "Hormones." Academic Press, Orlando.]

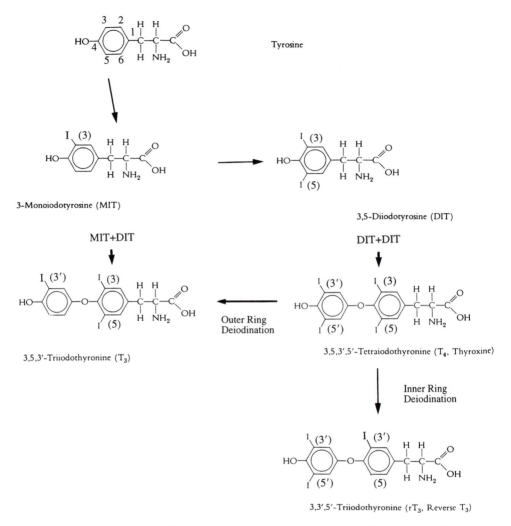

Figure 2-36.   Thyroid hormones and precursors. Synthesis of MIT, DIT, T_3, T_4, and rT_3.

the chemical structure of thyroxine was elucidated, its chemical name became **3,5,3′,5′-tetraiodothyronine** (Fig. 2-36). However, in 1952, J. Gross and Rosalind Pitt-Rivers discovered a second form of thyroid hormone lacking one iodide. This molecule was named **3,5,3′-triiodothyronine** or **triiodothyronine.** The abbreviation for thyroxine became **T_4,** emphasizing the number of iodides per molecule, whereas triiodothyronine became **T_3.** Many endocrinologists consider T_3 to be the biologically active form of thyroid hormone and T_4 to be a precursor. Although T_4 is not a propeptide, it is often termed a prohormone for T_3. This point is considered further in Chapter 7, and we refer to both of them here simply as the thyroid hormones.

A. Structure of Thyroid Hormones

The thyroid hormones are derived from two molecules of the amino acid tyrosine. The two amino acids are joined by elimination of the alanine side chain of one tyrosine and the attachment of its phenolic ring to the hydroxyl group extending from the phenolic group of the other tyrosine. The resulting chemical structure (Fig. 2-36) is a thyronine. The carbons of the recently added phenolic ring are designated by a prime ($'$) symbol that appears in their chemical names. This is an important designation that we will revisit later. Thyroid hormones are hydrophobic and hence poorly soluble in water but, like steroids, highly capable of readily passing through cell membranes.

B. Biosynthesis of Thyroid Hormones

Thyroid hormones are important developmental and metabolic regulators synthesized in the follicular cells of the thyroid gland under the influence of pituitary thyrotropin (TSH). The actual synthesis of thyroid hormones in the follicular cells involves, first, the synthesis of a large globular glycoprotein called **thyroglobulin.** This protein contains tyrosine residues that can be iodinated and then coupled to form T_3 or T_4. Details of this process are provided in Chapter 7 and only a general overview will be presented here. The process of coupling two tyrosines results in a thyronine. In most cases, T_4 is the major circulating thyronine and is converted peripherally to T_3 by a deiodinating enzyme located in the liver or in target cells. This enzyme is called 5′-deiodinase because it removes the 5′-iodide of the T_4 molecule to yield T_3. Another system of nomenclature calls it a type I deiodinase (see Chapter 7).

C. Transport of Thyroid Hormones in the Blood

Most of the circulating thyroid hormones (about 99%) are bound reversibly to plasma proteins. Binding to plasma protein by thyroid hormones is considered helpful for their transport in the blood, because T_3 and T_4 are somewhat hydrophobic and are not highly soluble in blood. However, others have argued that sufficient thyroid hormones could be transported free in the plasma to meet the demands of the animal. The relationship of bound and free hormone to their actions and pathological roles is controversial (see Chapter 7). Nevertheless, the binding proteins provide a ready reservoir of thyroid hormones that is already in the circulation and ready for immediate use.

Several different serum proteins are capable of binding and transporting thyroid hormones. In man, for example, about 75% of the bound hormones are linked to the α_2-globulins and 15 and 10%, respectively, are bound to prealbumin and albumin. Only a very small fraction ($<1\%$) are transported free in the blood. Being somewhat hydrophobic, free thyroid hormones are thought to cross cell membranes readily and are rapidly removed from the blood into target tissues or metabolized, especially by the liver and kidneys. There is evidence that thyroid hormones may enter cells by a carrier-mediated process.

D. Mechanism of Thyroid Hormone Action

The molecular mechanism of action for thyroid hormones appears to be similar to that described for estrogens. As previously stated, thyroid hormones enter target cells where they bind to nuclear receptor proteins. Following the binding to nuclear receptors, thyroid hormones initiate nuclear gene transcription resulting in synthesis of new proteins.

Nuclear receptors for thyroid hormones have been isolated and characterized. Typically, they have greater affinity for T_3 than for T_4, supporting the hypothesis that conversion of T_4 to T_3 is requisite for thyroid hormone action. Target cells for thyroid hormones are equipped with 5′-deiodinase to accomplish this conversion. Mitochondrial receptor proteins also have been demonstrated, and these mitochondrial receptors may be associated with some observed effects on mitochondrial protein synthesis and oxidative metabolism in target cells.

The interactions of thyroid hormones and nuclear DNA have proven to be even more complicated than what we described earlier for steroids. At least two major isoforms of nuclear **thyroid receptor proteins (TRs)** have been demonstrated in humans, mice, and rats, suggesting there may be a separate functional role for each type. In the rat, two forms, $TR\alpha$-1 and $TR\beta$-1, are found in all tissues, whereas another form, $TR\beta$-2, is limited to the pituitary, hypothalamus, and other brain areas. These receptors belong to the superfamily of receptor types that includes the steroid receptors and retinoic acid receptors.

Occupied TRs bind to **thyroid-responsive elements (TREs)** in or near specific genes in target cell nuclei, but may do this in a variety of ways. Monomers of occupied receptor may bind directly to TREs but they do not produce much transcription by the thyroid-responsive gene. Similarly, occupied TR homodimers bind for such a short time that they also produce little transcription. However, formation of a unique heterodimer between a monomer of TR and a **thyroid receptor auxiliary protein (TRAP)** produces maximal transcription following binding to a TRE. TRAPs are regular features of the nuclei of thyroid hormone target cells.

E. Metabolism of Thyroid Hormones

Thyroxine has several metabolic fates after being released from the thyroid gland. Approximately 33 to 40% is converted to T_3 by 5'-deiodination in the liver. This represents the source for most of the circulating T_3 in many species. About 15 to 20% of the circulating T_4 is deaminated and decarboxylated to form **tetraiodothyroacetate (TETRAC),** which is very soluble in water, has no physiological activity, and is excreted rapidly in urine or bile (Fig. 2-37). T_3 also may be converted similarly to form **triiodothyroacetate (TRIAC)** or the latter can be formed by deiodination of TETRAC. Some glucuronide conjugates of T_4 and to a lesser extent of T_3 also may be formed. All of these metabolized molecules are more soluble in water, cannot bind effectively to plasma-binding proteins or receptors, and are quickly eliminated via the urine.

Approximately one-half of the circulating T_4 is eventually converted by deiodination to a unique form of T_3 with the structure of 3,3',5'-triiodothyronine or **reverse T_3, rT_3** (Fig. 2-25). This conversion is accomplished by enzymatic 5-deiodination, which removes an iodide from the internal phenolic ring of a thyronine. Reverse T_3 has no biological activity, and it is degraded more rapidly than normal T_3. As a result of its rapid clearance, rT_3 levels in the blood are rather low. However, increases or decreases in circulating T_3 levels are always accompanied by reciprocal changes in rT_3 levels. The liver may continue to deiodinate T_3 and rT_3, removing the remaining iodides to produce **3,3'-diiodothyronine (T_2),** and then **monoiodothyronine, T_1.**

IV. Eicosanoids

The **eicosanoids** are small lipids derived from a common precursor, **arachidonic acid.** This group includes the prostaglandins, the leukotrienes, and the thromboxanes.

In the early 1930s, Maurice Goldblatt in England and U. S. von Euler in Sweden simultaneously discovered **prostaglandins (PGs).** Elucidation of these important regulatory

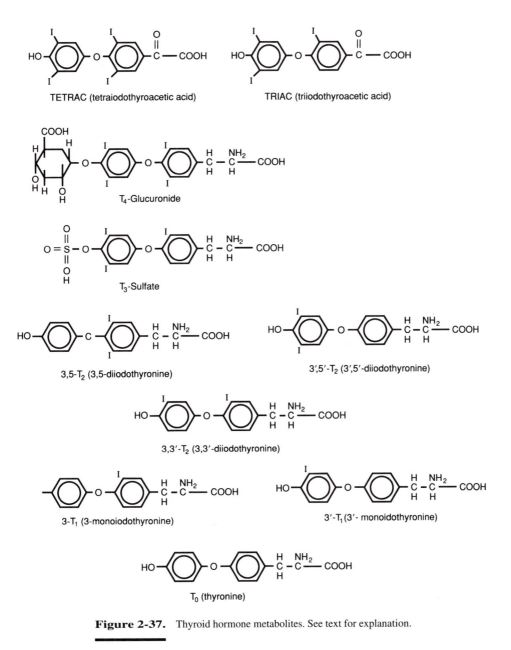

Figure 2-37. Thyroid hormone metabolites. See text for explanation.

compounds and their synthetic pathways resulted in the awarding of the 1982 Nobel Prize in Physiology or Medicine to three principal prostaglandin researchers: Sune Bergstrom, Bengt Samuellson, and John Vane.

Prostaglandins were named on the belief that their source in men was the prostate gland. Since then, they have been found in most tissues of both men and women. The PGs have many diverse actions including stimulation of smooth muscle contraction in the intestine and uterus, vasodilation (but may cause vasoconstriction in certain vessels), and modulation of central nervous system function. They also have been shown to stimulate synthesis of corticosteroids, testosterone, and a variety of specific enzymes. One PG ($PGF_{2\alpha}$) is believed to be the uterine luteolytic substance in certain mammalian species (see Chapter 11). Prostaglandins also reduce progesterone synthesis by the corpus luteum, induce ovulation and lactation in rodents, and may be involved in the induction of labor. Furthermore, they may induce inflammation and fever. The antiinflammatory and antipyretic (fever-decreasing) action of aspirin is due to its inhibition of PG synthesis. **Prostacyclin (PGI$_2$),** one form of prostaglandin, is a potent inhibitor of blood platelet aggregation and inhibits blood clotting.

Researchers discovered the **thromboxanes** during studies on prostaglandin metabolism and action. Thromboxane A2 causes translocation of free calcium ions to bring about changes associated with the shape of blood platelets. It is this change in platelet shape that allows platelets to aggregate and facilitate clotting. Other thromboxanes may be released from the platelet, causing local constriction of vascular smooth muscle. This paracrine effect might enhance clotting by reducing the diameter of the arterioles and slowing blood flow through the capillary beds of damaged tissues.

Leukotrienes are synthesized and released by white blood cells in response to injury. They contribute to inflammatory or allergic responses by causing contraction of vascular smooth muscle and by increasing vascular permeability. Increased levels of leukotrienes have been associated with allergic reactions, asthma, cystic fibrosis, septic shock, and a number of other disorders.

A. Chemical Structure of Eicosanoids

Prostaglandins are all related structurally to **prostanoic acid,** and they can be separated into six classes (A through F) on the basis of structural differences (Fig. 2-38). Prostacyclin (PGI$_2$) represents a unique class of prostaglandin called PGI. Bergstrom succeeded in elucidating the structures for 16 PGs in 1956. The most commonly occurring PGs are PGE$_1$, PGE$_2$, and $PGF_{2\alpha}$. The leukotrienes consist of at least 15 related compounds occurring in five structural classes, A, B, C, D, and E (Fig. 2-38). Thromboxanes are very short-lived molecules, some of which also are depicted in Fig. 2-38.

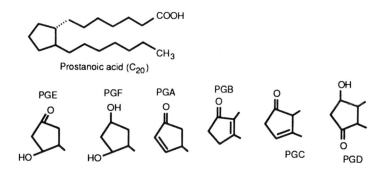

Figure 2-38. Prostaglandin structures. The basic chemical formula for the prostaglandins is that of prostanoic acid. These C$_{20}$ lipids are divided into classes (A, B, F, etc.) based on substitutions to the five-membered carbon ring. Modifications of the side chains result in different forms within a class, each designated by a subscript, e.g., $PGF_{2\alpha}$, PGE$_1$.

B. Biosynthesis of Eicosanoids

Arachidonic acid is synthesized from linolenic acid or diacylglycerol (DAG; see Section I,C,2). Once formed, arachidonic acid can be converted to any of the eicosanoids (Fig. 2-39). **Cyclooxygenase** is an enzyme that transforms arachidonic acid into endoperoxides, which are used to synthesize prostaglandins, prostacyclin, or thromboxanes. Drugs such as **aspirin** and **indomethacin** inhibit cyclooxygenase and block the syntheses of prostaglandins and thromboxanes. A separate enzyme, **5-lipoxygenase,** forms the leukotrienes from arachidonic acid. This enzyme is not inhibited by aspirin or indomethacin but can be inhibited by specific lipoxygenase inhibitors.

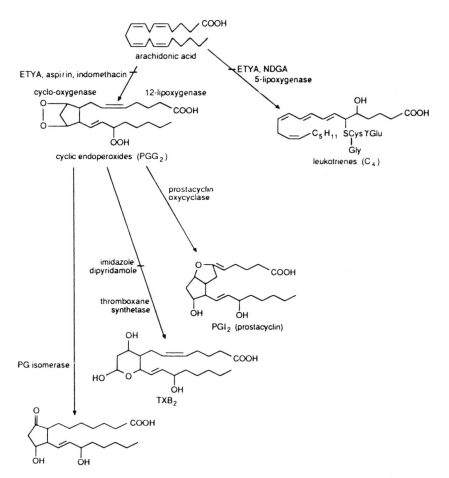

Figure 2-39.   Eicosanoid synthesis. The precursor for all eicosanoids, arachidonic acid, can be synthesized from diacylglycerol (DAG), which has dual roles as second messenger and eicosanoid precursor. Indomethecin, aspirin, and ETYA (eicosatetraynoic acid) inhibit the enzyme cyclooxygenase and block prostaglandin and thromboxane synthesis. ETYA and NDGA (nordihydroguaiaretic acid) block the enzyme 5-lipoxygenase and prevent leukotriene synthesis. ETYA is a modified form of arachidonic acid that competes for any enzyme that normally uses arachidonic acid as its substrate. [Modified from Bolander, F. F. (1989). "Molecular Endocrinology," Academic Press, San Diego.]

V. Other Important Regulators

In this section, we briefly describe a number of neural regulators and their chemistry to serve as background information for later discussions. Some limited information about their synthesis and metabolism is also provided.

A. Acetylcholine

Acetylcholine (ACh) was first demonstrated as a major neurotransmitter in the peripheral nervous system and only recently was its role in the brain clarified. Synthesis of ACh from its precursors, choline and acetate, is accomplished by an enzyme called **choline acetyltransferase (CAT).** After release into the synaptic space, ACh is degraded to choline and acetate by acetylcholinesterase (AChE) associated with the postsynaptic cell membrane. The presynaptic cells take up free choline for use in the resynthesis of ACh. The distribution of CAT immunoreactivity in the brain corresponds to the distribution of ACh receptors. In contrast, the distribution of AChE is much wider than that of either CAT or ACh receptors and therefore is an unreliable indicator of cholinergic synapses.

B. γ-Aminobutyric Acid (GABA)

The inhibitory neurotransmitter **gamma (γ)-aminobutyric acid (GABA)** is formed mainly from the amino acid glutamate by the enzyme **glutamate decarboxylase (GAD).** Glutamate may be synthesized either from glutamine or 2-oxoglutarate. GABA, an important central nervous system neurotransmitter, is inactivated by conversion to succinate or γ-hydroxybutyric acid. It may have a role as a neurohormone, too (see Chapter 4).

C. Interleukins

Certain white blood cells, such as lymphocytes and macrophages, secrete proteins called **interleukins** that operate in autocrine and paracrine fashion. For example, once a macrophage has identified a cell infected with a virus, it obtains and transfers the viral antigen complexed with a protein to a lymphocyte known as a helper T cell. The macrophage also secretes **interleukin 1 (IL-1)** that further activates the helper T cell. **Interleukin 2 (IL-2)** is a mitogen secreted by the helper T cell following its activation by IL-1, which stimulates another lymphocyte type called a B cell to divide and form plasma cells (antibody-synthesizing cells) and memory cells (responsible for long-term immunity). In autocrine fashion, IL-2 stimulates proliferation of the helper T cells, too.

There are more than a dozen recognized interleukins that may be involved in various aspects of the immune response system. Helper T cells also produce **interferon-γ,** which can transform macrophages and cause them to attack viral-infected cells more vigorously. The interleukins, interferons, and other factors secreted by cells involved with the immune response are known as **cytokines.** Recently, cytokines have been shown to be involved in other cellular processes. For example, they have been implicated in the endocrine mechanisms controlling release of certain pituitary hormones.

D. Miscellaneous Brain Neuropeptides

In recent years, numerous peptides have been identified in the brain that play important roles as neurotransmitters and neuromodulators. Some of these peptides also may function as neurohormones. Foremost among these are the **endorphins** and **enkephalins** that play important analgesic (pain-killing) roles. Other important peptides include substance P, VIP, galanin, neuropeptide Y (NPY), neuropeptide YY (PYY), peptide–histidine–isoleucine (PHI), neurotensin, cholecystokinin (CCK), angiotensin II (ANG-II), and calcitonin gene-related peptide (CGRP). The possible roles for many of these neuropeptides will be discussed in Chapter 4 in association with the regulation of pituitary gland functions.

E. Gaseous Chemical Regulators

Among the most surprising discoveries is the recognition of two well-known toxic gases as physiologically relevant, chemical regulators: **nitric oxide (NO)** and **carbon monoxide (CO)**. Failure to recognize them earlier in part is due to their ability to pass readily through cell membranes and the absence of discrete receptors on the cell membranes of target cells.

The first discovery establishing such regulatory substances was the identification of NO as the hypothesized **endothelium-derived relaxing factor (EDRF)** of endothelial cells lining blood arterioles. EDRF or NO diffuses into adjacent vascular smooth muscle cells to cause relaxation and allow dilation of the arterioles. These endothelial cells contain an enzyme, **nitric oxide synthetase (NOS),** which produces NO from the amino acid arginine. NOS has a heme group similar to that of hemoglobin acting as an electron acceptor in this reaction. Once formed, the NO quickly exits and enters adjacent smooth muscle cells where it interacts with intracellular enzymes. Because of its gaseous nature, NO can readily exit from cells and the body and special degrading systems may not be necessary. Soon after its discovery, NO released from certain pelvic nerves was identified as the causative factor producing penile erection through local effects on the vasculature of the penis. There also appear to be links between the ability of brain neurons to synthesize NO and the cellular basis for learning and memory.

Although best known for its role as an insidious poison, carbon monoxide may prove to be a physiological regulator, too. In 1987, CO was reported to cause dilation of blood vessels. The enzyme **heme oxygenase,** which breaks down heme, produces CO. Heme oxygenase is found in many cell types including neurons in certain brain regions. Preliminary studies suggest that CO readily enters smooth muscle cells, interacts with and activates the enzyme guanylyl cyclase, and produces a rise in cGMP. Elevated cGMP in turn promotes relaxation. As scientists look more carefully at these and other potential gaseous regulators, we may learn many new things about chemical regulation.

VI. Summary

Neurocrines and hormones may occur as peptides, proteins, modified amino acids (indole-amines, catecholamines, thyroid hormones), or lipids (steroids, eicosanoids). Peptides and proteins are synthesized as preprohormones and are modified to prohormones and then modified again to the final product. These processes usually occur intracellularly. Peptides are transported free in the blood or bound to plasma proteins. Most of the steroids and thyroid hormones as well as many amines are transported primarily bound to plasma proteins in equilibrium with a small quantity of free hormones.

Peptide, protein, and amine regulators bind to receptors located on the cell surface and typically produce their actions through production of second messengers. These mechanisms employ a variety of G proteins that influence levels of second messengers such as cAMP, IP$_3$, and DAG, which in turn may affect protein kinase activity or permeability of the cell membrane. Initial effects of these regulators usually affect membrane transport in (uptake) or out (secretion) of the cell or activate intracellular enzyme systems. Many of these membrane ligands also produce a delayed effect on protein synthesis.

Steroids and thyroid hormones enter the cytoplasm where they may either bind to cytosolic receptors (corticosteroids) or to nuclear receptors. Occupied receptors for these hormones act as transcription factors to affect nuclear mRNA production and subsequent protein synthesis. In a few cases, certain steroids may be ligands for receptors attached to the cell membrane.

Chemical regulators are metabolized in a variety of ways to inactive them so that activated cells can be inactivated again. Metabolism may involve inactivation by enzymes associated with the cell surface or by intracellular enzymes in the target cells or nearby cells. Reuptake of amines by presynaptic neurons is another method for inactivating the target cell. Steroids and thyroid hormones are often metabolized with addition of special components (e.g., sulfate, glucuronides) that increase their solubility in water and accelerate their excretion via the urine. Sensitivity of the target cells can also be influenced by increases (up-regulation) or decreases (down-regulation) in receptor number.

Peptides, proteins, indoleamines, catecholamines, thyroid hormones, and many other regulators are formed from one or more amino acids. The amino acid arginine is also the substrate for synthesis of the gaseous regulator, nitric oxide. Steroid hormones are all derived from cholesterol, and the eicosanoids have a common precursor, arachidonic acid.

Suggested Reading

Books

Bolander, F. F. (1994). "Molecular Endocrinology," 2nd Ed. Academic Press, San Diego.

Sherwood, N. M., and Hew, C. L. (1994). "Fish Physiology XIII. Molecular Endocrinology of Fish." Academic Press, San Diego.

Weintraub, B. D. (1995). "Molecular Endocrinology: Basic Concepts and Clinical Implications." Raven, New York.

Articles

Arimura, A. (1992). Receptors for pituitary adenylate cyclase-activating polypeptide: Comparison with vasoactive intestinal peptide receptors. *Trends Endocrinol. Metab.* **3,** 288–294.

Balla, T., and Catt, K. J. (1994). Phosphoinositides and calcium signaling: New aspects and diverse functions in cell regulation. *Trends Endocrinol. Metab.* **5,** 250–255.

Bar-Sagi, D. (1994). The Sos (Son of sevenless) protein. *Trends Endocrinol. Metab.* **5,** 165–169.

Baulieu, E.-E. (1991). The antisteroid RU 486: Its cellular and molecular mode of action. *Trends Endocrinol. Metab.* **2,** 233–239.

Blaustein, J. D., Olster, D. H., and Tetel, M. J. (1993). Heterogenous regulation of steroid hormone receptors in the brain. *Am. Zool.* **33,** 219–228.

Burwen, S. J., and Jones, A. L. (1987). The association of polypeptide hormones and growth factors with the nuclei of target cells. *Trends Biochem. Sci.* **12,** 159–162.

Cadena, R. L., and Gill, G. N. (1992). Receptor tyrosine kinases. *FASEB J.* **6,** 2332–2337.

Carson-Jurica, M. A., Schrader, W. T., and O'Malley, B. W. (1990). Steroid receptor family: Structure and functions. *Endocr. Rev.* **11,** 201–220.

Combarnous, Y. (1992). Molecular basis of the specificity of binding of glycoprotein hormones to their receptors. *Endocr. Rev.* **13,** 670–691.

Conti, M., Jin, S.-L. C., Monaco, L., Repaske, D. R., and Swinnin, J. V. (1991). Hormonal regulation of cyclic nucleotide phosphosdiesterase. *Endocr. Rev.* **12,** 218–234.

Delftos, L. J. (1991). Chromogranin A: Its role in endocrine function and as an endocrine and neuroendocrine tumor marker. *Endocr. Rev.* **12,** 181–188.

DeVivo, M., and Iyengar, R. (1994). G protein pathways: Signal processing by effectors. *Mol. Cell. Endocrinol.* **100,** 65–70.

Ferrara, N., Houck, K., Jakeman, L., and Leung, D. W. (1992). Molecular and biological properties of the vascular endothelial growth factor family of proteins. *Endocr. Rev.* **13,** 18–32.

Franklyn, J. A., and Sheppard, M. C. (1989). Thyroid and steroid hormone regulation of proto-oncogene expression. *Trends Endocrinol. Metab.* **1,** 35–39.

Guengerich, F. P. (1993). Cytochrome P450 enzymes. *Am. Sci.* **81,** 440–447.

Guiochon-Mantel, A., and Milgrom, E. (1993). Cytoplasmic–nuclear trafficking of steroid hormone receptors. *Trends Endocrinol. Metab.* **4,** 322–328.

Hammond, G. L. (1990). Molecular properties of corticosteroid binding globulin and the sex steroid binding proteins. *Endocr. Rev.* **11,** 65–79.

Hobkirk, R. (1993). Steroid sulfation: Current concepts. *Trends Endocrinol. Metab.* **4,** 69–74.

Ihle, J. N. (1994). Signaling by the cytokine receptor superfamily: Just another kinase story. *Trends Endocrinol. Metab.* **5,** 137–143.

Ing, N. H., and O'Malley, B. W. (1995). The steroid hormone receptor: Molecular mechanisms of action. In "Molecular Endocrinology: Basic Concepts and Clinical Implications" (B. D. Weintraub, ed.), pp. 195–215. Raven, New York.

Kelly, P. A., Djiane, J., and Edery, M. (1992). Different forms of the prolactin receptor: Insights into the mechanism of prolactin action. *Trends Endocrinol. Metab.* **3,** 54–59.

Kitamura, T., Ogorochi, T., and Miyajima, A. (1994). Multimeric cytokine receptors. *Trends Endocrinol. Metab.* **5,** 8–14.

Lazar, M. A. (1993). Thyroid hormone receptors: Multiple forms, multiple possibilities. *Endocr. Rev.* **14,** 184–193.

Lin, D., Sugawara, T., Straus, J. F., III, Clark, B. J., Stocco, D. M., Saenger, P., Rogol, A., and Miller, W. L. (1995). Role of steroidogenic acute regulatory protein in adrenal and gonadal steroidogenesis. *Science* **267,** 1828–1831.

Linder, M. E., and Gilman, A. G. (1992). G proteins. *Sci. Am.* **253**(7), 56–65.

Malbon, C. C., Rapiejko, P. J., and Watkins, D. C. (1988). Permissive hormone regulation of hormone-sensitive effector systems. *Trends Pharmacol. Sci.* **9,** 33–36.

Mathews, L. S. (1991). Molecular biology of growth hormone receptors. *Trends Endocrinol. Metab.* **2,** 176–180.

Mayer, E. A., and Baldi, J. P. (1991). Can regulatory peptides be regarded as words in a biological language? *Am. J. Physiol.* **261,** G171–G184.

McDonald, D. P. (1995). Unraveling the human progesterone receptor signal transduction pathway: Insights into antiprogestin action. *Trends Endocrinol. Metab.* **6,** 133–138.

McEwen, B. S. (1991). Steroid hormones are multifunctional messengers to the brain. *Trends Endocrinol. Metab.* **2,** 62–67.

Miller, W. L. (1988). Molecular biology of steroid hormone synthesis. *Endocr. Rev.* **9,** 295–318.

Moudgil, V. K. (1988). Steroid receptors in health and disease. *FEBS Lett.* **226,** 213–216.

Orti, E., Bodwell, J. E., and Munck, A. (1992). Phosphorylation of steroid hormone receptors. *Endocr. Rev.* **13,** 105–128.

Pestell, R. G., and Jameson, J. L. (1995). Transcriptional regulation of endocrine genes by second messenger signalling pathways. *In* "Molecular Endocrinology: Basic Concepts and Clinical Implications" (B. D. Weintraub, ed.), pp. 59–76. Raven, New York.

Pratt, W. B., Hutchison, K. A., and Scherrer, L. C. (1992). Steroid receptor folding by heat-shock proteins and composition of the receptor heterocomplex. *Trends Endocrinol. Metab.* **3,** 326–333.

Putney, J. W., Jr., and St. J. Bird, G. (1994). Calcium mobilization by inositol phosphates and other intracellular messengers. *Trends Endocrinol. Metab.* **5,** 256–260.

Putney, J. W., Jr., and Bird, G. St. J. (1993). The inositol phosphate-calcium signaling system in nonexciteable cells. *Endocr. Rev.* **14,** 610–631.

Rhodes, D., and Klug, A. (1993). Zinc fingers. *Sci. Am.* **254**(2), 56–65.

Rittmaster, R. S. (1993). Androgen conjugates: Physiology and clinical significance. *Endocr. Rev.* **14,** 121–132.

Robel, P., and Baulieu, E.-E. (1994). Neurosteroids: Biosynthesis and function. *Trends Endocrinol. Metab.* **5,** 1–8.

Roseler, W. J., Park, E. A., Klemm, D. J., Liu, J., Gurney, A. L., Vandenbark, G. R., and Hanson, R. W. (1990). Modulation of hormone response elements by promoter environment. *Trends Endocrinol. Metab.* **1,** 347–351.

Rosner, W. (1990). The functions of corticosteroid binding globulin and sex hormone-binding globulin: Recent advances. *Endocr. Rev.* **11,** 80–91.

Scammell, J. G. (1993). Granins: Markers of the regulated secretory pathway. *Trends Endocrinol. Metab.* **4,** 14–18.

Segre, G. V., and Goldring, S. R. (1993). Receptors for secretin, calcitonin, parathyroid hormone (PTH)/PTH-related peptide, vasoactive intestinal peptide, glucagonlike peptide I, growth hormone-releasing hormone and glucagon belong to a newly discovered G-protein-linked receptor family. *Trends Endocrinol. Metab.* **4,** 309–314.

Simpson, E. R., Mahendroo, M. S., Means, G. D., Kilgore, M. W., Hinshelwood, M. M., Graham-Lawrence, S., Amarneh, B., Ito, Y., Fisher, C. R., Michael, M. D., Mendelson, C. R., and Bulun, S. E. (1994). Aromatase cytochrome P450, the enzyme responsible for estrogen biosynthesis. *Endocr. Rev.* **15,** 342–355.

Spiegel, A. M., Shenker, A., Simonds, W. F., and Weinstein, L. S. (1995). G protein dysfunction in disease. *In* "Molecular Endocrinology: Basic Concepts and Clinical Implications" (B. D. Weintraub, ed.), pp. 297–318. Raven, New York.

Turgeon, J. L., and Waring, D. W. (1992). Functional cross-talk between receptors for peptide and steroid hormones. *Trends Endocrinol. Metab.* **3,** 360–365.

Vassart, G., Parmentier, M., Libert, F., and Dumont, J. (1991). Molecular genetics of the thyrotropin receptor. *Trends Endocrinol. Metab.* **2,** 151–156.

Visser, T. J., de Herder, W. W., Rooda, S. J. E. R., Rutgers, M., and van Buuren, J. C. J. (1990). The role of sulfation in thyroid hormone metabolism. *Trends Endocrinol. Metab.* **1,** 211–218.

Williams, G. R., and Brent, G. A. (1995). Thyroid hormone response elements. *In* "Molecular Endocrinology: Basic Concepts and Clinical Implications." (B. D. Weintraub, ed.), pp. 217–239. Raven, New York.

Yen, P. M., and Chin, W. W. (1994). New advances in understanding the molecular mechanisms of thyroid hormone action. *Trends Endocrinol. Metab.,* **5,** 65–72.

Zoeller, R. T. (1993). Molecular mechanisms of signal integration in hypothalamic neurons. *Am. Zool.* **33,** 244–254.

3

Approaches to
Endocrine Research

I. The Nature and Methods of Science

The basic method of scientific investigation has changed little over the years, but dramatic advances in observational, manipulative, and analytical tools during the last 50 years have resulted in a virtual explosion in our knowledge of animal biology that parallels similar events in physics, chemistry, engineering, and other scientific disciplines. For example, our observational powers have been increased several orders of magnitude by advances in the field of microscopy. Miniature radiotransmitters have made it possible to tag secretive animals such as rattlesnakes and follow their natural migrations without disturbing them. New techniques in chemistry such as high-performance liquid chromatography can be used to analyze volumes as small as a few microliters that may contain only a few nanograms or even picograms of an important molecule. From a tiny sample, we can determine the chemical structure of a molecule and, if it is a peptide construct, a gene that will direct the synthesis of large amounts of the molecule. The advent of computers and associated technology has not only automated many of our analytical procedures but has greatly augmented our ability to analyze information. Computers also can be used to simulate natural conditions and to construct models of natural phenomena that we can manipulate and use to predict events in nature.

A. The Scientific Method

The process used by animal biologists to investigate the lives and activities of animals is the **scientific method.** We seek facts called data (singular, datum) and organize them into hypotheses, theories, or laws that give these data order and meaning. Many scientists are engaged primarily in the processes of gathering data. These data may be accumulated through careful observation or by use of planned experiments. Creative scientists also take data and try to organize them to form generalizations. From a series of observations, the scientist might form a **hypothesis** (Gr. *hypothesis,* supposition) or a predictive statement, a statement of what may be the true explanation of certain phenomena. Until it is tested experimentally, it is only a working hypothesis that must be supported or rejected on the basis of experimental results. If the hypothesis does not hold, it must be rejected or possibly revised and retested. If supported, the scientist may choose to test it more vigorously or form additional hypotheses on the relationships observed.

Hypotheses must be testable through observation or experimentation. The proposal that life on earth originated from outer space would be difficult to test, as would the notion that dinosaurs became extinct because other animals ate their eggs. All scientific generalizations and hypotheses must be tested no matter how "self-evident" or "unlikely" they might appear. Hypotheses or theories that are contradicted by data obtained from valid testing procedures must be modified or discarded.

Testing of any hypothesis involves rigorous attention to details. Unless the test is reliable, the data obtained can neither contradict nor support the hypothesis or theory. Although it is possible to disprove hypotheses, it usually is not possible to prove one with a single observation or experiment. The data obtained may support the hypothesis, but since there are frequently many other ways one might test the hypothesis, it is not yet proven beyond all doubt.

The more ways scientists test the hypothesis, the more confident they become of its validity. When scientists discover that a preponderance of new data supports the generalization, it becomes a **theory** (Gr. *theoria,* speculation). Continued testing of the theory never stops. Every theory must be reexamined and modified if necessary as new data are accumulated. A scientific theory is not just an educated guess, as "theory" is often used by the nonscientist or as is suggested by the origin of the word. A scientific theory is an established concept based on accumulated data. Once a theory is accepted with certainty, it becomes a **law** or **principle.** However, even principles are still subjected to retesting and revision of their parts.

II. Controlled Experimental Testing

Science has accepted ways in which hypotheses or theories may be tested through observation or experimentation. There are techniques for making observations or for designing experiments and evaluating the result so that prejudice (bias) on the part of either investigator or subject is eliminated. Procedures are rigidly followed to control for inadvertent biases produced by the methodologies used, and an attempt is made to control other factors that might influence the results. This must be done so that scientists can be confident that the outcome of an experiment is a consequence of certain manipulations only. A **variable** is an event or condition that is subject to change. The scientist may allow or cause one or more variables to change while keeping all others constant. The changed or manipulated variable is the **independent variable** and those variables that are altered as a result of manipulating the independent variable are called **dependent variables.** One group of organisms serves as the **experimental group** subjected to the changed independent variable. The independent variable is unchanged for a second group known as the **control group.** One type of experimental control is a complete match to the manipulated animal minus only the factor being tested. If one group of rats is receiving injections of a drug or hormone dissolved in a solvent (called a **vehicle**), the appropriate control would be a second group of similar rats (matched for age, body weight, sex, history, etc.) receiving injections of an equivalent amount of vehicle without the drug or hormone (to control for effects of administering the injection). If a researcher wishes to investigate the effects of removal of some body organ, the appropriate control would not be an unoperated animal, because the anesthesia or the surgical procedure could influence the result. A more appropriate control would be a **sham-operated** animal, one that was also anesthetized and surgically disturbed but in which the gland in question was left intact. Looking at drug treatment of surgically altered animals requires a more complicated design (see Table 3-1).

Appropriate controls are often more difficult to establish in human subjects either because of moral and ethical issues or because of complicating psychological factors. It is well known, for example, that subjects expecting certain responses to treatments may respond differently than other subjects told to anticipate different responses. Consequently, in studies of the effects of a drug on the performance of long-distance runners, some subjects receive the real drug and the others are given an inactive substitute, or **placebo** (L., "I

Table 3-1. Comparison of Adequate and Inadequate Controls[a]

Treatment 1	Treatment 2 Adequate control	Treatment 2 Inadequate control	Treatment 3	Treatment 4
One independent variable: Surgery				
Surgically altered animal	Sham-operated animal	Unaltered animal		
One independent variable: Injection				
Animal injected with chemical regulator in vehicle (a vehicle is some medium)	Animal injected with vehicle only	Uninjected animal		
Two independent variables: Surgery and injection				
Surgery plus chemical regulator in vehicle	Sham operated plus chemical in vehicle		Surgery plus vehicle only	Sham operated plus vehicle only

[a] A simple experiment involving only one independent variable (either surgery or injection) and a more complicated experiment involving both independent variables. Why are some attempts to establish control groups labeled as inadequate? Note that although two treatment groups are all that is necessary when manipulating a single variable, four are needed for two variables. How many experimental groups would be needed if three independent variables were examined simultaneously (*hint:* $2^1 = 2$)?

will please"), administered in the same manner to control for psychological factors. In a **double-blind study,** neither the subjects nor the people administering the drug or placebo know which subjects are receiving which treatment. Another type of control is to switch treatments after a suitable period of observation, without telling the subjects, and then to observe the subjects for a second period. These approaches also can be useful when working with nonhumans.

A. Representative Sampling

An important method used by the biologist is the examination of a **representative sample.** It is rarely feasible to test all potential subjects in a population or all the individuals of one kind of animal, so the scientist must establish a representative sample to test. Again, care must be exercised to ensure that the subjects have not been identified with any biases that might influence the results and therefore reflect characteristics of the entire population. The size of the representative sample will be influenced by the availability of animals, expected variability in factors being measured, statistical test criteria, previous experience of the investigator, and so forth.

Since scientists work with what they discern to be representative samples, they need a reliable method of analyzing experimental results and comparing appropriate control groups to see if their results are valid and representative. At the end of an experiment, changes in dependent variables of the experimental group are compared to the same variables in the control group. The branch of mathematics called **statistics** is the scientific tabulation and treatment of data. Statistical treatments provide tests for determining if the results are highly reproducible (that is, what is the probability that the differences observed in a dependent variable between experimental and control group are significant).

Reports of experimental data must be examined carefully not only to see if statistical analyses were performed but whether the appropriate statistical test was performed. Thus, it is essential that all researchers have a background in statistical methods, not only for analyzing data but for designing experiments. The design of a study must reflect knowledge of what appropriate statistical tests are to be applied. Many published studies are difficult to interpret because of errors in experimental design and/or in the application of inappropriate statistical tests.

B. The Dose–Response Relationship

Biological systems are very sensitive to the quantity of available chemical regulators, and small changes in concentration may have profound effects on physiology and behavior. This influence of concentration on physiology and behavior is summarized in the **dose–response relationship.** We can use a graph to illustrate this by comparing the physiological or behavioral response (dependent variable) on the y axis with the dose of regulator (independent variable on the x axis (Fig. 3-1). Typically, there is a minimally effective dose necessary to produce any response. This is often called the **threshold dose.** Smaller doses producing no response are called **subthreshold doses.** Some systems show an **all-or-none response,** meaning that the effect, if produced at all, occurs at the maximal intensity and adding more regulator will not give a greater response (Fig. 3-1A). However, in most systems, increasing the dose produces a greater response up to some maximal or **optimal level** beyond which further increases in regulator actually produce a reduction in the optimal response (Fig. 3-1B). This reduction may be due to interfering effects of the regulator on

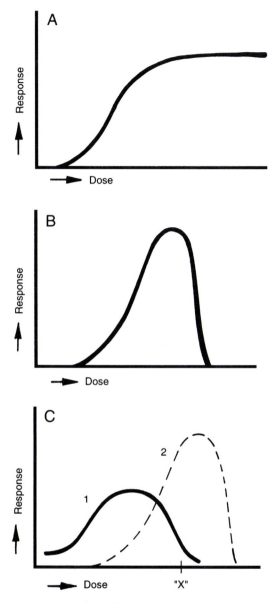

Figure 3-1. Dose–response relationships. (A) A system may respond to increasing dose until the process reaches a maximum. Additional regulator produces no change in the response over a wide range. (B) Relationship with optimal response to an "optimal" dose. A dose too low to produce a response is considered subthreshold. Doses above the "optimal" dose produce diminished responses and may indicate a toxicity to the tissue or animal. Such doses often are considered to be "pharmacological," although any level of a regulator that exceeds natural levels also could be termed pharmacological. Notice that when observing the response of a single dose you might not know whether you were above the optimal dose or below it. Consequently, when the data are not readily available, multiple doses should be employed to determine a useful dose–response relationship. (C) Species-specific or tissue-specific dose responses. An "optimal" dose for one tissue or one animal (1) might be either subthreshold or pharmacological for another (2).

other systems or may be simply a toxic action that decreases the ability of the system to respond. The lowest dose that produces a maximal response is called the **optimal dose.** Generally speaking, the natural physiological range for a normal regulator in an animal is between the threshold dose and the optimal dose. Notice that if you ran an experiment with only one dose of a hormone and measured a certain response, you could be using a physiological dose or a supramaximal dose (Fig. 3-1B). When designing an experiment that involves treatment with any regulator or other chemical such as antagonistic or agonistic drugs, it is essential to ascertain either from the literature or experimentally what the appropriate dose should be. Excessive doses may produce toxic effects via different pathways that influence your interpretation of the data.

When testing actions of a chemical substance on physiology and behavior, it is often difficult to separate direct actions from unanticipated actions on other tissues that might influence the results. Some drugs useful for studying one system in fact may produce toxic side effects of a **paradoxical nature** on other tissues. For example, the drug thiourea is an effective inhibitor of thyroid gland function. However, the response of the liver to thiourea produces effects opposite to those caused by surgical thyroidectomy, making interpretation of the effects of thiourea on the whole animal open to question.

The graphic illusion of the relationship between the concentration of a chemical substance and its action is called a **dose–response curve.** Comparison of the dose–response relationship in different tissues, in different individuals, or in different species may give very different results (Fig. 3-1C). Consequently, one must be very careful when using data based on one species for an experiment with another species or when extrapolating from doses observed in cell or tissue culture to the intact animal.

C. Occam's Razor and Morgan's Canon

An old principle of logic, called **Occam's razor,** should be invoked when interpreting data. Occam's razor states that if several explanations are compatible with the evidence, the simplest one should be considered the most probable one. Experience tells us that complicated explanations may not be necessary to explain a phenomenon when a simple one will do. A modification of Occam's razor, called **Morgan's canon,** was formulated in the late 1800s with respect to interpreting animal behavior. Morgan's canon says that when considering at what level a behavior might be controlled by the nervous system we should invoke the lowest level (that is, the least complicated) that will work. These are important principles to keep in mind as you continue your studies of physiology. This is not meant to imply that more complicated explanations are always incorrect but to emphasize that we do not have to make them more complicated than necessary.

D. Biological Rhythms

The release of hormones and other chemical regulators often occurs in bursts (**phasic secretion**) rather than at a continuous and constant rate (**tonic secretion**). Many of these bursts exhibit distinct, predictable diurnal, monthly, or seasonally cyclic patterns of secretion that are termed **biological rhythms** (Fig. 3-2). For example, maximal levels of testosterone in the blood of human males normally occur between 2 and 6 A.M. The monthly fluctuations for estradiol and progesterone levels in human female blood plasma recur with precise regularity. A predictable increase in androgens in the blood of the Italian frog, *Rana esculenta,* occurs every spring, signaling the onset of the spring breeding season. As one

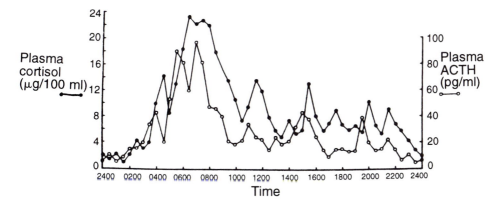

Figure 3-2. Diurnal pattern of plasma corticotropin (ACTH) and cortisol. More corticotropin (ACTH) produced in the pituitary gland is released in the early morning than at other times of the day. Since ACTH stimulates the adrenel cortex to secrete cortisol, the latter shows a matching rhythm in blood concentration. [From THE ADRENAL CORTEX by Vinson/Whitehouse/Hinson, © 1993. Adapted by permission of Prentice-Hall, Inc., Upper Saddle River, NJ.]

might expect, not only can hormone levels vary predictably, but the sensitivity of target cells may show diurnal or seasonal fluctuations. For example, the responsiveness of the pigeon crop to the stimulatory actions of the tropic hormone prolactin varies with time of day. Furthermore, the metabolic responses of killifish to a given dosage of prolactin varies with the season of the year in which the fish are treated.

The existence of biological rhythms suggests caution when interpreting data. Failure of a regulator to produce an effect when applied at 11:00 A.M. on September 15 may not mean that this regulator has no action in this animal but that the animal is not responsive to this dose at this time. If we had tried a different time of day or another time of the month or season of the year, we might have had different results. If the regulator had produced an effect, we still could not conclude that we would always see that effect.

The discovery of biological rhythms has opened many new areas of research and has opened old data to new interpretations. Today, experimental designs more frequently consider the times for treatment and the repeating of experiments with diurnal or seasonal treatments. Increasingly, experiments are conducted to understand what internal and/or external factors are responsible for regulating these rhythms.

III. Methods of Endocrine Analysis

In recent years, the development and application of new techniques have caused a revolution in endocrinological studies, resulting in many new and exciting discoveries. Some of these developments as well as older techniques are discussed briefly here.

A. Extirpation–Observation and Replacement–Observation

Early studies of hormone actions were limited to gross manipulations and observations because there were no precise ways to estimate levels of chemical regulators. Putative regulators were identified by first removing a gland or other organ (**extirpation**) and observing the effects of removal on the organism. Then an attempt was made to replace the lost tissue through transplants or grafts or by administering extracts of the tissue (**replacement**). If continued observations verified that the symptoms caused by extirpation were relieved by the replacement therapy, evidence for the existence of a chemical regulator was established.

A related early technique that evolved from the classical approaches involved the use of a **bioassay system,** a system employing animals or animal parts that could provide a quantitative or at least a qualitative estimate of the presence of a regulator in a tissue extract. Such bioassays made it possible not only to determine if a chemical fraction of a gland had the biological activity that represented the sought-after regulator, but the quantitative assays could provide an estimate of how much of the regulator was present. Through their use, bioassays made it possible to obtain seasonal data on fluctuations in chemical regulator levels. Furthermore, they could be used to isolate chemical regulators and eventually to obtain highly purified preparations. Bioassays made it possible to identify and test suspected agonists and antagonists, too, and bioassays were used to validate the methods now used to estimate regulator levels in tissues and body fluids.

Hormones and other chemical regulators operate at minute concentrations, typically in the ranges of micrograms, nanograms, or even picograms per milliliter of blood or extracellular fluid. This means 10^{-6}, 10^{-9}, or 10^{-12} g/ml (or 1 millionth, 1 billionth, or 1 trillionth of a gram, respectively). These natural levels are often referred to as "physiological" and should be considered when designing experiments to observe the actions of a regulator. Treatment levels that produce circulating levels in excess of natural or optimal levels are termed **pharmacological doses** (commonly in the milligram range). Interpretations of such pharmacological studies may be complicated with respect to understanding natural functions of regulators. Large doses greatly increase the probability of paradoxical or toxic side effects.

Until the advent of sensitive biochemical procedures that allowed us to measure actual hormone levels, endocrinology was limited to extirpation–replacement, "inject 'em and inspect 'em," or crude estimates based on bioassays. Many of these studies emphasized pharmacological doses that produced excessive and possibly unnatural responses. Some of the molecular approaches that have led to a revolution in our understanding of the nature of chemical regulation and produced an unprecedented explosion in data generation are described in the following sections.

B. Radioimmunoassay

Once purified molecules became available for experimentation, a number of techniques developed that allowed endocrinologists to routinely measure regulator levels in body fluids, localize specific molecules in a particular cell or part of a cell, and monitor these parameters under a multitude of experimental conditions. The first major technical breakthrough was the development of the **radioimmunoassay** (or **RIA**) by Rosalind Yalow and Solomon Berson. While studying the occurrence of insulin resistance in patients with diabetes mellitus, these investigators noted that these people had antibodies to insulin in their blood. In part because they had difficulty in convincing the scientific community of the validity of their observations, Yalow and Berson went ahead and showed that addition of excess insulin could displace radioactively labeled insulin from these antibodies. This demonstration, together with their fully developed mathematical treatment of these interactions, provided the foundations for the RIA, a technique that quickly revolutionized the entire field of endocrinology. Today, many RIAs are used routinely to measure all sorts of chemical regulators in nanogram and even picogram quantities, a feat never possible with bioassay systems. Because of his early death at age 54, Berson did not share the Nobel Prize awarded to Yalow in 1977 in recognition of the importance of their contribution to the entire field of physiology and medicine.

Development of any RIA relies, first, on the availability of a pure source of regulator that can be used to induce formation of highly specific antibodies against it. Second, one must be able to radioactively tag or label a quantity of the regulator. It is assumed that the antibody cannot distinguish between an unlabeled or "cold" molecule of the regulator and a labeled or "hot" one and will bind each ligand with equal affinity. A scientist simply sets up a balanced and carefully controlled competition between cold and hot regulators for the binding sites on antibody molecules. The competition is designed with a constant amount of antibody and constant amount of hot ligand in the reaction mixture. When no cold ligand is present, the amount of hot ligand bound to the antibody at the end of a prescribed period of time is defined as 100% binding. Then, additional mixtures are prepared with increasing quantities of cold ligand. As the quantity of cold ligand increases, the competition for antibody-binding sites increases in favor of the cold ligand so that less and less hot ligand is bound. Consequently, the percentage of hot ligand bound decreases with the addition of increasing amounts of cold ligand. The antibody–ligand complexes are then precipitated, leaving the unbound hot and cold ligands in the supernatant. By measuring the radioactivity of either the precipitate or the supernatant, the percentage of bound (or unbound; also called "free") hot ligand (dependent variable) can be plotted against the known concentrations of cold ligand added (independent variable; see Fig. 3-3). Radiation detection equipment (scintillation counter or gamma counter) is used to measure the amount of free or bound radioactivity. The plot of percent bound in Fig. 3-3 constitutes a standard curve from which the quantity of cold ligand in a sample of unknown quantity (the "unknown") can be estimated. A similar competition is set up with the same quantity of antibody and hot ligand plus a sample of the unknown solution in which we wish to determine the level of cold ligand. By determining the percent bound for the unknown sample (y axis), the quantity of cold ligand can be estimated by extrapolating to the x axis.

At first, this technique proved useful only for peptides, but soon it became possible to "trick" antibody-synthesizing cells into making specific antibodies against all manner of chemical substances. Hence, we can measure thyroid hormones and steroids with the same ease that we measure insulin or thyrotropin (TSH).

C. Immunoradiometric Assay

The **immunoradiometric assay (IRMA)** combines the use of radiolabeling with immunological precision to detect virtually all of the molecules of a particular hormone in a plasma sample. A similar method involves labeling the antibody with a fluorescing compound. These approaches have revolutionized the measurement of peptide hormones such as corticotropin (ACTH) and TSH. The IRMA employs two antibodies plus beads coated with the protein **avidin.** The first antibody, which is directed against the amino (or N) terminis of the hormone, is labeled with a radioactive tag. The second antibody is directed toward the carboxy (or C)-terminal end of the peptide hormone. In addition, the second antibody is combined with **biotin,** a molecule that binds specifically to avidin. The two antibodies form a sandwich around the hormone; the sandwich in turn is bound to the

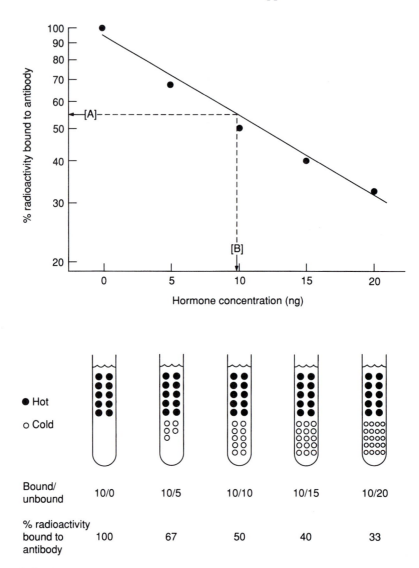

Figure 3-3. Standard curve for radioimmunoassay (RIA). A standard curve is prepared by placing the same amount of antibody and radiolabeled (hot) hormone into every test tube and adding a different known amount of unlabeled (cold) hormone to each tube, so that cold and hot hormone will compete for the same binding sites on the antibody. The greater the amount of cold hormone added, the lower will be the amount of radioactivity bound to the antibody. The antibody with its bound load of hot and cold hormone is precipitated from the solution and either the radioactivity of the supernatant or of the precipitate is counted. Thus, one can plot the percentage of hot hormone bound to the antibody against the amount of cold hormone in the solution to produce a standard relationship, or what is called a *standard curve*. If an additional tube is prepared with the same amounts of antibody and hot hormone but with an unknown amount of cold hormone (such as might be present in a blood sample), one can determine the percentage of hot hormone bound [A] and extrapolate from the standard curve to estimate how much cold hormone [B] was present in the blood sample.

avidin-coated bead (Fig. 3-4). When used with an excess of both antibodies, virtually all of the peptide hormone in the sample is captured, resulting in greater sensitivity than is pos-

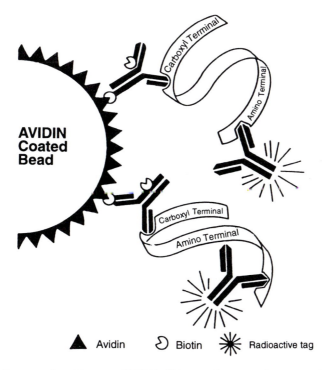

AVIDIN
Coated
Bead

Carboxyl Terminal

Amino Terminal

Carboxyl Terminal

Amino Terminal

▲ Avidin   ♘ Biotin   ✳ Radioactive tag

Figure 3-4. Immunoradiometric assay (IRMA). This procedure is an improvement over the standard RIA for measuring peptides and utilizes beads coated with the protein avidin, to which is bound antibodies to the C-terminal end of the peptide to be measured. A second antibody to the N terminis of the same peptide is also used to create a sandwich. With this approach, only intact peptides will be detected and not partially degraded ones. [Reprinted by permission from Findling, J. W., Engeland, W. C., and Raff, H. The use of immunoradiometric assay for the measurement of ACTH in human plasma. *Trends Endocrinol. Metab.* **1**, 283–287. Copyright 1990 by Elsevier Science Inc.]

sible with an RIA. Because the antibodies are directed against the terminal ends of the molecules, partially degraded fragments are not detected. Unfortunately, the highly specific nature of this detection procedure means that **heterologous assays,** i.e., assays performed in one species employing antibodies prepared against the hormone of another species, will be unacceptable unless exhaustive efforts are made to ensure validation and correlation with results using existing RIA methods.

D. High-Performance Liquid Chromatography

Chemists, while searching for techniques to improve the ability to separate closely related molecules from one another, developed a sophisticated separation system known as **high-performance liquid chromatography** or **HPLC.** This technique is a modification of column chromatography that operates under high pressures and relies on differential solubility of molecules in the solvents employed to wash the columns and the affinities of these same molecules for the substances of which the columns are made. A mixture of molecules in a particular solvent is applied to the column and the solvent collected in aliquots as it comes off at the bottom. If properly designed, all of one type of molecule will come off in the same

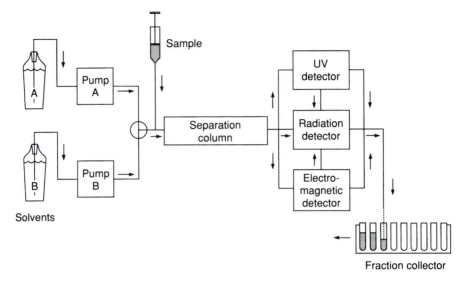

Figure 3-5. High-performance liquid chromatography (HPLC). A sample is applied to a separation column in a particular solvent. Additional solvents can be mixed and pumped through the separation column and, depending on the affinity of the sample components for the column and the solvents, the sample components will leave the column at different times and pass through some kind of detector. Some compounds, such as steroids, are best detected and quantified by their absorbance of ultraviolet light (UV) whereas others (e.g., catecholamines) can be quantified using electrochemical detectors. HPLC can be coupled to radiation detectors for identifying a labeled molecule or specific metabolic products of a radioactive precursor, and the radiation detector may be used in tandem with either UV or electrochemical detectors. Various fractions from the column may be recovered by using a fraction-collecting device.

aliquot. What aliquot contains a given type of molecule depends on the affinity of each type for the column versus the solvent. Special detectors allow investigators to identify quantitatively the presence of molecules in various aliquots. Such an approach allows investigators to quantitatively determine all of the steroids and steroid metabolites or biogenic amine neurotransmitters and their metabolites in a given sample simultaneously (see Fig. 3-5). With an RIA, a separate analysis must be done for each molecule you expect to find.

E. Immunohistochemistry

Another analytical use for antibodies is to localize particular regulators, synthesizing enzymes, or degrading enzymes in tissues, cells, or parts of cells. This approach, called **immunohistochemistry** (or immunocytochemistry), employs an antibody made in one species, say a mouse, against the specific molecule (antigen) you wish to localize in the brain of, say, a song bird. There are several variations of this technique available (see Fig. 3-6), but the basic procedure is similar. First, you make the mouse antibody (which would be a mouse gammaglobulin protein) and apply that to sections of the tissue placed on a microscope slide. Theoretically, the antibody will bind only to cells that contain the antigen. In another species, you might make an antibody against mouse gammaglobulin and conjugate that with an enzyme known as a peroxidase. This is also applied to the tissue sections. This second antibody will bind to the gammaglobulin, which has bound previously to the

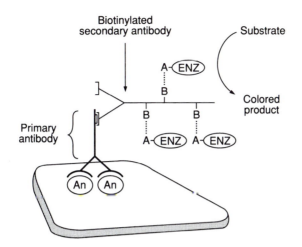

Figure 3-6. Immunocytochemistry. This method requires a pure source of antigen (An) to make a specific antibody (primary antibody) in, for example, a mouse. A secondary antibody is made against mouse immunoglobulins in, for example, a goat. The secondary antibody, which will recognize any mouse antibody, is complexed with several biotin molecules per secondary antibody molecule (i.e., biotinylated). Biotin has a high binding affinity for avidin. The primary antibody is applied to the cell or tissue section on a microscope slide followed by the biotinylated goat secondary antibody. Then a specific enzyme complexed to avidin is added. This step amplifies the response as several enzymes can be complexed to each secondary antibody. Addition of substrate for that enzyme and a chromagen yields a colored product that appears in the immediate vicinity of the antigen. This technique can be used with a fluorescing compound instead of an enzyme and can be adapted for use with the transmission electron microscope as well.

song bird antigen. Next, you add a substrate for the peroxidase enzyme that results in a colored product that will be localized in the cells containing the antigen–gammaglobulin–antigammaglobulin–peroxidase complex (Fig. 3-7).

There are other variations on this basic technique using different enzyme–substrate markers that allow one to examine more than one antigen in a single section. **Immunofluorescence** is a modification in which a compound is attached to the antigammaglobulin; this compound, rather than the peroxidase enzyme, will fluoresce under certain wavelengths of light (usually ultraviolet).

Another related approach is the **enzyme-linked immunoabsorbent assay (ELISA),** which is used to detect the presence of a specific molecule. Molecules of a specific regulator are coated onto walls of special microtiter plates (used commonly in immunological studies), where they compete with free molecules in plasma or a tissue extract for a specific antibody. The peroxidase–antiperoxidase method is then used to reveal the immobilized hormone–antibody complexes.

F. Techniques for Determining the Number and Characteristics of Receptors

For a protein to be a receptor for a chemical regulator, it must have certain properties. Because of the small number of regulator molecules usually present, a receptor must have a high **affinity** for the regulator; that is, a strong tendency to bind the regulator. Second, the

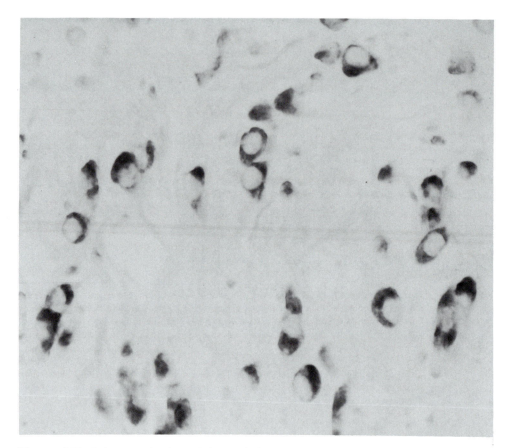

Figure 3-7. Immunocytochemistry of glucocorticoid receptors. Demonstration of immunoreactive glucocorticoid receptor-like material in neurons of the midbrain tegmentum of the brain of kokanee
salmon, *Oncorhynchus nerka kennerlyi.* The peroxidase procedure was used with a mammalian antibody prepared against the type-II mammalian glucocorticoid receptor to demonstrate
presence of immunoreactive material (dark staining) in the cytoplasm of these cells. Photograph courtesy of Laura L. Carruth, University of Colorado, Boulder.

receptor should have high **specificity** for the regulator and little tendency to bind other
molecules. A third feature of receptors is their low **capacity.** This means that all available
receptor sites are occupied at relatively low concentrations of regulator; that is, the receptor
is said to be **saturated.** Finally, the distribution of a putative receptor should correspond to
the known target tissues for the regulator and should be correlated with some biological
effect.

Many regulators may "stick" to proteins other than their specific receptors, especially
in cell or tissue homogenates, where the process of disrupting the normal cell architecture
may unmask potential binding sites normally not available to the regulator. Thus, it is
important to distinguish **specific binding** (high-affinity, low-capacity binding proteins)
from **nonspecific binding** (low-affinity, high-capacity binding proteins). These nonspecific
binding sites do not saturate unless huge doses of regulator are applied. To distinguish
between specific and nonspecific binding, we take advantage of the availability of radioactively labeled regulators to determine the total binding capacity of a cell or tissue homog-

enate. Then, by adding an excess of unlabeled regulator to another sample also containing the labeled regulator, a competition is set up for the small number of high-affinity, low-capacity sites (the true receptors), causing labeled regulator to be displaced from only the specific binding sites. Because the nonspecific binding sites have such a high capacity, there is little competition occurring there. This competition is done using a range of unlabeled concentrations, and the difference between total binding (homogenate sample without un-labeled excess) and nonspecific binding (homogenate with unlabeled excess) provides an estimate of specific binding (Fig. 3-8). However, since specific binding is a function of both the affinity of the receptors as well as the number of receptors, simply demonstrating spe-cific binding cannot be used for comparative purposes. Therefore, investigators compare the kinetics of the regulator–receptor binding, much like you would do for studies of sub-strate–enzyme interactions, to provide an estimate of both the affinity of a receptor for the specific regulator in question as well as an estimate of the number of receptors present. The most commonly used procedure for kinetic studies is the **Scatchard analysis,** a relation-ship first employed by G. Scatchard in 1949. The analysis of the experimentally derived data is similar to that done for enzymatic kinetics. Although this method involves certain assumptions and necessitates careful, repeatable laboratory procedures, it provides infor-mation on affinity (slope of the line), receptor number (y-axis intercept), and purity of the preparation (straight line vs. curvilinear plot).

A substance bound by a receptor can generally be termed a **ligand.** The chemical kinetics you have studied for reversible enzyme–substrate interactions are appropriate for describ-ing ligand–receptor interactions. Normally, a ligand [L] binds reversibly with its receptor [R] to form a **ligand–receptor complex** [LR]:

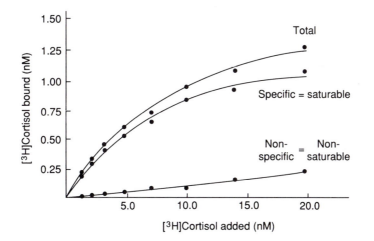

Figure 3-8. Specific and nonspecific binding by receptors: Determination of saturable (receptor or "specific") and nonsaturable (nonreceptor or "nonspecific") binding of cortisol to a liver cytosolic preparation using [³H] cortisol as the ligand. Nonsaturable binding is not greatly affected by the amount of ligand since it has a high capacity. This is determined by adding a large excess of unlabeled cortisol in an attempt to displace the saturable receptors (specific binding) while not affecting the nonsaturable (nonspecific) binding. The difference between the binding observed with and without the excess of unlabeled ligand is defined as the satur-able or specific binding. [After Chakraborti, P.K., and Weisbart, M. (1987). High-affinity cor-tisol receptor in the liver of the brook trout, *Salvelnus fontinalis* (Mitchell). *Can J. Zool.* **65,** 2498–2503.]

$$[L] \quad + \quad [R] \quad \rightleftharpoons \quad [LR]$$

free ligand   unoccupied receptor   bound ligand = occupied receptor.

The affinity or association of the ligand for the receptor is described by an **association rate constant, k_1,** and the tendency for LR to dissociate to L + R is described by the **dissociation rate constant, k_2.** At equilibrium,

$$k_1[L][R] = k_2[LR].$$

This can be expressed as

$$\frac{[L][R]}{[LR]} = \frac{k_2}{k_1} = K_d,$$

where K_d represents the dissociation constant. The reciprocal of the dissociation constant is the association constant K_a, and

$$K_d = \frac{1}{K_a}.$$

Therefore,

$$K_a = \frac{[LR]}{[L][R]}.$$

From this equation, you can see that the K_a is equal to the ratio of bound to free ligand, or [LR]/[L].

If a fixed number of target cells (in other words, a fixed number of receptors) is incubated in replicates with increasing concentrations of ligand in each replicate, the number of occupied receptors also will increase until all of the available receptors are complexed to ligand (saturation, or 100% bound). At saturation, the number of bound ligand molecules equals the number of available receptors while the ratio of bound ligand to free ligand approaches zero. If we plot the bound/free ratio for each replicate incubation against the bound concentration of ligand, the resulting theoretically straight line will intercept the y axis of our graph at a point that defines the total number of receptors or the receptor capacity. This graph (Fig. 3-9) is called a **Scatchard plot.** The slope of the line on the Scatchard plot is equal to the negative value of the K_a (or $-1/K_d$). Thus, by determining only the bound ligand and the ratio of bound ligand to free ligand (where the number of receptors is kept constant and only the concentration of free ligand varies) one can extrapolate the number of receptors and the association constant for that particular receptor.

Most Scatchard plots, however, do not yield a straight line but rather a downward curving line. The most common explanation for repeated observations of this sort is that one is dealing with a heterogeneous mixture of the high-affinity and low-affinity sites. Special mathematical methods are available to separate these binding sites and allow differentiation of the numbers of high- and low-affinity sites.

G. Reverse Hemolytic Plaque Technique

The **reverse hemolytic plaque technique** is sensitive enough to detect hormone release from a single cell and is based on antibody-directed, complement-mediated lysis of red blood cells (RBCs). This technique was developed from a well-established procedure for detecting bacterial or viral agents. Antigen that has an affinity for a particular antibody is

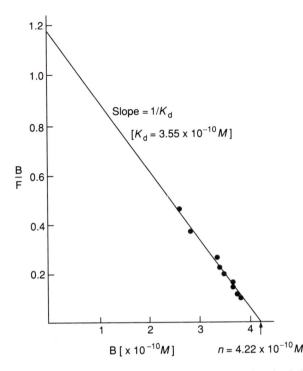

Figure 3-9. Scatchard plotting method for receptor determination. The Scatchard plot is described in text. B, Bound ligand; F, free ligand. [From Bolander, F. F. (1989). "Molecular Endocrinology." Academic Press, San Diego.]

coated on the surface of blood cells and a large number of these cells is mixed with a small number of dispersed hormone-secreting cells, such as pituitary cells; this mixture is then spread over the surface of a glass microscope slide. An antibody to the hormone is added, which can then complex with any hormone secreted by the pituitary cell as well as with the antibody antigen on the RBC. Complement from blood serum is added and the antibody–hormone–RBC complex facilitates the interaction of the RBCs with the complement, resulting in lysis of the RBCs. Thus, all RBCs in the vicinity of a secreting pituitary cell would be lysed, leaving a clear space, or *plaque,* on the slide.

H. Nucleic Acid Approaches in Endocrinology

The **polymerase chain reaction (PCR)** is another amplifying technique that allows the investigator to increase a small quantity of DNA to amounts that can be analyzed readily. The PCR can be used to create cDNA probes, to analyze DNA sequences, to screen for mutations of specific genes, and to amplify a specific mRNA using reverse transcription (making DNA from RNA instead of making RNA from DNA) coupled with PCR to amplify the product. cDNA probes can be prepared that bind to specific mRNAs and ascertain when a certain gene is turned on in a given cell. (cDNA is "copy" DNA made from RNA using a reverse transcription process.)

Transgenic animals are lines of animals produced by inserting multiple copies of a specific gene or genes into a fertilized egg. These genes may be responsible for producing a specific hormone such as growth hormone (GH). Insertion of GH genes into mice resulted in a phenotype characterized by the production of exceptionally large quantities of GH, and this trait was passed on to their offspring. In other cases, investigators may insert "reporter" genes that encode some unique proteins and are linked to a known hormone-responsive element. This reporter gene will only function once the appropriate transcription factor is present.

Blotting is another popular technique that can be used to identify gene activity at any of several levels. DNA extracts from a tissue or cell culture are subjected to electrophoresis in a gel, which spreads the DNAs out according to their size and nucleotide composition. Next, the DNA is transferred passively to a nitrocellulose filter that is laid over the gel. The DNA molecules migrate directly from their positions in the gel onto the filter. A radioactively labeled cDNA constructed for a particular gene can then be applied as a probe to the filter. The specific sequence of nucleotides in the cDNA probe will bind to the closest complementary sequence in the gel and thus indicate the presence of the transcription product of that gene. This procedure is called a **Southern blot** (named after its developer, E. M. Southern). A similar approach using a cDNA probe to identify a specific gene in an mRNA extract was named a **Northern blot,** apparently because mRNA is a mirror image of Southern's DNA. Later, identification of a specific polypeptide or protein gene product in similarly prepared blots from protein extracts became a **Western blot.** A specific protein or peptide is localized by using a labeled antibody probe. Finally (although other directional designations are still possible), the **Southwestern blot** was devised for identifying a transcription factor (protein) in a gel using a labeled DNA-recognition binding site probe.

Genetic engineering techniques provide another way to examine the actions of a gene regulator by constructing unique genes, inserting them into cells, and examining the actions of regulators. To simplify the determination of when a gene has been activated, a structural gene for a product that is easily measured is complexed to the regulatory unit of the test gene of interest. These **reporter genes** usually are genes producing certain enzymes. Once the nucleotide sequence of the test gene is known, the regulatory portion of that test gene can be complexed to a reporter gene that is easy to detect. Some common reporter genes encode luciferase, β-galactosidase, and chloramphenicol acetyltransferase (CAT). Numerous copies of these test gene–reporter gene complexes are inserted into cells. Addition of a hormone or other activator of the test gene will allow the hormone to bind to the regulator units of the normal gene plus the regulator unit complexed to the reporter genes, resulting in synthesis of the reporter enzyme. The occurrence of easily detected enzymatic activity following addition of substrate indicates successful activation. For example, hydrolysis of the substrate luciferin by luciferase yields light (this is the mechanism for bioluminescence observed in fireflies and many marine invertebrates). This system also is useful for studying factors that might prevent reporter gene activation.

Suggested Reading

Books

Beesley, J. E. (1993). Immuncytochemistry: A Practical Approach. IRL Press at Oxford Univ. Press, Oxford.

dePablo, F., and Scanes, C. G. (1993). "Handbook of Endocrine Research Techniques." Academic Press, San Diego.

The Endocrine Society (1996). "Introduction to Molecular and Celluluar Research," Syllabus. The Endocrine Society Press, Bethesda, MD.

Martin, B. M. (1994). "Tissue Culture Techniques: An Introduction." Birkhaeuser, Basel.

Articles

Barsano, C. P., and Baumann, G. (1989). Editorial: Simple algebraic and graphic methods for the apportionment of hormone (and receptor) into bound and free fractions in binding equilibria; or how to calculate bound and free hormone. *Endocrinology* **124**, 1101–1106.

Eidne, K. A. (1991). The polymerase chain reaction and its uses in endocrinology. *Trends Endocrinol. Metab.* **2**, 169–175.

Ekins, R. (1990). Measurement of free hormones in blood. *Endocr. Rev.* **11**, 5–46.

Findling, J. W., Engeland, W. C., and Raff, H. (1990). The use of immunoradiometric assay for the measurement of ACTH in human plasma. *Trends Endocrinol. Metab.* **1**, 283–287.

Nilson, J. H., Keri, R. A., and Reed, D. K. (1995). Transgenic mice provide multiple paradigms for studies in molecular endocrinology. In "Molecular Endocrinology: Basic Concepts and Clinical Implications" (B. D. Weintraub, ed.), pp. 77–94. Raven, New York.

Segre, G. V. (1990). Advances in techniques for measurement of parathyroid hormone: Current applications in clinical medicine and directions for future research. *Trends Endocrinol. Metab.* **1**, 243–247.

Shizuru, J. A., and Sarventnick, N. (1991). Transgenic mice for the study of diabetes mellitus. *Trends Endocrinol. Metab.* **2**, 97–104.

Stewart, T. A. (1994). Models of human endocrine disorders in transgenic rodents. *Trends Endocrinol. Metab.* **5**, 136–141.

Weiss, J., and Jameson, J. L. (1993). Perfused pituitary cells as a model for studies of gonadotropin biosynthesis and secretion. *Trends Endocrinol. Metab.* **4**, 265–270.

4

Organization of the Mammalian Hypothalamo– Hypophysial Axis

THE **hypothalamo–hypophysial axis** consists of the neurosecretory (NS) neurons forming the NS nuclei of the hypothalamic region of the brain and the **hypophysis** or **pituitary gland** (Fig. 4-1) plus the glands and other targets they directly control (Fig. 4-2). It represents the neuroendocrine link between the nervous system and the traditional endocrine system. It is a thoroughly integrated system with the hypothalamic NS neurons receiving innervations from ordinary neurons. Both these neurons and the NS neurons are responsive to direct and feedback effects of hormones present in the blood or cerebrospinal fluid (CFS).

The pituitary gland often has been called the "master gland" because its hormones control many diverse systems all essential for survival and reproduction. The discovery of the complex control of the pituitary by the hypothalamus has transferred our attention to the brain as being the "master gland," and the pituitary no longer is considered to be an autonomous or omnipotent regulator of physiology, but only another part of a major system for homeostatic regulation.

The embryonic vertebrate brain develops as four major regions known as the **telencephalon** (anteriormost), **diencephalon, mesencephalon,** and **rhombencephalon** (posteriormost). During subsequent development these regions differentiate into the major components of the adult brain (Fig. 4-3). The telencephalon becomes the **olfactory bulbs,** the **olfactory lobes,** and the **cerebral hemispheres.** The diencephalon differentiates into three regions: the dorsal **epithalamus,** the central **thalamus,** and the ventral **hypothalamus.** Most of the diencephalon becomes the thalamus, a major relay station between higher and lower portions of the brain. The floor or ventral portion of the diencephalon becomes the hypothalamus, containing various NS nuclei, which are sources for the neurohormones involved with regulation of pituitary function. The epithalamus is derived from the roof of the diencephalon and gives rise to the endocrine **epiphysial complex** that includes the **pineal gland** (see Chapter 6). The mesencephalon gives rise primarily to the **optic tectum,** and the rhombencephalon differentiates into the **cerebellum** and **medulla.**

The pituitary gland or hypophysis, which is located directly beneath the third ventricle of the brain, develops through fusion of a ventral growth or evagination from the diencephalon, the **infundibulum,** with an ectodermal sac known as **Rathke's pouch** (Fig. 4-4). The latter develops as an inward pocketing or invagination off the anterior roof of the oral cavity. The third ventricle is a cavity continuous with the other ventricles of the brain and the central canal of the spinal cord. It is filled with cerebrospinal fluid (CSF). Recent studies employing amphibian embryos have shown that the secretory cells of the adenohypophysis and the neurosecretory neurons of the hypothalamus have a common origin from the neural ridge of the embryo (Fig. 4-5). Thus, the connection between the infundibulum and Rathke's pouch appears to be a development that functionally reunites these cells again.

The term hypophysis is derived from *hypo,* under (the brain), + *physis,* growth. Its alternate name, pituitary gland, is derived from *pituita,* meaning slime or phlegm. The pituitary, located behind the nose and above the nasal and oral cavities in humans, was believed to be the source of phlegm, one of the four humors of the body proposed by Galen centuries ago. The other humors were blood, black bile, and yellow bile.

The mammalian hypothalamo–hypophysial axis will be discussed in this chapter, followed by that of nonmammalian vertebrates in Chapter 5. Although in an evolutionary sense such a discussion should begin with agnathan fishes and proceed through the other fishes and tetrapods, the mammalian system is better understood and provides the nomen-

A

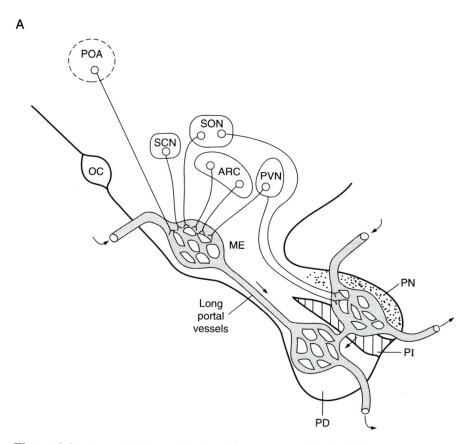

Figure 4-1. A generalized mammalian hypothalamo–hypophysial axis. (A) Hypothalamic region of brain showing several neurosecretory centers (POA, preoptic area; SCN, suprachiasmatic nucleus; ARC, arcuate nucleus; SON, supraoptic nucleus; PVN, paraventricular nucleus; OC, optic chiasm), the nuerohemal median eminence (ME) and pars nervosa (PN), as well as the glandular pars distalis (PD) and pars intermedia (PI). The PD, PI, and pars tuberalis (not shown) constitute the adenohypophysis. A distinct PI is not seen in some mammals. Long portal vessels connect the median eminence to the capillary bed of the adenohypophysis and short portal vessels may also connect the capillary beds supplying the PI to the PD. (B) Relationship between hypothalamic ns neurons through neurohormones released into the portal system and specific hormone-secreting cells in the pars distalis. See text for details.

clature with respect to structures, hormones, and functions that later were applied to nonmammalian systems. In fact, the entire field of endocrinology has branched in similar fashion from mammalian investigations, largely of a clinical orientation and motivation, to nonmammalian systems.

I. The Mammalian Hypophysis

The hypophysis of adult mammals is located ventral to the brain just posterior to the optic chiasma, and it remains attached to the hypothalamus by a stalklike connection (Fig. 4-1). It is separable into two regions, the **adenohypophysis** derived from Rathke's pouch and the

B

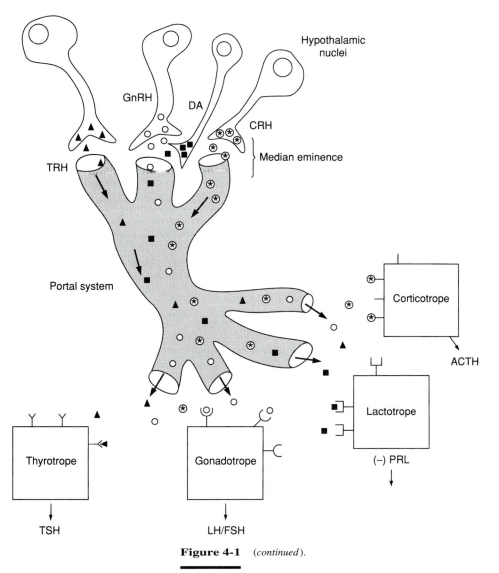

Figure 4-1 *(continued)*.

neurohemal **neurohypophysis,** which develops from the infundibulum. The anatomical terminology used here for all vertebrate pituitaries is based on Green (1951) and Purves (1961).

The adenohypophysis is an epithelial glandular structure (*adeno,* gland) and can be subdivided into three regions or pars (or bodies): the **pars anterior** (or **pars distalis**), the **pars tuberalis,** and the **pars intermedia.** These regions are distinguished by their cytological features as well as their anatomical relationships to the neurohypophysis. Two distinct subregions can be identified in the neurohypophysis: the more anterior pars eminens or **median eminence** and the **pars nervosa.** The two neurohemal regions of the neurohypophysis consist mostly of aminergic and peptidergic axonal endings mixed with blood capillaries

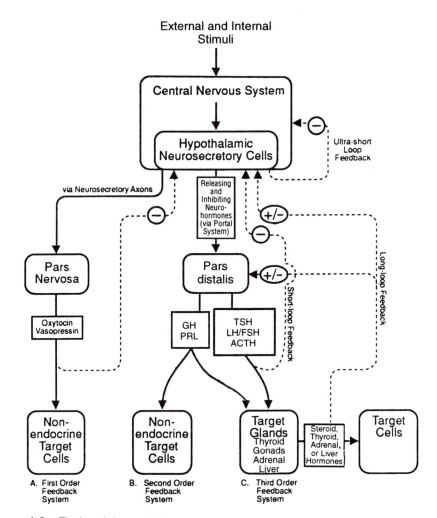

Figure 4-2. The hypothalamo–pituitary axes with major feedback loops. PRL is under strong inhibitory hypothalamic control and feedback loops may not be as important as for other tropic hormones. Release of TSH, LH, FSH, and ACTH is under stimulatory hypothalamic control and has well-defined feedback loops. GH release requires removal of inhibitory control and stimulation by a hypothalamic neurohormone. Some of the actions of GH may have feedback effects (e.g., levels of amino acids; see text).

and what are probably modified glial cells called **pituicytes.** An extensive vascular portal system, the **hypothalamo–hypophysial portal system,** develops between the median eminence of the neurohypophysis and the pars anterior of the adenohypophysis (Fig. 4-6). This portal system carries blood from the median eminence directly to the epithelial cells of the pars anterior (Fig. 4-1B).

The hypothalamo–hypophysial portal system forms a neurovascular link between the hypothalamus and the pituitary gland as described in the pioneering anatomical studies of Wislocki. Presumably, blood containing the neurohormonal regulators flows from the me-

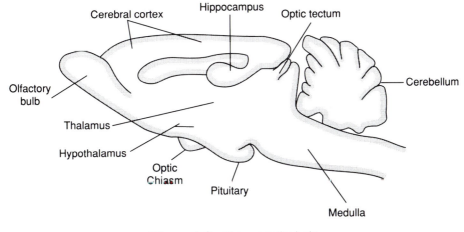

Figure 4-3. The mammalian brain.

dian eminence to the pars anterior, and the venous drainage from the latter carries pituitary tropic hormones into the general circulation. Additional data suggest that there may be significant blood flow from the adenohypophysis to the hypothalamus as well, and such **retrograde flow** may prove to be important in actions of pituitary hormones on the central nervous system.

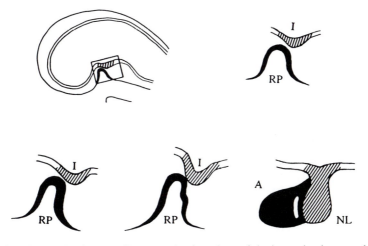

Figure 4-4. Pituitary development. Representation in embryo of the interaction between the infundibulum (I) extending down from the developing brain and contacting the evagination of Rathke's pouch (RP) from the embryonic pharynx resulting in formation of the pars nervosa (NL) from the infundibulum and the adenohypophysis (A) from Rathke's pouch. [Reprinted by permission of the publisher from Dubois, P. M., and ElAmraoui, A. Embryology of the pituitary gland. *Trends Endocrinol. Metab.* **6**, 1–7. Copyright 1995 by Elsevier Science Inc.]

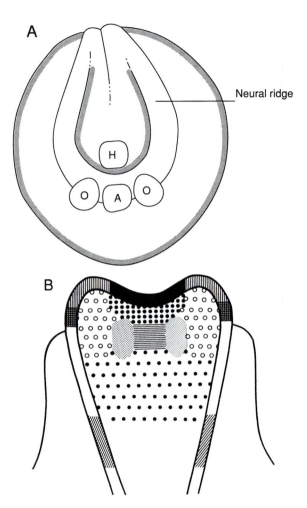

Figure 4-5. Anterior neural ridge origin for endocrine cells of hypothalamus and adenohypophysis. (A) Neurula of anuran amphibian embryo showing close proximity of origins for hypothalamus (H), adenohypophysis (A), and olfactory placodes (O). GnRH cells of the anterior hypothalamus have their origin from cells that migrate later from the olfactory placodes through the forebrain. Based on communications from K. Kawamura and S. Kikuyama. (B) Similar development in the avian embryo. ▮, Adenohypophysis; [●●●], hypothalamus; ▦, neurohypophysis; ▭, optic vesicles; [○ ○], telencephalon; [● ●], diencephalon; ▥, ectoderm of nasal cavity; ▤, olfactory placode; ▨, mesencephalic neural crest. [Reprinted by permission of the publisher from DuBois, P. M., and El Amraoui, A. Embryology of the pituitary gland. *Trends Endocrinol. Metab.* **6,** 1–7. Copyright 1995 by Elsevier Science Inc.]

Sympathetic fibers are known to innervate pituitary blood vessels and influence blood flow through the hypothalamo–hypophysial portal system. In addition, peptidergic fibers containing immunoreactive neuropeptides have been demonstrated, although their significance is unknown.

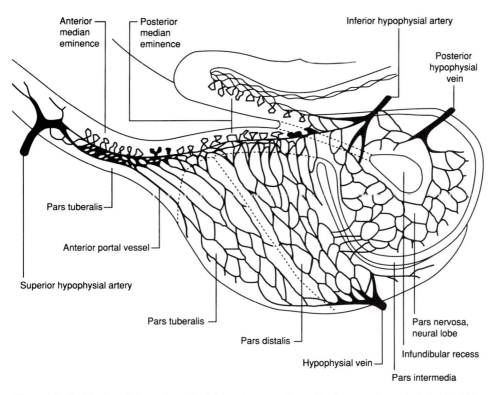

Figure 4-6.   The hypothalamo–hypophysial portal system of a cat. The long portal vessels drain blood from the median eminence to the pars distalis. Note that a few short portal vessels connect the pars intermedia and the pars distalis as well. [From Matsumoto, A., and Ishii, S. (1992). "Atlas of Endocrine Organs." Springer-Verlag, Berlin.]

A. Subdivisions of the Adenohypophysis

1. The Pars Tuberalis

The pars tuberalis develops from the lateral portions of Rathke's pouch and consists of a thin layer of cells projecting rostrally (anteriorly and dorsally) from the adenohypophysis. It is in contact with the median eminence of the neurohypophysis, and the blood vessels of the hypothalamo–hypophysial portal system pass near or through the pars tuberalis en route to the pars distalis.

The pars tuberalis is characteristic of all tetrapod vertebrates, but its function(s) has not been clarified. Evidence suggests that it is only an extension of the pars anterior related primarily to reproduction.

Structurally, the cells of the pars tuberalis are connected to the cerebrospinal fluid of the third ventricle in the brain through cellular processes originating in modified **ependymal cells** known as **tanycytes** (Fig. 4-7). Ependymal cells are epithelial cells that line the ventricles of the brain and surround the central canal of the spinal cord and form a protective layer that surrounds the nervous system. It has been suggested that tanycytes may selectively remove molecules, including various types of regulators, from cerebrospinal fluid and transfer them to cells of the pars tuberalis, causing the latter to release their stored

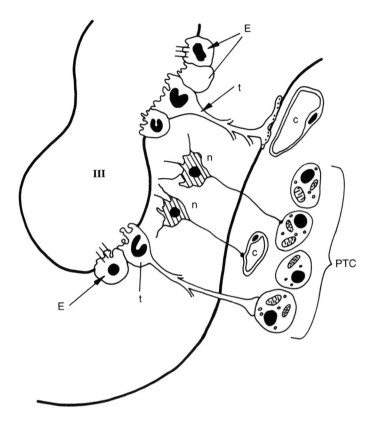

Figure 4-7. Relationship of modified ependymal cells lining the third ventricle to the pars tuberalis. Modified ependymal cells (E) or tanycytes (t) provide connections between the third ventricle (III) and capillaries (c) or pars tuberalis cells (PTC). Similar connections by neurons (n) also have been observed.

products. Although this is a highly speculative idea, such an interesting anatomical relationship demands some imaginative research to provide a better understanding of both tanycytes and the cells of the pars tuberalis.

2. Pars Intermedia

The portion of the adenohypophysis that makes contact with the pars nervosa of the neurohypophysis is defined as the pars intermedia. Indeed, differentiation of the pars intermedia occurs only if physical contact takes place with the infundibulum during development. Some mammals, such as whales, manatees, elephants, armadillo, pangolin, beaver, and adult humans, lack a pars intermedia.

The term **neurointermediate lobe** designates both the pars intermedia and the pars nervosa as an anatomical unit. In some species, the pars intermedia becomes separated from the remainder of the adenohypophysis by a cavity or cleft, accentuating the designation of a neurointermediate unit.

Only one glandular cell type appears in the mammalian pars intermedia, and it is supposedly responsible for secretion of the peptide hormone melanocyte-stimulating hormone or **melanotropin (MSH).** MSH stimulates skin cells, known as **melanocytes,** to synthesize a brown pigment, **melanin,** which causes increased deposition of pigment in the skin or hair. Most of those mammals lacking a pars intermedia lack hair and/or have few melanocytes.

3. Pars Anterior or Pars Distalis

The major portion of the adenohypophysis originally was designated as the pars anterior, and a variety of cellular types were identified there by selective staining procedures. In these animals lacking a true pars intermedia, the pars anterior was called the pars distalis. However, this latter term is now applied routinely as if it were synonymous with the pars anterior, especially since pars distalis was the appropriate term for adult humans and was so named in most textbooks. Because of extensive use of pars distalis in the literature for both anatomical designations, we will use the term "pars distalis" exclusively in this text to avoid confusion when the student encounters the modern literature. Likewise, the term "anterior lobe" is applied loosely to mean either the pars distalis or the pars anterior plus the pars intermedia. "Posterior lobe" often is used in place of neurointermediate lobe, but at other times this term indicates only the pars nervosa in animals lacking a pars intermedia. Because of their variable meanings, we will not use the terms "anterior" and "posterior" lobes.

Several cellular types in the pars distalis are responsible for secretion of the pituitary tropic hormones: **corticotropin,** or adrenocorticotropic hormone (**ACTH); thyrotropin,** or thyroid-stimulating hormone (**TSH); growth hormone** or somatotropin (**GH); prolactin (PRL);** and two **gonadotropins,** or gonadotropic hormones (**GTHs).** The two GTHS are **follicle-stimulating hormone** or follitropin (**FSH)** and **luteinizing hormone** or lutropin (**LH),** both of which are named for their effects in female mammals but have equally important and similar roles in males. All of these hormones are polypeptides or proteins. These six hormones, together with MSH from the pars intermedia, collectively are referred to as the tropic hormones of the adenohypophysis. A listing of tropic hormones, alternative names for them, their targets, and their general physiological roles are summarized in Table 4-1.

In addition, peptides known as **lipotropin (LPH)** and **β-endorphin** may be released from the adenohypophysis. Although not listed here as tropic hormones, they may perform endocrine functions. β-Endorphin is one of several peptides known as **endogenous opioid peptides (EOPs)** that bind to the same receptors as the drug morphine, an opioid or a derivative of opium.

B. Cellular Types of the Adenohypophysis

The cellular types in the pars distalis and the pars intermedia first were distinguished by utilizing special dyes in particular staining combinations. This differential uptake of dyes is due to the differential affinity of cytoplasmic granules for these dyes. Some cytoplasmic granules bind acidic dyes, and cells with these granules are termed acidophilic cells or **acidophils** (*philos,* love). Other granules bind basic dyes and the cells containing these granules are termed **basophils.** A pituitary basophil is defined as a cell that is stained by the aniline blue dye of the Mallory trichrome staining method. However, since aniline blue is in reality an acidic dye, some investigators have preferred to use the term **cyanophil** (*cyanos,* blue) to designate these cells. Cells that do not contain stainable cytoplasmic granules have been called **chromophobes,** although there is a current movement to call them **nongranulated cells.**

Table 4-1. Synonyms, Abbreviations, Cellular Source, Targets, and Actions for Mammalian Adenohypophysial Tropic Hormones

Name	Abbreviation[a]	Synonyms	Other abbreviations	Cellular source	One target	One action on target
Prolactin	PRL	Mammotropin, luteotropin, luteotropic hormone	LTH	Lactotrope	Mammary gland	Stimulates milk synthesis
Growth hormone	GH	Somatotropin, somatotropic hormone	STH	Somatotrope	Muscle	Stimulates incorporation of amino acids into protein
Corticotropin	ACTH	Adrenocorticotropic hormone, adrenocorticotropin		Corticotrope	Adrenal cortex	Stimulates synthesis and secretion of corticosteroids
Lipotropin	LPH	None		Corticotrope?	Adipose tissue?	Stimulates hydrolysis of fats to free fatty acids and glycerol
Melanotropin	MSH	Intermedin, melanocyte- or melanophore-stimulating hormone		Pars intermedia	Melanocyte, etc.	Stimulates synthesis of melanin pigment
Thyrotropin	TSH	Thyroid-stimulating hormone		Thyrotrope	Thyroid gland	Stimulates synthesis of thyroid hormone
Follicle-stimulating hormone	FSH	Follitropin		Gonadotrope	Gonad	Stimulates follicular development in females and spermatogenesis in males; estrogen in females
Luteinizing hormone	LH	Interstitial cell-stimulating hormone, lutropin	ICSH	Gonadotrope	Gonad	Stimulates androgen and progesterone synthesis in females and androgen secretion in males

[a] The names in the first column and the abbreviations are used throughout the text.

Table 4-2. Some of the Dyes Used in Cytological Observation of
Adenohypophysial Cells and Their Abbreviations

Dye or staining procedure	Abbreviation	Chemical specificity (if known)
1. Aldehyde fuchsin	AF	—
2. Alcian blue	AB	Disulfide bonds, mucopolysaccharides
3. Periodic acid–Schiff	PAS	Glycoproteins, mucopolysaccharides
4. Orange G	OG	—
5. Azocarmine	AZ	—
6. Lead hematoxylin	PbH	—
7. Iron hematoxylin	FeH	—

Various types of basophils or acidophils may be distinguished from one another in terms of their specific affinities for other dyes. A listing of dyes used to distinguish pituitary cellular types is found in Table 4-2, and a common nomenclature of the hormone-secreting cells of the adenohypophysis is provided in Table 4-3. Although cellular types now are identified routinely by immunocytochemistry, a knowledge of stainable features of these cells is helpful when interpreting the older mammalian literature and especially the comparative literature prior to the availability of specific antibodies for tropic hormones.

The electron microscope also has been used to characterize cellular types of the pars distalis on the basis of general cellular morphology and the size and shape of electron-dense cytoplasmic storage granules containing tropic hormones (Table 4-3). A combination of ultrastructural, tinctorial (staining with dyes), and immunocytochemical techniques leaves little doubt as to the cellular origins of the tropic hormones. Ultrastructural features of some pars distalis cells can be seen in Fig. 4-8.

1. Cytology of the Pars Distalis

Generally speaking, there is one cellular type responsible for synthesis and release of each tropic hormone. There are two different stainable acidophils, one responsible for secretion of GH and one for PRL. A weakly staining, basophilic cell is responsible for synthesis of ACTH. However, only two strongly basophilic cells are found, and they are responsible for

Table 4-3. Some Light and Electron Microscopic Features of Cellular Types in the
Mammalian Pars Distalis

Cellular type	Tropic hormone secreted	Stainability[a] for light microscope	In situ granule size (nm)
Thyrotrope (β basophil)	TSH	PAS(+), AF(+)	150
Gonadotrope (δ basophil)	FSH, LH	PAS(+), AF(−)	200
Corticotrope	ACTH	Weakly PAS(+), AF(+); maybe PbH(+)	200
Lactotrope (ϵ acidophil)	PRL	Azocarmine(+)	600–900
Somatotrope (α acidophil)	GH	OG(+)	Variable to 350

[a] See Table 4-2 for dye abbreviations.

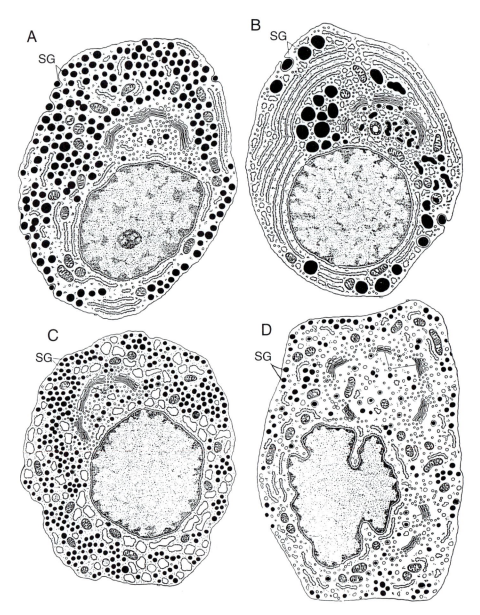

Figure 4-8. Cellular types in pars distalis drawn to the same scale. Note the obvious differences in secretory granule (SG) size that characterizes each of these cell types. A, Somatotrope; B, lactotrope; C, gonadotrope; D, corticotrope. [From Norman, A. W., and Litwack, G. (1992). "Hormones," Academic Press, San Diego.]

secretion of three glycoprotein tropic hormones: one secretes TSH and the other secretes LH and FSH. The two gonadotropins are produced in the same cell type based on immunochemistry, ultrastructure, and staining properties. In the following descriptions, the traditional mammalian designation for each of these five cellular types is given. The cells responsible for secreting tropic hormones are designated with the suffix "**trope.**"

The **thyrotrope** is the least common of the secretory cell types in the pars distalis. Thyrotropes have long cytoplasmic processes and contain spherical secretory granules. They occur primarily in the anterior-medial portion of the pars distalis and show little variations with sex or age.

Despite the chemical similarity of GTHs and TSH (see Section III,A), the **gonadotrope** is readily distinguishable from the thyrotrope. Gonadotropes represent about 15–20% of the cells and are distributed throughout the pars distalis cells. Two populations of spherical or slightly irregular secretory granules can be distinguished on the basis of size. At least three subtypes can be identified by differences in immunoreactivity. One subtype contains only FSH, one contains only LH, and the third contains both LH and FSH. It is not clear whether these represent separate clones of cells or different states of secretory activity.

Two acidophilic cell types are sources for GH and PRL, respectively (Fig. 4-9). The

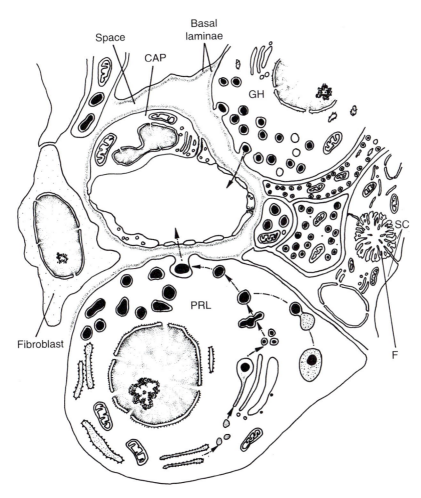

Figure 4-9. Somatotrope, lactotrope, and folliculostellate cells. A somatotrope (GH) and a lactotrope (PRL) are depicted adjacent to a capillary (CAP). Two folliculostellate cells (SC) form an extracellular follicle (F) at the far right.

somatotrope is the most abundant cell type, representing about 50% of the cells in the pars distalis, and is found mostly along the lateral margins of the pars distalis. The second acidophil is called a **lactotrope** and secretes PRL. The lactotropes account for between 10 and 25% of the cells in the pars distalis with the lower figure being common in men and nulliparous women (never having borne children). There are relatively few PRL-secreting cells in children. Lactotropes are distributed throughout the pars distalis, often found associated with gonadotropes. At least two lactotropes have been identified using ultrastructural criteria. One is very common, a sparsely granular cell with smaller spherical, oval, or irregular granules. The second type is uncommon, is densely granular, and typically occurs adjacent to capillaries. A third acidophilic cell type has been described, the **mammosomatotrope,** which secretes both GH and PRL, especially during pregnancy.

Corticotropin-secreting cells or **corticotropes** are intermediate in stainability between chromophobes and basophils. Corticotropes are located in a central wedge within the pars distalis and represent 10–15% of the total cells. They contain a variety of granules that are somewhat larger than those of thyrotropes and are immunoreactive for ACTH, LPH, and β-endorphin. Corticotropic cells also are immunoreactive for the protein **cytokeratin,** which occurs in the perinuclear area and is found in the pars distalis only in corticotropes. The number of corticotropes does not vary with age or sex but may vary markedly in a number of pathological states (see Chapter 9).

The sixth cellular type in the mammalian pars distalis is the nongranulated cell chromophobe that is not distinguished by selective staining techniques. Nongranulated cells may represent inactive, depleted, or undifferentiated cells, and some of them may differentiate into either basophils or acidophils, depending upon the stage of development, physiological conditions, or response to experimental manipulations. One type of nongranulated cell is the **null cell,** which has no special histologic, immunoreactive, or ultrastructural features other than a few small granules. Null cells may be the source of certain pituitary adenomas. A special type of nongranulated cell, the **follicostellate cell,** has been observed with the aid of the electron microscope in all vertebrates. These nongranulated cells exhibit **S-100 protein,** a characteristic of neuroglial cells in the brain, and S-100 protein is not found in any other cells of the adenohypophysis. The processes of these stellate (star-shaped) cells are very long and form a sort of network or reticulum between capillaries throughout the pars distalis. They are called "follicular" because of the way their stellate processes will sometimes surround or enclose small spaces (lumina or follicles; Fig. 4-9). Microvilli and sometimes cilia project into the follicular lumina. These stellate cells may perform a supportive or nutritional function. They are probably not the source of any tropic hormones, but there is evidence that they produce paracrine secretions (see Section II,B,2) that can influence tropic hormone release. Secretions from these cells in culture have been shown to attenuate release of GH, PRL, and LH following administration of substances that normally evoke their release.

2. Cytology of the Pars Intermedia

The **melanotrope** is the only glandular epithelial cell in the pars intermedia. The pars intermedia also contains stellate cells of unknown function that are interspersed among the MSH-secreting cells. In most mammals, the cleft separating the pars intermedia from the pars distalis is lined by ependymal-like cells called **epithelial cleft cells.** The cleft cells often are ciliated.

3. Cytology of the Pars Tuberalis

Two cellular types have been reported in the pars tuberalis of several mammals with three or possibly four cellular types occurring in the primate pars tuberalis. One cell type reacts specifically with antibody to pituitary LH and the second type specifically binds antibody to TSH. Occasionally, one or two rare cells are observed in primates that bind antibody to ACTH and GH. The majority of pars tuberalis cells are chromophobic in most mammals, but all are stainable types in humans. The pars tuberalis may represent a "fragment" of the pars distalis and it may function as an additional, but limited, source of tropic hormones.

C. Subdivisions of the Neurohypophysis

The mammalian neurohypophysis consists of two distinct neurohemal components. The median eminence is defined as the more anterior portion of the neurohypophysis that has a blood supply in common with the adenohypophysis, specifically the hypothalamo–hypophysial portal system. (Note that in some terminologies the median eminence is considered to be a subdivision of the hypothalamus.) An abundant but separate blood supply characterizes the pars nervosa (Fig. 4-6), which is that portion of the neurohypophysis in contact with the pars intermedia, when present, to form the neurointermediate lobe. Both the median eminence and the pars nervosa are composed of axonal tips of NS neurons originating in hypothalamic NS nuclei, capillaries, and pituicytes.

The function of pituicytes is unknown, but they could play a role as supportive elements or may be involved actively in storage and release of neurohormones from the neurohypophysis, similar to the role of stellate cells of the adenohypophysis. Pituicytes are possibly derived from ependymal or glial cells. Special aggregations of ependymal cells and capillaries form secretory structures, such as the **choroid plexus,** that produce CSF.

II. The Mammalian Hypothalamus

The NS nuclei of the hypothalamus and **preoptic area (POA)** produce neurohormones (Fig. 4-10) that regulate pituitary secretion. Axons from these nuclei travel either to the median eminence (Fig. 4-11) or to the pars nervosa. The neurohormones associated with the median eminence and adenohypophysis are termed **hypothalamic hypophysiotropic hormones** and are identified as either **releasing hormones** or **release-inhibiting hormones,** depending on whether they stimulate or inhibit tropic hormone release from the adenohypophysis (Table 4-4). These regulating hormones are mostly small peptides of as few as three to as many as 44 amino acids. At least one is considered to be a simple catecholamine (dopamine) derived from a single amino acid. The term "releasing hormone" is often used to refer to all of these neurohormones whether they stimulate or inhibit release of a tropic hormone.

The neurohormones associated with the pars nervosa are all very similar in structure (Fig. 4-12). Each is a peptide consisting of nine amino acid residues. The name "octapeptide" was used originally for these neurohormones because the two cysteine residues at positions 1 and 6 form a disulfide bridge to become cystine. Formation of this disulfide bond results in conversion of six amino acid residues into the characteristic five-amino acid ring structure with a side chain of three amino acids, hence the original name of octa-

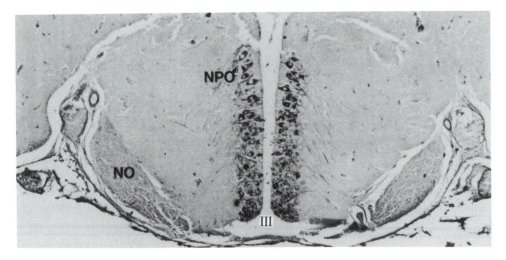

Figure 4-10. Neurosecretory neurons in the preoptic nucleus (NPO) of the rat. NO, optic nerve; III, third ventricle. [From Matsumoto, A., and Ishii, S. (1992). "Atlas of Endocrine Organs." Springer-Verlag, Berlin.]

peptide. However, because there are nine residues numbered 1 to 9, these peptides frequently are referred to as the **nonapeptide neurohormones.**

Four nonapeptides are stored in the adult pars nervosa and have been identified in the mammalian hypothalamus. These include the neutral nonapeptide **oxytocin (OXY),** as well as the basic nonapeptides **arginine vasopressin (AVP), lysine vasopressin (LVP),** and **phenypressin (PVP)** (see Chapter 6 for discussion of these neurohormones and their distribution in mammals and other vertebrates). All mammals have OXY plus one (usually) of the basic forms. So far, PVP has been demonstrated only in marsupial mammals.

The region of the mammalian brain that controls hypophysial function consists primarily of bilaterally paired NS nuclei, including the **anterior hypothalamic, suprachiasmatic,**

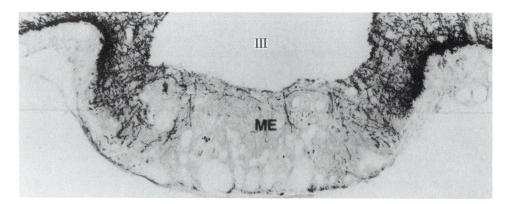

Figure 4-11. The rat median eminence. Immunoreactive GnRH fibers are stained in the lateral areas of the median eminence (ME). [From Matsumoto, A., and Ishii, S. (1992). "Atlas of Endocrine Organs." Springer-Verlag, Berlin.]

Table 4-4. Confirmed and Proposed Mammalian Hypothalamic
Hypophysiotropic Hormones

Name	Abbreviation	Alternate abbreviation	Chemical structure
Thyrotropin-releasing hormone	TRH	TSHRH, TRF	(Pyro)-E-H-P-NH$_2$
Gonadotropin-releasing hormone	GnRH	LRH, LHRH, LRF	(Pyro)-E-H-W-S-Y-G-L-R-P-G-NH$_2$
Somatostatin	SS	GHRIH, GRIH, SRIH, SRIF	A-G-C-K-N-F-F-W-K-T-F-T-S-C (S—S bridges)
Growth hormone-releasing hormone	GHRH	GRH, GRF	Peptide, amino acids
Prolactin release-inhibiting hormone (dopamine)	PRIH	PIF	Dopamine
Prolactin-releasing hormone	PRH	PRF	Same as TRH, VIP?
Corticotropin-releasing hormone	CRH	CRF	Peptide, 41 amino acids
Melanotropin release-inhibiting hormone	MSHRIH	MRIH, MIF	Dopamine
Melanotropin-releasing hormone	MSHRH	MRH, MRF	C-Y-I-Q-N-OH? | SH

ventromedial, dorsomedial, posterior hypothalamic, supraoptic, paraventricular, periventricular, and **arcuate nuclei** (Fig. 4-13). These nuclei are responsible for producing the hypothalamic hypophysiotropic neurohormones that regulate release of tropic hor-

Figure 4-12. Structure of neurohypophysial nonapeptide hormones in mammals. Phenypressin and mesotocin are found exclusively in marsupials. Lysine vasopressin is a variant of arginine vasopressin and is found in suiform eutherian mammals. The basic amino acids (Lys, Arg) at position 8 are associated with strong antidiuretic activity.

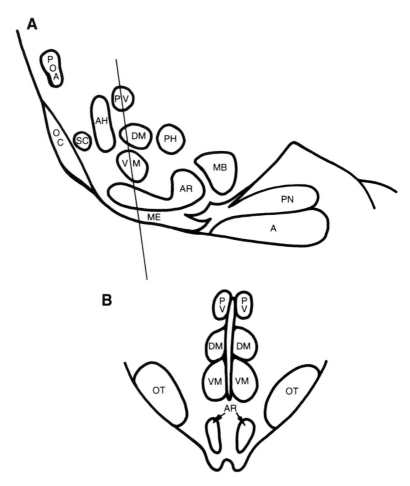

Figure 4-13. The hypothalamic neurosecretory centers. The nuclei depicted are as follows: AH, anterior
hypothalamic; AR, arcuate; DM, dorsomedial; PH, posterior hypothalamic; POA, preoptic
area; PV, paraventricular; SC, suprachiasmatic; VM, ventromedial. (A) Sagittal section of
brain and pituitary. A, adenohypophysis; OC, optic chiasm; MB, mammillary body; ME, median eminence; PN, pars nervosa. (B) Cross-section of brain at level indicated by the vertical
line in A. OT, optic tract.

mones from the hypophysis and for production of the nonapeptide neurohormones of the
pars nervosa.

Much of our initial knowledge about these NS centers was accumulated from observing
the effects of lesions or of localized electrical stimulation in the hypothalamus and POA as
well as from studies involving implants of crystalline hormones into these regions. Some
cautions for interpretation of data obtained from such studies are in order, however. The
use of disruptive lesions, for example, requires careful bilateral placement of comparable
lesions and leaves some uncertainty as to exactly what was destroyed by the lesions. Alterations in pituitary function following placement of lesions might involve destruction of the
NS neurons that elaborate a given hypothalamic hypophysiotropic hormone or may only
disrupt an NS tract. Furthermore, the lesion might have damaged non-NS neurons that
would normally modulate the activity of certain NS neurons. Damage to vascular elements

of the median eminence might also alter tropic hormone release patterns. Ideally, secretion of all tropic hormones should be monitored following placement of a particular lesion; yet, for practical reasons, this is rarely done. Usually only one (such as TSH) or, at most, two tropic hormone systems are examined, whereas others (e.g., PRL, LH, FSH, ACTH, and GH) are often ignored. Last, it is difficult to establish suitable controls for some of these procedures. In spite of such drawbacks, the use of these approaches in combination with immunocytochemical techniques has helped to establish the location of NS centers responsible for secreting each hypothalamic neuropeptide.

A. Hypothalamic Hypophysiotropic Hormones

Numerous studies have confirmed that the hypothalamus exerts a direct influence over functioning of the adenohypophysis. Microscopic observations indicate the absence of neural connections between the mammalian hypothalamus and the adenohypophysis. The presence of the hypothalamo–hypophysial portal system led to the establishment of what is now termed the **neurovascular hypothesis,** which states that hypothalamic neurohormones released into the portal circulation are responsible for controling tropic hormone release from the adenohypophysis. Severing the portal connections or transplanting the pituitary to some avascular site elsewhere in the body (i.e., an ectopic transplant) causes marked changes in the secretory pattern of the tropic hormones. Generally, such operations are followed by a marked reduction in circulating levels of TSH, FSH, LH, GH, and ACTH, whereas PRL and MSH levels increase. These observations led to the initial interpretation that release of the TSH, GH, ACTH, and the gonadotropins primarily is under stimulatory control (via releasing hormones) and that release of MSH and PRL normally is under inhibitory control (via release-inhibiting hormones). This initial interpretation has been modified as we have learned more about the many factors that influence tropic hormone release (see below).

If the severed blood vessels of the hypothalamo–hypophysial portal system are allowed to regenerate so that blood may again flow from the median eminence to the adenohypophysis, the normal secretory patterns for the tropic hormones resume. These latter observations support strongly the neurovascular hypothesis of hypothalamic control over tropic hormone release in mammals.

In recent years, many hypothalamic regulatory hormones have been proposed and their chemical identities established. Andrew Schally and Roger Guillemin shared a Nobel Prize for their initial isolation and characterization of **thyrotropin-releasing hormone (TRH)** and **gonadotropin-releasing hormone (GnRH),** respectively. Soon, with the advent of vastly improved biochemical techniques, other regulatory neurohormones were identified. Each hypothalamic regulatory hormone is named for the tropic hormone it was first shown to influence, and is designated according to whether it causes release (RH) or is release-inhibiting (RIH). The relative importance of releasing or release-inhibiting hormones for the various tropic hormones differs for each tropic hormone but is consistent for different mammalian species.

B. Control of Hypothalamic Hypophysiotropic Hormone Release

Release of hypothalamic regulatory hormones is influenced by neural activity, negative or positive feedback mechanisms, and direct actions of certain hormones. The following accounts represent an overview of the general pattern of control in mammals, but individual species may vary significantly from this pattern.

The predominant feedback loop involves production of blood-borne hormones or metabolites resulting from the actions of tropic hormones on specific peripheral target cells (Fig. 4-2). This feedback may alter the sensitivity of pituitary cells to hypothalamic RHs and RIHs or have effects on other neurons that innervate the hypothalamic NS neurons. Feedback effects of tropic hormones or even RHs and RIHs have been documented, but, for simplicity, we will focus our attention on the predominant feedback mechanisms described above. You will recall from Chapter 1 that most feedback is of the negative type, but enhancement of responsiveness or even positive feedback, such as the midcycle gonadotropin surge characteristic of female mammals, may occur.

1. Hypothalamic Factors

Peptides have been identified in hypothalamic neurons, and some of these may influence tropic hormone release, either directly or indirectly, through their actions as neurotransmitters, neurohormones, or neuromodulators. In addition, a variety of nonpeptide neurotransmitters (for example, acetylcholine, dopamine, norepinephrine, serotonin, GABA) affect tropic hormone release indirectly. The same neurotransmitter may stimulate release of one tropic hormone and inhibit release of another depending on the nature of their synaptic connections. Conversely, two or more different molecules might justifiably be termed specific releasing or release-inhibiting hormones for a particular tropic hormone. We will consider hypothalamic/POA agents and their multiple opportunities for interaction as we discuss the mechanisms involved in controlling the release of each tropic hormone (see Section IV).

Many pharmacological studies of nervous regulation of hypothalamic/POA NS centers have been conducted, employing neurotransmitters or drugs that either mimic or block the activity of various known neurotransmitters (see Chapter 2). Studies of this type have led to identification of the kinds of neurons that regulate release of individual hypothalamo–hypophysiotropic hormones. For example, application of dopamine to cultured pituitary cells with and without cocultured hypothalamic/POA tissue has made it possible to distinguish between the indirect stimulatory activity of dopamine on LH release via a neurotransmitter role in the hypothalamus/POA and its direct inhibitory action as a neurohormone on PRL release. Utilization of catecholamines and related drugs *in vivo* and *in vitro* also has contributed much to our understanding of neuronal regulation of hormone release.

The response of NS neurons to specific neurotransmitters and neuromodulators or of pituitary cells to neurohormones is determined by the presence of specific receptors for these substances on the cell membranes of the NS neurons. The inhibitory action of dopamine on PRL release mentioned above is accomplished through the binding of dopamine to receptors in the plasmalemma of the PRL-secreting cells. The ergot alkaloids, such as **ergocornine** and **ergocryptine,** can mimic the action of dopamine by binding to another receptor on the PRL cell membrane termed an α-receptor. Stimulation of this α-receptor by a PRL-RH might ordinarily evoke PRL release, but the use of these α-receptor blocking drugs (α-blockers) can completely inhibit hormone release even in the absence of the PRL-RIH.

2. Paracrine Factors in the Adenohypophysis

The discovery of the presence of many known regulatory peptides in cells of the adenohypophysis first suggested possible paracrine roles for tropic cell secretions and possibly paracrine secretions from some of the nongranulated cell types. Most of the supporting data for paracrine and sometimes autocrine functions for these peptides comes from culture systems in which the density and physical relationship of cell types may be very different from conditions within the pituitary gland. For example, renin, renin substrate, and angiotensin II (ANG-II) have been reported from gonadotropes in the rat and from lactotropes in humans. Release of PRL is stimulated in cultures of both rat and human cells by ANG-II. Although information is accumulating on possible paracrine factors affecting tropic hormone release (see suggested reading at end of chapter), *in vivo* studies have not confirmed any physiologically significant roles so far.

III. Tropic Hormones of the Adenohypophysis

The adenohypophysial tropic hormones are separable into three distinct chemical categories (Table 4-5). The hormones within each category exhibit considerable overlap in chemical structures (that is, amino acid sequences) and in some cases overlap in biological activities as well. Category 1 includes the glycoprotein hormones: TSH, FSH, and LH. Each of these hormones is composed of two polypeptide subunits, each of which contains specific carbohydrate moieties. GH and PRL constitute the category 2 tropic hormones. Both PRL and GH are fairly large, folded polypeptide chains, and they exhibit considerable structural and some functional overlap. Category 3 includes the smallest adenohypophysial peptides: ACTH, MSH, LPH, and β-endorphin. All of these category 3 molecules are similar chemically, have a common prohormone, and exhibit overlap in some of their biological activities.

In addition to the three categories of pituitary tropic hormones, certain tropic hormones of similar chemical structure and biological activity are produced in the placental mammals. As many as five tropic-like hormones are produced by the chorionic (fetal) portion of the placenta, including **chorionic gonadotropin (CG),** which is LH-like in both structure and function, and **chorionic somatomammotropin (CS),** which has some GH but mostly PRL-like activity. Both a **chorionic thyrotropin (CT)** and a **chorionic corticotropin (CC)** have been isolated from human placentas and may occur in other mammals. Pregnant mares

Table 4-5. Chemical Categories of Some Tropic and Placental Peptide/Protein Hormones

Category	Name	Site of synthesis
I	Thyrotropin (TSH)	Adenohypophysis: pars distalis
	Luteinizing hormone (LH)	Adenohypophysis: pars distalis
	Follicle-stimulating hormone (FSH)	Adenohypophysis: pars distalis
	Chorionic gonadotropin (CG)	Placenta
	Chorionic thyrotropin (CTSH)	Placenta
	Menopausal gonadotropin	Adenohypophysis: pars distalis
II	Growth hormone (GH)	Adenohypophysis: pars distalis; placenta
	Prolactin (PRL)	Adenohypophysis: pars distalis; placenta
	Chorionic sommatomammotropin (CS)	Placenta
III	Corticotropin (ACTH)	Adenohypophysis: pars distalis
	α-Melanotropin (α-MSH)	Adenohypophysis: pars intermedia
	β-Endorphin	Adenohypophysis: pars distalis and pars intermedia
	Chorionic corticotropin (CC)	Placenta

produce large quantities of a placental gonadotropin that has both FSH-like and LH-like properties. This glycoprotein hormone is termed **pregnant mare serum gonadotropin (PMSG).** The importance of these placental hormones in pregnancy is discussed in Chapter 11. A variant type of gonadotropin occurs in postmenopausal women, and is called **menopausal gonadotropin (MG).** Human MG is basically FSH-like and is produced by the postmenopausal adenohypophysis in large amounts in response to failure of the ovaries to produce adequate levels of hormones that normally cause negative feedback.

Much of our initial knowledge concerning structures and functions of tropic hormones initially came about as a result of the availability of pituitary glands and placentas from slaughtered domestic livestock. Initially, huge quantities of starting tissue were needed to yield 1 mg of pure hormone. Modern advances in biochemical techniques have reduced drastically the amount of tissue needed and have increased the efficiency of extraction and consequently the availability of purified tropic hormones from many animal sources. Since there are many structural variations in these polypeptides isolated from different animals, and some corresponding differences in biological activity, it has become increasingly important to designate the source of the hormone used in experimental studies. This is especially true when using mammalian hormones in nonmammals where a molecule that performs a particular function in mammals may provide different results in a nonmammal. Investigators who study mammalian tropic hormones usually designate the source of the hormone, such as bovine (cattle), ovine (sheep), porcine (pig), equine (horse), caprine (goat), and murine (rodent). An additional lower case letter preceding the abbreviation of a tropic hormone usually designates the species source. For example, bovine GH would be designated bGH, whereas GH prepared from humans would be hGH. This method of designating the source will have limited use as we prepare purified hormones from a greater number of species, but presently is a useful shorthand way to indicate source.

The activities of the various tropic hormones first were determined by bioassays (see Chapter 3). The classical bioassays for each tropic hormone are described below. These biological approaches are still used in the biochemical isolation and characterization of tropic hormones, especially in nonmammals and in such cases where purified hormones are not available. Once highly purified hormones became available, radioimmunoassays (RIA; see Chapter 3) were developed for the mammalian tropic hormones and are routinely employed to measure circulating levels.

There are, however, a number of drawbacks to widely employing RIAs for measurement of circulating hormone levels. Production of antibodies against hormones purified from pituitary glands may result in an antibody that reacts against some portion of a preprohormone or prohormone that is released from the cell rather than against the circulating biologically active form. Use of such an antibody might yield results that do not correlate with biological bioassay data since preprohormones and prohormones may not bind to the receptors on target cells. Some peptide hormones may occur in different sizes (number of amino acids) or have some molecules altered through conjugation to another molecule (e.g., acetylation) resulting in different forms with different biological potencies that all bind equally well to the antibody used in the RIA. The close similarities in structure among the various tropic hormones of a given category (for example, PRL, GH, and CS) may result in cross-reactivities to the antibody produced against only one hormone because all have gross structural similarities. The specificity of the antibody for the structure of the purified hormone antigen makes it difficult to use one antibody prepared against oFSH to estimate circulating levels of FSH in another species where differences in amino acid sequences of this species' tropic hormones might result in the cross-reactivity of the antibody to LH or TSH even when such cross-reactivity did not occur in sheep. These problems of variability and structural similarities make it absolutely essential that any RIA be validated in several ways (including by bioassay), especially if the species used for the antibody preparation is not phylogenetically close to the species in which it is being used to measure hormone levels. The development of the immunoradiometric assays of recent years has improved the selectivity of identifying specific molecules in plasma samples with less interference from closely related molecules or fragments (see Chapter 3).

A. Category 1 Tropic Hormones

All of the mammalian glycoprotein tropic hormones examined to date are composed of two peptide subunits (Fig. 4-14) with an assortment of carbohydrate moieties attached (Table 4-6). Molecular masses for these glycoproteins are about 32 kDa. The biological half-lives for TSH and LH are about 60 min whereas that of FSH is about 3 times longer (170 min). The longer half-life for FSH is attributed to differences in the carbohydrate components of these hormones.

Each glycoprotein tropic hormone consists of two subunits, an α **subunit** and a β **subunit.** The α subunit is identical in all three adenohypophysial glycoproteins as well as in chorionic gonadotropin. The β subunit is specific to each hormone and is responsible for its unique biological activity. Hence, radioimmunoassay procedures that employ antibodies made against β subunits are more accurate and show less cross-reactivity with other glycoprotein hormones.

Each glycoprotein subunit is synthesized as a separate prosubunit. Each prosubunit is coded by a different gene, modified posttranslationally, and eventually the α and β subunits are coupled to form a heterodimer. Although the α subunits of category 1 tropic hormones are nearly identical in amino acid composition, there is considerable variation among the β subunits of LH, FSH, and TSH. The α and β subunit genes for hLH are located on different chromosomes (6 and 19, respectively). Both β subunits for hLH and hCG occur on chromosome 19, suggesting that the βhCG gene arose by a relatively recent duplication of the βLH gene that occurred about the time mammals evolved from reptiles. Considerable overlap occurs between β subunits of LH and CG, which have very similar biological activities.

The carbohydrate components represent 15 to 30% of the molecular weight of the glycoprotein subunits and resulting heterodimers. Glycoprotein hormones also show consid-

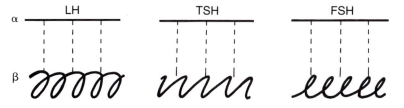

Figure 4-14. Generalized structure of pituitary glycoprotein hormones. The α subunit is common to all three hormones but the β subunit of each is unique and is responsible for the type of biological activity of the heterodimer. Subunits alone are without significant activity.

Table 4-6. Percent Carbohydrate Composition of Subunits Isolated from Pituitary
Gonadotropins and Thyrotropins

Carbohydrate	α Subunits			β Subunits		
	TSH-α	LH-α	FSH-α	TSH-β	LH-β	FSH-β
Hexose	7.3	8.3	2.8	3.0	4.3	13.8
Hexosamine	13.3	8.6	2.3	6.9	5.8	7.1
Sialic acid	—	—	0.8	—	—	3.7

erable specificity in their carbohydrate composition (Table 4-6). For example, FSHs contain larger quantities of sialic acid than do the others, and the sialic acid is largely associated with the FSH β subunit. Sialic acid protects FSH from rapid degradation by the liver. Treatment of FSH with the enzyme neuraminidase selectively removes sialic acid, reduces the biological activity of FSH, and allows it to be degraded more rapidly.

It is relatively easy with chemical procedures to dissociate these glycoproteins into their respective subunits. These separated subunits have little if any biological activity when administered to animals. Furthermore, it is possible to recombine the dissociated subunits and restore full biological activity. Any α subunit combined with any β subunit results in a fully active hormone characteristic of the source of the β subunit. Thus, when an α subunit isolated from TSH is combined with a β subunit from FSH, a glycoprotein with FSH activity and no TSH activity results.

1. LH: Actions and Bioassays

Synthesis of androgens in both males and females is caused by LH (or by structurally similar CG) action on the testes and ovaries. LH acts through a G protein-based, cAMP second-messenger system. Gamete release (sperm release or spermiation in males and ovulation in females) also is under the control of LH. Furthermore, LH causes formation of the corpus luteum in females from the ruptured ovarian follicles remaining after ovulation (see Chapter 11). LH also may be responsible for progesterone synthesis and release by the corpus luteum in some species.

Since three pituitary tropic hormones were named for their actions in females (PRL, FSH, and LH), an effort was mounted some years ago to rename LH for its action on the androgen-producing cells that occur between the seminiferous tubules of the testis rather than for inducing corpus luteum formation (luteinization) in the female. Hence, it was suggested that LH be named the **interstitial cell-stimulating hormone (ICSH)** for its action on the steroidogenic interstitial cell (also called the Leydig cell) of the testis. (There also is an LH-sensitive interstitial cell in the ovary; see Chapter 11.) Although the use of ICSH occurs sporadically in the literature, the name of LH has prevailed and is used today by most endocrinologists.

Numerous bioassays have been developed for quantitatively measuring LH activity (e.g., Table 4-7). One of the first bioassays was the spermiation response observed following injection of pituitary extracts or purified LH into frogs or toads. At one time, this bioassay was widely employed by doctors to detect the presence of CG in urine of women and confirm early pregnancy. The structural and functional similarity of hCG and hLH to amphibian gonadotropins is the basis for this clinical test. A spermiation test, however, responds to highly purified FSH, too, and cannot be considered a specific bioassay for

Table 4-7. Relative Effectiveness of Purified Ovine Gonadotropin (FSH and LH) in Some Gonadotropin Bioassays

Bioassay	Minimal effective dose (μg/ml or μg/injections[a])		Relative potency
	FSH	LH	
A. Testis weight maintenance in hypox lizard (*Anolis carolinensis*)	0.01	20.0	FSH>>LH
B. *In vitro* ovulation of frog ovary (*Xenopus laevis*)	>200.0	0.5	LH>>>FSH
C. Spermiation response by frog (*Hyla regilla*)	0.5	1.0	FSH≅LH

[a] Injection for bioassays A and C.

LH-like hormones. Pregnancy is now determined from urine samples with a simple and highly sensitive immunoassay involving an antibody produced specifically against hCG.

A popular early bioassay for LH was the **ovarian ascorbic acid depletion test (OAAD),** which was based upon a quantitative reduction of ascorbic acid in ovarian tissue following administration of LH. The degree of depletion if proportional to the dose of LH, and this bioassay is not affected by FSH. The relationship of ascorbic acid to hormone secretion is not understood, however, and this rather cumbersome assay is seldom used. The pigment response to LH by feathers of the African weaver finch, *Euplectes franciscanus,* was another highly specific bioassay for LH or CG, but it too is cumbersome and is not employed routinely.

The ovulatory response of amphibian ovaries *in vitro* to LH or CG (ovulation) has resulted in development of several similar bioassays. Administration of pituitary LH or progesterone stimulates ovulation in isolated fragments of anuran or urodele amphibian ovaries. Luteinizing hormone causes progesterone synthesis in follicular cells surrounding the oocyte, and it is the local progesterone binding to oocyte receptors that induces the ovulatory event. This *in vitro* ovarian bioassay also can be for detection of progesterone. Although the system is somewhat responsive to a few other steriods, it is relatively unresponsive to FSH and other tropic hormones.

2. FSH: Actions and Bioassays

Whereas the major actions for LH are stimulation of androgen synthesis and gamete release, FSH is primarily involved with gamete preparation: follicle development in females and spermatogenesis in males. In females, FSH also stimulates the conversion of androgens into estrogens. FSH may stimulate estrogen synthesis in the testis, but there is little if any contribution to circulating levels (see Chapter 11). Like LH, FSH binds to a membrane receptor and stimulates cAMP production as a second messenger.

The major bioassay for FSH prior to development of RIAs was the increase in testis or ovarian weight in hypophysectomized rats following injection of pituitary extracts or purified preparations of tropic hormones. Development of any of the androgen-sensitive sex accessory structures (see Chapter 11) in the test animals would be evidence of LH contamination. Another specific bioassay for FSH is the maintenance of testis weight in male lizards, *Anolis carolinensis,* following hypophysectomy (see Chapter 5, Table 5-3). The lizard testis regresses rapidly following hypophysectomy and normally is unresponsive to LH. Hence, this bioassay is very specific for small quantities of highly purified FSH.

3. TSH: Actions and Bioassays

TSH operates via a cAMP-dependent mechanism to increase synthesis of thyroid hormones, cause release of stored thyroid hormones, and secondarily increase iodide uptake by thyroid cells. Stimulation of thyroid gland function can be quantified cytologically by means of the TSH dose-dependent stimulation of epithelial height of the thyroid follicles in hypophysectomized animals. However, this bioassay requires several days to obtain results of these tissue changes, and a more rapid bioassay measures the amount of radioactive iodide (radioiodide) accumulated by thyroid follicles following administration of purified TSH or pituitary extracts. This bioassay can be performed *in vitro* as well as *in vivo*. Epithelial cell height and radioiodide accumulation indicate the degree of TSH stimulation and are also proportional to circulating TSH levels. However, these measurements provide no consistent information concerning rates of thyroid hormone synthesis and release (see Chapter 7 for more details of TSH-stimulated thyroid events).

The human has been shown to produce variant TSHs, one of which is associated with a pathological condition known as Graves' disease. Normal TSH has a biological half-life of about 0.25 hr, and maximal radioiodide uptake in thyroids of hypophysectomized rats is observed 4 hr after TSH administration. The so-called **long-acting thyroid-stimulator (LATS)** in Graves' disease has a biological half-life of 7.5 hr and causes maximal radioiodide uptake 12 hr after administration. LATS is not a product of the pituitary but is an aberrant immunoglobulin (see Chapter 7).

B. Category 2 Tropic Hormones

Two pituitary tropic hormones, GH and PRL, plus the placental tropic hormone, CS, comprise category 2. Multiple copies of the genes for hGH and hCS are found on chromosome 17, whereas the hPRL gene is on chromosome 6. Prolactin and GH are large single polypeptide hormones of similar structure (Fig. 4-15) and molecular mass (about 22 to 23 kDa). Human CS is extremely similar to both hGH and PRL, although in other mammals CS is more like PRL than GH. There is an 85% homology between hGH and hCS as well as considerable overlap with hPRL, hence the name "somatomammotropin." CS also is known as placental lactogen, but this older name does not reflect its GH-like actions. It is estimated that duplication of the GH gene and evolution of the CS gene occurred between 85 and 100 MYBP. This is a relatively recent event compared to the separation of the GH and PRL genes about 400 MYBP.

The human placenta also produces the pituitary forms of GH and PRL. Placental PRL accumulates in amniotic fluid. A smaller (16-kDa) variant of pituitary PRL has been isolated from the rat placenta. A 20-kDa variant of pituitary hGH has been demonstrated during the second half of pregnancy. A similar molecule is produced in the pituitary by alternative processing of mRNA from the normal GH gene, but does not appear in the circulation.

Both GH and PRL appear in the circulation as monomers as well as dimers (e.g., "big" GH) or oligomeres (eg., "big–big" GH). All of these forms are measureable by RIA, but the monomers have greater biological activity. In addition, GH, at least in humans and rabbits, is known to interact with a plasma **GH-binding protein,** complicating further the picture of circulating levels. Because of the heterogeneity of category 2 hormones, all references to GH or PRL in future discussions will be to the normal monomeric forms unless indicated.

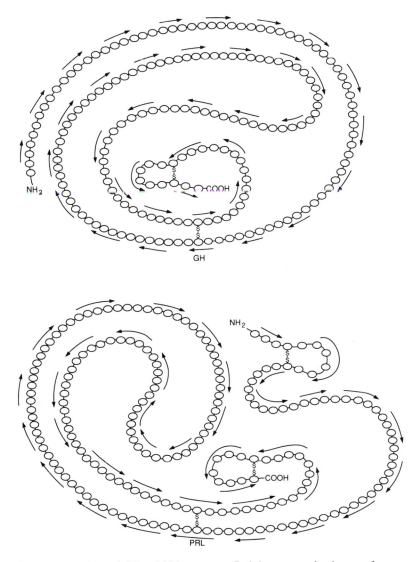

Figure 4-15. Comparison of GH and PRL structures. Both hormone molecules are of comparable size, exhibit considerable overlap in amino acid sequences, and may have similar actions when tested at higher doses in some systems. There are two disulfide bonds responsible for folding of the polypeptide chain in GH and three in PRL.

Both GH and PRL produce a number of common effects on osmoregulation (renal function, intestinal fluid absorption), selective tissue growth (prostate gland, sebaceous gland), lactation, and other processes. These common actions are related to their structural similarity. PRL and hCS, unlike GH, only have weak effects on body growth and metabolism.

Membrane receptors for GH and PRL are monomeric proteins that span the cell membrane only once and lack tyrosine kinase activity (see Chapter 2). Both PRL and GH receptors are similar to one another and neither receptor activates adenylate cyclase. The exact mechanism of action for GH and PRL is unknown. The extracellular domain of the hGH receptor is identical to the serum hGH-binding protein mentioned earlier, suggesting it may be a hydrolysis product from the membrane receptor.

1. GH: Actions and Bioassays

Growth hormone is often termed a protein anabolic hormone because it stimulates incorporation of amino acids into proteins and has a negative effect on nitrogen excretion. The classical bioassay for GH is the histological measurement of the thickness of the epiphysial cartilage in the tibia of the hypophysectomized rat or mouse following treatment with pituitary extracts or purified molecules. Currently RIA techniques are employed routinely to measure blood and pituitary content of GH in mammals and the tibia bioassay is only infrequently invoked.

Growth hormone represents about one-half of the total hormone content of the human adenohypophysis. It has been characterized as a protein composed of 191 amino acids (MW 21.5 kDa) having a biological half-life in blood of 20 to 40 min. hCS also consists of 191 amino acids, of which 161 are identical to those in hGH; yet, as mentioned earlier, hCS has rather low GH activity. Human GH has been synthesized in the laboratory.

Crude GH preparations consist of a collection of protein isohormones. Each form is thought to have its own actions, and collectively they produce all the effects normally attributed to pituitary GH activity. The gene responsible for synthesis of the 21,500-Da form of GH has been cloned and inserted successfully into the genome of mice. Growth rates of mice with the inserted genes and growth rates of their offspring are about twice that of normal mice. This approach to GH therapy has important implications for future treatment of GH-based growth deficiencies in humans.

Growth hormone stimulates absorption of amino acids and protein synthesis, especially by skeletal muscle cells. It cooperates with insulin to channel utilization of amino acids, fatty acids, and carbohydrates following a meal (see Section IV,C,3). Furthermore, GH becomes an important regulator of blood glucose and amino acid utilization during short-term and long-term starvation.

Circulating levels of hGH are highest during the period of maximal growth (ages 2 to 17 years). A daily secretory rhythm becomes established at about 4 years of age and continues throughout adult life. This pattern of GH secretion is both irregular and spontaneous, depending upon the physiological state of the individual, but episodes of GH release are frequently correlated with the onset of deep sleep.

Optimal growth-promoting actions of GH are obtained in hypophysectomized animals only when thyroid hormones are administered together with GH. This relationship between thyroid hormones and GH has been described as a synergism, that is, the growth response elicited by combined therapy with thyroid hormones and GH is greater than predicted by adding together the responses obtained with each hormone administered alone. Either thyroid hormones or GH will reinitiate some growth in hypophysectomized animals, but complete resumption of normal growth requires combined therapy. Furthermore, intact animals that exhibit thyroid deficiencies grow slowly and abnormally (see Chapter 7).

Thyroid hormones may influence synthesis of GH in intact rats, but their actions pertinent to hypophysectomized animals are peripheral. Thyroid hormones maintain a "responsive state" in the target cells so that they are more sensitive to GH (see Chapter 7 for more details of this "permissive" effect).

The effects of steroids on growth are complex. Androgens and estrogens can increase the responsiveness of human tissues to hGH, but to a lesser extent than do thyroid hormones. The mechanism of this steroid effect is not understood. Steroids, especially androgenic ones, have important effects on amino acid and carbohydrate metabolism unrelated to the roles of GH (see Chapter 14). Androgens are known to stimulate protein synthesis and hypertrophy of skeletal muscle, and estrogens increase protein synthesis in the uterus. Conversely, the increase in androgens and estrogens at puberty causes cessation of proliferation of the epiphyseal plates in the long bones of the appendicular skeleton and renders these tissues unresponsive to GH. This results in a permanent cessation of growth in stature.

Direct metabolic actions of GH on protein synthesis, amino acid transport, and lipolysis have been reported in several tissues. However, the growth effects of GH are mediated indirectly by the production of two peptide regulators in the liver or, in some cases, directly in target tissues. These peptides were first called **sulfation factors** because of effects on incorporation of sulfate into cartilage during GH-stimulated cartilage growth, a phenomenon that could not be invoked by direct application of GH to cartilage cells *in vitro*. Later, they became known as **somatomedins** since they mediated the actions of the somatotropic hormone, GH. These peptide growth stimulators are structurally related to insulin and have some insulin-like activity in addition to their growth-promoting actions. Consequently, they are known as **insulin-like growth factors (IGF-I, IGF-II).** A bioassay for IGFs has been developed using costal cartilage from pigs *in vitro*. The liver, then, can be considered a target endocrine gland for GH, since it synthesizes and releases IGFs into the circulation. IGFs are transported in the blood while complexed to specific plasma **IGF-binding proteins.** In other target tissues, IGFs synthesized by GH target cells may be necessary for producing GH-linked effects in these targets. Levels of both IGF-I and IGF-binding proteins are depressed in hypothyroid patients and elevated in hyperthyroid patients, indicating thyroid hormones can affect GH actions through actions at the liver.

In adult mammals, IGF-I is the principal IGF, and it binds to a receptor that is very similar to the receptor for insulin and can even bind insulin weakly. In contrast, IGF-II primarily is a fetal growth factor with a unique receptor unlike those for insulin and IGF-I. IGF-II has only weak insulin-like activity. Both IGFs are powerful mitogens (i.e., they stimulate cellular division or mitosis). Levels of hIGF-I increase at about 6 to 8 years of age and peak during puberty. Lower, constant levels of IGF-I are characteristic of adults. hIGF-II apparently is a fetal growth factor and may be produced under the influence of hCS.

2. Prolactin: Actions and Bioassays

Prolactin consists of a single chain of 199 amino acids (23 kDa) and, like GH, occurs in multiple isohormones. It has been shown to produce a variety of distinctive actions in animals, including effects associated with reproduction, growth, osmoregulation, and the integument (Table 4-8). Furthermore, PRL may produce synergistic actions with ovarian, testicular, thyroid, and adrenal hormones. The best-known action for PRL is the lactogenic effect on the mammary gland of female animals for which the hormone was named. Prolactin stimulates DNA synthesis, cellular proliferation, and the synthesis of milk proteins (casein and lactalbumin), free fatty acids, and lactose by the mammary gland. A similar effect is produced by CS from the placenta. Evidence suggest an IGF-like peptide is synthesized in the liver under the influence of PRL, which is a necessary synergist with PRL

Table 4-8. Prolactin Actions in Mammals

Actions related to reproduction
    Mammary development and lactation
    Preputial gland size and activity
    Synergism with androgen on male sex accessory glands
    Luteotropic in rodents
    Fertility in dwarf mice
    Increased testis cholesterol
    Increased androgen binding in human prostate
    Stimulation of glucuronidase activity in rodent testis
    Parental behavior
    Decreased copulatory activity in male rabbits
    Advanced puberty in rats
    Vaginal mucification in rats
    Antiovulatory and antiluteinizing actions in rats
    Relaxation of uterine cervix in rats
    Reduced catabolism of progesterone by rat uterus
    Inhibition of myometrial contractions
    Increased estradiol binding by rat uterus
    Decreased GTH release
Actions related to growth and development
    Mammary development
    Sebaceous and preputial gland growth
    Hair growth
    Erythropoietic actions
    Renotropic actions
    Spermatogenic actions
    Male sex accessory development
Actions related to water and electrolyte balance
    Lactation
    Increased Na$^+$ retention at renal level
    Corticotropic
Actions on integumentary structures
    Mammary development and lactation
    Sebaceous and preputial gland size and activity
    Hair maturation
Actions on steroid-dependent targets or synergisms with steroids
    Mammary growth (ovarian steroids)
    Milk secretion (corticosteroids)
    Sebaceous and preputial gland secretion (gonadal and cortical
        steroids)
    Growth and secretion of male sex accessory glands (androgens)
    Luteotropic action (estrogens?)
    Renal Na$^+$ reabsorption (aldosterone?) and renotropic action
        (androgens)
    Spermatogenesis (androgens)
    Advanced puberty (gonadal steroids)
    Hair growth (androgens, corticosteroids)
    Vaginal mucification in rats (estrogen and progesterone)

for these actions. This factor has been named **synlactin,** but it has not been isolated and identified, yet.

In some species (e.g., rat, sheep), PRL may influence the synthesis of progesterone by the corpus luteum. This action was responsible for the older name for prolactin, luteotropic hormone or LTH. There is evidence in male mammals for effects of PRL on certain sex accessory structures. These reproductive actions of PRL are discussed in Chapter 11 with respect to the overall regulation of reproduction in mammals.

Like the situation for GH, PRL actions on the mammary gland and possibly on other targets require an interaction with additional hormones. Estrogens favor cell proliferation and growth of the mammary gland, making the mammary more responsive to PRL. Glucocorticoids also potentiate the actions of PRL in all species examined, although the nature of this interaction is unclear. PRL may stimulate production of small peptides by mammary cells, which in turn may be mediators of PRL action. Progesterone inhibits PRL actions on the mammary gland and can block lactogenesis. One hypothesis suggests progesterone competes for glucocorticoid binding and/or blocks gene activation by glucocorticoids. The stimulatory actions of insulin on the mammary gland may be related to its IGF-like activity rather than its being a true interactor with PRL.

Several specific bioassays for PRL have been reported utilizing animals representing four major vertebrate classes (teleosts, amphibians, birds, and mammals):

1. The sodium-retaining bioassay in the cichlid teleost, *Sarotherodon (Tilapia) mossambicus*
2. The xanthophore-expanding bioassay in the teleost *Gillichthyes mirabilis*
3. The red-eft water drive performed in the newt *Notophthalmus viridescens*
4. The crop sac assay performed in the domestic pigeon
5. The *in vitro* mouse mammary gland assay

The extremely similar structures of GHs and PRLs within and among species make measurements by RIAs for either PRL or GH difficult, and extreme caution should be used when interpreting RIA data, especially the use of antibodies to mammalian PRL to measure levels in nonmammals. There is less overlap in their biological actions, and bioassays are especially useful for characterizing prolactin molecules from different sources. We discuss only the pigeon and mouse bioassays here. Bioassays employing the other vertebrate animals are described in Chapter 5.

a. The Pigeon Crop Sac Bioassay

The pigeon crop sac bioassay is a precise quantitative, dose-related bioassay yielding a standard curve from which relative activities of unknown preparations can be assessed. The bioassay is performed on 6-week-old pigeons that have not reached sexual maturity. The pigeon crop sac is a bilateral extension of the esophagus that in brooding birds produces a cytogenous secretion that is fed to the hatchlings. For this bioassay, the crop sac first is "primed" by injection of the birds with a small amount of mPRL (usually ovine). Then, on one side of the crop, the test solution containing either a known or unknown amount of mPRL or pituitary extracts is applied by subcutaneous injection directly over the crop surface. On the opposite side, a control solution (usually the same saline vehicle used for the test solution) is similarly applied. After removal of the crop sac several days later, a given area of the crop epithelium is scraped off around each control or test application site, dried, and weighed. The weight of the dried crop sac epithelial area is plotted against the known dose of mPRL used so that a standard curve is produced. The PRL content of unknowns can be identified by extrapolation from the standard curve.

Positive results have been obtained in the crop sac bioassay with lungfish pituitaries as well as pituitaries from all tetrapod groups, but most piscine PRLs will produce only an

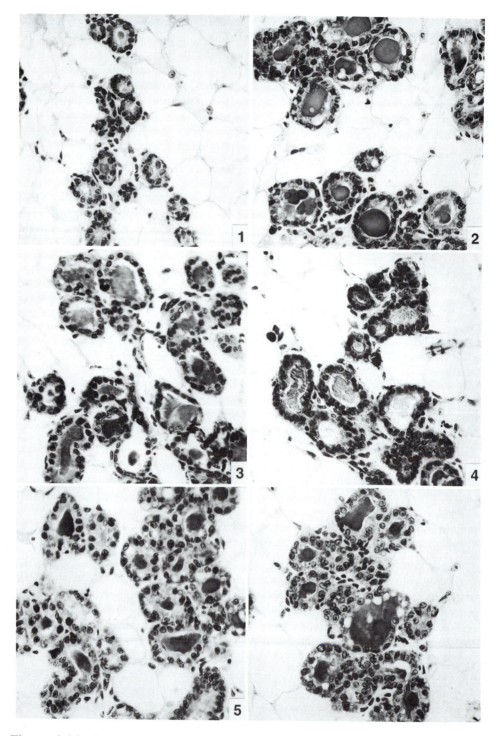

Figure 4-16. *In vitro* bioassay in mouse mammary gland for PRL. Secretion rating (SR) is accomplished using a visual scale [SR = 1 (lowest activity) to 5 (highest).] (1) Control explant. Alveoli not distended and containing little secretion; SR = 1. (2) Effect of extract from

atypical response or no response at all. Although this bioassay is very consistent and sensitive, it has the major disadvantage of the time involved in performing it.

b. Mammary Gland in Vitro Bioassay

For this bioassay, mammary gland explants from pseudopregnant or midpregnant mice (or rabbits) are cultured in a precise medium that includes some other hormones (e.g., glucocorticoids, insulin). The addition of PRL or pituitary homogenates to the culture medium causes cytological changes that can be quantified according to a numerical index (Fig. 4-16). These cytological changes are correlated with the ability of PRL to stimulate milk synthesis. Like the crop sac bioassay, mammotropic (lactogenic) activity is present in the pituitaries of all tertrapod species tested in the mouse mammary bioassay, but piscine pituitary preparations yield minimal responses, if any (see Chapter 5, Table 5-6). Like the crop sac bioassay, the mammary gland response is not a rapid assessment.

C. Category 3 Tropic Hormones

This category comprises several hormones derived from the same precursor, a prohormone known as **proopiomelanocortin** or **POMC,** and includes ACTH, MSH, LPH, and the EOP, β-endorphin. POMC-related peptides are found in cells of the brain, the pars distalis, and the pars intermedia (when present). In the brain, two additional prohormones, **proenkephalin** and **prodynorphin,** are the sources for the neural EOPs, the **enkephalins** and the **dynorphins.** All of the category 3 molecules are produced by variations in posttranslational processing of POMC, and consequently exhibit considerable overlap in their biological activities due to possession of similar amino acid sequences (see Fig. 4-17).

In corticotropic cells of the pars distalis, POMC is cleaved to produce ACTH (39 amino acids), an N-terminal **16K fragment** (MW 16,000) with no known biological activity, and a large form (91 amino acids) of LPH known as β-**lipotropin** or β-**LPH,** representing the C-terminal portion of POMC. In turn, β-LPH may be cleaved to a 58-amino acid fragment called γ-**LPH** (consisting of residues 1–58 of β-LPH) and β-**endorphin** (residues 61–91 of β-LPH). In melanotropes of the pars intermedia, ACTH is cleaved further to yield α-**MSH** (residues 1–13 of ACTH) and a **corticotropin-like peptide** called **CLIP** (residues 18–39 of ACTH).

In the brain, proenkephalin is the precursor for either the opioid pentapeptide **Met-enkephalin** or the opioid pentapeptide **Leu-enkephalin,** which differ only at position 5. **Dynorphin** is a larger opioid peptide occurring in two forms (13 or 16 amino acids) derived from prodynorphin (also known as proenkephalin B). There are two smaller partly homologous peptide versions of the dynorphins called α- and β-**neoendorphins.** The enkephalins, dynorphins, and endorphins all have the same four or five N-terminal amino acids that apparently allow them to bind to similar receptors.

---

amphibian pituitary; SR = 5. (3) Effect of extract from reptilian pituitary; SR = 4. (4) Effect of extract from avian pituitary; SR = 3. (5) Effect of guinea pig pituitary; SR = 4. (6) Purified ovine prolactin; SR = 5. [From Nicoll, C. S., Bern, H. A., and Brown, D. (1966). Occurrence of mammotrophic activity [prolactin] in the vertebrate adenohypophysis. *J. Endocrinol.* **34,** 343–354.]

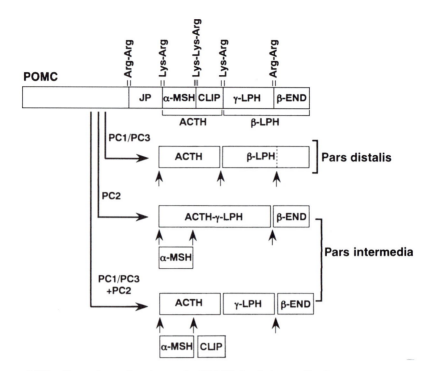

Figure 4-17. Fates of proopiomelanocortin (POMC) in pituitary cells. Corticotropes in the pars distalis process POMC with proteolytic enzymes (PC1/PC3) to yield primarily ACTH. The enzyme PC2 required to release α-MSH is confined to pars intermedia cells. Note that there are two scenarios possible for producing α-MSH. PC2 also cleaves ACTH to α-MSH and corticotropin-like peptide (CLIP). β-Endorphin production predominates in pars intermedia cells but also can occur in corticotropes. Amino acid residues that constitute cleavage sites are shown along the top edge of POMC (Arg-Arg, etc.)

The amide of ACTH and β-endorphin may be acetylated by an interaction with *N*-acetyltransferase and acetyl-coenzyme A. This reaction occurs posttranslationally in the secretory granules of melanotropes. Thus, a high proportion of α-MSH and β-endorphin is acetylated prior to release.

1. Corticotropin

Mammalian ACTH has been purified from several mammalian sources (bovine, porcine, ovine, human). In all species, it has 39 amino acids in a single peptide chain with a molecular weight of about 4500. Amino acids 1–23 of ACTH have full biological activity, 1–19 have 80% of full activity, but the fragment 1–16 has very little biological activity. Amino acids 24–39 are obviously outside of that region of the molecule responsible for its biological activity. The structures of the active fragments of several mammalian ACTHs are provided in Table 4-9.

The classical bioassay for ACTH has been the **adrenal ascorbic acid depletion test (AAAD),** which is similar to the OAAD bioassay described for LH. The lowest effective dose of purified ACTH in this bioassay is 0.2 mU or about 1.2×10^{-9} g (about 1 ng). The adrenal cortex secretes certain corticosteroids in response to ACTH stimulation. The

Table 4-9. Structural Variation of Active Portion of Mammalian ACTHs

Source of ACTH	1 25	26	27	28	29	30	31	32	33 39
Porcine	Ser .... Asn	Gly	Ala	Glu	Asp	Glu	Leu-	Ala	Glu .... Phe
Ovine	Ser .... Asp	Gly	Ala	Glu	Asp	Glu	Ser	Ala	Gln .... Phe
Bovine	Ser .... Asn	Gly	Ala	Glu	Asp	Glu	Ser	Ala	Gln .... Phe
Human	Ser .... Asn	Gly	Ala	Glu	Asp	Glu	Ser	Ala	Glu .... Phe

AAAD test is rarely used today and has been replaced by measurement of either circulating ACTH or corticosteroids with RIA.

2. Melanotropin

The name of this tropic hormone stems actually from its actions in amphibians and reptiles where it causes dispersion of melanin pigment granules contained within a specialized cell, the **melanophore,** located in the skin (see Chapter 5). Dispersal of these melanin granules causes the animal to appear darker. In mammals, the epidermal melanin-containing cell is a **melanocyte.** Not only does this cell synthesize melanin under the influence of MSH but it extrudes it into the extracellular compartment where it can be accumulated in **keratino-cytes** (keratin-containing cells). Animals that change from a white "winter coat" to a brown "summer coat" employ the services of MSH to stimulate melanin production for the summer coat. Hypophysectomy of the short-tailed weasel during the winters causes the summer coat to be white. Treatment of hypophysectomized weasels with either MSH or ACTH (which has inherent MSH-like action) is sufficient to cause regrowth of the normal brown summer coat.

The standard bioassays for MSH have been developed in amphibians and reptiles utilizing their ability to adapt to dark or light backgrounds. One bioassay involves measurement (using a reflectometer) of the quantity of light reflected from an isolated piece of skin removed from a light-adapted frog. A similar bioassay employs isolated pieces of skin from the light background-adapted *Anolis carolinensis,* the American chameleon. A rapid all-or-none response (darkening) occurs *in vitro* in the presence of MSH.

The adenohypophyses of all vertebrates tested by bioassay possess MSH activity, including birds that lack a pars intermedia. As mentioned above, α-MSH and CLIP are released following the hydrolysis of ACTH in melanotropes. α-MSH usually is acetylated prior to release, which slows its degradation and increases its biological activity. It is not possible to distinguish MSH activity caused by α-MSH or by ACTH, which contains the entire α-MSH sequence at its N-terminal end, binds readily to MSH receptors, and hence has innate MSH-like activity. The physiological roles for α-MSH are not known in birds or in most mammals. A physiological role has not been established for MSH in fishes where most of their pigmentary changes apparently are under dual sympathetic–parasympathetic innervation, allowing for rapid responses.

There is a common heptapeptide core (seven amino acids) in α-MSH, ACTH, and LPH, and this similarity in structure is reflected clearly by overlapping biological activity. Corticotropin has considerable MSH activity, and LPH has rather low ACTH activity but strong MSH action. Melanotropin has both weak ACTH and LPH activity. These overlaps in function affect interpretation of bioassayable data on MSH activity, i.e., which peptide is measured by the bioassay?

Three separate MSH molecules have been identified in extracts of mammalian pituitaries, but only α-MSH appears to be released from POMC *in vivo* and occurs in the circulation. Hydrolysis of the 16K fragment can yield γ-MSH (18 amino acids). β-MSH (22 amino acids) is a hydrolysis product of γ-LPH. It is assumed that only α-MSH is a true melanotropin and that β-MSH and γ-MSH are artifacts of extraction.

3. Lipotropin

In addition to its role as a precursor for endorphins, LPH has been proposed as an adenohypophysial hormone that stimulates lipolysis in adipose tissue (that is, hydrolysis of fats to free fatty acids and glycerol). A standardized bioassay for LPH activity involves culturing mouse or rabbit epididymal (testis) fat pads and measuring the release of glycerol or free fatty acids or both into the culture medium. Another bioassay technique for LPH employs measurement of the inhibition of incorporation of ^{14}C-labeled acetate into lipid following addition of LPH to the culture medium. This procedure is, in effect, a measurement of lipogenesis, which is inversely related to lipolysis. Lipotropin, presumably of pituitary origin, has been identified in the systemic circulation, but levels of circulating LPH have not been linked to changes in lipid metabolism, leaving open the question of any physiological role for LPH.

4. The Endorphins and Enkephalins

Morphine is an opiate analgesic (pain-killing) drug that binds to specific receptors in the central nervous system. Scientists postulated that there also would be endogenous compounds that produce analgesic opiate-like (morphine) effects on the central nervous system. A search for endogenous analgesics has resulted in identification and chemical characterization of two groups of biologically active opiate-like peptides. The larger peptides are known as endogenous morphines or endorphins, so named because they are endogenous and produce morphine-like effects. These include the dynorphins, β-endorphin, and some C-terminal hydrolysis products of β-endorphin that are acetylated at the N-terminal end. These alterations in β-endorphin markedly reduce its analgesic properties. The opioid pentapepetides called enkephalins (meaning "in the head") also bind to opioid receptors. The distribution of endorphins in the pituitary and central nervous system parallels that observed for ACTH and LPH, indicating they are products of POMC hydrolysis. The enkephalins and dynorphins are produced from a different but related prohormone and are localized in other neurons.

Painful stimuli elevate levels of endorphins and enkephalins in the CSF, and they appear to exhibit the features required for endogenous opiate-like agents. There is little doubt that these compounds probably function as neuromodulators or neurotransmitters within the central nervous system related to their morphine-like actions. The action of morphine, a nonpeptide, is blocked by closely related molecules such as **naloxone** (Fig. 4-18). The effects of endorphins also are blocked by naloxone, implying closeness in mechanisms of action for morphine and the endorphins. Three types of opioid receptors have been identified with differing affinities for the various opioids (see Table 4-10). In addition to their involvement with pain preception, endorphins influence release of neurotransmitters affecting tropic hormone release and can inhibit oxytocin release (see Fig. 4-19). The possible roles for endorphins and enkephalins as regulators of behavior, especially as related to painful stimuli, and their possible endocrine implications represent one of the most ex-

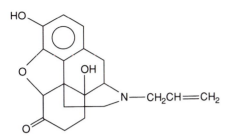

Naloxone

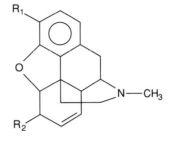

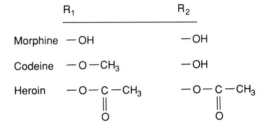

Figure 4-18. Chemical structures of opiate drugs and the opiate antagonist, naloxone. Three common opiates (morphine, heroin, codeine) differ according to the groups attached to the complex carbon ring at R_1 and R_2.

citing areas of neuroendocrinology to appear since the discovery of the first hypothalamo–hypophysiotropic hormone.

IV. Regulation of Tropic Hormone Secretion in Mammals

For each of the tropic hormones, we will describe the nature and actions of the various RHs and RIHs involved, what the major feedback loops are, and how other neural factors may influence secretion of a tropic hormone. In addition, we will develop an overall scheme that summarizes or models each tropic hormone regulatory system.

Table 4-10. Affinity of Opioids for Major Opioid Receptor Types

| Opioid | Receptor types | | |
	μ Receptor (mu)	δ Receptor (delta)	κ Receptor (kappa)
Morphine	High	Low	None
Naloxone (antagonist)	High	Low	None
β-Endorphin	High	Low	None
Enkephalins	Low	High	None
Dynorphin	None	Low	High

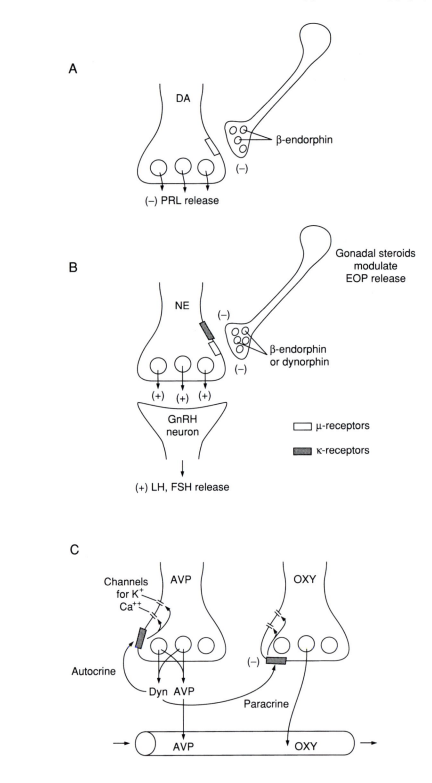

A. Regulation of Thyrotropin Secretion

Release of TSH is under stimulatory control as evidenced by the lack of TSH release following disruption of the portal vessels or explanting of the adenohypophysis. The major factor controlling release is a tripeptide known as thyrotropin-releasing hormone (TRH) or thyroliberin. The factors controlling TSH release are summarized in Fig. 4-20C.

1. Thyrotropin-Releasing Hormone (TRH)

The tripeptide TRH (Table 4-4) was one of the first hypothalamic regulatory hormones to be identified chemically. It is found in the NS neurons of many nuclei but in highest concentration in the paraventricular nucleus (PVN), which sends TRH-immunoreactive fibers to the median eminence. TRH appears in the hypophysial blood following electrical stimulation of the appropriate regions of hypothalamus, and causes release of TSH *in vivo* and *in vitro* from the adenohypophysis. The cAMP second-messenger system is responsible for mediating the action of TRH on the thyrotrope. It appears that this tripeptide is the endogenous hypothalamic hypophysiotropic TRH of mammals.

Extrahypothalamic TRH also is present in other brain regions, the spinal cord, the pineal gland, and the neurohypophysis, as well as in some other tissues. The common occurrence of TRH outside the hypothalamus and its presence in extrahypothalamic regions of the nervous system of mammals, nonmammalian vertebrates, and even invertebrates has led to the suggestion that TRH may also function as a neuromodulator or neurotransmitter. Administration of synthetic TRH causes depression of firing in certain brain neurons, and pituitary-like TRH receptors have been demonstrated in many brain areas.

The biological half-life for TRH in peripheral blood is very short (2 min in mice), apparently because peptidases in the blood rapidly inactivate TRH. Were it not for modifications of both the C- and N-terminal amino acids that slow peptidase degradation of TRH, this tripeptide would be destroyed even more rapidly (a pyroglutamate on the N-terminal end and an amidated C-terminal end). However, it needs to travel only a short distance through the portal blood to reach the thyrotropes so its rapid rate of degradation in peripheral blood is trivial.

2. Other Neural Factors Affecting TSH Secretion

Both stimulatory and inhibitory neural control of TRH release occurs in mammals. Experimental studies have demonstrated that TRH neurons in the paraventricular nucleus are innervated by dopaminergic and norepinephrine-secreting neurons. Stimulation of nonrepinephrine-secreting neurons can evoke release of TRH, whereas DA is an inhibitor of TRH release. The stimulatory role of NE in humans is supported but not confirmed. Serotoninergic fibers have been shown to inhibit TRH release, although some studies indicate a stimulatory role for 5-HT. Effects of 5-HT in humans are based largely on use of

Figure 4-19. Endogenous opioid peptides (EOPs) as neuromodulators. EOPs affect hormone release in at least three ways. They can (A) increase PRL release by blocking dopamine (DA) release, (B) inhibit norepinephrine (NE) stimulation of GnRH release, or (C) prevent oxytocin (OXY) release when arginine vasopressin (AVP) release occurs in the pars nervosa. [Modified from R. E. Brown. "An Introduction to Neuroendocrinology." © Cambridge University Press, 1994. Reprinted with permission of Cambridge University Press.

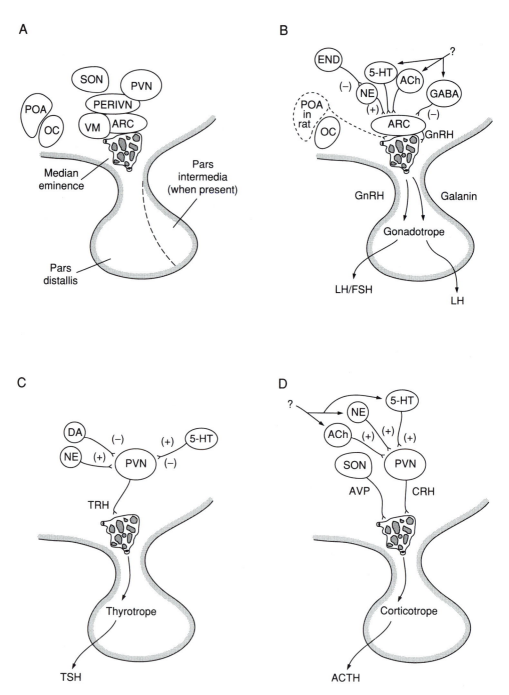

Figure 4-20. Hypothalamic control of tropic hormone secretion. Note that possible widespread inhibitory effects of somatostatin other than its actions on GH release are not included. Many of the ordinary neurons depicted actually have their cell bodies outside of the hypothalamus (e.g., all NE-secreting neurons). See text discussions of these factors. (A) General location of neurosecretory nuclei; (B) control of GTH release; (C) control of TSH release; (D) control of ACTH release; (E) control of GH release; (F) control of PRL release; (G) control of MSH release. AMG,

pharmacological agents known to influence serotonergic neurons or 5-HT receptors. These studies are somewhat controversial in interpretation, and no clear role has been established.

Somatostatin can function as a TSH-inhibiting hormone, but its physiological role as

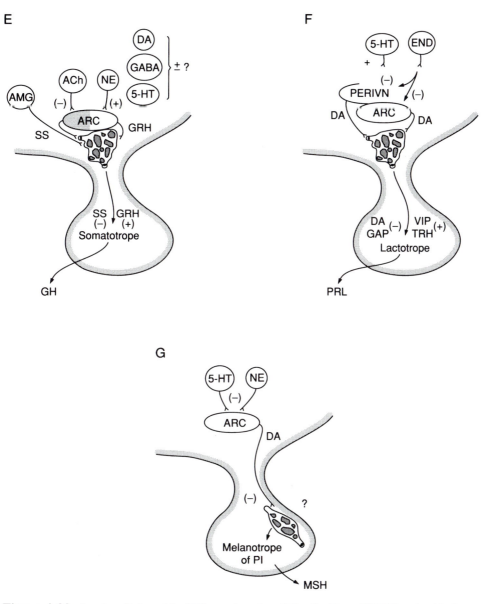

Figure 4-20 (*continued*). Amygdala; ARC, arcuate nucleus; OC, optic chiasm; PERIVN, periventricular nucleus; POA, preoptic area; PVN, paraventricular nucleus; SON, supraoptic nucleus; ACh, acetylcholine; DA, dopamine; END, endogenous opiates; GABA, γ-aminobutyric acid; 5-HT, serotonin; NE, norepinephrine; AVP, arginine vasopressin; CRH, corticotropin-releasing hormone; GAP, GnRH-associated peptide; GnRH, gonadotropin-releasing hormone; GRH, GH-releasing hormone; SS, somatostatin; TRH, thyrotropin-releasing hormone; VIP, vasoactive intestinal peptide.

a TRIH has not been confirmed. Release of somatostatin can be inhibited by adrenergic α_2-agonists, which is in keeping with the known stimulatory action of norepinephrine on TSH release.

3. Feedback Effects on TSH Secretion

Feedback of thyroid hormones occurs primarily at the thyrotopic cells in the pars distalis and reduces their sensitivity to TRH. Additional evidence suggests that feedback inhibition of the hypothalamic dopaminergic and serotonergic neurons by thyroid hormones is also important.

Studies with other neurotransmitters (e.g., acetylcholine, histamine, GABA) and related pharmacological agents are too few and the results too diverse to provide any clear picture of their roles, if any, in controlling TSH release.

B. Regulation of Gonadotropin Secretion

Release of the gonadotropins (GTH or Gn) is under stimulatory control similar to that described for thyrotropin. Both LH and FSH release are caused by a decapeptide (10 amino acids; Table 4-4) called gonadotropin-releasing hormone (GnRH). Because earlier studies had focused on the actions of this peptide on release of LH, it was first named luteinizing hormone-releasing hormone (LHRH) or luliberin. However, its dual action causing release of both LH and FSH implies that GnRH is a more appropriate name. Although there is evidence for a hypothalamic FSHRH, it is generally accepted that endogenous GnRH causes release of both gonadotropins. Neural and neurohormonal factors affecting GTH secretion are summarized in Fig. 4-20 B.

1. Gonadotropin-Releasing Hormone (GnRH)

Endogenous GnRH originates in NS neurons located from the preoptic area to the medial basal hypothalamus, depending on the species. In rats and mice, most of the GnRH-immunoreactive cells sending fibers to the median eminence are located in the POA, anterior to the optic chiasm. In the guinea pig and human, the medial basal hypothalamus (mainly the arcuate nucleus) is the major residence for GnRH neurons. GnRH immunoreactive cells also have been identified in the olfactory region, and studies support the hypothesis that at least some of the hypothalamic GnRH cells migrated from the olfactory region into the hypothalamic area (see below).

GnRH cells also contain other regulatory molecules including the **delta sleep-inducing peptide (DSIP)** and the peptide **galanin.** Nothing is known about the importance of DSIP in GnRH neurons but galanin is released along with GnRH and is itself a releaser of LH. Galanin has no effect on FSH release, however. Furthermore, galanin content varies in GnRH neurons of females during the reproductive cycle. Synthesis of galanin peaks just prior to ovulation and may participate in the generation of the midcycle surge in LH that brings about ovulation and corpus luteum formation (see Chapter 11).

a. GnRH Cells

Neurons that secrete gonadotropin-releasing hormone (GnRH) do not form a discrete, compact nucleus but develop an interconnected network that produces an endogenous, synchronized, pulsatile pattern of GnRH release. This network is the **GnRH pulse generator.** Special GnRH neurons called **GT-1 neurons** (a special cell line derived from transgenic mice) exhibit an oscillatory pattern of GnRH release *in vitro* and, together with some related cell lines, have provided a model system for studying the GnRH pulse generator.

GT-1 neurons synthesize and exhibit pulsatile release of GnRH but do not carry any biochemical glial cell markers. The absence of glial markers supports the contention that these cultures consist only of neurons. The pulsatile nature of their secretion implies an endogenous oscillator controlling this behavior and experiments suggest an autocrine role for GnRH to control its own release. From these experiments, researchers have concluded that operation of the endogenous oscillator depends on Ca^{2+} influx through **voltage-sensitive calcium channels (VSCCs)** and an autocrine positive feedback of released GnRH, which causes additional GnRH release.

Ultrashort negative feedback has been established for GnRH on normal GnRH-secreting neurons of the rat hypothalamus both *in vivo* and *in vitro*. The mechanism for this auto-feedback is not clear, but it could occur through recurrent collateral fibers of GnRH neurons that synapse on their own cell body or dendrites. GT-1 neurons do have GnRH receptors and binding of GnRH to these receptors is associated with a rapid, dose-dependent increase in intracellular Ca^{2+}. This results in a two-phase response involving an initial phospholipase C-mediated, inositol trisphosphate (IP_3)-dependent mobilization of Ca^{2+} (see Chapter 2) and a sustained entrance of Ca^{2+} through VSCCs. The phospholipase C system activates other internal mechanisms involving phospholipase D and diacylglycerol (DAG) that sustain activation of protein kinase C (see Fig. 4-21).

Initial binding of GnRH activates additional GnRH release (positive feedback) but soon is followed by inhibition and loss of spontaneous pulsatility (negative feedback). These dual autocrine actions of GnRH result in regular pulsatile episodes of GnRH release into the medium. Additional mechanisms for activating GnRH release in GT-1 cells have been demonstrated and these may influence the *in vivo* responses of GnRH-secreting neurons. Among the agents that can affect Ca^{2+} mobilization in GnRH cells are the peptides called **endothelins** and catecholamines (dopamine, norepinephrine, and epinephrine). Receptor channels for glutamate and GABA are also present on GnRH neurons and may be important modulators of the hypothesized endogenous oscillator controlling pulsatile GnRH release. The estrogen-induced, preovulatory GnRH surge also may be related to this autocrine positive feedback mechanism.

b. GnRH

Examination of numerous mammalian species has shown the structure of GnRH has been highly conserved although a number of variants are known in nonmammals (see Chapter 5). GnRH operating through a cAMP-dependent mechanism causes release of both GTHs, FSH, and LH, from the adenohypophysis, although a greater amount of LH release is always observed following administration of synthetic GnRH. Since GnRH was first synthesized, approximately 1400 analogs have been created, ranging in action from analogs that are even more potent stimulatory agents than native GnRH to inhibitors of gonadotropin release that can block the action of endogenous GnRH. One super-releaser has approximately 150 times the potency of native GnRH. All of the releasing analogs liberate both FSH and LH. Prolonged treatment with most GnRH agonistic analogs results in downregulation of GnRH receptors, however, and chronic administration causes a reduction in gonadotropin release.

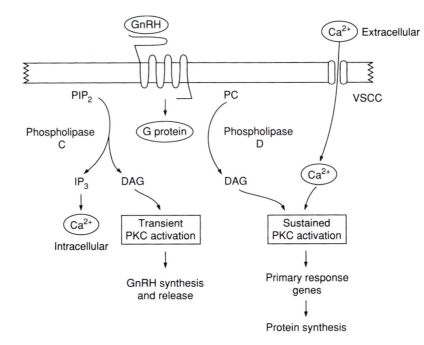

Figure 4-21. A scheme for the autocrine action of GnRH. The GnRH receptor is linked to a G-protein mechanism for activating phosphokinase C (PKC). Proposed modulatory influences of glutamate and GABA receptors have been omitted for simplicity. This two-phase response involves a transient effect of intracellular calcium on PKC activity and a sustained phase maintained by influx of extracellular calcium through voltage-sensitive calcium channels (VSCC). PIP_2, phosphoinositol diphosphate; PC, phosphatidylcholine; respective substrates for IP_3 and DAG or DAG only. Other abbreviations can be found in the text. [Modified by permission of the publisher from Stojilkovic, S. S., Krsmanovic, L. Z., Spergel, D. J., and Catt, K. J. Gonadotropin-releasing hormone neurons. Intrinsic pulsatility and receptor-mediated regulation. *Trends Endocrinol. Metab.* **5,** 201–209. Copyright 1994 by Elsevier Science Inc.]

   The half-life of GnRH is very short, for it is rapidly degraded in peripheral plasma. The short-term success of several potent synthetic analogs of GnRH appears to be related to their relative resistance to degradation by peptidases.

   Like TRH, GnRH has been identified immunologically in extrahypothalamic nervous tissue and in the pineal gland of some species. Similarly, GnRH, like TRH, can cause depression of neural function in the central nervous system. Thus, GnRH may play a role as a neuromodulator or neurotransmitter in addition to its involvement with gonadotropin release.

   Studies of early amphibian development showed that the precursors of adenohypophysial cells as well as certain hypothalamic neurons have their origins in portions of the embryonic neural ridge (Fig. 4-5). From these studies, we have learned in both amphibians and mammals that one of these neural ridge derivatives, the **nasal placode,** is the origin of certain GnRH neurons of the hypothalamus as well as olfactory neurons that eventually make up the olfactory, terminal, and vomeronasal nerves. Nasal placodes are bilateral structures responsible for development of both left and right olfactory tracts. The olfactory neurons migrate first from the nasal region and contact the forebrain, inducing formation of the olfactory bulbs. The NS neurons follow an extracelluar matrix laid down by olfactory neurons along their migration route and are identifiable by their secretion of GnRH. The degree of migration of these NS cells in different species probably explains the variability in GnRH distribution. Ablation (surgical destruction or extirpation) of both nasal placodes in amphibians results in the absence of GnRH cells identified as immunoreactive for mammalian GnRH (mGnRH) in certain regions of the brain. Unilateral ablation results in absence of GnRH cells only on the side of the ablation. Similarly, a genetically based abnormality known as **Kallmann's syndrome** in humans results in a failure for nasal placode cells to migrate and produces patients with anosmia and hypogonadism.

2. Other Factors Affecting Gonadotropin Secretion

Secretion of GnRH is under the stimulatory control of NE-secreting neurons in both rats and primates. Activity of these NE-secreting neurons can be depressed by other neurons secreting β-endorphin. The only other neurotransmitter with distinct stimulatory actions on gonadotropin release is **glutamate,** an excitatory amino acid neurotransmitter. The actions of cholinergic, serotonergic, dopaminergic, and GABAnergic factors on GnRH and gonadotropin release are based largely on pharmacological studies and are considered controversial due to the variety of reported effects.

3. Feedback Effects on FSH and LH Secretion

Gonadal steroids generally have negative feedback effects on gonadotropin release at the level of the gonadotropes, reducing their sensitivity to GnRH. Estrogen-concentrating neurons in the POA of rats release GABA, which exerts presynaptic inhibition on NE-secreting neurons, lowers GnRH output from the hypothalamus, and may represent another pathway for negative feedback operating on gonadotropin release.

Selective negative feedback on FSH secretion occurs at the gonadotrope, reducing FSH output but not LH output in response to GnRH. This feedback is accomplished through production of a peptide by the gonads known as **inhibin.** The predominence of LH in the midcycle surge of gonadotropin at ovulation in females is due in part to this selective action of inhibin. Further discussion of inhibin and related peptides can be found in Chapter 11.

Positive feedback also occurs in the gonadal axis of female mammals. Pituitary gonadotropins stimulate the synthesis of estrogens, especially estradiol. In rats, estradiol binds to neurons in the preoptic area and causes GnRH neurons to release GnRH, which causes release of more gonadotropin, further elevating estradiol levels. Positive feedback by steroids has been confirmed on the pituitary gonadotrope as well as on brain neurons. Regardless of its site of action, positive feedback is responsible for the midcycle surge of LH that stimulates ovulation. Rupture of the ovarian follicle causes a reduction in estradiol synthesis, and the secretion of progesterone by the postovulatory follicle (corpus luteum) and reinstates negative feedback.

Gonadotropes have endothelin receptors, which when occupied can cause fluctuations in Ca^{2+} movements similar to those produced by GnRH. Endothelins are peptides composed of 21 amino acids (Fig. 4-22) and named for their original source, the endothelial cells of the pig aorta. There are three closely related endothelin peptides (endothelin-1,

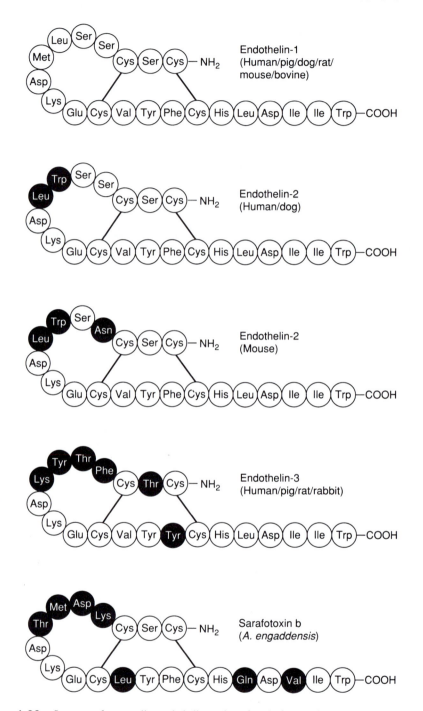

Figure 4-22. Structure of mammalian endothelins and sarafotoxin from snake venom. The vasoactive peptide, endothelin-1 was first isolated from pig aorta and may play an important modulatory role in neuroendocrine and endocrine systems. The black circles represent amino acid substitutions that differ from endothelin-1. [Modified from Masaki, T. Endothelins: Homeostatic and compensatory actions in the circulatory and endocrine systems. *Endocr. Rev.* **14,** 256–268, 1993. © The Endocrine Society.]

endothelin-2, and endothelin-3) that are similar to the peptide **sarafotoxin S6B** isolated recently from snake venom. Endothelins bind to two distinct receptor subtypes (ET_A and ET_B) coupled to G-protein mechanisms. (ET_A receptors preferentially bind endothelin-1 and endothelin-2 whereas ET_B receptors do not discriminate among the endothelins. Originally described as potent cardiovascular vasoconstrictors, their actions on neuroendocrine and endocrine systems indicate a more varied role. Immunoreactive endothelins have been demonstrated in paraventricular and supraoptic neurons that are connected to the pars nervosa, in extrahypothalamic neurons and glial cells (astrocytes), the adenohypophysis, the placenta, corpora lutea of the ovary, the thyroid gland, and the parathyroid glands. The actions of endothelins on gonadotropes involve voltage-sensitive calcium channels suggesting that endothelins and GnRH may activate a common intracellular pathway, although each operates through different receptors.

C. Regulation of Growth Hormone Secretion

Although initial studies involving transection of the hypothalamo–hypophysial portal system or explant of the pituitary implied that release of GH was under stimulatory hypothalamic control, we now know it is under both inhibitory and stimulatory control by neurohormones. Secretion of GH is episodic; that is, it occurs in bursts separated by longer intervals of lowered release. For example, in rats, GH is secreted at intervals of about 3 hr. In humans, maximal episodes of GH release occur during sleep (Fig. 4-23). Neural and neurohormonal factors influencing GH secretion are summarized in Fig. 4-20E.

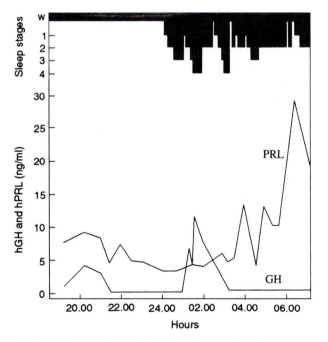

Figure 4-23. Diurnal pattern of GH and PRL release. [From Baulieu, E., and Kelly, P. (1992). "Hormones: From Molecules to Disease," p. 206. Chapman & Hall, London.]

1. Growth Hormone-Releasing and GH Release-Inhibiting Hormones

Release of GH, *in vivo,* is controlled primarily by a GH release-inhibiting peptide (GHRIH), also called **somatostatin (SS).** However, even in the absence of somatostatin, a releasing hormone known as **somatocrinin (GHRH)** or somatoliberin is necessary to stimulate GH release. The episodic bursts of GH release occur during periods of lowered somatostatin levels and elevated somatocrinin.

The periventricular nucleus of the anterior hypothalamus and the amygdala of the limbic system are the sources for most of the somatostatin-containing nerve terminals in the median eminence. This tetradecapeptide (14 AAs; Table 4-4) is a strong inhibitor of GH release. Somatostatin appears to work through a GTP-dependent G_i protein that prevents the elevation of cAMP necessary to provoke GH release (see Chapter 2).

Somatostatin also occurs in extrahypothalamic nervous tissue (both brain and spinal cord). A possible neuromodulator/neurotransmitter function in the central nervous system has been suggested, similar to that for TRH and GnRH. It is also present in the mucosa of the stomach where it locally inhibits release of the gastric hormone, gastrin (see Chapter 13), and in the pancreatic islets where it inhibits both glucagon and insulin release (see Chapter 14). A larger form of somatostatin (28 AAs) was extracted first from the gut and pancreatic tissues and later was shown to exist in nerve endings of the median eminence along with the tetradecapeptide. Studies have since shown that somatostatin-28 is released in equal or greater amounts than somatostatin-14 into portal blood. Furthermore, pituitary receptors show stronger binding affinity for the larger peptide. Both forms of somatostatin apparently are produced from the same prohormone.

Somatocrinin is associated mainly with the arcuate nucleus from which GHRH-immunoreactive fibers extend to the median eminence. GHRH occurs in the same neurons that produce other neural regulators. Some neurons in the arcuate nucleus contain DA and neurotensin as well as GHRH. Others exhibit GHRH colocalized with DA and galanin.

A potent peptide releaser of GH was first isolated and characterized chemically from a human pancreatic tumor, and the same gene was later found to be active in the hypothalamus. In human, porcine, bovine, ovine, and caprine mammals, somatocrinin is composed of 44 amino acid residues and is related chemically to the glucagon-secretin family of peptides (see Chapter 2). In the rat, GHRH has only 43 amino acids and differs considerably in amino acid composition from the other mammalian GHRHs. The significance of this deviation in rat GHRH structure is unknown.

2. Other Neural Factors Affecting GH Secretion

There is no shortage of investigations that have reported the effects of pharmacological and/or disruptive techniques on GH secretion. However, there are considerable conflicting data, and it is difficult to determine whether effects observed are direct or indirect. Furthermore, it is often difficult to ascertain whether an effect is mediated through control of somatostatin or somatocrinin release. For example, an increase in GH secretion could represent either a decrease in somatostatin release, an increase in somatocrinin release, or both.

Factors that have been shown to elicit GH release under certain conditions include NE, DA, 5-HT, GABA, **neuropeptide Y (NPY),** opiates, **vasoactive intestinal peptide (VIP),** thyroid hormones, glucocorticoids, and estrogens. Inhibition of GH secretion has been reported following treatment with GABA, 5-HT, substance P, neurotensin, large quantities of glucocorticoids, and androgens. From this brief summary, it seems clear that catecholamines are stimulatory. Serotonin is usually stimulatory, probably acting through effects on NE-secreting neurons. The dichotomous actions of GABA could be due to its ability to inhibit release of somatostatin and GHRH under different conditions. Substance P and neurotensin increase somatostatin release, explaining their prevention of GH secretion.

Some of the confusion as to roles of various regulators on GH release stems from species differences. For example, stress, which is usually associated with elevated glucocorticoids, generally increases GH secretion in primates, inhibits GH secretion in rodents, and has no effect in domestic ungulates. Moderate exercise also causes GH release in humans, and partial or total food deprivation also increases GH in primates as well as in domestic ungulates.

3. Feedback Effects on GH Secretion

Many of the actions of GH on body growth are mediated by two peptides produced in the liver and secreted into the general circulation under the direction of GH rather than by GH itself. These peptides are insulin-like growth factor I (IGF-I) and insulin-like growth factor II (IGF-II). IGF-I is the predominent form produced in adults whereas IGF-II is primarily a fetal factor. Plasma IGF-I reduces release of GH through negative feedback.

GH produces metabolic effects including elevation of glucose and free fatty acid levels as well as decreases in plasma amino acids. These metabolic products can affect GH release, but it appears they play no major physiological role. Elevated glucose can decrease GH output and insulin-induced hypoglycemia elevates GH release. High levels of free fatty acids also depress GH levels, and, conversely, depressed fatty acid levels elevate GH release. Both glucose and free fatty acid feedback effects are consistent with the reduction in both following insulin injection.

GH release is also sensitive to one amino acid, arginine. Infusion of arginine causes a marked increase in GH release. However, other amino acids are not very effective stimulators of GH release, and no elevation in circulating GH is observed following consumption of a high-protein meal. Hence, the physiological importance of amino acid feedback is questionable.

D. Regulation of Prolactin Secretion

Prolactin has the distinction of being the only pituitary tropic hormone for which there is evidence for two distinct RHs and RIHs, although most authorities favor only one of each. The primary control over PRL release, as mentioned earlier, is inhibitory in mammals, and the pituitary releases PRL when freed either surgically or chemically from hypothalamic control. This is in marked contrast to the situation with GH where both the absence of somatostatin and the presence of GHRH are necessary to elicit secretion. Consequently, ectopic or explanted pituitaries spontaneously release large quantities of PRL.

In lactating mammals, suckling or electrical stimulation of the teat sends neural input to the hypothalamus and evokes release of the nonapeptide oxtocin into the circulation, causing contraction of cells lining the ducts of the mammary glands and milk ejection. This **neuroendocrine reflex** also stimulates a delayed release of PRL, which stimulates the mammary gland to replace lost milk. Neural and neurohormonal factors affecting PRL release are summarized in Fig. 4-20 F.

1. Prolactin-Releasing Hormone (PRH)
and Release-Inhibiting Hormone (PRIH)

The consensus is that mammalian prolactin release-inhibiting hormone (PRIH) is the catecholamine neurotransmitter, dopamine, which is a potent inhibitor of PRL release *in vivo* or *in vitro* (Fig. 4-24). Dopamine is released from hypothalamic neurons in the arcuate nucleus and periventricular hypothalamus as a neurohormone. It is elevated in the hypothalamo–hypophysial portal system under physiological conditions where PRL release is inhibited.

Two additional PRIH candidates have been proposed. The first is GABA, which has been shown to block PRL release *in vivo* and *in vitro* by direct action on lactotropes. However, large amounts of GABA are required to inhibit PRL release and such levels are not found in either portal or peripheral blood samples, suggesting that GABA may not be a physiological PRIH.

There is also support for a peptide PRIH that can inhibit PRL release. This peptide PRIH is the 56-amino acid C-terminal fragment of the GnRH prohormone. It remains to be demonstrated, however, that this **GnRH-associated peptide (GAP)** is a physiological regulator.

Like the situation for GH, there is evidence for an endogenous factor or factors that stimulate PRL release. This PRL-releasing hormone (PRH) at first was thought to be the tripeptide already named TRH, since administration of synthetic TRH stimulates PRL release as well as TSH release at least under some conditions. However, the neuropeptide VIP stimulates PRL release during suckling when TRH is ineffective. Administration of TRH to suckling mice causes elevated TSH release but no increase in PRL release such as caused by VIP. Furthermore, TSH release is never observed during normal suckling. Any decision as to the importance of TRH for PRL release in other mammals awaits the results of similar studies.

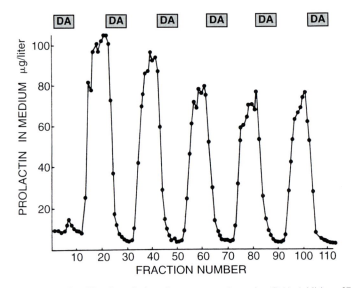

Figure 4-24. Response of perifused rat pituitary lactotropes to dopamine (DA). Addition of DA immediately depresses the spontaneous release of PRL normally seen *in vitro*. [From Martini, L. and Besser, G. M. (1977). "Clinical Neuroendocrinology." Academic Press, New York.]

2. Other Neural Factors Affecting Prolactin Secretion

Several neuropeptides evoke PRL release under experimental conditions, including **neurotensin, epidermal growth factor (EGF),** and EOPs. The opioids apparently block the activity of DA-secreting neurons that normally prevent PRL release. Somatostatin also can inhibit PRL release. GABA, which can directly inhibit PRL release at the lactotrope, can also stimulate PRL release through actions on hypothalamic neurons. It is not clear whether these actions of somatostatin and GABA have any physiological significance.

3. Feedback and Other Chemical Effects on Prolactin Secretion

The major feedback loop established in rats involves PRL feeding back directly on hypothalamic dopaminergic neurons. PRL also increases GABA synthesis and release into the portal circulation, supporting its inhibitory action on the lactotrope as part of the feedback mechanism rather than as a true PRIH.

Estrogens moderate PRL release in rats and, to a lesser extent, in sheep by enhancing the sensitivity of lactotropes to hypothalamic PRHs. This effect may involve an estrogen-induced reduction in their sensitivity to DA by altering receptor levels and/or intracellular second-messenger systems. Estradiol is concentrated by certain GABA-secreting neurons that decrease their activity and possibly release some PRH-secreting neurons from tonic GABA inhibition.

Gonadotropes release the octapeptide **angiotensin-II (ANG-II)** as one response to GnRH. ANG-II then binds to specific receptors on nearby lactotropes. This potentially paracrine action on the lactotrope may have some importance for PRL release but confirmatory studies are needed.

E. Regulation of Corticotropin Secretion

Corticotropin secretion by corticotropes of the pars distalis is primarily under stimulatory neuropeptide regulation by the neuropeptide **corticotropin-releasing hormone (CRH)** or corticoliberin. Evidence has been presented that suggests the existence of an inhibitory hypothalamic factor, but this is not confirmed. Although ACTH and MSH are synthesized by hydrolysis of the same preprohormone in the melanotrope and the corticotrope (see Section III,A,3), there are many differences in the regulation of their release. Secretion of ACTH stimulates cortisol (e.g., humans) or corticosterone (e.g., rats) or both (e.g., deer) from the adrenal cortex, and the circadian rhythm of ACTH secretion is reflected in a similar rhythm for corticosteroid release. In humans, there is a gradual decrease in ACTH plasma levels with the onset of sleep to a minimum at about midnight (Fig. 3-2). Then, a series of bursts of ACTH release occurs, culminating in maximal plasma levels of ACTH and cortisol at about 6 A.M. Following these peaks, there is a gradual decline in both hormones during the daylight hours with occasional bursts of secretion. Release of ACTH may occur following a mid-day meal but not an evening meal. Neural and neurohormonal factors affecting ACTH release are summarized in Fig. 4-20D.

1. Corticotropin-Releasing Hormone (CRH)

The amino acid composition of CRH consists of 41 amino acid residues, and it releases ACTH from the corticotrope *in vivo* and *in vitro*. The highest hypothalamic levels in rats and humans are in the paraventricular and arcuate nuclei. In rats, CRH occurs in the highest concentration in the median eminence, which exceeds concentrations in other brain regions by 10 to 100 times. CRH occurs in other hypothalamic nuclei as well as in other regions of the brain. In rats and humans, AVP and CRH have been colocalized in the same neurons.

Human and rat CRHs are identical peptides and oCRH differs from these at seven residues. There is considerable homology (at 20 positions) in CRH to **sauvagine,** a 40-residue peptide isolated from skin of the frog, *Phylomedusa sauvagii,* as well as overlap with the structure of a hypotensive agent, **urotensin I,** isolated from the spinal cord and urophysis of teleostean fishes.

ACTH release also is stimulated by the nonapeptide **arginine vasopressin (AVP),** and there is good evidence to support a physiological role for AVP in enhancing the responsiveness of corticotropes to CRH. During stress (see Chapter 9), the ratio of secreted AVP to CRH increases from 2:1 to as high as 9:1. Apparently, AVP released from the median eminence acts directly upon corticotropes to markedly increase their sensitivity to CRH and prevents shutdown of ACTH secretion via corticosteroid feedback on CRH release (see Section III,E,3). CRH-like bioactivity in the supraoptic nucleus is due primarily to the presence of AVP and not CRH, whereas the major CRH-like bioactivity in the paraventricular nucleus is due to CRH.

Although the Battleboro strain of rats lacks the ability to synthesize AVP, these rats still show a stress response. However, extracts from Battleboro hypothalami exhibit only 20% of the CRH activity demonstrated in normal rats. Addition of AVP restores ACTH-releasing activity to 100%.

CRH is also a potential inhibitor of LH and GH release at the level of the hypothalamus. These effects do not involve opioid neurons but the exact mechanism is unknown. CRH can stimulate GHRH release and inhibit somatostatin release, which would explain its effect on GH. The reduction in LH release when CRH is elevated may provide the mechanism whereby stress shuts down the reproductive axis.

2. Other Neural Factors Affecting ACTH Secretion

In addition to CRH and AVP, release of ACTH has been observed following administration of several classical neurotransmitters (norepinephrine, epinephrine, serotonin, acetylcholine) and a variety of neuropeptides (OXY, VIP, PHI, ANG-II). Although all of these compounds have been reported within nerve fibers of the median eminence, none of them appears to be of major importance in influencing ACTH release.

GABA has direct inhibitory actions on rat corticotropes that have receptors for GABA. Data concerning the action of GABA on ACTH secretion in humans are conflicting and no generalizations are yet possible.

3. Feedback Effects on ACTH Secretion

Negative feedback occurs through the actions of glucocorticoids both on CRH neurons in the hypothalamus and corticotropes in the pars distalis. However, during chronic stress, as mentioned previously, the output of AVP, which is not subject to glucocorticoid feedback, increases. As a result, the sensitivity of corticotropes to CRH is enhanced. This action helps to sustain prolonged elevation of glucocorticoids in the face of negative feedback on CRH release during chronic stress.

F. Regulation of Melanotropin Secretion

Melanotropin is released in large amounts when the pituitary is transplanted to an ectopic site or placed *in vitro,* indicating chronic inhibition of MSH release is the normal condition. Although evidence exists for stimulatory factors regulating MSH release, it seems likely that removal of inhibitory controls is most important for eliciting MSH release. A summary of neural and neurohormonal factors affecting MSH release is provided in Fig. 4-20G.

The predominant form of circulating MSH is acetylated or nonacetylated α-MSH, a peptide consisting of the first 13 amino acids of ACTH. α-MSH is actually produced by hydrolysis of ACTH synthesized within the melanotrope.

1. Melanotropin-Releasing Hormone (MRH) and Release-Inhibiting Hormone (MRIH)

Dopamine has been demonstrated to be a physiological melanotropin release-inhibiting hormone (MRIH) in numerous species. It is not clear whether α-MSH release is controlled by a neurohormone or by direct innervation. Several peptides have been shown to possess intrinsic MRIH activity, but the exact chemical identity of the endogenous neurohormone is not known. Extrahypothalamic MRIH activity has been demonstrated, and at least one of the synthetic releasers of α-MSH, a potential MRH, is an antidepressant when administered to humans.

2. Feedback Effects on α-MSH Secretion

There does not appear to be any direct feedback on MSH release. Stimulation of melanocytes by α-MSH does not release anything into the circulation that would be a feedback candidate. Perhaps the strong negative control of α-MSH release precludes the requirement for negative feedback.

V. The Hypothalamic Nonapeptide Neurohormones

The hypothalamic nonapeptide neurohormones are synthesized in the supraoptic and paraventricular hypothalamic nuclei. Most of these neurons project their axons to the pars nervosa, although some travel to the median eminence. Nonapeptides stored in the pars nervosa can be released directly into the general circulation in response to neural stimulation. The targets for these hormones are located at considerable distances from the pars nervosa (for example, kidney, mammary gland, uterus).

Two neurohypophysial nonapeptide hormones (usually AVP and OXY) are present in the pars nervosa of most adult mammals. Lysine vasopressin (LVP), a variant of AVP, is produced by members of the mammalian order Suina, which includes peccaries, domestic pigs, and the hippopotamus. Most of the species in this group exhibit both AVP and LVP in addition to OXY. Phenypressin (PVP), a unique vasopressin-like molecule, replaces vasopressin in the pars nervosa of marsupials.

Fetal mammals secrete a molecule that at first appears to be a hybrid of AVP and OXY, being composed of the side chain of AVP with the ring structure of OXY. This nonapeptide is known as **arginine vasotocin (AVT)** and is characteristic of adult nonmammalian vertebrates (Fig. 4-12; see also Chapter 6). The pineal gland of at least some adult mammals also contains AVT.

The vasopressins (AVP, LVP) and PVP function as antidiuretic agents, increasing the ability of the kidneys to reabsorb water from the glomerular filtrate. At higher doses, vasopressins cause vasoconstriction and can elevate blood pressure (a pressor effect). This action may increase glomerular filtration and water excretion. AVT produces similar actions in nonmammals and has been suggested to play an osmoregulatory role in fetal mammals.

As mentioned earlier, OXY stimulates contraction of myoepithelial cells lining the ducts of the mammary glands and causes ejection of milk. The stimulatory action of OXY on smooth muscle of the uterus is related to the induction of labor and the birth process. OXY also stimulates contractions in oviducts as well as in the vas deferens (sperm duct) of males.

VI. Clinical Aspects of the Hypothalamo–Hypophysial System

A. Disorders of the Hypothalamus

The most common clinical disorder of the hypothalamus is **diabetes insipidus.** The patient produces an abnormally large volume (3 to 8 liters/day) of dilute urine. This diuresis is most commonly due to the absence of AVP, which normally controls water reabsorption in the kidney (see Chapter 6 for more details on the action of AVP). Forty to 50% of such patients are **idiopathic** (i.e., denoting a disease of unknown cause) and exhibit no other evidence of neuroendocrine dysfunction. Some of these patients have been identified at autopsy as having degeneration of the supraoptic and paraventricular nuclei. About 15% of the cases are related to the presence of tumors within the brain, producing pressure on the supraoptic and paraventricular nuclei, which indirectly reduces secretion of AVP. Physical damage (such as a brain lesion) or certain infections (e.g., encephalitis) account for the remainder. **Nephrogenic diabetes insipidus** is the result of a failure of the kidney tubules to respond to normal or above normal levels of AVP.

The **syndrome of inappropriate antidiuresis (SIAD)** is caused by excessive AVP release. High levels of AVP result in excessive water retention, elevated blood pressure, and reduced urine production. Drugs such as demeclocycline block the action of AVP on the kidney and are used to treat this condition. However, long-term use of such drugs induces nephrogenic diabetes insipidus. Medications that control AVP release from the pars nervosa often are not uniformly effective and can influence other hypothalamic functions.

Tumors within the central nervous system are responsible for a number of other disorders. One of the more dramatic consequences is **sexual precocity** (accelerated sexual maturation; see Chapter 11 for a technical description of precocity). Precocity is much more common in males. About one-fourth of precocity cases are correlated with the presence of a pineal tumor, and 95% of these occur in males. A number of other cases of sexual precocity are associated with hypothalamic tumors, most of which also are found in males. One type of tumor, associated with precocity, the **harmatoma,** occurs in the posterior hypothalamus. It consists of masses of partially disoriented glial and ganglion cells or of normal cells located in abnormal sites. Harmatomas may secrete GnRH, which could explain their effects on early sexual maturation. Other causes for precocity are listed in Chapter 11.

Several other pathologies have been identified including the following:

1. **Hyponatremia:** Low blood sodium may be correlated with a number of factors including brain carcinoma, basal skull fractures, meningitis, and encephalitis.
2. **Hypernatremia:** Excessive levels of blood sodium may occur as a result of aneurysms in the brain, pineal tumors, and so forth. This condition is not accompanied by fluid imbalance.

B. Disorders of the Adenohypophysis

Acromegaly is a spectacular disorder of GH regulation and has been publicized heavily. When it occurs in children, it can lead to gigantism. This is a rare disorder, however, affecting from 3 to 40 individuals per million people in the United States each year. It is a well-known disorder because acromegaly was the first disorder of the pituitary gland to be recognized and because it produces giants. Acromegaly is caused by overproduction of GH due either to the absence of adequate somatostatin to suppress GH release or by the absence of negative feedback to suppress release. Growth hormone-secreting tumors release GH autonomously. Approximately half of acromegalic patients are deficient in one or more additional pituitary hormones, usually the gonadotropins. Not only do such patients exhibit excessive growth, but body proportions become distorted. When acromegaly develops in adults, only the body proportions become distorted since growth in stature has ceased and cannot be reinitiated with any quantity of GH. Cartilage tends to proliferate in joints resulting in abnormally proportioned hands and elongate jaws. The nose and ears, whose supporting tissue is cartilage, enlarge markedly causing distortion in appearance. There also are marked effects on other systems, for example, excessive sweating and secretion of sebum by the skin, enlargement of the heart, and hypertension may develop. Life expectancy is shortened considerably. Use of excessive amounts of GH by people to augment muscle growth in order to improve physical performance can produce similar effects.

In contrast, lack of sufficient GH during early life can cause short stature. However, other explanations for subnormal growth are known. Furthermore, it is important to distinguish between short stature with normal body proportions and dwarfism where the individual exhibits distorted features. Laron-type dwarfism occurs in humans with normal GH levels but very low IGF blood levels. Due to possession of a mutant allele, the liver of these patients lacks GH receptors. GH-binding proteins are absent from the blood as well. African pygmies lack the binding proteins for GH and show the normal pattern for growth, but the inability to maintain normal circulating levels of GH causes growth to be much less than normal.

Pituitary chromophobe adenomas are the most common source of pituitary-related problems. (Any benign or noncarcinogenic glandular tumor can be termed an adenoma.) They rarely secrete any hormones (occasionally GH and seldom TSH or ACTH), and their effects are usually due to pressure on the brain or optic chiasm caused by growth of the adenoma. Most patients experience severe headaches and visual disturbances (even blindness can result). Sometimes the production of one or more pituitary hormones may be reduced. Other adenomas may secrete excessive amounts of prolactin (prolactinomas). TSH, ACTH, or gonadotropins may be products of pituitary adenomas, and their clinical impacts are discussed in Chapters 7, 9, and 11, respectively. Surgical removal of the adenoma is the most common treatment and is highly successful.

Partial or total **hypopituitarism** refers to selective or total absence of pituitary hormones. These defects may reside in the adenohypophysis itself (primary disorder) or be due to hypothalamic dysfunction (secondary disorder).

There has been considerable interest in the role(s) of EOPs on mental disturbances such as depression and schizophrenia. Clinical studies, however, have yielded mixed results, and it is not clear whether mental illness is associated with EOPs.

VII. Summary

The adenohypophysis produces tropic hormones and consists of a pars distalis, a pars intermedia, and a pars tuberalis. The embryonic origin of the adenohypophysis (Rathke's pouch) is generally assumed to be from oral ectoderm, but studies suggest a neural origin similar to the neurohypophysis and hypothalamus. The major portion of the adenohypophysis consists of the pars distalis, which produces GH, PRL, GTHs, (FSH, LH), ACTH, TSH, and endorphins. The posterior portion of the adenohypophysis is the pars intermedia, which is responsible for synthesis of α-MSH. A number of mammals lack a pars intermedia and α-MSH may not be an important pituitary product in those species. The pars tuberalis contains some stainable cell types, but its physiological importance is not clear.

The neurohypophysis forms from the infundibulum and consists of an anterior neurohemal area, the median eminence, and a more posterior neurohemal structure, the pars nervosa. The median eminence stores the neurohormones produced in the hypothalamus that regulate tropic hormone release from the adenohypophysis. The hypothalamo–hypophysial portal system connects the median eminence to the adenohypophysis. The pars nervosa has no common blood supply with the adenohypophysis. It is responsible for storage of nonapeptide neurohormones (usually AVP and OXY) produced in the hypothalamus until they are released into the general circulation.

There are three categories of tropic hormones based on chemical structure. Category 1 includes the glycoproteins, LH, FSH, and TSH. Category 2 includes the large peptides, GH and PRL. Category 3 includes the very similar smaller peptides, ACTH, MSH, and LPH. Structural similarities suggest that LPH may be related to the endorphins, which are linked to opiate actions on the mammalian central nervous system. Placentas of mammals produce up to four tropic-like hormones, including CG (LH-like), CS (PRL-like and GH-like to some degree), CT (TSH-like) and CC (ACTH-like), as well as PRL and GH. Pregnant mare serum gonadotropin is a chorionic-type GTH with both FSH- and LH-like properties. Menopausal gonadotropin is FSH-like and is produced by the pituitary of postmenopausal women.

The NS nuclei controlling tropic hormone release are generally located in the preoptic area or the ventral region of the hypothalamus. Aminergic and peptidergic neurons terminate in the median eminence where they release the hypothalamo–hypophysiotropic regulating hormones into the portal circulation. These NS neurons are innervated by regular neurons (e.g., aminergic, cholinergic, serotonergic), which influence their release. These regulating neurons may be responsive to hormones or other factors as well.

Thyrotropin release is stimulated by TRH secreted primarily by neurons in the paraventricular nucleus. TRH release can be increased by NE and E and inhibited by 5-HT and DA. Somatostatin also blocks TSH release. However, thyroid hormones feedback primarily on the thyrotropes with only minor effects in the hypothalamus.

Gonadotropin release (LH and FSH) is induced by hypothalamic GnRH, which in turn is stimulated by NE but inhibited by opioid-secreting neurons. Location of the GnRH neurons is species-specific and no generalizations are possible. Negative feedback occurs generally through actions of gonadal steroids on hypothalamic neurons or on gonadotropes. Inhibin secreted by the gonads feeds back specifically to limit FSH release from the gonadotrope. Positive feedback by estrogens may occur primarily at either the hypothalamus (e.g., rats) or the pituitary gonadotrope (e.g., primates).

Growth hormone release is under strong inhibitory control by hypothalamic somatostatin but somatocrinin (GHRH) is necessary to evoke release in the absence of somatostatin. Somatostatin comes primarily from the periventricular nucleus and the amygdala of the limbic system. Somatocrinin is produced in the arcuate nucleus. Many neural and NS factors can affect GH release, but the pathway and physiological importance of most observations are unclear. Catecholamines are always stimulatory. Many actions of GH are mediated by IGFs and the latter form the primary negative feedback on GH release. Metabolites (glucose, free fatty acids, arginine) can affect GH release, but their physiological role is uncertain.

Prolactin release, like GH, is under strong inhibitory control by dopamine from the arcuate and periventricular nuclei acting as a neurohormone. It also can be inhibited by GAP, although a physiological role for GAP is not clear. Unlike the situation for GH, no PRH seems necessary to get PRL release in the absence of DA inhibition. During suckling, however, hypothalamic VIP is an important stimulant. Although TRH has been shown to induce PRL release, it may not be a physiological releaser. Negative feedback occurs through direct actions of PRL on hypothalamic neurons.

Corticotropin secretion is stimulated by CRH, and AVP enhances the sensitivity of corticotropes to CRH. Corticotropes also produce β-LPH from which they release β-endorphin. β-LPH and ACTH are produced from the same preprohormone. CRH is produced primarily in the paraventricular and arcuate nuclei and greatest levels are observed in the median eminence. AVP in the median eminence comes primarily from the supraoptic nucleus. Glucocorticoids provide negative feedback directly on corticotropes and on hypothalamic CRH neurons. In chronic stress, glucocorticoid feedback is in part overcome by elevated AVP.

Release of melanotropin, like PRL, is inhibited by dopamine, probably acting as a neurohormone (MRIH), and no MRH is necessary. Melanotropes first produce ACTH from a prohormone and then hydrolyze it to α-MSH and CLIP prior to release. Like corticotropes, they also release β-endorphin into the general circulation. There is little evidence for negative feedback on α-MSH release.

NS neurons in the supraoptic and paraventricular nuclei produce the nonapeptides AVP and OXY, which are stored in the pars nervosa. A few mammals make LVP as well as AVP and marsupials produce PVP instead of AVP. Fetal mammals produce AVT instead of AVP and OXY.

Suggested Reading

Books

Brown, R. E. (1994). "An Introduction to Neuroendocrinology." Cambridge Univ. Press, New York.

Colmers, W. F., and Wahlestedt, C. (1993). "The Biology of Neuropeptide Y and Related Peptides." Humana Press, Totowa, NJ.

Imura, H. (1994). "The Pituitary Gland," 2nd Ed., Comprehensive Endocrinology Revised Series. Raven, New York.

Melmed, S. (1995). "The Pituitary." Blackwell, Cambridge, MA.
Motta, M. (1991). "Brain Endocrinology," 2nd Ed., Comprehensive Endocrinology Revised Series. Raven, New York.
Muller, E. E., and Nistico, G. (1989). "Brain Messengers and the Pituitary." Academic Press, San Diego.
Nemeroff, C. B. (1992). "Neuroendocrinology." CRC Press, Boca Raton, FL.
North, W. G. (1993). "The Neurohypophysis: A Window on Brain Function," Vol. 689. N.Y. Acad. Sci., New York.

Articles
General/Miscellaneous

Beck-Peccoz, P., Persani L., and Faglia, G. (1992). Glycoprotein hormone α-subunit in pituitary adenomas. *Trends Endocrinol. Metab.* **3,** 41–45.
Besedovsky, H. O., and Del Ray, A. (1996). Immune-neuro-endocrine interactions: Facts and hypotheses. *Endocr. Rev.* **17,** 64–102.
Brann, D. W., and Mahesh, V. B. (1992). Excitatory amino acid neurotransmission: Evidence for a role in neuroendocrine regulation. *Trends Endocrinol. Metab.* **3,** 122–126.
Deschepper, C. F. (1991). The renin-angiotensin system in the pituitary gland. *Trends Endocrinol. Metab.* **2,** 104–107.
Dubois, P. M., and El Amraouci, A. (1995). Embryology of the pituitary. *Trends Endocrinol. Metab.* **6,** 1–7.
Gershon, M. D. (1993). Development of the neural crest. *J. Neurobiol.* **24,** 141–145.
Green, J. D. (1951). The comparative anatomy of the hypophysis with special reference to its blood supply and innervation. *Am. J. Anat.* **88,** 225–311.
Houben, H., and Denef, C. (1990). Regulatory peptides produced in the anterior pituitary. *Trends Endocrinol. Metab.* **1,** 398–403.
Merchenthaler, I. (1991). Current status of brain hypophysiotropic factors: Morphologic aspects. *Trends Endocrinol. Metab.* **2,** 219–226.
Pearse, A. G. E., and Takor Takor, T. (1976). Neuroendocrine embryology and the APUD concept. *Clin. Endocrinol.* **5,** Suppl. 229s–244s.
Purves, H. D. (1961). Morphology of the hypophysis related to its function. *In* "Sex and Internal Secretions" (W. C. Young, ed.), 3rd Ed., pp. 161–239. Boilliere, Tindall and Cox, London.
Rawlings, S. R., and Hezareh, M. (1996). Pituitary adenylate cyclase-activating polypeptide (PACAP) and PACAP/vasoactive intestinal polypeptide receptors: Actions on the anterior pituitary gland. *Endocr. Rev.* **17,** 4–29.
Said, S. I. (1991). Vasoactive intestinal peptide: Biologic role in health and disease. *Trends Endocrinol. Metab.* **2,** 107–112.
Schwartz, J., and Cherny, R. (1992). Intercellular communication within the anterior pituitary influencing the secretion of hypophysial hormones. *Endocr. Rev.* **13,** 453–475.
Spangelo, B. L., and MacLeod, R. M. (1990). The role of immunopeptides in the regulation of anterior pituitary hormone release. *Trends Endocrinol. Metab.* **1,** 408–412.
Steele, M. K. (1992). The role of brain angiotensin II in the regulation of luteinizing hormone and prolactin secretion. *Trends Endocrinol. Metab.* **3,** 295–301.
Swabb, D. F., Hofman, M. A., Lucassen, P. J., Purba, J. S., Raadsheer, F. C., and Van de Nes, J. A. P. (1993). Functional neuroanatomy and neuropathology of the human hypothalamus. *Anat. Embroyl.* **187,** 317–330.

Gonadotropins and Releasing Hormones

Barbieri, R. L. (1992). Clinical applications of GnRH and its analogues. *Trends Endocrinol. Metab.* **3,** 30–34.
Beitins, I. Z., and Padmanabhan, V. (1991). Bioactive follicle-stimulating hormone. *Trends Endocrinol. Metab.* **2,** 145–151.
Gharib, S. D., Wierman, M. E., Shupnik, M. A., and Chin, W. W. (1990). Molecular biology of the pituitary gonadotropins. *Endocr. Rev.* **11,** 177–198.
Knobil, E. (1992). Remembrance: The discovery of the hypothalamic gonadotropin-releasing hormone pulse generator and of its physiological significance. *Endocrinology* **131,** 1005–1006.

Krsmanovic, L. Z., Stojilkovic, S. S., and Catt, K. J. (1996). Pulsatile gonadotropin-releasing hormone release. *Trends Endocrinol. Metab.* **7**, 56–59.

Petit, C. (1993). Molecular basis of the X-chromosome-linked Kallmann's syndrome. *Trends Endocrinol. Metab.* **4**, 8–13.

Rissman, E. F. (1996). Behavioral regulation of gonadotropin-releasing hormone. *Biol. Reprod.* **54**, 413–419.

Schwanzel-Fukuda, M., Jorgenson, K. L., Bergen, H. T., Weesner, G. D., and Pfaff, D. W. (1992). Biology of normal luteinizing hormone-releasing hormone neurons during and after their migration from olfactory placode. *Endocr. Rev.* **13**, 623–634.

Stojilkovic, S. S., Krsmanovic, L. Z., Spergel, D. J., and Catt, K. J. (1994). Gonadotropin-releasing hormone neurons: Intrinsic pulsatility and receptor-mediated regulation. *Trends Endocrinol. Metab.* **5**, 201–209.

Tsai, P.-S., and Weiner, R. I. (1996). Regulation of gonadotropin-releasing hormone neurons by basic fibroblast growth factor. *Trends Endocrinol. Metab.* **7**, 65–68.

TSH and Releasing Hormones

Magner, J. A. (1990). Thyroid-stimulating hormone: Biosynthesis, cell biology, and bioactivity. *Endocr. Rev.* **11**, 354–385.

Tixier-Vidal, A., and Faivre-Baumann, A. (1992). Ontogeny of thyrotropin-releasing hormone biosynthesis and release of hypothalamic neurons. *Trends Endocrinol. Metab.* **3**, 59–64.

GH, PRL, and Releasing Hormones

Amselem, S., Duquesnoy, P., and Goosens, M. (1991). Molecular basis of Laron dwarfism. *Trends Endocrinol. Metab.* **2**, 35–40.

Barkum, A. L. (1992). Acromegaly. *Trends Endocrinol. Metab.* **3**, 205–210.

Baumann, G. (1991). Growth hormone heterogeneity: Isohormones, variants, and binding proteins. *Endocr. Rev.* **12**, 424–449.

Baxter, R. C. (1993). Circulating binding proteins for the insulinlike growth factors. *Trends Endocrinol. Metab.* **4**, 91–96.

Ben-Johnathan, N., and Lie, J.-W. (1992). Pituitary lactotrophs: Endocrine, paracrine, juxtacrine, and autocrine interactions. *Trends Endocrinol. Metab.* **3**, 254–258.

Corpas, E., Harman, S. M., and Blackman, M. R. (1993). Human GH and human aging. *Endocr. Rev.* **14**, 20–39.

Devsa, J., Lima, L., and Tresguerres, J. A. F. (1992). Neuroendocrine control of growth hormone secretion in humans. *Trends Endocrinol. Metab.* **3**, 175–182.

Dieguez, C., and Casanueva, F. F. (1995). Influence of metabolic substrates and obesity on growth hormone secretion. *Trends Endocrinol. Metab.* **6**, 55–59.

Frawley, L. S. (1994). Role of the hypophyseal neurointermediate lobe in the dynamic release of prolactin. *Trends Endocrinol. Metab.* **5**, 107–112.

Herrington, A. C. (1994). New frontiers in the molecular mechanisms of growth hormone action. *Mol. Cell. Endocrinol.* **100**, 39–44.

Korbonits, M., and Grossman, A. B. (1995). Growth hormone releasing peptide and its analogues: Novel stimuli to growth hormone release. *Trends Endocrinol. Metab.* **6**, 43–49.

LeRoith, D., Adamo, M., Werner, H., and Roberts, C. T., Jr. (1991). Insulinlike growth factors and their receptors as growth regulators in normal physiology and pathologic states. *Trends Endocrinol. Metab.* **2**, 134–139.

Lewis, U. J. (1992). Growth hormone: What is it and what does it do? *Trends Endocrinol. Metab.* **3**, 117–121.

Siddiqui, R. A., McCutcheon, S. N., Blair, H. J., Mackenzie, D. D. S., Morel, P. C. H., Brier, B. H., and Gluckman, P. D. (1992). Growth allometry of organs, muscles and bones in mice from lines divergently selected on the basis of plasma insulin-like growth factor-I. *Growth Devel. Aging* **56**, 53–60.

Sinha, Y. N. (1992). Prolactin variants. *Trends Endocrinol. Metab.* **3**, 100–106.

POMC Derivatives

Childs, G. V. (1991). Multipotential pituitary cells that contain adrenocorticotropin (ACTH) and other pituitary hormones. *Trends Endocrinol. Metab.* **2**, 112–117.

Mains, R. E., and Eipper, B. A. (1990). The tissue-specific processing of pro-ACTH/endorphin: Recent advances and unsolved problems. *Trends Endocrinol. Metab.* **1,** 388–394.

Margioris, A. N. (1993). Opioids in neural and nonneural tissues. *Trends Endocrinol. Metab.* **4,** 163–168.

Orth, D. N. (1992). Corticotropin-releasing hormone in humans. *Endocr. Rev.* **13,** 164–191.

Plotsky, P. M., Cunningham, E. T., and Widmaier, E. P. (1989). Catecholaminergic modulation of corticotropin-releasing factor and adrenocorticotropin secretion. *Endocr. Rev.* **10,** 437–458.

Solomon, S. (1993). Corticostatins. *Trends Endocrinol. Metab.* **4,** 260–264.

Endothelins

Battistini, B., D'Orleans-Juste, P., and Sirois, P. (1993). Endothelins: Circulating plasma levels and presence in other biologic fluids. *Lab. Invest.* **68,** 600–628.

Macrae, A. D., and Bloom, S. R. (1992). Endothelin: An endocrine role. *Trends Endocrinol. Metab.* **3,** 153–157.

Masaki, T. (1993). Endothelins: Homeostatic and compensatory actions in the circulatory and endocrine systems. *Endocr. Rev.* **14,** 256–268.

Stojilkovic, S. S., and Catt, K. J. (1992). Neuroendocrine actions of endothelins. *Trends Pharmacol. Sci.* **13,** 385–391.

5

Comparative Aspects of the Hypothalamo–Hypophysial System in Nonmammalian Vertebrates

THE FOLLOWING account describes the major features of the hypothalamo–hypophysial axes of nonmammalian vertebrates as contrasted to the description given for mammals in Chapter 4. Hormones described in Chapter 4 that also appear in nonmammals are not re-defined, although their names appear with abbreviations the first time they are used. If the reader is not familiar with nonmammalian vertebrates, a brief discussion of the evolution-ary relationships and importance of various vertebrate taxonomic groups listed here can be found in Appendix A.

The hypophysial system is unique to the vertebrate chordate animals, and no invertebrate chordate, or for that matter any other invertebrate deuterostome phylum, exhibits an endo-crine gland closely integrated with the brain. However, there is a hint of a homologous system in the primitive cephalochordate called amphioxus (*Branchiostoma*). A shallow epithelial groove located in the oral cavity of *Branchiostoma* appears in close proximity to the simple dorsal nervous system of this animal (Fig. 5-1). It is known as **Hatschek's pit**

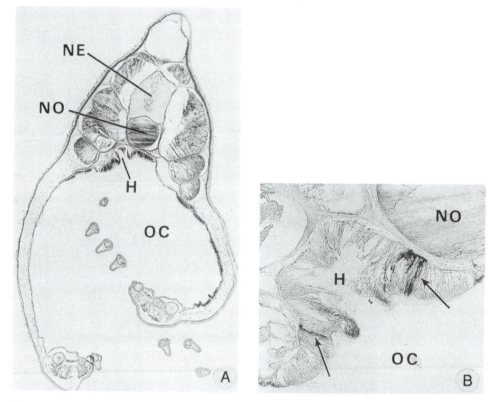

Figure 5-1. Hatschek's pit in the cephalochordate, amphioxus. (A) Cross-section through the oral cav-ity (OC) of an adult animal, showing two basic chordate features: the dorsal nerve cord (NE) and the notochord (NO). Hatschek's pit (H) showing immunoreactive substance P (dark staining) appears on the dorsal pharyngeal surface similar to the location of Rathke's pouch in vertebrate embryos. (B) Enlargement of Hatschek's pit, showing immunoreactive metenkephalin-like material (arrows). [From Nozaki, M., and Gorbman, A. (1992). The question of functional homology of Hatschek's pit of amphioxus (*Branchiostoma belcheri*) and the vertebrate adenohypophysis. *Zool. Sci.* **9**, 387-395.]

and has been found to react positively to antibodies against substance P, Met-enkephalin, cholecystokinin (CCK), and mammalian luteinizing hormone (LH). Claims for immuno-reactivity to gondadotropin-releasing hormone (GnRH) have not been confirmed, however. In another group of invertebrate chordates, the ascidians or tunicates, we find the so-called neural gland, which is also associated with the oral cavity. Although no immunohistochem-ical studies have been reported at the time of this writing, the embryological association of all three structures (pituitary, Hatschek's pit, and neural gland) with the oral epithelium, and indeed the nasal placodes, is suggestive of homology. These observations, of course, do not prove homology between these two orally derived structures but are certainly sup-portive of such a relationship.

With few exceptions, the brain structures responsible for secretion of neuropeptides are paired areas. However, in discussing the presence of these secretory areas or nuclei and their related neuropeptides, it is often cumbersome to refer always to their paired nature. In the following accounts, the reader should assume all nuclei are paired or bilateral even though they may be referred to in the singular.

I. The Fishes

The piscine hypothalamo–hypophysial system is separable into the same major divisions as that of mammals: **hypothalamus, neurohypophysis,** and **adenohypophysis** (Fig. 5-2). Some marked differences from mammals occur in fishes as well. There is no pars tuberalis in fishes, although a possibly homologous structure, the pars ventralis, occurs in sela-chians (chondrichthyeans). Although there is no median eminence and no hypothalamo–hypophysial portal system in hagfishes and teleosts, both are present in lampreys and sharks as well as in nonteleostean bony fish groups. The pars distalis typically is differentiated into two subregions or zones, each with its special cellular types. The pars intermedia is intimately interdigitated in most fishes with the pars nervosa of the neurohypophysis to form a **neurointermediate lobe.**

Finally, posterior to the neurointermediate lobe in cartilaginous fishes and most bony fishes is a unique structure formed from the floor of the diencephalon. This structure is probably derived from ependymal cells and is known as the **saccus vasculosus.** It is espe-cially well developed in some groups, but its function is unknown. Although the saccus vasculosus is possibly not an endocrine gland, it is a prominent feature that evolved early among jawed fishes and is useful in comparing early vertebrate phylogenies.

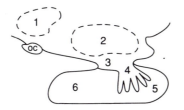

Figure 5-2. Generalized hypothalamo-hypophysial axis in fishes. 1, Preoptic area; 2, hypothalamus; 3, median eminence (when present); 4, pars nervosa; 5, pars intermedia; 6, pars distalis. Areas 4 and 5 together constitute the neurointermediate lobe. OC, Optic chiasm.

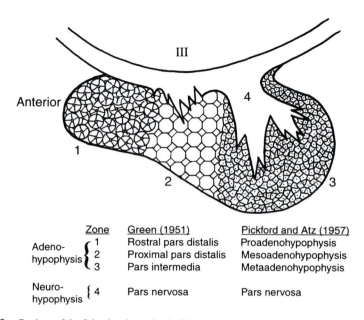

	Zone	Green (1951)	Pickford and Atz (1957)
Adeno-hypophysis {	1	Rostral pars distalis	Proadenohypophysis
	2	Proximal pars distalis	Mesoadenohypophysis
	3	Pars intermedia	Metaadenohypophysis
Neuro-hypophysis {	4	Pars nervosa	Pars nervosa

Figure 5-3. Regions of the fish adenohypophysis. The terminologies of Green (1951) and Pickford and Atz (1957) are compared for three histologically distinct regions (1, 2, and 3). Only the Green nomenclature is used in text.

Two different terminologies have been proposed for the subregions of the piscine adenohypophysis. The nomenclature proposed by Green is used here in preference to the alternative system proposed by Pickford and Atz, because the Green system is similar to mammalian terminologies (Fig. 5-3). In this discussion, each term is defined initially, followed by the terms proposed by Pickford and Atz (italicized and in parentheses).

There are three distinct zones recognized in the adenohypophysis by both schemes. The most anterior and rostral (dorsal) portion of the piscine adenohypophysis often consists of follicles of cells and is termed the **rostral pars distalis** (*proadenohypophysis*). The remainder of the pars distalis comprises the **proximal pars distalis** (*mesoadenohypophysis*). The third region is the **pars intermedia** (*metaadenohypophysis*).

Each of these regions of the adenohypophysis is readily distinguished cytologically and generally each contains different cellular types that produce different tropic hormones. The cellular types found in each region and the hormones they are thought to produce are summarized in Table 5-1. Distribution of cellular types are shown in Fig. 5-4. Although alternative schemes have been proposed, the classification scheme for pituitary cell types used here is the same one described for mammals in Chapter 4.

For our discussions of hypothalamic control of hypophysial functions, we focus on two regions in which the majority of neurosecretory (NS) neurons reside (Fig. 5-5). The first area is located just anterior and dorsal to the optic chiasma and is called the **preoptic area** as in mammals. This region marks the telencephalic and diencephalic boundary and usually is considered part of the telencephalon. Endocrinologists often include it when they are discussing the hypothalamic control of pituitary function as though it were part of the diencephalon. The second region includes the hypothalamus proper. With the exception of

Table 5-1. Comparative Cytology of the Pars Distalis from Representatives of Different Vertebrate Groups

Hormone	Vertebrate group	Cellular type	Alternate names	Ultrastructural determination of cytoplasmic granule size (nm)
TSH	Chondrichthyes: Selachians	Type 1 basophil	Type I	90–120
	Osteichthyes			
	Holosteans	Type 1 basophil (amphophil)		—
	Teleosts	Type 1 basophil	δ-Basophil	400
	Amphibia: Anurans	Type 1 basophil		150–400
	Reptiles	Type 1 basophil		300–400 × 200–250
	Birds	Type 1 basophil	δ-Basophil	50, 100, 200
	Mammals	Type 1 basophil	β-Basophil	150
GTH	Chondrichthyes: Selachians	Type 2 basophil	Type V, VI	100–700
	Osteichthyes			
	Holosteans	Type 2 basophil		
	Teleosts	Type 2 basophil	β- and γ-Basophils	60–160; 80–240
	Amphibia: Anurans	Type 2 basophil		Polymorphous to 900
	Reptiles	Type 2 basophil		150–270; 600–800
	Birds	Type 2 basophil	β-Basophil; γ-basophil	120–200 120–400
	Mammals	Type 2 basophil	δ-Basophil	200
ACTH	Chondrichthyes: Selachians	Type 3 basophil	Type II	140
	Osteichthyes			
	Holosteans		Acidophil	
	Teleosts	Type 3 basophil	ε-Cell	110–250
	Amphibia: Anurans	Type 3 basophil		100–200
	Reptiles	Type 3 basophil (amphophil)		—
	Birds	Type 3 basophil	ε-Cell	150–300
	Mammals	Type 3 basophil		200
PRL	Chondrichthyes: Selachians	Type 1 acidophil	Type IV	263
	Osteichthyes			
	Holosteans	Type 1 acidophil		—
	Teleosts	Type 1 acidophil	η-Cell	Polymorphic; 170–350
	Amphibia: Anurans	Type 1 acidophil		180–500
	Reptiles	Type 1 acidophil		—
	Birds	Type 1 acidophil	η-Cell	Polymorphic; 250–300
	Mammals	Type 1 acidophil	ε-Acidophil	Polymorphic; 600–900
GH	Chondrichthyes: Selachians	Type 2 acidophil	Type III	200
	Osteichthyes			
	Holosteans	Type 2 acidophil		—
	Teleosts	Type 2 acidophil	α-Cell	—
	Amphibia: Anurans	Type 2 acidophil		180–250
	Reptiles	Type 2 acidophil		310
	Birds	Type 2 acidophil	α-Acidophil	250–300
	Mammals	Type 2 acidophil	α-Acidophil	350

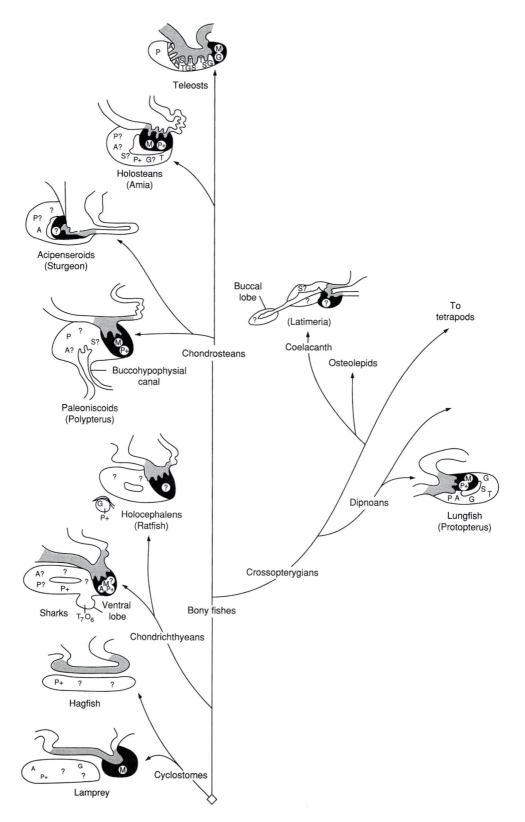

Teleosts

Holosteans
(Amia)

Acipenseroids
(Sturgeon)

Paleoniscoids
(Polypterus)

Buccohypophysial
canal

Chondrosteans

Buccal
lobe

(Latimeria)

Coelacanth

Osteolepids

To
tetrapods

Dipnoans

Lungfish
(Protopterus)

Holocephalens
(Ratfish)

Sharks

Ventral
lobe

Bony fishes

Crossopterygians

Chondrichthyeans

Hagfish

Lamprey   Cyclostomes

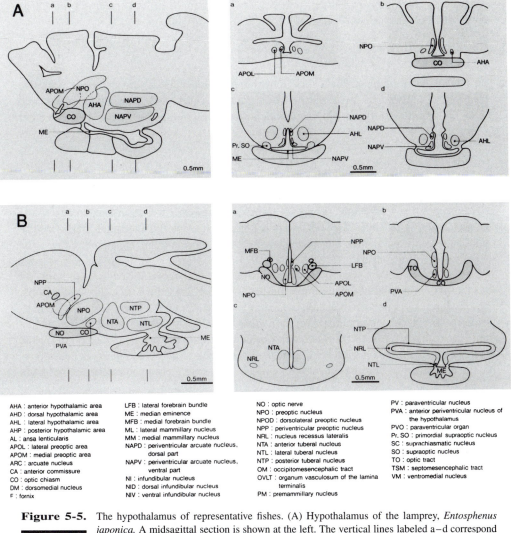

AHA : anterior hypothalamic area
AHD : dorsal hypothalamic area
AHL : lateral hypothalamic area
AHP : posterior hypothalamic area
AL : ansa lenticularis
APOL : lateral preoptic area
APOM : medial preoptic area
ARC : arcuate nucleus
CA : anterior commissure
CO : optic chiasm
DM : dorsomedial nucleus
F : fornix

LFB : lateral forebrain bundle
ME : median eminence
MFB : medial forebrain bundle
ML : lateral mammillary nucleus
MM : medial mammillary nucleus
NAPD : periventricular arcuate nucleus,
dorsal part
NAPV : periventricular arcuate nucleus,
ventral part
NI : infundibular nucleus
NID : dorsal infundibular nucleus
NIV : ventral infundibular nucleus

NO : optic nerve
NPO : preoptic nucleus
NPOD : dorsolateral preoptic nucleus
NPP : periventricular preoptic nucleus
NRL : nucleus recessus lateralis
NTA : anterior tuberal nucleus
NTL : lateral tuberal nucleus
NTP : posterior tuberal nucleus
OM : occipitomesencephalic tract
OVLT : organum vasculosum of the lamina
terminalis
PM : premammillary nucleus

PV : paraventricular nucleus
PVA : anterior periventricular nucleus of
the hypothalamus
PVO : paraventricular organ
Pr. SO : primordial supraoptic nucleus
SC : suprachiasmatic nucleus
SO : supraoptic nucleus
TO : optic tract
TSM : septomesencephalic tract
VM : ventromedial nucleus

Figure 5-5. The hypothalamus of representative fishes. (A) Hypothalamus of the lamprey, *Entosphenus japonica.* A midsagittal section is shown at the left. The vertical lines labeled a–d correspond to the cross-sectional views at the right. (B) Hypothalamus of the teleost *Anguilla japonica,* arranged like that of the lamprey in (A). [Modified from Matsumoto, A., and Ishii, S. (1989). "Atlas of Endocrine Organs: Vertebrates and Invertebrates." Springer-Verlag, Berlin.]

the preoptic nucleus of hagfishes, the NS nuclei of the hypothalamo–hypophysial system are all paired structures. Other telencephalic and mesencephalic structures may produce similar peptides or may be targets for fibers secreting peptides usually associated with the

Figure 5-4. Comparative anatomy of the fish pituitary. The neurohypophysis is indicated by shaded areas and the pars intermedia is shown in black. Approximate distributions of cell types secreting tropic hormones are indicated by letters. A, Corticotropin; G, gonadotropins; M, melanotropin; P, prolactin; P+, somatolactin; S, growth hormone or somatotropin; T, thyrotropin; ?, unidentified. [Modified from Schreibman, M. P. (1986). Pituitary Gland. *In* "Vertebrate Endocrinology: Fundamentals and Biomedical Implications. Volume 1, Morphological Considerations" (P. K. T. Pang and M. P. Schreibman, eds.), pp. 11–56. Academic Press, Orlando, Florida.]

hypothalamo–hypophysial system, but they are not thought to be central players in the regulation of endocrine function.

A. Agnathan Fishes

The jawless or agnathan fishes (Gr., *a,* without; *gnathos,* jaws) are the most primitive of vertebrates. They are separable into two groupings: the very primitive hagfishes and the evolutionarily divergent lampreys, which are more similar in many respects to other fishes.

1. The Myxinoids (Hagfishes)

The Atlantic and Pacific hagfishes possess the most primitive hypothalamo–hypophysial system among the vertebrates (Figs. 5-4 and 5-5). The hagfish system lacks many of the features characteristic of other piscine and tetrapod groups. Furthermore, hagfishes are much more primitive in many respects than even their closest living agnathan relatives, the lampreys (Petromyzontidae).

The hagfish brain has several NS regions. Anterior and dorsal to the optic chiasm, we find the **preoptic nucleus,** which appears to produce NS products that are stored in the neurohypophysis, a neurohemal area comparable to the mammalian pars nervosa. There is, however, no anterior neurohemal region in the Atlantic hagfish comparable to the median eminence, although a very primitive anterior neurohemal area has been described for the Pacific hagfish and has been termed a median eminence. Attempts to demonstrate immunologically reactive neuropeptides (e.g., GnRH, substance P, and some other peptides) in the hagfish brain or neurohypophysis have been unsuccessful.

The origin of the adenohypophysis of hagfishes appears to be from endoderm rather than from ectoderm, presenting an additional puzzle with respect to the origin of the pituitary. Furthermore, it raises the question of possible homology of the hagfish adenohypophysis to that of other vertebrates and supports the viewpoint that hagfishes are aberrant vertebrates, are not on the mainline evolutionary pathway, and cannot readily be compared to other vertebrates.

The hagfish adenohypophysis is not differentiated into subregions; that is, there is no pars distalis or pars intermedia. The adenohypophysis is composed primarily of nonstainable cells and rare periodic acid–Schiff reagent positive [PAS(+)] basophils or an occasional acidophil. Electron micrographs of the hagfish adenohypophysis show rare granular cells with cytoplasmic granules 100–200 nm in diameter. These granular cells are believed to represent the two rare stainable cellular types identifiable with the light microscope. When hagfish adenohypophysial tissue is cultured *in vitro,* no observable changes take place in either granular or nongranular cells.

Certain cells of the myxinoid adenohypophysis exhibit cytological modifications where they make contact with the neurohypophysis. These altered cells collectively are termed **modified adenohypophysial tissue,** and it has been proposed that this apparent induction by neurohypophysial tissue may represent phylogenetically the origin of the pars intermedia. In other vertebrates, only the pars intermedia develops following contact of the presumptive adenohypophysis with the neurohypophysis.

Bioassays of hagfish pituitaries for prolactin (PRL) activity have proven negative, and anti-ovine PRL antibody does not bind to hagfish adenohypophysial cells. Hypophysectomy of Pacific hagfish produces no convincing alterations in either thyroid or gonadal tissue. Bioassayable corticotropin (ACTH)-like and thyrotropin (TSH)-like activities have been demonstrated in pituitary extracts of a Pacific hagfish, *Eptatretus stouti*. Gonadotropin (GTH) activity has been demonstrated in the pituitary of *Eptatretus burgeri,* a shallow water, seasonally breeding hagfish. Obviously, the hagfishes demonstrate a very primitive stage of hypothalamo–hypophysial interaction.

2. Lampreys (Petromyzontids)

In the lampreys, the more anterior rostral pars distalis is composed of basophils and non-stainable cells (Fig. 5-4). These same cellular types as well as carminophils (PRL) are found in the proximal pars distalis, which is located between the rostral pars distalis and the pars intermedia. Only the melanotrope occurs in the pars intermedia.

The distinct pars nervosa and the pars intermedia form a well-developed neurointermediate lobe. Peptidergic neurons terminate in the pars nervosa where the single nonapeptide neurohormone arginine vasotocin (AVT) is stored. A second anterior neurohemal region is associated with the pars distalis and has been referred to as a median eminence. However, this structure is devoid of portal blood vessels, so it is unlikely that it functions as a median eminence. Studies do show that materials can diffuse readily from the brain into the adenohypophysis via this putative median eminence.

There are more NS nuclei identifiable in the lamprey brain than have been demonstrated in their relatives, the hagfishes. Two major NS centers occur in the lamprey brian anterior and dorsal to the optic chiasm: the **medial** and **lateral anterior preoptic areas** as well as the preoptic nucleus. Both GnRH and PRL have been identified in neurons of the preoptic area using immunocytochemistry. GnRH occurs in olfactory regions, too. Two additional areas are found in the hypothalamus: the **anterior hypothalamic area** and the **dorsal** and **ventral periventricular arcuate nuclei** (see Fig. 5-5).

B. Chondrichthyean Fishes

1. Sharks, Rays, and Skates (Selachians)

The selachian (or elasmobranch) hypothalamo–hypophysial system possesses two anatomical features not found in agnathan fishes. The **pars ventralis** represents a fourth subdivision of the adenohypophysis unique to selachians. It is located ventral to the proximal pars distalis, to which it is connected by a stalk (Fig. 5-4). Localization of GTH and TSH activity in the pars ventralis has prompted some investigators to suggest that the pars ventralis is homologous to the pars tuberalis of the tetrapod adenohypophysis. The terms *rostral* and *proximal* may be somewhat misleading when applied to the selachian pars distalis because of the presence of the ventral lobe, which is probably homologous to part of the proximal pars distalis of other fishes. Consequently, these regions are sometimes designated as the rostral, median, and ventral lobes of the pars distalis. The distribution of cellular types in the selachian pituitary is summarized in Table 5-1.

The second feature appearing for the first time in selachians is the saccus vasculosus derived from the hypothalamic ependyma and located immediately posterior to the neurointermediate lobe. It is not as well developed as its homolog in bony fishes, but it does possess the unique **coronet** cellular type characteristic of the saccus vasculosus of bony fishes (see Fig. 5-6). Immunoreactive **melanin-concentrating hormone (MCH)** has been colocalized with melanotropin (α-MSH) in the saccus vasculosus of the shark, *Scyliorhinus*.

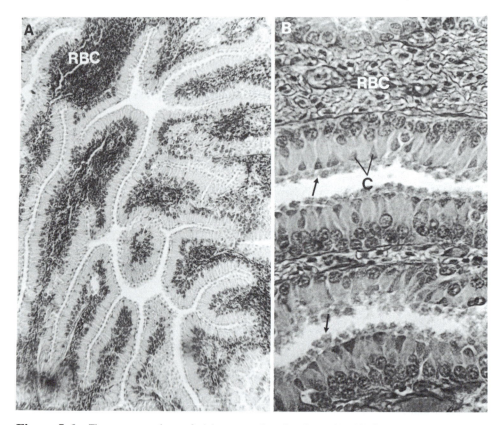

Figure 5-6. The saccus vasculosus of rainbow trout, *Oncorhynchus mykiss*. (A) Coronet cells and ependymal secretory cells. (B) Higher magnification showing mucoid secretory products (arrows) of saccus vasculosus cells. C, Coronet cells; RBC, red blood cells.

The neurointermediate lobe shows extensive interdigitation of neural and endocrine cells. The pars nervosa stores some unique nonapeptides in addition to AVT (see Chapter 6). These hormones are presumably produced in the preoptic nucleus. The pars intermedia contains only one cellular type associated with MSH activity.

There is a well-developed median eminence with a hypothalamo–hypophysial portal system connecting it to the pars distalis. The median eminence consists of anterior and posterior neurohemal areas. The posterior region receives both peptidergic and aminergic NS axons and appears to be linked by portal vessels to the proximal pars distalis but not to the ventral lobe. There is considerably less NS material in the anterior neurohemal area, which appears to be connected by capillaries to the rostral pars distalis. It is tempting to suggest that the median eminence has differentiated to increase the efficiency of delivering hypothalamic neurohormones to specific adenohypophysial cells.

2. Ratfishes (Holocephalans)

The hypothalamo–hypophysial system of the ratfishes is readily subdivided cytologically into rostral pars distalis, proximal pars distalis, and pars intermedia. Although several cellular types have been demonstrated with selective staining procedures, no experimental studies have verified which cells produce which tropic hormones.

Like selachians, holocephalans possess a unique region associated with the adenohypophysis, called the **pharyngeal lobe.** This structure is located in the roof of the mouth outside the cranium, as though a portion of Rathke's pouch had stayed behind and had not become incorporated into the adenohypophysis proper. The pharyngeal lobe consists of follicles, and may be homologous to the follicular rostral pars distalis of bony fishes. On the other hand, it may be homologous to the ventral lobe of the selachian and the buccal lobe of the coelacanth pituitary (a primitive bony fish discussed below).

The ratfish neurohypophysis includes a prominent median eminence connected to the rostral pars distalis and the proximal pars distalis by a hypothalamo–hypophysial portal system. Somatostatin, thyrotropin-releasing hormone (TRH), and GnRH immunoreactivities are present in the hypothalamus. The pars nervosa is mingled with the pars intermedia of the adenohypophysis to form a typical neurointermediate lobe. A well-developed saccus vasculosus is also present.

C. Bony Ray-Finned Fishes (Actinopterygians)

Four taxonomically distinct groups of ray-finned fishes are recognized traditionally by most authorities. Three of these (polypterans, chondrosteans, and holosteans) are considered to be rather primitive whereas the teleosts are recognized as the most highly evolved of these fishes.

1. Polypteran Fishes (*Polypterus* and *Calamoichthyes*)

The primitive African polypterid fishes possess all the typical piscine features including rostral and proximal divisions of the pars distalis, a neurointermediate lobe, and a saccus vasculosus, and they retain as adults a connection between the hypophysis and the mouth cavity called the **buccohypophysial canal** (*bucca,* cheek). This canal is believed to be a remnant originally connecting Rathke's pouch to the oral cavity (Fig. 5-4). The buccohypophysial canal or duct is lined with a weakly staining cell type that does not appear to be associated with production of any tropic hormones.

The pars nervosa contains AVT and an oxytocin-like nonapeptide neurohormonal product of the preoptic nucleus known as **isotocin (IST).** These two neurohormones are characteristic of all ray-finned fishes (see Chapter 6). The hypothalamo–hypophysial portal system with a defined median eminence is well developed. Peptidergic fiber tracts travel from the preoptic nucleus via the median eminence to terminate in the pars nervosa. Aminergic neurons that terminate in the median eminence probably originate in the preoptic region, too.

2. Chondrostean Fishes (Sturgeon, Spoonbill)

There is less information concerning the hypothalamo–hypophysial system of chondrostean fishes than for any other ray-finned bony fishes. There is no buccohypophysial canal in chondrosteans, but a hypophysial cavity is present. The cavity separates the pars distalis and pars intermedia and may be homologous to the buccohypophysial canal and the hypophysial cleft described in some mammals. The pars distalis consists of a rostral zone and a proximal zone (rostral and proximal pars distalis). Numerous follicles occur throughout the pars distalis, and their lumena, which are filled with a basophilic colloidal material, are considered to be remnants of the hypophysial cavity. The entire pars distalis may be homologous to the proximal pars distalis of teleostean fishes, although the follicles themselves may be homologous to the follicles of the teleostean rostral pars distalis. If this interpretation proves correct, the terms rostral pars distalis and proximal pars distalis may not be applicable to the two zones of the chondrostean pars distalis.

The pars intermedia of the sturgeon is large and closely associated with the pars nervosa to produce a typical neurointermediate lobe. The pars nervosa is basically hollow and is similar to the saccus vasculosus. Both peptidergic and aminergic fibers have been reported in the pars nervosa.

The hypothalamus contains a well-developed preoptic nucleus that provides peptidergic fibers to the pars nervosa. The **nucleus lateralis tuberis (NLT)** is a new nucleus in bony fishes and consists of peptidergic and aminergic neurons. There is a well-developed median eminence consisting of aminergic axonal endings from the nucleus lateralis tuberis. The median eminence is separated from the pars distalis by a connective tissue sheath so that no neurons penetrate the pars distalis. A hypothalamo–hypophysial portal system possibly conducts neurohormones from the median eminence to the pars distalis, and immunoreactive corticotropin-releasing hormone (CRH) has been demonstrated in the median eminence.

3. Holostean Fishes (Gars and the Bowfin)

Adult holostean fishes do not exhibit a buccohypophysial duct nor is there any hypophysial cleft, although a transient hypophysial cleft does appear during development of the pituitary. The adenohypophysis consists of a follicular rostral pars distalis, a proximal pars distalis, and a pars intermedia, which interdigitates with the pars nervosa to form a neurointermediate lobe.

The preoptic nucleus has separated into two distinct portions, the dorsal **pars magnocellularis** made up of larger NS neurons and the ventral **pars parvocellularis** of smaller cells. Peptidergic fibers from the preoptic nucleus pass to the pars nervosa where nonapeptide hormones (AVT and IST) are stored (see Chapter 6). Aminergic fibers also appear in the pars nervosa and are believed to come from the nucleus lateralis tuberis.

The median eminence is connected to the pars distalis by a well-developed portal system. In addition, there are a limited number of peptidergic and aminergic axons that penetrate the pars distalis. However, only aminergic fibers are associated with the median eminence of the bowfin, *Amia calva*. Here, in these near relatives of the teleostean fishes, is the modest beginning of a shift from neurovascular control to neuroglandular control of the adenohypophysis directly by neurons of the hypothalamus as is seen among the teleosts.

4. Teleostean Fishes

The hypothalamo–hypophysial systems of several teleostean species have been studied in detail. In view of the vast adaptive radiation that teleosts have undergone, it is not surprising to find considerable variation in this sytem. Although only a few species have been examined, they have been examined in greater detail than other piscine groups, and the various cellular types have been identified and their secretions determined. There is a strong tendency for localization of cellular types in particular regions of the adenohypophysis, which has aided scientists in determining what hormone each cell type produces. Unlike the situation with mammals, two kinds of GTH-secreting cells have been demonstrated in the teleostean proximal pars distalis. There are two GTHs in salmonids; one that is like follicle-

Table 5-2. Presence of Neuropeptides in Neurons Contacting Secretory Cells of the Adenohypophysis of the Sea Bass, *Dicentrarchus labrax*[a,b]

Region	Secretory cell type	Peptides present
Rostral pars distalis	ACTH cells	SS, GHRH, CRH, AVT, IST, SP, NT, GAL
	PRL cells	GAL
Proximal pars distalis	GH cells	SS, GHRH, AVT, IST, CCK, SP, NPY, GAL
	GTH cells	GnRH, IST, GHRH, AVT
	TSH cells	GHRH, CRH, AVT, IST, SP, GAL
Pars intermedia	MSH cells	GHRH, CRH, MCH, AVT, IST, SP
	PAS(+) cells	GHRH, CRH, MCH, AVT, IST, SP

[a] See text for explanation of peptide abbreviations.
[b] After Moons, L., Cambré, M., Ollevier, F., and Vandescande, F. (1989). Immunocytochemical demonstration of close relationships between neuropeptidergic nerve fibers and hormone-producing cell types in the adenohypophysis of the sea bass (*Dicentrarchus labrax*). *Gen. Comp. Endocrinol.* **73,** 270.

stimulating hormone (FSH-like) that controls spermatogenesis and oogenesis, and a second one that is LH-like and responsible for final gamete maturation and ovulation or sperm release. Other teleosts also may exhibit two GTHs, but some appear to have one LH-like GTH.

Teleosts exhibit extensive innervation of the pars distalis by preoptic and hypothalamic neurons. There is no distinct portal system and hence no true median eminence in most species examined, and the control of tropic hormone release is neuroglandular by direct peptidergic or aminergic innervation rather than neurovascular as found in many primitive fishes and in the tetrapods (see Table 5-2).

The pars intermedia is intimately associated with the pars nervosa of the neurohypophysis to form a neurointermediate lobe. Direct innervation from the pars nervosa is responsible for controlling MSH release from the pars intermedia. In addition to PbH(+) MSH-secreting cells, a PAS(+) cell type, the **PIPAS cell,** is present in the pars intermedia of teleosts with the exception of the salmonids. A similarly staining cell also occurs in the pars intermedia of some primitive bony fishes (i.e., *A. calva* and *Polypterus*).

The diversity of teleosts makes generalizations about the distribution of neuropeptides very difficult. Furthermore, development and extensive application of immunohistochemistry to the teleost brain has resulted as it did for mammals in the discovery that neuro peptides may appear in many neurons that are not associated directly with the activity of the hypothalamo–hypophysial system. For example, GnRH neurons may be found in the anteriormost olfactory regions of the brain as well as in the midbrain, and the amount of thyrotropin-releasing hormone (TRH) present in the telencephalon typically is much greater than in the hypothalamus. We will describe the anatomy of the brain of the Japanese eel, *Anguilla japonica,* as a representative but not necessarily typical teleost (Fig. 5-5).

The preoptic area located anterior to the optic chiasm sports several specialized areas in the eel. In addition to the preoptic nucleus, there are also medial and lateral preoptic areas are the anterior periventricular nucleus of the anterior hypothalamus. The NLT is separated into anterior, posterior, and lateral parts that are principally responsible for secretion of TRH, MCH, as well as growth hormone-releasing hormone (GHRH), GnRH, CRH, and AVT. Immunoreactive α-MSH also has been demonstrated in the NLT of carp and rainbow trout. Magnocellular and parvocellular NS neurons of the preoptic nucleus have been reported to contain either GHRH, GnRH, somatostatin (SS), CRH, or AVT. Axons from most of these neurosecretory neurons follow well-defined tracts to the pars distalis or the neurointermediate lobe of the pituitary, whereas others may connect with a variety of other brain structures.

Two types of CRH and CRH-secreting cells have been described in the common sucker, *Catostomus commersoni.* CRH-containing cells in the preoptic nucleus react best to antisera for mCRH and often contain the nonapeptide AVT. In contrast, the axonal processes of CRH cells in the NLT react to antibodies for sauvagine and urotensin-I and do not contain AVT. Experimental studies suggest that the preoptic CRH colocalized with AVT is responsible for controlling release of ACTH whereas CRH in the NLT controls synthesis. These observations also suggest an ACTH-releasing role for AVT that is similar to the role for arginine vasopressin (AVP) in mammals. A similar pattern for CRH distribution has been reported for several species of eel (*Anguilla* spp.). The complexity of connections in this system is exemplified by the observations that in the sea bass, *Dicentrarchus labrax,* the corticotropic cells apparently are innervated by a variety of neurons containing AVT, CRH, neurotensin, GHRH, or SS.

D. Bony Fishes: The Lobe-Finned Fishes (Sarcopterygians)

The lobe-finned fishes are a primitive group of bony fishes that is mostly extinct. Although they once thrived in the ancient oceans and radiated in many directions, only two lobe-finned groups have living representatives. The dipnoans or lungfishes are represented by three genera restricted to three continents (*Protopterus,* tropical Africa; *Lepidosiren,* tropical South America; *Neoceratodus,* temperate and subtropical Australia). The crossopterygians are thought to be very close to the fishes that gave rise to amphibians. There is one living crossopterygian, the coelacanth, *Latimeria chalumnae.*

The hypothalamo–hypophysial system of *Neoceratodus* is similar to that of primitive bony fishes, including possession of a neurointermediate lobe. This lungfish is closely related to the primitive and extinct Devonian lungfishes. There is a follicular rostral pars distalis containing cells that secrete corticotropin (ACTH) and prolactin (PRL). In contrast, the African and South American lungfishes are of more recent origin, and we find their pituitary systems are similar to those of tetrapods, especially amphibians. For example, there is less regionalization of cellular types in the adenohypophysis and no neurointermediate lobe. However, this similarity with amphibians could be the result of parallel or convergent evolution, since *Protopterus* and *Lepidosiren* are considered to be more distantly related to the ancestors of the tetrapods than is *Neoceratodus.* In general, hypothalamic regulation in lungfishes appears to be neurovascular as in primitive actinopterygians, rather than neuroglandular as in teleosts. The pars tuberalis, a tetrapod feature, is missing in all living lungfishes.

The hypothalamo–hypophysial system of the coelacanth is much more piscine-like than it is tetrapod-like. The anatomy of the adenohypophysis is much like that of the selachians. There is a **buccal lobe** in the pars distalis, which, based on the cell types present there, may be homologous to the ventral lobe of selachians. The pars distalis may be subdivided into rostral and proximal portions with stainable cellular types appearing that are distributed similarly to those of teleosts. The pars intermedia forms a typical neurointermediate lobe as in other bony fishes. Unlike teleosts, there is a distinct median eminence connected by a well-developed hypothalamo–hypophysial portal system to the adenohypophysis, particularly in the proximal region. However, some NS axons appear to penetrate the proximal pars distalis similar to the situation described for the holostean fish, *A. calva.*

Although the sarcopterygian fishes represent descendants of fishes ancestral to the tetrapods, extant members of this group provide no transitional stages with respect to anatomical features of the hypothalamo–hypophysial system. However, the distribution and structure of MSH and the structure of PRL isolated from the Australian lungfish are more like tetrapods than like other bony fishes.

II. The Tetrapod Vertebrates: Anatomical Considerations

The tetrapod neuroendocrine systems exhibit most of the characteristics described for mammals and lack the major features that characterize the fishes. The median eminence and pars nervosa are well-developed, distinct neurohemal structures. There is no remnant or suggestion of a saccus vasculosus in tetrapods. The amphibian pars distalis shows a tendency for cellular regionalization and the pars distalis of reptiles and birds consists of two somewhat distinct zones. However, these zones are not anatomically separated as are the rostral and proximal zones in fishes. The pars tuberalis is a consistent tetrapod feature and appears to have secretory cells. No functional role for the pars tuberalis has been established for any tetrapod.

The tetrapod hypothalamus becomes progressively differentiated into more distinct nuclei from amphibians to reptiles and then to birds and mammals, although they are never as distinct as they appear in drawings. Although some direct control of the pars intermedia is present in amphibians, there is a predominance of neurovascular control rather than direct neural control of adenohypophysial function in all tetrapods.

A. Amphibians

The amphibian hypothalamo–hypophysial axis has most of the features characteristic of the tetrapod system (Fig. 5-7), including the presence of a pars tuberalis (Fig. 5-8). The prominent preoptic area (Fig. 5-9) contains several NS centers including the lateral and medial preoptic areas and the preoptic nucleus we saw in the fishes. GnRH, GnRH-associated peptide (GAP), GHRH, SS, AVT, TRH, CRH, and another oxytocin-like nonapeptide, **mesotocin (MST),** have been reported in the preoptic region. In addition, a separation of the posterior preoptic nucleus is named the **suprachiasmatic nucleus (SCN)** and a separate, nearby **ventromedial nucleus** is described in some species. AVT has been reported in the suprachiasmatic nucleus of bullfrogs. The **infundibular nucleus** located in the basal hypothalamus supplies aminergic and peptidergic fibers to the median eminence. This nucleus is separable into **dorsal (NID)** and **ventral (NIV)** components and is homologous to at least part of the major hypophysiotropic region of the mammalian hypothalamus. The relationship between the infundibular nucleus and the nucleus lateralis tuberis of fishes is uncertain. Immunoreactive-like TRH, SS, neuropeptide Y (NPY), and **atrial natriuretic peptide (ANP)** have been identified in the NID, whereas TRH, α-MSH, and NPY occur in the NIV. Pituitary adenylyl cyclase-activating peptide (PACAP) has been demonstrated in the preoptic area as well as in the NID and NIV and PACAP-immunoreactive fibers appear in the median eminence, suggesting a role in controlling pituitary function. Experimental studies have demonstrated that PACAP stimulates cAMP production and calcium mobilization in anuran pars distalis cells.

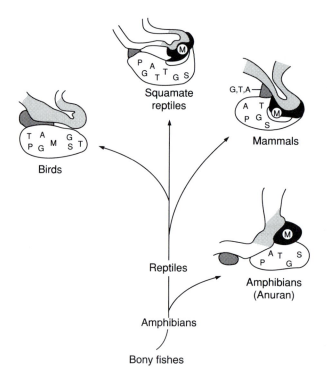

Figure 5-7. Comparative anatomy of the tetrapod pituitary. See Fig. 5-4 for explanation of abbreviations. [Modified from Schreibman, M. P. (1986). Pituitary gland. *In* "Vertebrate Endocrinology: Fundamentals and Biomedical Implications. Volume 1, Morphological Considerations" (P. K. T. Pang and M. P. Schreibman, eds.) pp. 11–56. Academic Press, Orlando, Florida.

The pars nervosa of the neurohypophysis receives peptidergic fibers originating in the magnocellular neurons of the preoptic nucleus. At least two nonapeptide hormones (AVT and MST) are stored in this well-developed neurohemal organ (see Chapter 6). TRH is colocalized with MST in the pars nervosa and has been shown to cause MSH release. A tract of ANP-reactive axons travels from the dorsal hypothalamus through the median eminence to enter the pars nervosa. AVT-secreting cell bodies in the preoptic area and axonal endings in the pars nervosa also contain GHRH. Neurons containing immunoreactive MCH project from the anuran preoptic nucleus to the pars nervosa. These neurons also contain α-MSH, although α-MSH neurons originating in the NIV do not contain MCH. There is no tendency for development of a neurointermediate lobe, nor is a saccus vasculosus or any comparable structure found in amphibians.

The adenohypophysis consists of pars tuberalis, pars intermedia, and pars distalis. On the basis of ultrastructural comparison of cytoplasmic granules and other features, what appear to be two separate granular cellular types have been observed in the anuran pars tuberalis (Fig. 5-8). There appears to be a neural pathway extending from the ependymal lining of the third ventricle to the pars tuberalis, but no functional correlations have been reported.

The amphibian pars intermedia has a poor vascular supply, but it is directly innervated by aminergic neurons thought to originate in the caudal aminergic nuclei of the hypothalamus. Release of α-MSH appears to be under direct neural control of aminergic neurons but

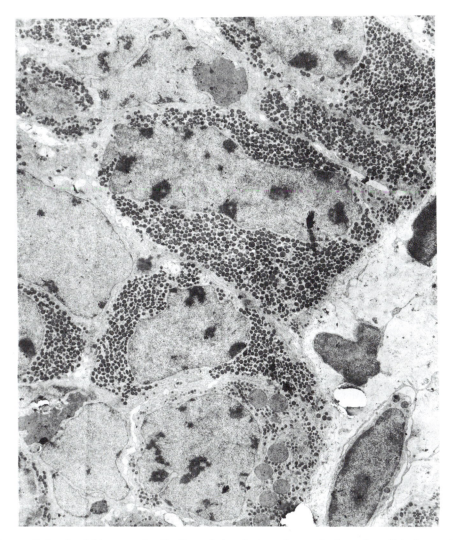

Figure 5-8. Amphibian pars tuberalis. Transmission electron micrograph of granular cells in the pars tuberalis of the frog, *Rana pipiens*. Two cell types can be identified on the basis of granule size. (Photograph courtesy of K. T. Fitzgerald, University of Colorado, Boulder.)

is also affected by peptidergic fibers. NPY- and ANP-immunoreactive neurons enter the pars intermedia of anurans, and NPY inhibits release of α-MSH whereas ANP is stimulatory.

The pars distalis is not separable into discrete regions although there is a tendency for some regionalization of cellular types (Fig. 5-10). Urodele amphibians exhibit greater regionalization than do anuran species. Careful cytological studies have been performed on many amphibian species, often coupled with experimental manipulations, bioassays, or correlations with specific life history events. Unfortunately, there has been considerable disagreement among researchers in this area with respect to establishment of specific sources for the various tropic hormones. The cytology of the pituitary of *Rana temporaria* represents the basic amphibian condition (Table 5-1).

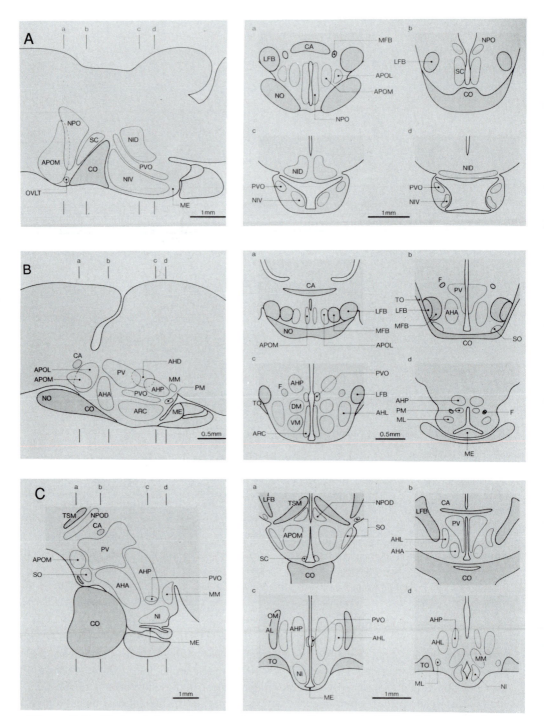

Figure 5-9. The hypothalamus of representative nonmammalian tetrapods. (A) Hypothalamus of the bull-frog, *Rana catesbeiana*. (B) Hypothalamus of the snake, *Elaphe conspicillata*. (C) Hypothalamus of the Japanese quail, *Coturnix coturnix japonicus*. See Fig. 5-5 for abbreviations. [Modified from Matsumoto, A., and Ishii, S. (1989). "Atlas of Endocrine Organs: Vertebrates and Invertebrates." Springer-Verlag, Berlin.]

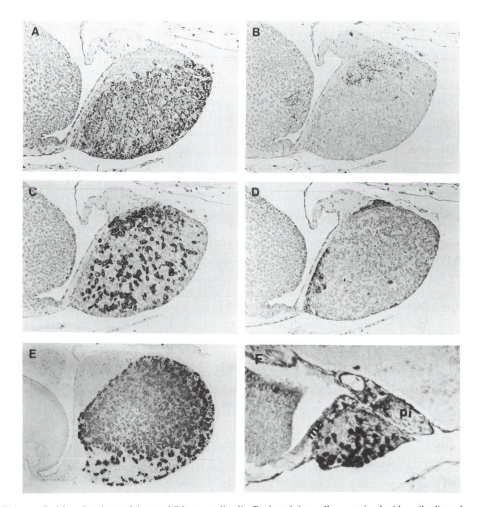

Figure 5-10. Cytology of the amphibian pars distalis. Dark-staining cells are stained with antibodies selective for different pituitary hormones. (A) Immunoreactive PRL-secreting cells; (B) immunoreactive GH-secreting cells; (C) immunoreactive βLH-secreting cells; (D) immunoreactive ACTH-secreting cells; (E) immunoreactive βFSH-secreting cells; (F) immunoreactive βTSH-secreting cells. (A–D) are from the Japanese newt, *Cynops pyrrhogaster,* (E) is from adult bullfrog, *Rana catesbeiana,* and (F) is from postmetamorphic *Bufo calamita.* (A–D) are courtesy of Shigeyasu Tanaka and Sakae Kikuyama, Waseda University, Tokyo. (E) is reprinted with permission from Tanaka, S., Park, M. K., Hayashi, H., Hanaoka, Y., Wakabayashi, K., and Kurosumi, K. (1990). Immunocytochemical localization of the subunits of glycoprotein hormones (LH, FSH, and TSH) in the bullfrog pituitary gland using monoclonal antibodies and polyclonal antiserum. *Gen. Comp. Endocrinol.* **77**, 88–97, Academic Press, San Diego. (F) is reprinted with permission from Garcia-Navarro, S., Malagon, M. M., and Gracia-Navarro, F. (1988). Immunohistochemical localization of thyrotropic cells during amphibian metamorphosis: A sterological study. *Gen. Comp. Endocrinol.* **71**, 116–123, Academic Press, San Diego.

B. Reptiles

The reptilian hypothalamus is depicted in Fig. 5-9. In reptiles, there is no preoptic nucleus but the medial and lateral preoptic areas are still present. Replacing the preoptic nucleus are the separate **supraoptic nucleus** and the **paraventricular nucleus.** These nuclei con-

sist of peptidergic NS neurons that terminate in the pars nervosa and produce the octapeptide neurohormones AVT and MST. CRH has been colocalized with AVT in some neurons
of the paraventricular nucleus. The hypothalamus is separated into **anterior, dorsal,** and
posterior hypothalamic areas, paraventricular organs, periventricular nucleus, and
both **ventral** and **dorsal medial nuclei.** The infundibular nucleus, also called the **arcuate
nucleus,** is considered to be homologous to the ventral infundibular nucleus of amphibians
and the arcuate (infundibular) nucleus of birds as well as to the major hypophysiotropic
area of the mammalian hypothalamus. Aminergic and peptidergic NS fibers (including
GnRH-containing neurons) originate in this nucleus and terminate in the median eminence.
Immunoreactive MCH is present in neurons of the periventricular and lateral hypothalamic
nuclei of a turtle, a lizard, and two snake species, but these fibers project to other brain
regions and probably do not influence the pituitary.

The adenohypophysis is well developed in most reptiles, and it consists of a pars distalis,
a pars tuberalis, and a pars intermedia (Fig. 5-7). The primitive reptilian condition probably
is best represented by the Rhynchocephalia (*Sphenodon punctatus*), Chelonia (turtles), and
Crocodilia. The pars distalis appears as two distinct regions reminiscent of the condition in
bony fishes. There is a rostral or **cephalic lobe** and a **caudal lobe** in the pars distalis, and
the distribution of cellular types in these lobes is similar to that described for fishes (Table 5-
1). The pars tuberalis is well developed in the Rhynchocephalia, Chelonia, and Crocodilia,
but it is greatly reduced and sometimes absent in lizards. In adult snakes, the pars tuberalis
is completely absent. There is no explanation for this apparent disappearance of the pars
tuberalis in squamates.

Although the intermediate lobes of many lizards and burrowing snakes are markedly
reduced or absent, some reptiles have well-developed intermediate lobes. It is especially
elaborate in certain chelonians, crocodilians, snakes, and anoline lizards.

C. Birds

The avian hypothalamo-hypophysial system differs from that of other tetrapods (and fishes)
in that the pars intermedia is absent in adults of all species. A well-developed pars tuberalis
is present, and the pars distalis consists of cephalic and caudal lobes homologous to those
described for reptiles (Fig. 5-7).

Unlike mammals, the primary capillaries of the portal system in birds lie superficially or
in grooves on the surface rather than penetrating the pars tuberalis to form the complex
vascular bed seen in mammals. This anatomical difference allows an investigator (with a
steady hand) to sever the portal connection with the pars distalis without interrupting fiber
tracts.

The hypophysiotropic region of the hypothalamus supplies aminergic and peptidergic
fibers to the median eminence. The region anterior and dorsal to the optic chiasm is occupied by the medial anterior preoptic area; a dorsal preoptic nucleus; and suprachiasmatic,
paraventricular, and supraoptic nuclei (Fig. 5-9). As in reptiles and mammals, the supraoptic nucleus and the paraventricular nucleus are responsible for secretion of the nonapeptide neurohormones (see Chapter 6). Immunoreactive oCRH also has been localized in the
paraventricular nucleus. The hypothalamus itself includes the anterior and lateral hypothalamic areas, paraventricular organ, and an infundibular or arcuate nucleus.

In some birds, i.e., the pigeon, the Japanese quail, and the white-crowned sparrow, the
median eminence is separable into an anterior neurohemal and a more posterior neurohemal area similar to that described for selachians. Each of these areas has its own portal

connection to the pars distalis, which consists of a cephalic and a caudal lobe. It is not certain how widespread this phenomenon of two median eminentia is among other avian species. However, such potential regionalization of both the median eminence and the pars distalis could represent a mechanism to increase the efficiency of delivery of hypothalamic neurohormones to cellular types regionalized in various parts of the pars distalis (Table 5-1). This neurovascular specialization is analogous to the system of direct innervation of pars distalis cells that was observed in teleosts.

III. The Tropic Hormones of Nonmammalian Vertebrates

The same categories of tropic hormones are found in nonmammals as were discussed for mammals in Chapter 4. Category 1 includes the glycoprotein hormones LH, FSH, and TSH. Growth hormone (GH) and PRL constitute hormones of category 2, and the derivatives of **proopiomelanocortin (POMC)** represent category 3 [ACTH, MSH, corticotropin-like peptide (CLIP), endorphins].

A. Gonadotropins

Purified GTHs isolated from nonmammalian species are similar to those of mammals, and several of them have been partially characterized (Table 5-3). Although more GTHs are being isolated each year, no clear phylogenetic pattern has emerged.

Mammalian GTHs have been examined extensively for activity in lower vertebrates, and many reviews of this work are available for fishes, amphibians, reptiles, and birds (see the

Table 5-3. Relative Potencies of Follicle-Stimulating Hormone and Luteinizing Hormone Purified from Different Vertebrates and Tested in the *Anolis* and *Xenopus* Gonadotropin Bioassays[a]

Hormone	Relative potency	
	Anolis bioassay	*Xenopus* bioassay
Ovine FSH	100.0[b]	<0.005
Ovine LH	0.04	2.0[c]
Snapping turtle FSH	3.0[b]	0.002
Snapping turtle LH	0.13	1.8[c]
Chicken FSH	4.3[b]	<0.0015
Chicken LH	0.85	0.08[c]
Bullfrog FSH	7.0[b]	0.004
Bullfrog LH	1.2	0.33[c]

[a] Based on Licht, P., Papkoff, H., Farmer, S. W., Muller, C. H., Tsui, H. W., and Crews, D. (1977). Evolution of gonadotropin structure and function. *Recent Prog. Horm. Res.* **33**, 169–248.
[b] *Anolis* bioassay for FSH: Maintenance of testis weight in hypophysectomized lizard.
[c] *Xenopus* Bioassay for LH: *In vitro* ovulation.

Suggested Reading list at the end of this chapter). Generally, the hormonal control of reproduction involves two GTHs with LH-like and FSH-like activities. The apparent failure to find two GTHS in some vertebrate groups (e.g., squamate reptiles) may be due to derived secondary reliance on only one GTH with repression of synthesis of the other. Specific activity for both mammalian FSH and LH in nonmammals may be related to biological half-life differences between the two hormones when they are examined *in vivo* and emphasizes the need for good *in vitro* bioassays in a variety of vertebrates.

It has been proposed that the most primitive glycoprotein hormone was an LH-like molecule. Later, the LH gene duplicated and became modified to produce a TSH-like hormone. Follicle-stimulating hormone presumably diverged even later from TSH. This scheme was suggested from studies of purified β-subunit structure for these mammalian hormones. Subunit structures are beginning to appear for teleostean and other vertebrate GTHs, and a more critical examination of this hypothesis may be possible soon.

Category 1 tropic hormones have now been isolated from teleosts and all of the tetrapod groups. However, they have not been isolated from jawless fishes or sarcopterygian fishes, and limited information is available for the chondrichthyean fishes.

1. Gonadotropins in Chondrichthyean Fishes

Gonadotropin activity is present in both the proximal pars distalis and the ventral lobe of selachians. Antibody to mammalian GTHs binds to cells of the ventral lobe. Bioassay of the selachian proximal pars distalis reveals the presence of an LH-like GTH that will stimulate oocyte maturation in the clawed frog *Xenopus laevis*.

2. Gonadotropins in Bony Fishes

The first observations on teleostean GTHs suggested a single LH-like GTH in teleosts was present. Mammalian FSH appeared to be inactive in teleosts and mammalian LH was able to support the entire reproductive process. Furthermore, antibody directed against mammalian LH bound to salmon gonadotropic basophils but antibody against mammalian FSH did not. Gonadotropins purified from salmon, carp, and rainbow trout were attributed high LH-like activity when tested with the frog spermiation bioassay, as did pituitary extracts prepared from lungfishes. This spermiation reaction, however, subsequently has been shown to respond to highly purified FSH as well as LH and should not be interpreted as evidence for the exclusive LH nature of teleost gonadotropin. Furthermore, removal of sialic acid, a carbohydrate in higher concentration in mammalian FSH than in LH, reduced the activity of salmon GTH prior to its assay in the lizard testicular bioassay (see Section III,A,4), suggesting it is FSH-like. In contrast, crude extracts of salmon pituitaries had no effect on chicken Sertoli cells, which are reported to be highly sensitive to mammalian FSH.

Analyses of salmon gonadotropin preparations has revealed the presence of two glycoprotein molecules termed **GTH-I** and **GTH-II.** The pituitary content of GTH-I is much greater than that of GTH-II (e.g., 100 times greater in juvenile coho salmon) and is also present in greater amounts in the blood of salmon undergoing vitellogenesis or spermatogenesis. Although both hormones show equal potencies in stimulating steroidogenesis *in vitro,* GTH-II apparently is released primarily during the final step of gonadal maturation when it stimulates synthesis of the **oocyte maturational steroid hormone (17,20β-dihydroxy-4-pregnen-3-one).** Immunohistochemical studies have demonstrated that GTH-I and GTH-II are produced in separate pituitary cell types and functionally resemble mammalian FSH and LH, respectively.

A nonglycoprotein GTH preparation that stimulated uptake from the blood of protein precursors utilized in yolk synthesis by growing oocytes (i.e., vitellogenesis) has been prepared from pituitaries of a few teleostean species. This "GTH" does not consist of subunits like other GTHs, appears to be limited in its distribution among the teleosts, and is of uncertain significance.

3. Gonadotropins in Amphibians

Two distinct GTHs have been isolated and purified from the bullfrog, the tiger salamander *Ambystoma tigrinum,* and the leopard frog, *Rana pipiens.* Purified bullfrog LH and FSH both stimulated spermatogenesis and spermiation in bullfrogs, but only bullfrog LH elevated plasma androgen levels. *Ambystoma* LH was not as effective as bullfrog LH in raising plasma androgen levels in bullfrogs. Conversely, *Ambystoma* LH was more effective than bullfrog LH in *Ambystoma,* although the latter was considerably more effective than *Ambystoma* FSH.

Bullfrog LH is more effective than ovine LH in stimulating progesterone synthesis in amphibians. Bullfrog LH also stimulates reptilian and avian thyroids, emphasizing the structural closeness of LH and TSH molecules and their actions.

These studies reveal that nonmammalian GTHs can be expected to differ markedly in structure and function both when compared to their mammalian counterparts and among more closely related species. Caution must be applied before assuming an LH of species *X* will be more LH-like in its action when tested in species *Y,* or that a molecule labeled as an LH can even be called that when applied to another species, in which it might induce FSH-like effects. This diversity in action may be due in part to variability in receptor structures that might recognize different regions of the GTH or TSH molecules or cue on subtle changes in shape due to variations in amino acid sequences.

4. Gonadotropins in Reptiles

Studies employing injections of mammalian hormones into squamate reptiles have emphasized a role for FSH-like GTHs but suggest no role for LH-like hormones. Earlier data might be interpreted as a consequence of a very short biological half-life for injected mammalian LH or that mammalian FSH might be similar enough to both a squamate FSH and LH to possess both activities. Purification of gonadotropic activity from squamate reptiles, however, has verified that only one FSH-like molecule is present, although both FSH-like and LH-like GTHs have been isolated from chelonians and crocodilians. The presence of only one molecular species of GTH in squamates does not rule out the possibility that this molecule has intrinsic FSH-like and LH-like activities. Since both FSH- and LH-like GTHs are found in other reptiles, the squamate condition is probably secondarily derived and does not represent a primitive condition.

There is great variability in the responses of reptilian tissues to GTHs purified from different vertebrate groups. Testicular weight maintenance in hypophysectomized *Anolis carolinensis* is readily maintained by FSH-like hormones purified from mammals, birds, reptiles, or amphibians, but this testicular parameter is very insensitive to LH-like gonadotropins from any source. Nevertheless, plasma androgen levels in hypophysectomized *A.*

carolinensis may be elevated by either FSH-like or LH-like GTHs from almost any source. Similar effects on plasma levels of androgens or on androgen synthesis by minced testes *in vitro* have been observed for crocodilians. The situation in chelonians is not clear as testes of some species do not respond to LH-like hormones *in vitro,* whereas at least one species (*Chrysemys picta*) does respond to *in vivo* injections of mammalian LH but not as well as it responds to mammalian FSH. The reptilian ovary, like the testis, exhibits broad sensitivity to mammalian and other FSHs and LHs. FSHs are generally the most effective GTHs in reptiles, although there are some conflicting reports regarding this point.

5. Gonadotropins in Birds

Birds conform to the typical mammalian pattern of two GTHs, LH and FSH, although avian species are rather insensitive to mammalian FSH and LH preparations. Two separate GTHs have been extracted and partially purified from domestic galliform birds. However, studies of the duck support the existence of only one GTH.

Chicken LH stimulates interstitial cells in the testis of chickens or Japanese quail, and the seminiferous tubules are stimulated by chicken FSH. Mammalian LH and avian LH are reported to be more effective than FSHs in stimulating androgen production by chicken interstitial cells *in vitro.* Some conflicting reports suggest that avian FSH and LH are more nearly equal in their actions on minced pigeon or chicken testes, however. Reptilian LH is more effective than reptilian FSH in stimulating androgen production by the avian testis. Purified amphibian GTHs (both LH-like and FSH-like) are equally effective, although their overall activity in birds is low.

B. Thyrotropins

Mammalian TSHs have been shown to stimulate thyroid function in all vertebrates examined with the exception of the hagfishes. Similarly, pituitary extracts from most nonmammals exhibit TSH-like activity when tested in mammals. The specific actions of TSH on nonmammalian thyroid glands and the functions of thyroid hormones in nonmammals are similar to those for mammals (see Chapters 4 and 7). Although little is known about the structures of nonmammalian TSHs, it is apparent that there is great similarity to those of mammals. Purified bullfrog TSH is thyrotropic in both anurans and urodeles but is ineffective on thyroids of reptiles and birds.

As mentioned earlier, bullfrog LH exhibits thyrotropic activity in fishes, reptiles, and birds. Highly purified mammalian LH also has intrinsic TSH activity in teleosts. The presence of TSH-like activity intrinsic to these LH molecules supports the suggestion that evolution of TSH-like receptors capitalized on this intrinsic feature and later duplication and modification of the gene may have led to the origin of the modern TSH gene.

C. Category II Tropic Hormones: Prolactin and Growth Hormone

Prolactin has so many different actions in vertebrates that it is almost impossible to describe them all. Indeed, people have suggested a variety of names for this hormone that might be more descriptive, such as versitilin, ubiquitin, panaceanin, and miscellanin. Many of the reported actions of PRL (see Table 5-4), however, may be grouped into a few overlapping, general types: 52% of the reported actions are related to growth phenomena; 38% are

Table 5-4. Prolactin Actions in Nonmammalian Vertebrates

Actions related to reproduction
  Teleosts
    Skin mucous secretion (e.g., discus "milk")
    Reduction of toxic effects of estrogen
    Growth and secretion of seminal vesicles
    Parental behavior (nest building, fin fanning, buccal incubation of eggs)
    Maintenance of brood pouch in male seahorse
    Gonadotropic
  Amphibians
    Water drive (prior to reproduction)
    Secretion of oviductal jelly
    Spermatogenic and/or antispermatogenic
    Ovulation
    Stimulation of cloacal gland development
  Reptiles
    Antigonadotropic
  Birds
    Production of crop "milk"
    Formation of brood patch
    Antigonadal
    Premigratory restlessness
    Parental behavior
    Synergism with steroids on female reproductive tract
    Suppression of sexual phase of reproductive cycle
Actions related to growth and development
  Teleosts
    Proliferation of melanocytes
    Growth of seminal vesicles
    Renal glomerular growth, tubule stimulation and proliferation
  Amphibians
    Tail and gill growth
    Limb regeneration
    Proliferation of melanophores
    Structural changes accompanying water drive
    Brain growth in tadpoles
    Cloacal gland development
    Ultimobranchial stimulation
  Reptiles
    Tail regeneration
    Skin sloughing
  Birds
    Proliferation of pigeon crop sac mucosa
    Epidermal hyperplasia in brood patch
    Feather growth
    Development of female reproductive tract
Actions related to water and electrolyte balance
  Cyclostomes
    Electrolyte metabolism in hagfish
  Teleosts
    Survival of hypophysectomized euryhaline freshwater species
    Restoration of water turnover in hypophysectomized *Fundulus kansae*
    Restoration of plasma Na^+ and Ca^{2+} in hypophysectomized eels when given with cortisol
    Skin, buccal, and gill mucous secretion
    Reduced gill Na^+ efflux (reduced permeability)

continues

Table 5-4   (*continued*).

Actions related to water and electrolyte balance (cont.)
Teleosts
Reduced gill permeability to water
Inhibition of gill Na^+/K^+-ATPase
Renotropic action (increased glomerular size)
Increased urinary water elimination and decreased salt excretion
Stimulation of renal Na^+/K^+-ATPase
Decreased water absorption and increased Na^+ absorption in flounder bladder
Decreased salt and water absorption from eel gut
Amphibians
Skin and electrolyte changes associated with water drive, metamorphosis
Sodium and water transport across toad bladder
Restoration of plasma Na^+ in hypophysectomized newts
Possible hypercalcemia in toads
Reptiles
Restoration of plasma Na^+ levels in hypophysectomized lizard
Birds
Stimulation of nasal (orbital) salt gland secretion
Actions on integumentary structures
Teleosts
Reduced gill Na^+ efflux
Reduced gill permeability to water
Inhibition of gill Na^+/K^+-ATPase
Restoration of water turnover in hypophysectomized *Fundulus kansae*
Skin, buccal, and gill mucous secretion
Melanogenesis and proliferation of melanocytes (synergism with MSH)
Dispersal of yellow pigment in cutaneous xanthophores (?)
Maintenance of brood pouch in male seahorse
Amphibians
Skin changes associated with water drive
Proliferation of melanophores
Effects on toad bladder
Skin yellowing in frogs
Reptiles
Epidermal sloughing
Birds
Production of crop "milk"
Formation of brood patch
Stimulation of feather growth
Stimulation of nasal gland secretion
Actions on steroid-dependent targets or synergisms with steroids on targets
Cyclostomes
Electrolyte metabolism in hagfish
Teleosts
Na^+ retention by gills (corticosteroids)
Na^+ retention by kidney (corticosteroids)
Salt and water movement in gut (corticosteroids)
Synergism with androgens on catfish seminal vesicles
Dispersal of yellow pigment in xanthophores (corticosteroids) (?)
Maintenance of brood pouch in male seahorse (corticosteroids)
Amphibians
Stimulation of oviductal jelly secretion (estrogens and progestogens)
Na^+ transport across anuran bladder (aldosterone)
Water-drive structural changes (sex steroids)
Spermatogenesis (androgens)
Cloacal gland development (androgens)

continues

Table 5-4   (*continued*)

Actions on steroid-dependent targets or synergisms with steroids on targets (cont.)
  Reptiles
    Restoration of plasma Na⁺ levels in hypophysectomized lizard (corticosteroids)
    Antigonadotropic actions (sex steroids?)
  Birds
    Formation of brood patch (synergism with ovarian or testicular steroids)
    Parental behavior (possible progesterone synergism)
    Synergism with estrogens and progestogens on female reproductive tract
    Stimulation of nasal (orbital) gland secretion (corticosteroids)
    Stimulation of feather growth (sex steroids in some species)
    Antigonadotropic actions (sex steroids?)

related to reproduction; 25% involve water and electrolyte balance; 26% are concerned with integumentary structures; and 35% concern actions synergistic with other hormones. The most primitive role for PRL may be related to osmotic–ionic balance, with the other actions being acquired later among vertebrates through coevolution of hormone structure and receptors. Certainly, the major role for PRL in teleostean fishes is related to osmotic regulation in fresh water, and even in humans PRL release can be induced following alterations in blood osmotic pressure. Regardless of the multiplicity of roles or even the establishment that it may be an osmoregulatory hormone, the name "prolactin" for this hormone is here to stay.

Mammalian GHs are effective in most nonmammals, and most nonmammalian preparations exhibit GH activity in mammals. The differences among vertebrate GHs are exemplified by their varied effectiveness when tested in a single system such as the rat tibia bioassay (Table 5-5). Comparative studies are hampered further by the structural similarity of GH and PRL. For example, PRL can have clear growth-promoting actions in larval amphibians. Structural analyses of GHs from about 20 fishes and an equal number of tetrapods suggests a molecular phylogeny that recognizes a distinct dichotomy with the teleosts and other bony fishes diverging from the tetrapods (Fig. 5-11). Unfortunately, this scheme lacks in-

Table 5-5. Potencies of Growth Hormone Determined in the Rat Tibia Assay[a]

Species	Potency[b]
Sturgeon	0.40
Tilapia	0.05
Bullfrog	1.15
Leopard frog	0.36
Snapping turtle	0.24
Sea turtle	0.12
Ostrich	0.94
Duck	0.16
Marsupial	0.03
Mammal	1.00

[a]From Chester-Jones, I., Ingleton, P. M., and Phillips, J. G. (1987). "Fundamentals of Comparative Vertebrate Endocrinology." Plenum Press, New York.
[b]Bovine GH standard = 1.0. Marsupial GHs were nonparallel.

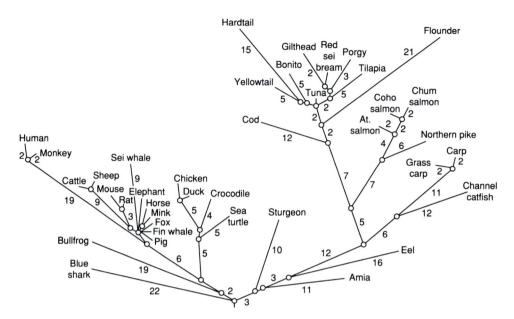

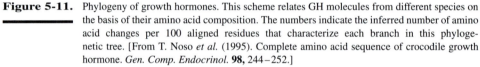

Figure 5-11. Phylogeny of growth hormones. This scheme relates GH molecules from different species on the basis of their amino acid composition. The numbers indicate the inferred number of amino acid changes per 100 aligned residues that characterize each branch in this phylogenetic tree. [From T. Noso *et al.* (1995). Complete amino acid sequence of crocodile growth hormone. *Gen. Comp. Endocrinol.* **98**, 244–252.]

formation on lungfishes and the coelacanth and therefore can provide only limited information concerning the evolution of tetrapod GH. Reptilian and avian GHs form a cluster separate from mammals and the crocodilians and birds seem to form a subcluster. Primate GHs are distinctly mammalian but form a subcluster.

Prolactin provides an excellent example for considering the evolution of endocrine systems. There have been evolutionary changes in the structure of the PRL molecule as evidenced by the failure of fish PRLs to work in avian and mammalian bioassays, although mammalian PRLs retain piscine activity (Table 5-4). The antigenic portion of piscine PRLs is similar to mammals as evidenced by the binding of mammalian PRL antibodies to the lactotropes of the piscine rostral pars distalis. Furthermore, new PRL target tissues have evolved (for example, the crop sac of birds and the mammary glands of mammals) that possess receptors specific for the "newer" portions of the molecules. This is evidenced by the failure of the piscine PRLs to activate responses in avian and mammalian tissues. It is unfortunate that there are no bioassays employing reptilian tissues and that purified reptilian PRLs are not available to fill in this part of the evolutionary changes that appear to be supported by studies with the bioassays of the other major groups.

1. Category II Hormones in the Fishes

a. Prolactin

Immunochemical studies suggest that the pituitaries of hagfish (agnathans) contain a molecule that binds antibody prepared against mammalian PRL, but only negative bioassays have been reported for hagfish pituitaries. However, lamprey pituitaries do contain bioassayable PRL, based on responses observed in the xanthophore-expanding *Gillichthyes* bioassay (see below). Similarly, testing of pituitary extracts from selachians in the *Gillichthyes* bioassay indicates PRL activity in the rostral pars distalis. No bioassays of holocephalan pituitaries have been reported. Several teleostean PRLs and lungfish PRL have been purified.

As mentioned previously, PRL plays an important role in osmotic regulation in freshwater teleosts. Teleostean PRLs have little biological activity in amniote bioassays, but they work well in amphibians and teleosts (see Table 5-4). Curiously, purified PRLs have been prepared from five species of teleosts, and all chemically resemble mammalian GH more closely than mammalian PRL. However, both highly purified tilapia GH and PRL can stimulate synthesis of milk proteins in cultured rabbit mammary gland cells. Two similar PRL molecules (isohormones) have been isolated from both tilapia and salmon, but the significance of this observation is not clear.

Some investigators have proposed that since the structures of teleostean and most other piscine PRLs are different from those of amphibians and the amniotes, they should be given a separate name. Thus, **paralactin** is sometimes used to refer to the piscine PRL-like substance in the rostral portion of the pars distalis. However, lungfish PRL biochemically is more like tetrapod PRL than other piscine PRLs.

b. Fish Bioassays for Prolactin

The **sodium-retaining bioassay** in fishes stresses the osmoregulatory actions of PRL in teleostean fishes, and particularly its effect on sodium uptake across the gill and resultant alterations in plasma sodium levels. Only some pituitaries from teleostean species, however, will produce a measurable response in this bioassay, and all other piscine and other nonmammalian preparations are inactive. Purified mammalian PRLs, curiously, work very well in this bioassay. This observation supports the hypothesis that the osmoregulatory portion of the piscine PRL molecule has been retained in the structure of mammalian PRL with the addition of new structural modifications that are responsible for crop sac and mammary gland actions.

This bioassay was developed for fish PRL first in hypophysectomized guppies, *Poecilia latipinna.* Later, it was modified to use intact, seawater-adapted tilapia, *Sarotherodon mossambicus.* Adaptation of these fishes to seawater almost eliminates PRL from the circulation and alleviates the need for hypophysectomy.

The **xanthophore-expanding bioassay** is performed in the gobiid fish *Gillichthyes mirabilis,* the long-jawed mudsucker. It is an all-or-none bioassay for determination of the smallest dose of purified hormone or the greatest dilution of pituitary homogenate or extract that will cause local yellowing in 50% of the fish tested following injection of the test material beneath the preopercular skin. The bioassay does have the distinct advantage, however, of being extremely sensitive to piscine (including agnathan) and amphibian pituitary homogenates. For example, teleostean pituitaries are more than 100,000 times more effective in the assay than is oPRL that is used as a standard. As little as 1/100th of a tiger salamander pituitary gives a positive response in the xanthophore-expanding bioassay, whereas an entire salamander pituitary of the same size is required to produce a minimal response in the pigeon crop sac bioassay.

c. Growth Hormone in the Fishes

The presence of a growth hormone has not been established for cyclostome fishes. However, growth hormone activity can be demonstrated in the selachian pars distalis with the rodent tibia bioassay (see Chapter 4). Insulin-like growth factor (IGF) activity also has been demonstrated for two sharks, *Squalus acanthias* and *Mustelus canis,* although no definite link to GH has been established. No data on holocephalans are available.

Mammalian GHs are very active in promoting growth of fishes, and polypterid, chondrostean, holostean, and teleostean pituitaries exhibit GH activity in the mammalian tibia bioassay (see Chapter 4). Furthermore, by means of immunochemical techniques employing anti-rat GH antibody, GH immunoactivity has been demonstrated in the pituitaries of chondrostean, holostean, and teleostean fishes. Growth hormones isolated from fishes show great variability from species to species. Two variants of GH showing 94% homology in terms of amino acid sequence have been identified in chum salmon.

As described for mammals, there is a marked influence of thyroid hormones on growth in teleosts. Thyroid hormones accelerate growth in fishes, and thyroid hormone treatment restores growth rates to normal in radiothyroidectomized rainbow trout. Although it has not been demonstrated conclusively, it is reasonable to assume a synergistic relationship between thyroid hormones and GHs similar to that reported for mammals. Androgens also produce a positive effect on growth of salmonid fishes, suggesting a protein anabolic action for these steroids that is probably independent of the action of GH.

GH secretion correlates directly with induction of IGF-I mRNA in livers of growing coho salmon, and similar relationships between GH and IGF-I levels have been reported for rainbow trout, long-jawed mudsucker (*Gillichthyes*), and Japanese eels (*Anguilla*). Transfer of juvenile coho salmon prematurely into seawater decreases IGF-I levels and growth although GH levels increase. In coho salmon, GH appears to be important for seawater adaptation although it does not seem to be important in adaptation to seawater by young tilapia. IGF-binding proteins are described for several species including coho salmon and *Gillichthyes*. These observations suggest that involvement of GH and IGF-I levels in teleostean growth is similar to that in mammals.

d. Somatolactin in Teleosts

Madeleine Olivereau in France correlated the cytological appearance of PIPAS cells in the pars intermedia of teleosts with low-calcium environments and named them **calcium-sensitive cells.** A new teleostean hormone chemically similar to GH and PRL has been isolated by M. Ono and associates from the pars intermedia of Japanese flounder (*Paralichthys olivaeus*) and by Marilyn Rand-Weaver and coworkers from the Atlantic cod, *Gadus morhua.* The new hormone was named **somatolactin (SL)** and was shown by immunocytochemistry in several teleosts to reside in the PIPAS cells. In salmonids, such as the rainbow trout (*Oncorhynchus mykiss*), which lacks PIPAS cells, there are cells in the pars intermedia that bind antibodies to SL. These SL-immunoreactive cells are stimulated cytologically by placing rainbow trout in low-calcium environments and are cytologically repressed when the fish are in high-calcium environments. SL-containing cells in all teleosts appear to be involved in some manner with calcium balance.

Rising plasma levels of SL also have been correlated with smolting of seaward-migrating salmonids and with gonadal maturation. Plasma SL peaks at spawning in Pacific salmon (*Oncorhynchus keta, O. kisutch, O. nerka*). Furthermore, SL stimulates steroidogenesis in ovarian or testicular fragments isolated from coho salmon although SL is less potent in this regard than is GTH-I. This possible steroidogenic role for SL is similar to effects reported for GH and PRL in several teleosts and suggests the action of these latter hormones may be a reflection of their chemical similarity to SL.

2. Category II Hormones in the Amphibians

a. Prolactin

Prolactins from most amphibians cross-react with antibody to rat growth hormone, and PRL is thought by some to be the larval GH of amphibians. Tiger salamanders were the only animals tested whose PRL did not cross-react with rat GH antibody. In addition to possible influences on larval growth, PRL is antimetamorphic in both anurans and urodeles, that is, it blocks metamorphosis from the aquatic larva to the semiterrestrial or terrestrial juvenile form. Nevertheless, PRL also is essential for normal metamorphosis (see Chapter 8). Prolactin induces water-drive behavior in newts and salamanders (see below) and influences secondary sexual characters associated with breeding. Integumentary effects of PRL have been observed on the skin of newts and salamanders. Prolactin also affects water balance in anurans and urodeles as it does in fishes. Japanese workers have isolated amphibian PRLs and developed specific radioimmunoassays for studying possible functions of PRL in activities such as breeding migrations.

b. An Amphibian Bioassay for Prolactin

Prolactin induces a second metamorphosis or **"water drive"** in newts that is characterized by migration of the juvenile terrestrial form, the eft, back to water where breeding will occur. Water-drive behavior is accompanied by a series of physiological and morphological changes as well, but these are not related to the bioassay per se. Following injection of PRL, the efts will leave the dry areas of their laboratory containers and submerge themselves in water. No other hormone has been found to induce water-drive behavior. This bioassay has the disadvantage of being an all-or-none response. Pituitary preparations for all classes of vertebrates except the class Agnatha, possess water-drive-inducing activity (Table 5-6). Treatment with mammalian PRL stimulates locomotor activity in tiger salamanders that may be a component of water-drive behavior. Curiously, terrestrial anuran amphibians that migrate to water for breeding do not exhibit elevated PRL levels. The failure for PRL to elicit water-drive behavior in anurans suggests that the PRL-based water drive is an exclusively urodele phenomenon.

Table 5-6. Bioassayable Prolactin in Vertebrate Pituitaries

Vertebrate group	Gillichthyes[a] xanthophore-yellowing response	Tilapia Na+-retaining response	Red-eft water drive	Pigeon crop sac assay	Mouse mammary in vitro
Agnatha	+	−	−	−	−
Chondrichthyes	+	−	+	−	−
Chondrostei	+	−	?	−	−
Holostei	+	−	?	−	−
Teleostei	+	+, −	+	−	−[b]
Dipnoi	+	−	+	+	+
Amphibia	+	−	+	+	+
Reptilia	+	?	+	+	+
Aves	+	?	+	+	+
Mammalia	+	+, −	+	+	+

[a] This assay system may not be specific for prolactin (see text).
[b] Purified tilapia PRL stimulates casein synthesis in the rabbit mammary gland.

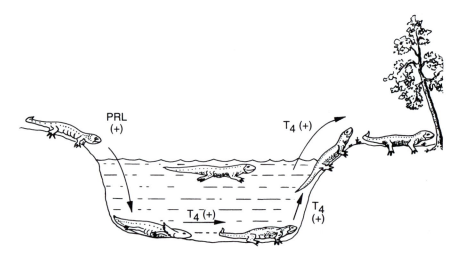

Figure 5-12. Land and water drives in urodele amphibian life histories. Prolactin (PRL) is associated with increased locomotor behavior and a preference for seeking out water, associated usually with reproductive behaviors. Increased thyroid activity is associated with a terrestrial preference, which probably causes the newly metamorphosed animal or postreproductive animal to leave the water. Similar relationships may occur in anurans (see text). T_4, Thyroxine.

The antithesis of water drive is the return to land by newts and salamanders as well as by frogs and toads after breeding. This **"land-drive"** behavior appears to be a consequence of elevated thyroid hormones, and experimentally elevated thyroid hormones prevent breeding migrations of anurans to water. Antibodies against PRL as well as treatment with thyroxine (T_4) can influence substrate preference in urodeles (Fig. 5-12).

c. Growth Hormone

The assessment of GHs in amphibians is complicated by the observation that PRL may be a larval growth hormone, whereas GH per se may operate only in transformed or metamorphosed individuals. Both GH and PRL have been isolated from frogs (*R. catesbeiana* and *R. pipiens*). The large number of similarities in the physical properties of amphibian and mammalian GHs and the similarities in amino acid composition suggest that there has been considerable conservatism expressed in the evolution of tetrapod GHs. Purified frog GHs are not as effective as bovine GH in the rat tibia bioassay, however (Table 5-5).

Cross-reaction occurs between antibody prepared against rat GH and the PRLs prepared from most amphibian species. Of the amphibian species examined to date, only the tiger salamander appears to produce a GH distinctly different from mammalian PRL. Nevertheless, antibody prepared against bullfrog PRL does not cross-react immunologically with bullfrog GH. This observation supports the interpretation that the inability to distinguish a distinct tropic hormone, such as PRL, in a nonmammal with antibody prepared against some mammalian tropic hormone, such as GH, does not mean the endogenous hormones (PRL and GH) are not separate and distinct molecules.

IGF activity has been demonstrated in *Bufo woodhousei* and *X. laevis* but no link to either PRL or GH has been established. This is an important area that demands more attention in all nonmammalian groups.

3. Category II Hormones in the Reptiles

a. Prolactin

The reptiles are the only major vertebrate group in which a good bioassay has not been developed, and little work has been done with PRL in reptiles. Mammalian PRL stimulates growth in juvenile snapping turtles, *Chelyldra serpentina,* and lizards, *Lacerta sicula sicula.* Appetite is also stimulated in *Lacerta* by PRL. A possible effect on water balance has been reported in turtles in which PRL influenced glomerular filtration. Purified reptilian PRLs give positive responses in other vertebrate bioassays (Table 5-6), and this PRL activity appears to be confined to the cephalic (rostral) lobe of the pars distalis. Considerable research remains to be done with respect to PRL in reptiles.

b. Growth Hormone

Mammalian GH, like PRL, stimulated growth in juvenile snapping turtles and in the lizard *Lacerta* although the sites of action for GH and PRL in the lizard appear to be different. Growth hormone stimulated appetite in *Lacerta,* as reported for PRL, and also produced a marked increase in growth of the digestive tract (splanchnomegaly), an action reported for PRL in other vertebrate groups (Table 5-4).

Purified GH has been prepared from the caudal lobe of adult snapping turtle pituitaries and sea turtles. Many characteristics of these GHs are similar to those of other tetrapods. Turtle GHs are very effective in the rat tibia bioassay.

4. Category II Hormones in the Birds

a. Prolactin

Prolactin plays several essential roles in avian reproduction. In some birds, a ventral portion of the body becomes defeathered and highly vascularized. This region is called a *brood patch* and is used by the bird for incubating the eggs. The brood patch facilitates transfer of heat from parent to incubating egg. Estrogens synergize with PRL in formation of the brood patch. PRL is also involved in fat deposition and induction of migratory restlessness in birds. This restless behavior, or **zugunruhe,** appears just prior to the actual migration of birds.

The amino acid composition and electrophoretic properties of chicken PRL are similar to those of mammalian PRLs. Chicken PRL appears to be chemically and functionally distinct from chicken GH.

b. Growth Hormone

Growth hormone activity has been demonstrated in pituitaries of chickens and turkeys by means of the mouse tibia bioassay. However, antibovine GH antibody apparently does not cross-react with chicken pituitary extracts, nor does purified GH from duck pituitaries cross-react with rat GH antibody. These observations suggest unique differences in bird GHs, which are probably more similar to reptilian than to mammalian GHs.

D. Category III: The Proopiomelanocortin Group

As in mammals, ACTH-, MSH-, CLIP-, and endorphin-like peptides are all derived from the same precursor, POMC. In some species, lipotropin (LPH) has been reported but will not be included in these discussions since no biological role for it is known other than as a precursor for endorphins (see Chapter 4). In corticotropes [PbH($+$)] of nonmammals, POMC is hydrolyzed to yield ACTH and β-endorphin whereas in the melanotrope, the end products are MSH, CLIP, and β-endorphin. Furthermore, the occurrence of N-acetylation of MSH and β-endorphin is similar to that in mammals.

1. Corticotropin Actions in the Fishes

Little is known about ACTH actions in the more primitive fish groups and not much more is known about its actions in teleosts. Treatment of lampreys with mammalian ACTH causes cytological changes in the interrenal cells that are suggestive of stimulation. ACTH has been purified from shark pituitaries, but its effects on interrenal tissue have not been studied. Secretion of ACTH in teleosts is correlated with enhancement of cortisol secretion from the interrenal gland (homologous to the adrenal cortex of mammals).

2. Corticotropin Actions in Nonmammalian Tetrapods

Activity of the steroidogenic enzyme 3β-hydroxy-Δ^5-steroid dehydrogenase is enhanced in the bullfrog interrenal following treatment with ACTH. Blood levels of corticosterone are elevated by ACTH treatment and lowered by hypophysectomy in several anuran species. Similar responses to ACTH and hypophysectomy are reported for lizards, turtles, and birds. In addition, bird interrenals also respond to PRL and GH as well as to ACTH. In all tetrapods examined, ACTH seems to work through a cAMP-dependent mechanism as it does in mammals.

3. Melanotropin Actions in Nonmammals

Many vertebrates show changes in pigmentation or pigmentary patterns that are correlated with environmental factors or particular events in their life histories. Rapid pigmentary responses are under direct neural control, and the slower responses are generally the result of endocrine control or a combination of neural and endocrine control.

Vertebrates possess a variety of specialized, pigmented cells referred to as **chromatophores** (see Table 5-7). These cellular types singly or in special combinations are the bases for color patterns associated with the integument. **Physiological color changes** involving displacement of pigment granules within a pigment cell are characteristic of many fishes, amphibians, and some reptiles. All vertebrates exhibit increases in the number of chromatophores and/or increases in the amount of pigment contained within or in the vicinity of the chromatophores. This type of color change is termed **morphological color change.**

Dopamine has proven to be the most potent inhibitor of MSH release, and several investigators have proposed that dopamine is a hypothalamic MSH release-inhibiting hormone (MRIH) in nonmammals as well as in mammals (see Chapter 4).

a. Pigmentation Patterns

One of the most important chromatophores of lower vertebrates is the dermal **melanophore,** which contains melanin granules concentrated in special organelles called **mela-**

Table 5-7. Vertebrate Chromatophores

Chromatophore	Organelle	Pigment	Color
Melanophore	Melanosomes	Melanins	Brown, black (yellow, red)
Iridophore (guanophore, leukophore)	Reflecting platelets	Guanine, adenine, hypoxanthine, uric acid	—
Xanthophore	Pterinosomes	Pteridines	Yellow, orange
	Carotenoid vesicles	Carotenoids	Yellow, orange, red
Erythrophore	Pterinosomes	Pteridines	Red, orange
	Carotenoid vesicles	Carotenoids	Yellow, orange, red

nosomes (Fig. 5-13). Melanophores differ from **melanocytes,** which deposit their melanin products extracellularly (Fig. 5-14). When melanosomes are concentrated around the nucleus of the melanophore, the skin appears lighter than when they are dispersed throughout the cytoplasm. The degree of concentration of melanosomes is inversely related to the darkness of the skin (Fig. 5-15). Coloration patterns in some vertebrates may be determined by the distribution of types of chromatophores and by the relationship of dermal melan-

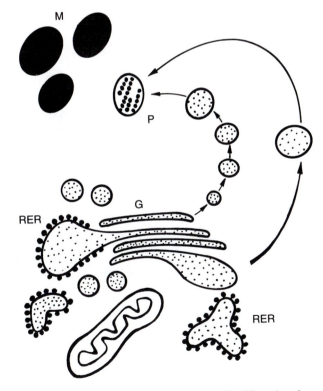

Figure 5-13. Production of melanosomes. Premelanosomes (P) differentiate from Golgi vesicles (G). They exhibit tyrosinase activity and synthesize melanin by polymerizing tyrosine. Once melanization of these organelles is complete, they are called melanosomes (M). RER, Rough endoplasmic reticulum.

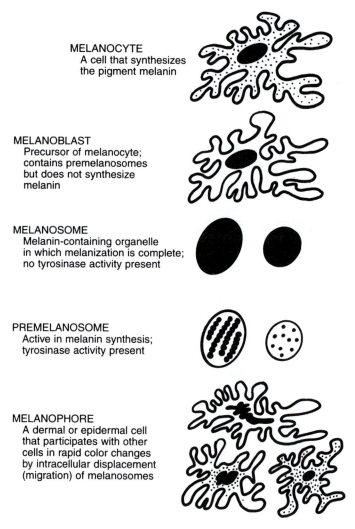

MELANOCYTE
A cell that synthesizes
the pigment melanin

MELANOBLAST
Precursor of melanocyte;
contains premelanosomes
but does not synthesize
melanin

MELANOSOME
Melanin-containing organelle
in which melanization is complete;
no tyrosinase activity present

PREMELANOSOME
Active in melanin synthesis;
tyrosinase activity present

MELANOPHORE
A dermal or epidermal cell
that participates with other
cells in rapid color changes
by intracellular displacement
(migration) of melanosomes

Figure 5-14. Melanophores and melanocytes. The terminology presented here is based on the Sixth International Pigment Cell Conference held in 1966. Melanophores can disperse their melanosomes uniformly in the cytoplasm or concentrate them to varying degrees around the nucleus.

ophores to other chromatophores. For example, dark spots on the skin of the frog *R. pipiens* are due to concentrations of melanocytes and extracellular deposition of melanin, whereas adjacent regions of the skin that may vary from light green to black are occupied exclusively by dermal melanophores and other chromatophores. The green color is produced by interactions of other chromatophores overlying the melanophores, which are masked when the melanosomes are dispersed throughout the cytoplasm of the melanophore.

The major target for MSH is the dermal melanophore, and in a few cases, other chromatophores may be affected. The actual mechanism of how melanosomes migrate in and out of the stellate processes of the melanophore is not completely understood. It appears that microtubules are essential for aggregation of the melanosomes, and these may be ac-

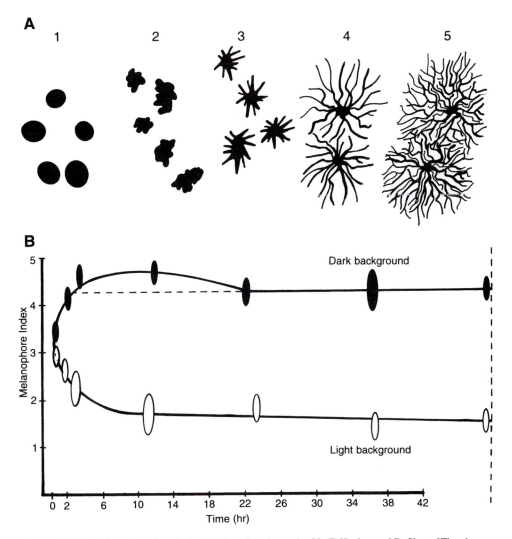

Figure 5-15. The melanophore index. (A) Based on the work of L. T. Hogben and D. Slome [The pigmentary effector system. VI. The dual character of endocrine coordination in amphibian colour change. *Proc. R. Soc. London* **108,** 10–53 (1931)], this index defines five stages of melanophore dispersion throughout the cytoplasm of a melanophore, from the least (1) to the most dispersed stage (5). The degree of dispersion is directly proportional to the amount of MSH present in the system. (B) Changes in melanophore index of an anuran during adaptation to dark or light backgrounds. Based on Novales, R. R. (1974). Actions of melanocyte-stimulating hormones. *In* "Handbook of Physiology, Section 7, Endocrinology," Vol. 4, Pt. 2, pp. 347–366. Williams & Wilkins, Baltimore, Maryland.

tivated by MSH through the second messenger cAMP. Any experimental treatment known to interfere with microtubule formation, such as application of the drug **colchicine,** blocks melanosome aggregation. Ultrastructural observations, however, in some amphibian melanophores do not support the presence of microtubules oriented properly to cause melanosomes to disperse, and the role of microtubules needs further clarification.

b. Melanotropin in Agnathan Fishes

Hagfishes (Myxinoidea) show no ability to adapt to different backgrounds, suggesting no role for MSH with respect to background adaptation. However, hypophysectomy of lampreys (Petromyzontidae) produces permanent paling, presumably due to loss of MSH. Exogenous MSH causes melanosome dispersion in lampreys. Nothing is known about hypothalamic factors controlling MSH release in cyclostomes.

c. Melanotropin in Chondrichthyean Fishes

Sharks adapt to background changes very slowly, requiring up to 100 hr to achieve maximal adaptation. Hypophysectomy abolishes this ability to "background adapt," whereas ectopic transplants of the neurointermediate lobe cause permanent darkening regardless of the background on which the shark is placed. The dogfish shark darkens following injection of either α- or β-MSH. Melanophore responses to MSH are paralleled by similar responses in xanthophores.

Both aminergic and peptidergic fibers penetrate into the dogfish pars intermedia. It has been hypothesized that the peptidergic fibers control synthesis of MSH, and that aminergic fibers are responsible for controlling MSH release. The pars intermedia of the holocephalan *Hydrolagus colliei* also exhibits NS innervation, although the type of innervation has not been determined. It is not clear what mechanism predominates in controlling release of MSH in these fishes.

d. Melanotropin in Bony Fishes: Teleosts

The pigmentary responses of fishes are under the control of direct aminergic innervation of melanophores and other chromatophores (Tables 5-8 and 5-9), and MSH does not play a major role in physiological color changes. The melanophores of most species do not

Table 5-8. Drugs Causing Melanosome Dispersion When Injected Subcutaneously in the Opercular Area of the Fish *Scopthalmus aquosus*[a]

Drug	MED[b] (μg)
Serotonin blockers	
LSD-25	0.8
Dibenamine	0.2
Adrenergic blockers	
Dihydroxyergotamine	0.6
Propranolol	6.0
Depressants	
Pentobarbital sodium	80.0
Phenothiazine ataractics	
Chlorpromazine	0.8
Cholinergic substances	
Acetylcholine	No effect up to 500 μg

[a]Modified from Scott, G. T. (1972). The action of psychoactive drugs on pigment cells of lower vertebrates. *In* "Pigmentation: Its Genesis and Biologic Control." (V. Riley, ed.), pp. 327–342. Appleton-Century-Crofts, New York.

[b]Minimum effective dose (three or four fish responding out of five fish tested).

Table 5-9. Drugs Causing Melanosome Aggregation When Injected Subcutaneously in the Opercular Area of the Fish *Scopthalmus aquosus*[a]

Drug	MED[b] (μg)	After pyrogallol[c] treatment (μg)
Epinephrine	0.04	0.00010
Isoproterenol	0.08	0.00002
Dopamine	0.02	0.02
Dopa	Inactive	Inactive
Phenylephrine	0.0002	0.0004
Melatonin	0.06	0.06
5-HT	0.1	2.0

[a] Modified from Scott, G. T. (1972). The action of psychoactive drugs on pigment cells of lower vertebrates. *In* "Pigmentation: Its Genesis and Biologic Control" (V. Riley, ed.), pp. 327–342. Appleton-Century-Crofts, New York.

[b] MED, Minimum effective dose (three or four fish responding out of five fish tested).

[c] Pyrogallol is an inhibitor of the enzyme catechol *O*-methyltransferase, which inactivates epinephrine and norepinephrine.

respond to either hypophysectomy or exogenous MSH, whereas others show only a limited response to MSH following denervation of the melanophores. Dispersion of pigments in melanophores of goldfish *Carassius auratus* and erythrophores of the European minnow *Phoxinus phoxinus* occurs following treatment with MSH. In contrast, melanophore dispersion can be induced only in denervated melanophores of the killifish *Fundulus heteroclitus* because of the presence of overriding neural stimuli in intact fish.

MSH may play some role in morphological color changes in teleosts. Treatment of xanthic goldfish (goldfish normally lacking any melanophores) with MSH stimulates melanophore differentiation and melanogenesis. Hypophysectomy of *F. heteroclitus* causes a reduction in the number of melanophores, and treatment with MSH results in an increase in the number of melanophores. A conflicting report suggesting melanophore differentiation in xanthic goldfish is caused by ACTH and not by MSH has yet to be resolved.

Melanin-concentrating hormone (MCH) was isolated first from teleostean pituitaries before it was demonstrated in mammals. It is a potent cyclic heptadecapeptide (17 amino acids) that causes contraction of teleostean melanophores but, in fact, produces weak melanosome dispersion in amphibian and reptilian melanophores. MCH also blocks ACTH and MSH release from teleostean pituitaries.

e. Melanotropin in Amphibians

Direct control of amphibian melanophores is under endocrine regulation, and there is no evidence for direct innervation of amphibian melanophores with the possible exception of limited neural control in *R. pipiens*. However, release of MSH from the pars intermedia is under neural control in most amphibians.

There have been reports of dual "innervation" of the pars intermedia in amphibians of a different type than in the cartilaginous fishes. Adrenergic fibers have been shown to penetrate the pars intermedia of *R. temporaria* and *Bufo arenarum,* and two types of adrenergic fibers synapse with pars intermedia cells of *R. pipiens*. No innervation of the pars intermedia is observed in the African clawed frog *X. laevis*, however. Dual control of MSH release as suggested for some anurans may operate through the presence of α- and β-adrenergic receptors in the plasmalemma of the pars intermedia cells. The catecholamines dopamine, norepinephrine, and epinephrine, as well as α-receptor agonists such as phenylephrine, all inhibit MSH release from the pars intermedia. Isoproterenol, a catecholamine agonist, stimulates MSH release by activating β-adrenergic receptors. This effect is blocked by antagonists of β receptors, such as propranolol. Known antagonists of α receptors (for example, dibenamine and dihydroergotamine) block the inhibitory actions of the catecholamines.

Certain cautions must be employed when one is attempting to interpret the actions of various drugs on MSH release from the pars intermedia. Epinephrine, for example, inhibits MSH release when applied at higher doses (10^{-5} M) but stimulates MSH release when applied at lower doses (10^{-6} or 10^{-7} M). Apparently at higher doses epinephrine saturates both α and β receptors, but the α effect predominates. At lower doses, epinephrine is more readily bound to the β receptors so that the β effect predominates, and MSH is released. The observation that a certain compound stimulates release of MSH does not necessarily imply a normal *in vivo* role for that substance or for the type of innervation it might suggest. For example, the actions of the other catecholamines under experimental conditions may only reflect an influence on the normal mechanism whereby dopamine controls MSH release.

Pigment organelles of another pigment cell, the **xanthophores,** are normally in an expanded state in most amphibians. In the tree frog *Hyla arenicolor*, however, the xanthophores are normally in an aggregated condition, and they may be dispersed by the application of MSH. An endogenous role for MSH on xanthophore pigment organelles has not been confirmed, nor is it clear how widespread this phenomenon might be.

The aggregation of reflecting platelets in the **iridophores,** another special chromatophore, of *R. pipiens* skin is stimulated by either cAMP or MSH. These iridophores possess α- and β-adrenergic receptors, and stimulation of the α receptors produces dispersion of the reflecting platelets of the iridophore. These data suggest a possible role for catecholamines in iridophore regulation.

As mentioned earlier, MCH is present in the anuran brain but no melanin-concentrating action has been demonstrated. In fact, MCH caused melanophore expansion in several studies. Additional studies are needed to elucidate the physiological importance of MCH.

f. Melanotropin in Reptiles

Reptiles exhibit a variety of mechanisms for regulating dermal pigment cells, including neural and endocrine mechanisms. Unlike the well-established pattern of adrenergic innervation in most amphibians, no innervation of the pars intermedia has been observed at either the light or electron microscopic level in several lizard species. The primitive tuatara of New Zealand, *Sphenodon punctatus*, also exhibits no innervation of the pars intermedia, implying this may be a basic pattern that was established in the earliest of reptiles. Innervation of pigment cells does occur in some reptiles, and pigmentary control of dermal melanophores may be neural, endocrine, or both.

Color changes in the true chameleon *Chameleo pumilis* are under direct neural control, and MSH has no effect on the skin chromatophores. Melanophores of the closely related species *Chameleo jacksoni*, however, respond to both MSH and ACTH. The opposite extreme is found in the American chameleon, *Anolis carolinensis*, in which there is no general neural control over melanophore responses, and melanosome dispersion can be readily induced *in vitro* by application of MSH or cAMP to isolated pieces of *Anolis* skin. However, there is a patch of skin near the eye that darkens in response to catecholamines. Pretreatment with α-adrenergic-blocking agents obliterates the response of *Anolis* melanophores to MSH. Horned lizards (*Phrysonoma* spp.) exhibit both neural and hormonal regulation of melanophores.

Neither xanthophores nor iridophores of *A. carolinensis* show any response to MSH. It is not known whether any non-melanin-containing chromatophores of other reptiles show any regulatory control.

g. Melanotropin in Birds

The observation that feather pigments (including melanin) are under the control of gonadal, thyroidal, and gonadotropic hormones seems to be related to the loss of the pars intermedia and MSH in birds. The only reported action for MSH in birds possibly is related to a developmental action. Embryonic implants of chicken pituitaries cause formation of black feathers where normally only white feathers would develop. This effect can be mimicked by treatment with either β-MSH or ACTH.

4. Endorphins in Nonmammals

β-Endorphin is produced by the pituitaries of all jawed vertebrates but is absent in agnathans. When present, endorphins are usually N-acetylated except in the cartilaginous dogfishes, although it is N-acetylated in the more primitive ratfish. No physiological role has been established for the pituitary endorphins of nonmammals.

IV. Some Comparative Aspects of Hypothalamic Control of Adenohypophysial Function in Nonmammals

Numerous studies have documented the distribution of mammalian-like neuropeptides in the brains of nonmammals (Table 5-10), but relatively little is known about the functional roles for these neuropeptides. In general, there is more plasticity in nonmammals with respect to the actions of the peptides that bear specific names for their roles in mammals. When making comparisons, it is important to bear in mind that hypothalamic control in teleosts is neuroglandular rather than neurovascular as in tetrapods.

A. Category I: The Glycoprotein Hormones

The focus in studies of glycoprotein hormone regulation has been on the gonadotropins rather than on thyrotropin. In fact, it has been and continues to be a problem to determine what neuropeptide(s) is(are) the endogenous releaser for TSH in many cases. Consequently, the picture is more complete for GTH release, where decapeptides similar or identical to mGnRH are active releasers of GTHs.

1. Control of Gonadotropin Release

Many of the cells that secrete GnRH apparently are derived from the embryonic neural ridge destined to give rise to the nasal olfactory epithelium (see Chapter 4) and migrate posteriorly from the nasal region into the preoptic area of the brain. These GnRH-secreting

Table 5-10. Localization of Neuropeptides in the Nonmammalian Vertebrate Brain

Group	Neuropeptide[a]	Brain location[b]													
		Telen-cephalon	POA	POR	PON	SON	PVN	OVLT	SCN	NLT	PVO	PERIV	IFN	NID	NIV
Teleosts	GHRH				+					+					
	TRH			+	+					+					
	GnRH	+		+	+					+					
	cfGnRH				+										
	ACTH									+					
	MCH									+					
	CRH				+					+					
	CRH + AVT				+										
	AVT				+										
	α-MSH														
Lungfish	TRH	+								+					
Amphibians	GnRH				+										
	GAP														
	CRH				+										
	AVT				+										
	MST				+										

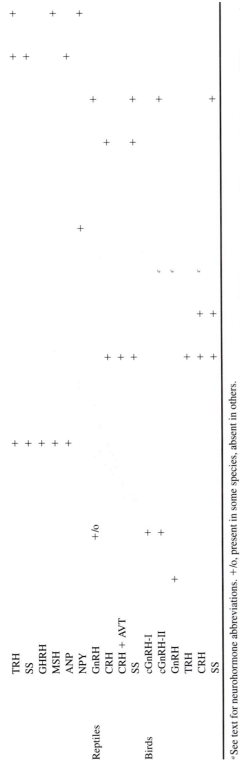

The table lists neurohormones for Reptiles and Birds:

Reptiles: TRH, SS, GHRH, MSH, ANP, NPY, GnRH, CRH, CRH + AVT, SS

Birds: cGnRH-I, cGnRH-II, GnRH, TRH, CRH, SS

[a]See text for neurohormone abbreviations. +/o, present in some species, absent in others.

[b]POA, Preoptic area; SCN, suprachiasmatic nucleus; POR, preoptic recess; NLT, nucleus lateralis tuberis; PON, preoptic nucleus; PVO, periventricular organ; SON, supraoptic nucleus; PERIV, periventricular nucleus; PVN, paraventricular nucleus; IFN, infundibular nucleus; NID, nucleus infundibularis dorsalis; NIV, nucleus infundibularis ventralis; OVLT, organum vasculosum laminae terminalis.

[c]Fibers entering this area but not originating there.

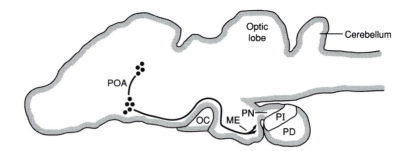

Figure 5-16. GnRH in the toad brain. GnRH-immunoreactive cell bodies apparently reach the preoptic area (POA) by migrating into the brain from the olfactory epithelium. Fibers project from the POA to the median eminence (ME), where they release GnRH into the portal vessels connected to the pars distalis (PD). PI, Pars intermedia; PN, pars nervosa; OC, optic chiasm. [Redrawn from Y. Jokura and A. Urano (1986). Extrahypothalamic projection of luteinizing hormone-releasing hormone fibers in the brain of the toad, *Bufo japonicus. Gen. Comp. Endocrinol.* **62,** 80–88.]

cells send axonal processes to the median eminence (Fig. 5-16). In addition, some GnRH cells as well as the other neuropeptide-secreting cells of the hypothalamus had their origin in a different region of the neural ridge and developed into the hypothalamic neurosecretory neurons.

Several variants of the decapeptide mGnRH are well known (Table 5-11), and purified GnRHs from many sources are effective at eliciting GTH release in teleosts, amphibians, reptiles, and birds. More variants can be expected as researchers examine more species, especially among the teleosts. mGnRH has been isolated from sheep, pigs, rats, mice, and humans as well as from amphibians. Two variants were found in chickens and were named **chicken I (cGnRH-I)** and **chicken II (cGnRH-II).** Salmon proved to have a unique decapeptide **(sGnRH)** as did catfish **(cfGnRH),** dogfish shark **(dfGnRH),** and lamprey **(lGnRH).** Many species exhibit more than one form of GnRH (see Fig. 5-17).

Although lGnRH is a form of GnRH found in the most primitive living vertebrate ex-

Table 5-11. Amino Acid Composition of Decapeptide Vertebrate Gonadotropin-Releasing Hormone Molecules[a]

Source	Position									
	1	2	3	4	5	6	7	8	9	10
Lamprey I	E	H	Y	S	L	E	W	K	P	G-NH$_2$
Lamprey III	E	H	W	S	H	D	W	K	P	G-NH$_2$
Dogfish	E	H	W	S	H	G	W	L	P	G-NH$_2$
Catfish	E	H	W	S	H	G	L	N	P	G-NH$_2$
Salmon	E	H	W	S	Y	G	W	L	P	G-NH$_2$
Sea bream	E	H	W	S	Y	G	L	S	P	G-NH$_2$
Chicken I	E	H	W	S	Y	G	L	Q	P	G-NH$_2$
Chicken II	E	H	W	S	H	G	W	Y	P	G-NH$_2$
Mammal	E	H	W	S	Y	G	L	R	P	G-NH$_2$

[a] Single-letter amino acid code is available in Appendix C. There is absolute conservation at positions 1, 2, 4, 9, and 10. Strong homology is seen at positions 3 and 6, with only lampreys showing variations. In marked contrast, there is little correspondence at position 8, which shows extensive variability.

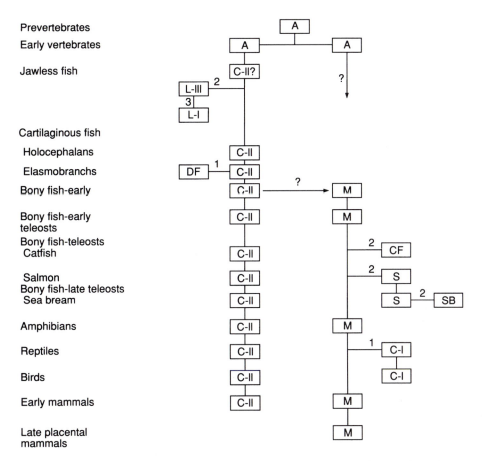

Figure 5-17. Proposed evolution of GnRH molecules in vertebrates. This hypothetical scheme focuses on the existence of an ancestral GnRH (A) that was present in the earliest vertebrates, and which may have undergone an earlier duplication resulting in the chicken II form (C-II) and a mammalian GnRH (M); or the latter may have arisen from C-II in the early bony fish. Names for vertebrate GnRHs: C-I, chicken I; CF, catfish; DF, dogfish shark; L-I, L-III, lamprey forms; S, salmon; SB, sea bream. [From Sherwood, N. M., Parker, D. B., McRory, J. E., and Lescheid, D. W. (1994). Molecular evolution of growth hormone-releasing hormone and gonadotropin-releasing hormone. *In* "Fish Physiology" (N. M. Sherwood and C. L. Hew, eds.), Vol. 13, pp. 3–66. Academic Press, San Diego.]

amined to date, the most widely distributed GnRH is cGnRH-II. This form is found in the chondrichthyean fishes, teleosts, lungfishes, amphibians, reptiles, birds, and marsupial mammals (see Fig. 5-17). cGnRH-II is more prevalent in extrahypothalamic regions and another form of GnRH (usually sGnRH, mGnRH, or cGnRH-I) often is associated with the median eminence and is responsible for release of pituitary GTHs. In amphibians, reptiles, and birds, cGnRH-II has been observed as far posteriorly as the medulla and anteriorly in the olfactory regions.

Frequently, more than one GnRH may be a GTH releaser when applied exogenously (Table 5-12), but usually only one is distributed appropriately to release GTHs. The other GnRH (usually cGnRH-II) probably functions as a neurotransmitter or neuromodulator.

Table 5-12. Effects of Neural and Endocrine Factors on Gonadotropin Release in Nonmammalian Vertebrates

Factor[a]	Effect	Vertebrate group[b]			
		Teleosts	Amphibians	Reptiles	Birds
Mammalian GnRH	(+) Plasma androgens	×		×	×
Salmon GnRH	(+) GTH release	×	×	×	
Lamprey GnRH	(+) GTH release	×	○	○	
Dogfish GnRH	(+) GTH release	×			
Chicken GnRH-I	(+) GTH release	×	×	×/○	×
Chicken GnRH-II	(+) GTH release	×		×	
AVT	(+) GTH slightly			×	
DA, DA agonists	(−) Inhibit GTH release		×		
Arachidonic acid	(+) GTH release	×			
E_2	(+) Plasma androgens	×			
E_2	(−) GTH release			×	
T	(+) GTH release				×
T	(−) GTH release			×	
P_4, E_2	Enhanced GnRH stimulation of GTH release				×
VIP	(+) GTH release			○	×
TRH	(+) GTH release			○	
Ovine CRH	(+) GTH release			○	
GHRH	(+) GTH release			○	
Morphine	(+) GTH release				○
NPY	(+) GTH release	×			
Serotonin	(+) GTH release	×			

[a] E_2, Estradiol; T, testosterone; P_4, progesterone. See text for other abbreviations.
[b] ×, Effect; ○, no effect; blank, not tested.

Receptors for GnRH are found in several brain regions, and cGnRH-II may control certain aspects of sexual behavior.

Judy King and Robert Millar in South Africa as well as Nancy Sherwood and coworkers in Canada have identified and studied the various forms of GnRH in vertebrates. Their work has led to a proposed scheme for the evolution of GnRH forms involving an early gene duplication and subsequent variation of one gene and relative conservatism of the other (Fig. 5-17). This scheme is based in part on the observation that cGnRH-II is so widely distributed and in part on the structures of the various decapeptides that are found with cGnRH-II. Although it may be premature to lock tenaciously onto this scheme until more species are examined, this pattern is certainly plausible. Caution must be applied, however, whenever phylogenetic schemes are based on only a fragment of a gene product, and such analyses should include the much larger GnRH-associated peptide (GAP).

Numerous neuropeptides, amine neurotransmitters, and steroids have been shown to alter GTH release in vertebrates (Table 5-12). Some of these amines operate at the level of the GnRH neurons in the brain. For example, dopamine directly inhibits release of GTHs from the pituitary in teleosts.

2. Control of Thyrotropin Release

Although mTRH is an effective releaser of TSH in mammals and is present in the brain of all nonmammalian vertebrates examined so far, it has no effect on TSH release in fishes

Table 5-13. Effects of Neural and Endocrine Factors on Thyrotropin Release in Nonmammalian Vertebrates

Factor[a]	Effect	Chon-drichthyes	Amphibians Urodeles	Amphibians Anura	Reptiles	Birds
TRH	Increases plasma T_4		O/×[c]	×[c]	×	
TRH	Stimulates TSH release			×[c]/O	×	×
Ovine CRH	Stimulates TSH release			×	×	×
GnRH	Stimulates TSH release	×		×	O	
Mammalian GHRH	Stimulates TSH release			O	×	
DA	Stimulates TSH release				O	
NE	Stimulates TSH release				O	
VIP	Stimulates TSH release				×	
Sauvagine	Stimulates TSH release				×	
Urotensin I	Stimulates TSH release				×	

[a] See text for abbreviations.
[b] ×, Effect; O, no effect; blank, not tested.
[c] Observed only in metamorphosed adults.

and larval amphibians. Release of TSH in amphibians, reptiles, and birds is more responsive to mCRH than TRH, and several studies suggest CRH is the "TRH" of larval amphibians, too. Thus, the molecule we call "TRH" in mammals is probably not the "TRH" of nonmammals.

Experimental evidence suggests that the control of TSH release is less specific in nonmammals. Thyrotropes of nonmammalian tetrapods release TSH not only in response to TRH and CRH but also in response to mGnRH, mGHRH, VIP, and the CRH-like peptides, sauvagine and urotensin I (see Table 5-13).

Teleostean fishes represent an even greater anomaly in the control of TSH release. Not only are they unique among the vertebrates in having neuroglandular control rather than neurovascular control over TSH release, but TSH release is under inhibitory control. The possible significance of this condition is discussed in Chapter 8.

B. Category II: Growth Hormone and Prolactin

Release of both category II tropic hormones is very similar to what has been observed in mammals. Notable exceptions are the neural control of prolactin release exhibited in teleosts and the fact that prolactin release in birds is entirely under stimulatory control.

1. Control of Growth Hormone Release

Release of GH is altered predictably by mammalian hypothalamic neuropeptides (Table 5-14). Treatment of teleosts, amphibians, reptiles, and birds with mGHRH causes release of GH. Similarly, somatostatin inhibits GH release, at least in teleosts, reptiles, and birds. Several other neuropeptides have been shown to cause GH release including TRH (in amphibians, reptiles, birds), GnRH, and NPY (teleosts). The actions of catecholamines on GH release do not show any consistent pattern and the response varies greatly among the vertebrates. The response to TRH and GHRH in birds can be blocked by IGF treatment, suggesting a negative feedback effect of IGFs may act on avian GH cells as in mammals.

Table 5-14. Effects of Neural and Endocrine Factors on Growth Hormone Release in Nonmammalian Vertebrates

Factor[a]	Effect on growth hormone release	Nonmammalian vertebrate group[b]			
		Teleosts	Amphibians	Reptiles	Birds
Mammalian GHRH	Stimulates GH release	×	×	×	×
TRH	Stimulates GH release		×	×	×
Dogfish shark GnRH	Stimulates GH release	×			
Salmon GnRH	Stimulates GH release	×			
VIP	Stimulates GH release			×	
DA	Stimulates GH release	×		○	
NPY	Stimulates GH release	×			
NE	Stimulates GH release			○/×	
NE	Inhibits GH release	×			
IGF	Suppresses response to TRH and GHRH				×
SS	Inhibits GH release	×		×	×
Morphine	Inhibits GH release				×
Serotonin	Inhibits GH release	×			

[a] See text for abbreviations.
[b] ×, Effect; ○, no effect; blank, not tested.

Soon after GH release was demonstrated in the goldfish following intraperitoneal injections of synthetic hGHRH, immunologically and chromatographically similar molecules were isolated from fishes. Apparently, the prohormone that yields GHRH in fish also produces a peptide very much like the pituitary adenylate cyclase-activating polypeptide (PACAP) described for mammals (see Chapter 4). However, in mammals, PACAP is produced by a separate gene. The mammalian PACAP precursor consists of PACAP and a GHRH-like peptide, suggesting it is homologous to the fish GHRH gene. Immunoreactive hGHRH-like material also is present in amphibians, reptiles, and birds, and hGHRH stimulates GH release in chickens, turtles, and frogs. Furthermore, a PACAP-like peptide has been isolated from the frog, *R. esculenta.* The relationship of GHRH-like and PACAP-like molecules in these vertebrate groups will be of special interest as we try to unravel the evolution of these genes.

2. Control of Prolactin Release

Observations on the effects of neural factors on PRL release in nonmammals is summarized in Table 5-15. As in mammals, the major control over PRL release seems to be inhibitory. Dopamine inhibits PRL release in teleosts and amphibians as well as do VIP and hGAP. The latter peptide is also a PRL release inhibitor in mammals. The inhibitory action of VIP is somewhat unexpected in light of its stimulatory role in birds and mammals. **Peptide histidine isoleucine (PHI)** is a VIP-like peptide that also inhibits PRL release in teleosts. However, PHI can stimulate TSH release in amphibians.

Stimulation of PRL release by TRH occurs in nonmammals as reported for mammals. The major control of prolactin release in birds is stimulatory and appears to be under the direct control of TRH. If this proves to be a universal role for TRH in nonmammals, its abundant presence in the brains of nonmammals may be explained, at least in part.

Table 5-15. Effects of Neural and Endocrine Factors on Prolactin Release
in Nonmammalian Vertebrates

Factor[a]	Effect on prolactin release	Nonmammalian vertebrate group[b]			
		Teleosts	Amphibians	Reptiles	Birds
Dopamine	Inhibits PRL release	×	×		
Apomorphine (DA agonist)	Inhibits PRL release	×			
Somatostatin	Inhibits PRL release	×			
VIP	Inhibits PRL release	×	×		
PHI	Inhibits PRL release	×			
	Stimulates PRL release		×		
GAP	Inhibits PRL release	×	×		
TRH	Stimulates PRL release	×	×	×	×
pVIP	Stimulates PRL release				×

[a] See text for abbreviations.
[b] ×, Effect; blank, not tested.

C. Category III: The Proopiomelanocortin Group

Although the several peptides of the POMC group have a common precursor, the regulation of ACTH and MSH release is entirely different as described for mammals (Chapter 4). Little is known about the control of endorphin release.

1. Control of Corticotropin Release

Synthetic oCRH is an effective releaser of ACTH in teleosts, amphibians, and birds (Table 5-16). The distribution of CRH immunoreactivity in reptiles is supportive of a similar role there, although definitive experiments have not been reported yet. In teleosts, the CRH-like peptides sauvagine and urotensin I also can evoke ACTH release, but these CRH-like peptides so far have proven ineffective in amphibians. As in mammals, AVT or AVP also can stimulate ACTH release in amphibians and birds, but the details of this action have not been substantiated. The appearance of immunoreactive AVT-like material, together with CRH in the median eminence of bullfrog larvae at metamorphosis, supports an attendant releasing role.

Table 5-16. Effects of Neural and Endocrine Factors on Corticotropin Release
in Nonmammalian Vertebrates

Factor[a]	Effect	Nonmammalian vertebrate group[b]			
		Teleosts	Amphibians	Reptiles	Birds
Ovine CRH	Elevates plasma corticoids	×	×		×
Sauvagine	Elevates plasma corticoids	×	○		
Urotensin-I	Elevates plasma corticoids	×	○		
AVT/AVP	Elevates plasma corticoids		×		×
DA antagonists	Elevates plasma corticoids				×
Morphine	Elevates plasma corticoids	×			
	Inhibits ACTH release	×			
ANGI-I	Elevates plasma corticoids	○			

[a] See text for abbreviations.
[b] ×, Effect; ○, no effect; blank, not tested.

Table 5-17. Effects of Neural and Endocrine Factors on Melanotropin Release in Nonmammalian Vertebrates

Factor[a]	Effect on melanotropin release	Nonmammalian vertebrate group[b]		
		Teleosts	Amphibians	Reptiles
TRH	(+) MSH release	×	×	×/○
Ovine CRH	(+) MSH release	×	○	
Sauvagine	(+) MSH release	×	○	
Urotensin I	(+) MSH release	×	○	
MST/AVT	(+) MSH release		○	
MCH	(−) MSH release	×		
NPY	(−) TRH-induced MSH release		×	
DA	(−) MSH release	×		
Naloxone	(+) DA, which (−) MSH		×	

[a] See text for abbreviations.
[b] ×, Effect; ○, no effect; blank, not tested.

2. Control of Melanotropin Release

Studies of MSH release by hypothalamic factors are almost confined to teleosts and amphibians (see Table 5-17). In fishes, MSH release is caused by treatment with TRH or CRH as well as by sauvagine and urotensin-I. Both MCH and DA block MSH release in teleosts. Treatment with TRH causes MSH release in amphibians and sometimes in reptiles, but CRH-like peptides are ineffective.

V. Possible Role for the Epiphysial Complex in Hypothalamic Function

The epiphysial complex (see also Chapter 6) includes the pineal gland, which is believed to secrete its products into the blood and possibly into the cerebrospinal fluid. The pineal gland secretes the amines melatonin and serotonin and possibly some small peptides, including AVT. Numerous studies have documented antigonadal actions of pineal extracts or melatonin in fishes, amphibians, reptiles, birds, and mammals (see Chapter 6). In addition, effects on thyroid function have been noted in some of these groups (see Chapter 6). Melatonin enhances certain aspects of the immune response system in mammals, but this role has not been confirmed for nonmammals. It is possible that these pineal actions are mediated via effects on the hypothalamus, or the specific sites of action may be at the target endocrine glands themselves.

Melatonin blocks pigment dispersal in amphibian dermal melanophores through a direct antagonism of the action of MSH. This effect in amphibians appears to be a unique action for melatonin.

VI. Summary

There is a progressive evolution of hypothalamic centers in vertebrates evident from examination of different vertebrates. The pattern of regulation has branched within the fishes, with the most recent fishes (teleosts) exhibiting predominantly neuroglandular control of tropic hormone release. Primitive fishes, including the chondrichthyeans, and tetrapods have specialized in neurovascular control with the development of a distinct median eminence and a hypophysial portal blood system. The distinct regionalization of tropic cell types in the fish adenohypophysis is less marked in tetrapods. The adenohypophysis of fishes is usually separable into a rostral pars distalis, a proximal pars distalis, and a pars intermedia. The pars tuberalis of tetrapods is absent in fishes but a unique ependymal structure is evident, the saccus vasculosus. Lungfishes lack both a pars tuberalis and the saccus vasculosus.

All of the mammalian tropic hormones have a molecular counterpart among the nonmammals, and many nonmammalian functions can be stimulated by mammalian tropic hormones. The established occurrence of at least one hormone from each category in all gnathostomes (jawed vertebrates) suggests that early in vertebrate evolution three cellular types differentiated [basophil, acidophil, and possibly PbH(+) cells, respectively], and each began elaboration of one of three types of molecules: glycoproteins (TSH, LH, and FSH), large peptides (GH and PRL), and the POMC-related peptides (ACTH, MSH, and endorphins). These three primitive molecular types attained functional significance as tropic hormones and gave rise via amino acid substitutions, modified cleavage of prohormones, or both, to the additional hormones that characterize each category.

Nonmammals produce the same tropic hormones as do mammals and there is considerable homology among them. Nevertheless, there is clear evidence of changes in the functions of these molecules related both to structural changes in the tropic hormones themselves and alterations in receptors such that new functions have evolved in some cases. A new hormone, somatolactin, has been isolated from the pars intermedia of teleosts, which may be involved in calcium regulation and/or reproduction.

Regulation of the release of tropic hormones in nonmammals follows the mammalian pattern for the most part. Many of the same neuropeptides are present in the brains of nonmammals as described for mammals but they show greater variations in their effects, often causing release of different tropic hormones than implied by their mammalian names.

Examination of the actions of tropic hormones in nonmammals has resulted in the development of numerous bioassays that have proven useful for quantifying levels during purification of tropic hormones. Furthermore, these bioassays are critical for measuring biological activity for a particular tropic hormone when purified sources from either mammalian or nonmammalian animals are not available. Caution must be exercised when comparing molecules of different species, especially when making comparisons across larger taxa.

Suggested Reading

Books

Bagnara, J. T., and Hadley, M. E. (1973). "Chromatophores and Color Change: The Comparative Physiology of Animal Pigmentation." Prentice-Hall, Englewood Cliffs, NJ.

Chester-Jones, I., Ingleton, P. M., and Phillips, J. G. (1987). "Fundamentals of Comparative Vertebrate Endocrinology." Plenum, New York.

Holmes, R. L., and Ball, J. N. (1974). "The Pituitary Gland, A Comparative Account." Cambridge Univ. Press, Cambridge.

Articles

Andersen, A. C., Tonon, M.-C., Pelletier, G., Conlon, J. M., Fasolo, A., and Vaudry, H. (1992). Neuropeptides in the amphibian brain. *Int. Rev. Cytol.* **138,** 89–210.

Ball, J. N. (1981). Hypothalamic control of the pars distalis in fishes, amphibians, and reptiles. *Gen. Comp. Endocrinol.* **44,** 135–390.

Barry, J. (1979). Immunohistochemistry of luteinizing hormone-releasing hormone-producing neurons of the vertebrates. *Int. Rev. Cytol.* **60,** 179–221.

Bern, H. A. (1983). Functional evolution of prolactin and growth hormone in lower vertebrates. *Am. Zool.* **23,** 663–671.

Dores, R. M., McDonald, L. K., Goldsmith, A., Deviche, P., and Rubin, D. A. (1993). The phylogeny of enkephalins: Speculations on the origins of opioid precursors. *Cell. Physiol. Biochem.* **3,** 231–244.

Fontaine, M., and Oliverau, M. (1975). Aspects of the organization and evolution of the vertebrate pituitary. *Am. Zool.* **15** (Suppl. 1), 61–81.

Gonzalez, G. C., Belenky, M. A., Polenov, A. L., and Lederis, K. (1992). Comparative localization of corticotropin and corticotropin releasing factor-like peptides in the brain and hypophysis of a primitive vertebrate, the sturgeon *Acipenser ruthenus L. J. Neurocytol.* **21,** 885–896.

Green, J. D. (1951). The comparative anatomy of the hypophysis with special reference to its blood supply and innervation *Am. J. Anat.* **88,** 225–311.

Hansen, G. N. (1983). Cell types in the adenohypophysis of the primitive actinopterygians, with special reference to immunocytochemical identification of pituitary hormone producing cells in the distal lobe. *Acta Zool. Suppl.* **1983,** 1–87.

Harrisson, F. (1988). Facts and hypotheses concerning the function of non-granulated cells in the adenohypophysis of vertebrates. *BioEssays* **8,** 168–171.

King, J. A., and Millar, R. P. (1992) Evolution of gonadotropin-releasing hormones. *Trends Endocrinol. Metab.* **3,** 339–346.

Lederis, K., Fryer, J. N., Okawara, Y., Schronrock, Chr., and Richter, D. (1994). Corticotropin-releasing factors acting on the fish pituitary: Experimental and molecular analysis. *In* "Fish Physiology, Molecular Endocrinology" (N. M. Sherwood and C. L. Hew, eds.), pp. 67–100. Academic Press, San Diego.

Lovejoy, D. A., Fischer, W. H., Ngamvongchon, S., Craig, A. G., Nahorniak, C. S., Peter, R. E., Rivier, J. E., and Sherwood, N. M. (1992). Distinct sequence of gonadotropin-releasing hormone (GnRH) in dogfish brain provides insight into GnRH evolution. *Proc. Natl. Acad. Sci. U.S.A.* **89,** 6373–6377.

Nozaki, M., and Gorbman, A. (1992). The question of functional homology of Hatschek's pit of amphioxus (*Branchiostoma belcheri*) and the vertebrate adenohypophysis. *Zool. Sci.* **9,** 387–395.

Peter, R. E. (1986). Vertebrate neurohormonal systems. *In* "Vertebrate Endocrinology: Fundamentals and Biomedical Implications. Volume 1, Morphological Considerations" (P. K. T. Pang and M. Schreibman, eds.), pp. 57–104. Academic Press, San Diego.

Pickford, G. E., and Atz, J. W. (1957). "The Physiology of the Pituitary Gland of Fishes." New York Zoological Society, New York.

Reiner, A. J. (1992). Neuropeptides in the nervous system. *In* "Biology of the Reptilia" (C. Gans and P. S. Ulinski, eds.), Vol. 17, pp. 588–739. Univ. of Chicago Press, Chicago.

Sawyer, C. H. (1978). History of the neurovascular concept of hypothalamo–hypophysial control. *Biol. Reprod.* **18,** 325–328.

Schreibman, M. P. (1986). Pituitary gland. *In* "Vertebrate Endocrinology: Fundamentals and Biomedical Implications. Volume 1, Morphological Considerations" (P. K. T. Pang and M. Schreibman, eds.), pp. 11–56. Academic Press, Orlando, FL.

Sherwood, N. M., Parker, D. B., McRory, J. E., and Lescheid, D. W. (1994). *In* "Fish Physiology, Molecular Endocrinology" (N. M. Sherwood and C. L. Hew, eds.), Vol. XIII, pp. 3–66. Academic Press, San Diego.

Specker, J. L., Ingleton, P. M., and Bern, H. A. (1984). Comparative physiology of the prolactin cell. *In* "Prolactin Secretion: A. Multidisciplinary Approach" (F. Mena and C. M. Valverde-R, eds.), pp. 17–30. Academic Press, New York.

Wallace, M. (1984). The molecular evolution of prolactin and related hormones. *In* "Prolactin Secretion: A Multidisciplinary Approach" (F. Mena and C. M. Valverde-R., eds.), pp. 1–16. Academic Press, New York.

6

Neurohormones
of the Pars Nervosa
and the Epithalamus

IN THIS chapter, we consider in some detail the hypothalamic nonapeptide neurohormones stored in the pars nervosa of the neurohypophysis (see Chapters 4 and 5) as well as some neurohormones produced by a derivative of the epithalamus, the **pineal gland** or **epiphysis.** The epithalamus represents the roof of the diencephalic portion of the brain, which has differentiated into a variety of secretory structures in vertebrates as well as into photo-receptors (Fig. 6-1). The pineal gland, one of these secretory structures, is also linked to an important hypothalamic nucleus, the **suprachiasmatic nucleus (SCN).** The pineal and the nearby **parapineal** are known as the **epiphysial complex.** Two additional prominent dorsal evaginations of the brain occur in the epithalamus: the **paraphysis** and the **dorsal sac.**

First, we will examine the nonapeptides of the pars nervosa in mammals. Next, we

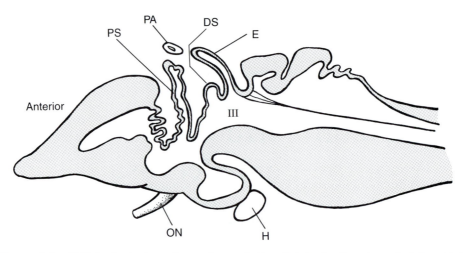

Figure 6-1. Epithalamic structures. Among the evaginations that develop from the roof of the epithalamus are the epiphysis cerebri or pineal (E), the parietal or parapineal organ (PA), the dorsal sac (DS), and the paraphysis (PS). ON, Optic nerve; H, hypophysis.

will consider the role of the pineal gland in mammals. Last, we will consider these two systems in nonmammals.

I. The Hormones of the Mammalian Pars Nervosa

In 1895, Oliver and Shafer reported a rapid increase in blood pressure following administration of pituitary gland extracts to anesthetized mammals. Later, this **pressor effect** was localized to the pars nervosa and the active principle was named **vasopressin** for its pressor effect on the vasculature. Pars nervosa extracts were also found to cause contraction of uterine smooth muscle, and this **uterotonic effect** was shown to reside in oxytocin. The ability of pars nervosa extracts to cause **milk ejection** from mammary glands also was due to the presence of oxytocin. Much later, oxytocin was found to induce a **depressor effect** (a reduction in blood pressure) in birds. All of these actions led to development of specific bioassays for vasopressin and oxytocin.

The pressor effect of vasopressin also caused diuresis through an increase in kidney blood pressure and glomerular filtration rate. Smaller doses of vasopressin that failed to produce an elevation in blood pressure were found to have the opposite effect on the kidney: antidiuresis. This **antidiuretic effect** is now believed to be the more important role for vasopressin in a physiological sense and is the basis for its alternative name, **antidiuretic hormone.** Vasopressin and possibly oxytocin may function as neurotransmitters or neuromodulators in other parts of the nervous system. Following its release into the hypothalamo–hypophysial portal system, vasopressin may be a physiological modulator that facilitates corticotropin (ACTH) release (see Chapter 4).

A. Chemistry of Mammalian Nonapeptides

Both vasopressin and oxytocin consist of nine amino acid residues organized into a ring structure of six residues joined through the disulfide bonds of two cysteine residues and

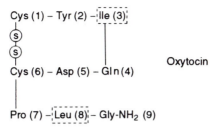

Figure 6-2. Oxytocin and arginine vasopressin. Comparison of primary structures of the common mammalian nonapeptides associated with the pars nervosa.

a side chain of three amino acids (Fig. 6-2). Vasopressin differs from oxytocin by having a phenylalanine at position 3 in the ring instead of isoleucine (Table 6-1) and an arginine at position 8 in the side chain (instead of leucine). This form of vasopressin is known as **arginine vasopressin (AVP).** The presence of the basic arginine makes AVP a basic molecule as contrasted to the neutral oxytocin nonapeptide. **Lysine vasopressin (LVP)** is a

Table 6-1. Structure of Neurohypophysial Nonapeptide Hormones in Vertebrates

Hormone	Amino acid sequence								
	1	2	3	4	5	6	7	8	9
Oxytocin	Cys -	Tyr -	Ile	- Gln	- Asn -	Cys -	Pro -	Leu	- Gly - NH₂
Mesotocin	Cys -	Tyr -	Ile	- Gln	- Asn -	Cys -	Pro -	Ile	- Gly - NH₂
Seritocin	Cys -	Tyr -	Ile	- Gln	- Ser -	Cys -	Pro -	Ile	- Gly - NH₂
Isotocin	Cys -	Tyr -	Ile	- Ser	- Asn -	Cys -	Pro -	Ile	- Gly - NH₂
Glumitocin	Cys -	Tyr -	Ile	- Ser	- Asn -	Cys -	Pro -	Gln	- Gly - NH₂
Valitocin	Cys -	Tyr -	Ile	- Gln	- Asn -	Cys -	Pro -	Val	- Gly - NH₂
Aspargtocin	Cys -	Tyr -	Ile	- Asn	- Asn -	Cys -	Pro -	Leu	- Gly - NH₂
Asvatocin	Cys -	Tyr -	Ile	- Asn	- Asn -	Cys -	Pro -	Val	- Gly - NH₂
Phasvatocin	Cys -	Tyr -	Phe	- Asn	- Asn -	Cys -	Pro -	Val	- Gly - NH₂
Arginine vasopressin	Cys -	Tyr -	Phe	- Gln	- Asn -	Cys -	Pro -	Arg	- Gly - NH₂
Lysine vasopressin	Cys -	Tyr -	Phe	- Gln	- Asn -	Cys -	Pro -	Lys	- Gly - NH₂
Phenypressin	Cys -	Phe -	Phe	- Gln	- Asn -	Cys -	Pro -	Arg	- Gly - NH₂
Arginine vasotocin	Cys -	Tyr -	Ile	- Gln	- Asn -	Cys -	Pro -	Arg	- Gly - NH₂

Table 6-2. Biological Activity of Neurohypophysial Peptides[a,b] in Various Bioassays

Hormone	Activity units[c]				
	Uterotonic (rat)	Depressor (chicken)	Milk ejection (rabbit)	Pressor (rat)	Antidiuretic (rat)
Oxytocin	450	450	450	5	5
Mesotocin	291	502	330	6	1
Isotocin	145	310	290	0.06	0.18
Glumitocin	10	—	53	0.35	0.41
Valitocin	199	278	308	9	0.8
Aspargtocin	107	201	298	0.13	0.04
Arginine vasotocin	120	300	220	255	260
Arginine vasopressin	17	62	69	412	465
Lysine vasopressin	5	42	63	285	260

[a] Based on data from Acher, R. (1974). Chemistry of the neurohypophysial hormones: An example of molecular evolution. *In* "Handbook of Physiology, Sec. 7, Endocrinology, Vol. 4, Part 1," pp. 119–130. Williams & Wilkins, Baltimore, MD.
[b] For distribution of these hormones among the vertebrates, see Table 6-3.
[c] Units are expressed in international units per micromole of pure synthetic substance; 1 mg of synthetic oxytocin = 500 USP units.

variant of AVP with a different basic amino acid, lysine, at position 8 (Table 6-1). LVP occurs in pigs, hippopotamus, peccaries (all in the mammalian order Suiformes), one mouse, and several marsupials. Another variant of AVP has another phenylalanine at position 2 as well as at position 3. This molecule, called **phenypressin (PVP),** occurs among marsupials. These basic molecules (AVP, LVP, PVP) tend to have good pressor and antidiuretic properties and weak uterotonic, milk-ejecting, and depressor activity (see Table 6-2). Oxytocin is strong in the latter bioassays, but has weak pressor and antidiuretic activity.

In fetal mammals, a nonapeptide molecule is found with the ring structure of oxytocin (isoleucine at position 3) and the basic side chain of AVP (arginine at position 8; see Table 6-1). The molecule was named **arginine vasotocin (AVT)** and it has good activity in all of the nonapeptide bioassays (see Table 6-2). No physiological role for AVT has been determined in fetal mammals, and it is absent in the pars nervosa of adults.

Nonapeptides are synthesized as part of larger propeptides. Early studies established the presence of additional peptides in pars nervosa extracts that lacked biological activity in the nonapeptide bioassays. These peptides were thought to play some carrier function to bring the nonapeptides to the secretory granules and were named **neurophysins.** The first stains specific for staining the secretions of the pars nervosa, allowing us to trace the neurosecretory neurons to their source in the hypothalamus, were actually staining the neurophysins. We now recognize two distinct propeptides, **prooxyphysin** (also called prooxytocin) and **propressophysin** (provasopressin) as the prohormones for oxytocin and the vasopressins, respectively (Fig. 6-3). When prooxyphysin is hydrolyzed, it yields oxytocin plus **neurophysin I,** a peptide of 92 amino acids. Hydrolysis of propressophysin yields vasopressin, **neurophysin II** (93 amino acids), and a short, unnamed glycopeptide (39 amino acids).

B. Hormones of the Nonmammalian Pars Nervosa

With the exception of the jawless agnathan fishes (cyclostomes), nonmammals, like mammals, typically produce two nonapeptides: one basic nonapeptide and a neutral one. In

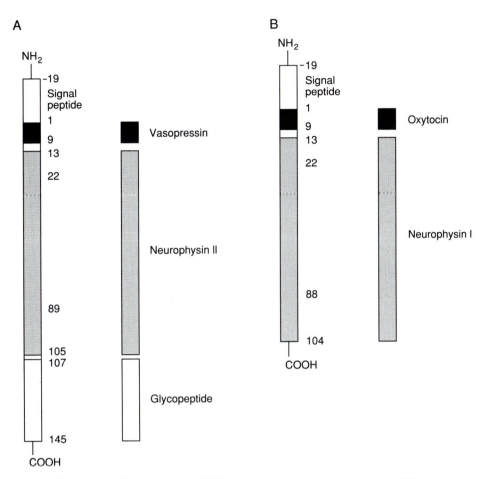

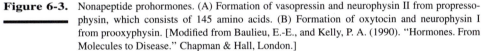

Figure 6-3. Nonapeptide prohormones. (A) Formation of vasopressin and neurophysin II from propressophysin, which consists of 145 amino acids. (B) Formation of oxytocin and neurophysin I from prooxyphysin. [Modified from Baulieu, E.-E., and Kelly, P. A. (1990). "Hormones. From Molecules to Disease." Chapman & Hall, London.]

virtually every case, the basic peptide is AVT, and this nonapeptide is the only one found in the cyclostomes. The vertebrate nonapeptides and their structures are provided in Table 6-1. Their distribution among the vertebrates and biological activities in the five standard bioassays are provided in Tables 6-2 and 6-3.

Numerous attempts to develop meaningful phylogenetic trees based on the distribution of the nonapeptides have met with limited success. One reason lies in the attempt to construct a phylogeny based on only a small fraction of the parent molecule, which is the direct gene product. Consequently, much of the genetic information has been overlooked in these schemes. However, a limited phylogenetic scheme based on their prohormones has been proposed (Fig. 6-4).

A second problem is the relatively small number of species in which the neurohypophysial nonapeptides have been identified with absolute certainty. The proliferation of forms in the chondrichthyean fishes suggests that we may eventually find a greater variety

Table 6-3. Neurohypophysial Nonapeptide Hormones in Vertebrates

Nonapeptide hormone	Where found
Arginine vasotocin	Agnatha, Chondrichthyes, Osteichthyes, Amphibia, Reptilia, Aves
Glumitocin	Chondrichthyes: Skates
Aspargtocin	Chondrichthyes: Sharks
Valitocin	Chondrichthyes: Sharks
Isotocin (ichthyotocin)	Osteichthyes: Actinopterygii
Mesotocin	Osteichthyes: Dipnoi, Amphibia, Reptilia, Aves, marsupials
Seritocin	Amphibia: Anura
Arginine vasopressin	Most mammals
Lysine vasopressin	Suiform mammals, marsupials
Phenypressin	Marsupials
Oxytocin	Mammals

of these molecules in much larger taxa such as the teleostean fishes or the reptiles than have been identified to date. Nevertheless, certain generalizations are still valid, and it is very likely that AVT represents the most primitive (and possibly ancestral) form of nonapeptide from which the others evolved. It also can be argued that the functional

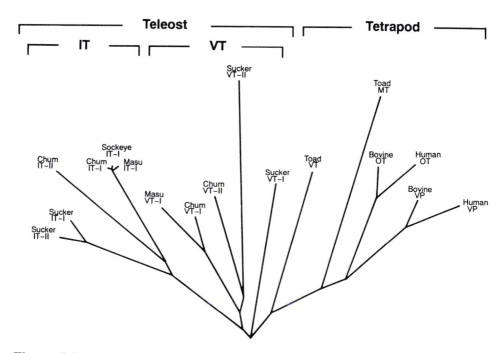

Figure 6-4. Phylogeny of nonapeptide prohormones, based on 119 amino acid residues in the prohormones. IT, Isotocin; VT, arginine vasotocin; MT, mesotocin; VP, vasopressin; OT, oxytocin. Two distinct gene groups have been identified in fishes for isotocin (IT-I and IT-II) and vasotocin (VT-I and VT-II). [From Urano, A., Kubokawa, K., and Hiraoka, S. (1994). Expression of the vasotocin and isotocin gene family in fish. *In* "Fish Physiology: Molecular Endocrinology of Fish" (N. M. Sherwood and C. L. Hew, eds.), pp. 101–132. Academic Press, San Diego.]

portion of the gene product might be more strongly influenced by selective forces than are the nonhormonal portions.

C. The Biological Actions of Vasopressin and Oxytocin

As we have already discussed, the main physiological role of vasopressin appears to be its antidiuretic action on the kidney, with a secondary role in elevating blood pressure through effects on vascular smooth muscle. Oxytocin causes contraction of reproductive tract smooth muscles in both males and females and contraction of myoepithelial cells lining the ducts of the mammary gland in females.

A brief examination of the structures of the nonapeptides reveals the bases for their biological activities. Antidiuretic and pressor activities apparently require a basic amino acid at position 8, and these actions are enhanced by phenylalanine at position 3. Peptides with neutral amino acids at position 8 exhibit predominantly oxytocin-like actions and appearance of isoleucine at position 3 enhances that effect (compare Tables 6-1 and 6-2).

As already mentioned, vasopressin and oxytocin may act as neurotransmitters or neuromodulators in the central nervous system and AVP affects release of ACTH from the pituitary. Furthermore, a number of metabolic actions for both neurohormones have been reported, but it is not clear how important these putative roles are.

1. Antidiuresis and Blood Vascular Effects: Vasopressins

In the kidney, blood pressure determines the glomerular filtration rate at which water and solutes are filtered through capillary tufts known as glomeruli and enter the nephrons as the glomerular filtrate. Normally, the filtrate lacks plasma proteins and the cellular components of the blood, and, initially, the concentrations of solutes such as Na^+ and glucose in the filtrate are identical to those in blood plasma. Numerous mechanisms operate to return solutes and most of the water back to the blood vascular system. If something interferes with the reabsorption process, a larger than normal volume of urine will be produced (i.e., diuresis). Should more reabsorption occur than normal, more fluid is reabsorbed and anti-diuresis results. Regulation of water reabsorption is critical to maintaining normal blood volume and blood pressure. Likewise, any marked changes in blood volume and concomitant changes in blood pressure will have effects on glomerular filtration rate and may lead to antidiuresis or diuresis.

Vasopressins act on the cells lining the collecting duct portion of the kidney tubules by increasing production of cAMP, which brings about changes in microtubules and micro-filaments that lead to an increase in permeability of these cells to water. Because of the high salt concentration maintained in the extracellular fluids of the medullary portion of the kidney, water readily moves from the lumen of the collecting duct, through the cells lining the kidney, and into the medullary extracellular compartment. From here, water moves osmotically into the blood, resulting in an increase in total blood volume and an increase in blood pressure. This reuptake of water from the glomerular filtrate reduces the volume of urine formed and elevates its solute concentration.

An increase in blood osmotic concentration and/or a decrease in blood pressure triggers release of vasopressin, which causes increased water reabsorption, antidiuresis, and a homeostatic decrease in blood osmotic concentration and/or elevation of blood pressure. Similarly, an increase in blood pressure and/or a decrease in blood osmotic concentration represses vasopressin secretion and causes diuresis with a corresponding drop in blood volume and pressure. Thus, vasopressin is important in the minute-to-minute regulation of blood volume and pressure as water uptake occurs in the intestine.

A secondary action of vasopressin occurs in the brain, where it stimulates thirst. Consumption of water will also add fluid to the blood vascular system and will increase blood pressure.

Chronic high blood pressure can cause the release of **atrial natriuretic peptide (ANP)** from the heart into the general circulation. ANP primarily produces its effects at the level of the kidney and at the adrenal cortex (see Chapter 9), accelerating sodium loss and hence promoting diuresis. In addition, ANP inhibits release of vasopressin and further accelerates diuresis in an attempt to compensate homeostatically for the high blood pressure.

Ethyl alcohol has an inhibitory effect on vasopressin release. Hence, consumption of alcoholic beverages produces a diuresis by increasing blood volume, blood pressure, and glomerular filtration rate, as well as through reduction in the efficiency of the reabsorption of water in the kidney. This inhibition may be prolonged so that excessive dehydration occurs, contributing to production of severe headache.

High doses of vasopressin cause contraction of arteriole smooth muscle and elevate blood pressure. This, in turn, increases the glomerular filtration rate sufficiently to override the normal antidiuretic action of the hormone and produces a net diuresis. Secretion of sufficient vasopressin to bring about arteriole constriction and attendant diuresis probably occurs following severe drops in blood pressure such as following hemorrhage. Neural regulation as well as other endocrine regulators of blood pressure, such as the angiotensins (see Chapter 9), may be more important regulators than vasopressin, however.

The hypothalamus also is the source of AVP that potentiates ACTH release from the adenohypophysis, but, in this case, it is released directly into the hypothalamo–hypophysial portal system and does not come from the pars nervosa. This effect on ACTH release apparently is very important in prolonging the stress response (see Chapter 9).

2. Uterotonic and Milk-Ejection Activities: Oxytocin

The contraction of uterine smooth muscle caused by oxytocin was almost immediately recognized as a methodology for artificially inducing labor in women, and it has been employed extensively for this purpose as well as to stimulate clamping down of the uterine muscles after birth to reduce postpartum bleeding. It was many years, however, before it was proven that oxytocin actually participates in the natural induction of labor in humans and other mammals.

Oxytocin not only produces rhythmic contractions in the female reproductive tract, but it is responsible for contraction of the vas deferens and epididymis during ejaculation. The sensation of orgasm, which involves rhythmic contractions of reproductive smooth muscle in both men and women, is induced by oxytocin as well.

In women, cows, and other female mammals, release of milk from the mammary glands is also induced by oxytocin. Exposure of mammary glands to estrogens causes development of the glands and the myoepithelial cells that line the ducts. Prolactin stimulates milk synthesis (see Chapter 4), and oxytocin causes contraction of the myoepithelial cells to cause milk ejection.

Release of oxytocin is brought about through a neuroendocrine reflex. Suckling of the newborn on the nipple sends neural impulses to the brain, which reach the hypothalamus and direct release of oxytocin from the pars nervosa into the general circulation. Sufficient oxytocin in the circulation also can stimulate oviduct contractions leading, in the mother, to a pleasurable sensation during suckling.

Like vasopressin, oxytocin has been shown to have effects on tropic hormone release. However, no distinct role for oxytocin as a releasing hormone has been verified.

II. The Epiphysial Complex

Almost all vertebrates exhibit one or two epithalamic structures that constitute the epiphysial complex (Fig. 6-5). The components of this complex are the pineal organ and a more anterior projection, the parapineal organ. In fishes, amphibians, and some reptiles (lizards), these organs are basically saclike diverticula that are more or less open to the third ventricle of the brain. They consist of a basal portion composed of sensory and ependymal (supportive) cells and may have an attached stalk with a distal end vesicle that contacts the dorsal brain case. These structures possibly arose in primitive fishes as a pair of diverticula that for some reason later changed positions relative to one another. A well-developed parapineal organ has been retained only in cyclostomes and lizards, whereas the pineal organ is found in all vertebrate groups with the exception of crocodilians. In anamniotes, as well as

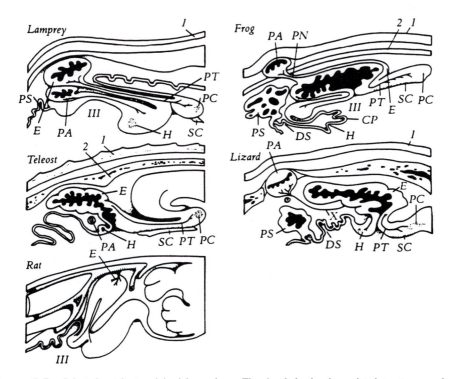

Figure 6-5. Selected vertebrate epiphysial complexes. The pineal gland and associated structures are shown for a lamprey, teleost fish, frog, lizard, and rat. 1, Skin; 2, skull; III, third ventricle; CP, choroid plexus; DS, dorsal sac; E, epiphysis cerebri or pineal organ; H, habenular commisure; PA, parietal or parapineal organ; PC, posterior commissure; PN, pineal nerve; PS, paraphysis; PT, pineal tract; SC, subcommissural organ; X, parietal nerve.

in lizards, the pineal organ has retained its sensory functions. However, in other reptiles as well as birds and mammals, the pineal appears to be only an endocrine structure and often is termed simply the pineal gland.

The anteriormost evagination that actually develops from the telencephalon is known as the paraphysis. The paraphysis is best seen in amphibians as a highly vascularized, saclike diverticulum (Fig. 6-5); it may function similarly to the choroid plexus in producing cerebrospinal fluid. The dorsal sac arises in most vertebrates as an epithalamic evagination just posterior to the paraphysis but anterior to the epiphysial complex. It is especially prominent in the ganoid fishes (chondrosteans, holosteans) and becomes less conspicuous in teleostean fishes. In most vertebrates the dorsal sac contributes to formation of the choroid plexus, and these structures may be indistinguishable in the adult.

The pineal complex is connected to an adjacent ependymal structure, the **subcommissural organ** of Dendy (SCO). The ependymal cells of the SCO produce an aldehyde fuchsin-positive secretion rich in disulfide bonds and cysteine. This secretion is similar to that observed in the pineal ependyma. The major secretory product of the SCO is a noncellular fiber that in some species extends into the central canal of the spinal cord for its entire length. This structure is known as **Reissner's fiber.** Its significance is not clear. The SCO and its Reissner's fiber have been described in vertebrates from cyclostomes to mammals. Once it was supposed that Reissner's fiber was involved in regulating posture through tension produced in it by flexion of the body. This tension presumably operated through influences of Reissner's fiber on pressure-sensitive neurons. A more plausible suggestion is the possibility that Reissner's fiber contributes to formation of cerebrospinal fluid. Formation and dissolution of the fiber into the cerebrospinal fluid have been documented as a temperature-dependent process in the frog, *Rana esculenta.* Reissner's fiber also binds biogenic amines (epinephrine, norepinephrine) present in the fluid in both *R. esculenta* and in mammals (cow, cat), suggesting still another role. Studies with mammals and reptiles imply the existence of a functional relationship among the pineal complex, the SCO, and the adrenal cortex, but the nature of that relationship remains somewhat obscure. Cytological activation of the SCO in the lizard *Lacerta s. sicula* has been correlated positively with seasonal activities of adrenal cortical cells and of testicular steroidogenic cells. The actual role or roles for the SCO and its secretory products must await further research, but preliminary data would suggest it is somehow related to activity of the pineal complex.

The human pineal gland was described by Galen during the second century as a structural (supportive) element within the brain. Much later, in 1646, René Descartes described it as a small gland in the brain in which "the soul exerted its function more particularly than in any other part." It was, however, three centuries later that scientists determined what it was that the soul was doing through the pineal gland. McCord and Allen in the early 1900s observed that pineal extracts caused blanching (lightening of the skin) of amphibian larvae by causing a concentration of melanin within the melanophores.

A. The Pineal Gland and Melatonin

Many years after the observation that pineal extracts caused the skin of frogs to lighten, Lerner and coworkers succeeded in isolating and characterizing the active skin-lightening agent, 5-methoxyl-*N*-acetyltryptamine or **melatonin** (see Chapter 2, Fig. 2-5). Since that time, a number of biologically active indolamines and related compounds have been isolated from pineal tissue, including **serotonin (5-hydroxytryptamine, or 5-HT),** *N*-acetyl-

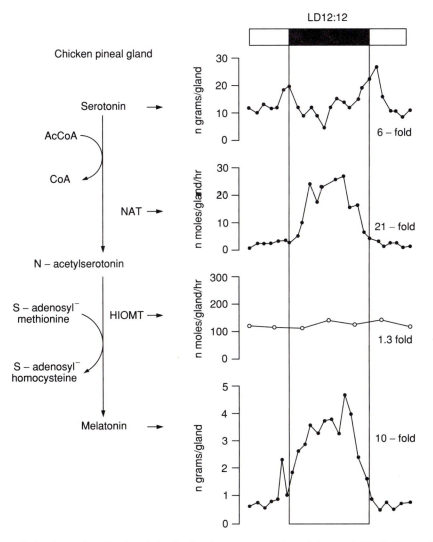

Figure 6-6. Serotonin and melatonin levels related to enzyme activity and photoperiod. Both the quantity of melatonin and activity of the rate-limiting enzyme, *N*-acetyltransferase (NAT), in the pineal gland of chickens exhibit increases during the scotophase. The enzyme hydroxyindole-*O*-methyltransferase (HIOMT) does not exhibit a diurnal rhythm. Levels of serotonin appear to be correlated with onset of both the scotophase and the photophase. Similar observations have been made in mammals. [From Binkley, S. A. (1990). "The Clockwork Sparrow: Time, Clocks, and Calendars in Biological Organisms." Prentice-Hall, Englewood Cliffs, New Jersey.]

serotonin, 5-methoxytryptophol (5-MTP), and **5-hydroxytryptophol (5-HTP).** Both melatonin and serotonin are secreted by the pineal and each has its unique actions.

The initial substrate for synthesis of the indoleamines is the amino acid tryptophan, which is converted to serotonin and then to melatonin (Fig. 6-6). Two enzymes have been studied extensively with respect to their role in regulating melatonin synthesis: ***N*-acetyltransferase (NAT)** and **hydroxyindole-*O*-methyltransferase (HIOMT).** NAT converts 5-HT to *N*-acetylserotonin, which in turn is converted to melatonin by HIOMT.

The conversion of 5-HTP to 5-MTP is also catalyzed by HIOMT. Another enzyme complex, **monoamine oxidase (MAO),** acts on 5-HT, converting it to its inactive metabolite, **5-hydroxyindoleacetic acid (5-HIAA).** Thus, two enzymes, NAT and MAO, compete for the same substrate, 5-HT, and their relative activities could represent mechanisms for regulating melatonin synthesis.

Early observations suggested that HIOMT was the rate-limiting enzyme for melatonin synthesis, and several studies describe correlations between HIOMT activity and melatonin synthesis. More recent studies, however, have established NAT as the rate-limiting enzyme in melatonin synthesis.

Both mammalian and avian pineals are innervated by sympathetic fibers, and variations in norepinephrine synthesis are correlated with pineal functions. Circadian rhythms in tyrosine hydroxylase activity, the rate-limiting enzyme for norepinephrine synthesis, have been reported in rats, and the rhythm associated with this enzyme correlates positively with observed rhythms in HIOMT activity and NAT activity.

Melatonin has been shown to produce a rather limited number of cellular effects. The major enzymes influenced by melatonin are those invoked in steroid and amine transformations, for example, 5α-reductase and MAO, respectively. The actions of melatonin on MAO activity and consequent effects on tissue 5-HT levels may explain effects observed on pancreas, pituitary, testes, and the pineal itself following administration of melatonin.

B. Functions of the Pineal Gland in Mammals

A central action for the pineal gland is the regulation of endogenous rhythms, a role probably related to the primitive role of the epiphysial complex as a photoreceptive organ. The mammalian pineal gland also has been implicated as an inhibitor of reproductive and thyroid activity. The pineal has come under scrutiny as a potential regulator of aging and of the immune system. Some actions of melatonin are summarized in Table 6-4.

1. The Pineal Gland and Endogenous Rhythms

Melatonin levels in the blood exhibit a distinct diurnal rhythm, being greater at night than during the day in almost every species examined. This circadian rhythm usually persists

Table 6-4. Summary of Nonreproductive Actions of Melatonin and the Pineal in Mammals

Target	Description of action
Melanophores (melanocytes)	Melatonin implants in weasels, *Mustela erminea,* cause them to grow white coats (typical of winter) in the spring instead of brown coats
Hair	Melatonin inhibits hair growth of intact or pinealectomized mice
Connective tissue	Pinealectomy reduces permeability of subcutaneous connective tissue
Adrenal cortex	Pineal substance, adrenoglomerulotropin, claimed to stimulate aldosterone release Pineal may alter release of ACTH from adenohypophysis
Parathyroid	Pinealectomy of rat caused hypertrophy of parathyroids, which was reduced by administration of pineal extract or melatonin
Cardiovascular system	Vasopressor activity reported for pineal extracts, probably due to presence of AVT
Immune response	Chronic administration of pineal extracts caused leukocytosis, lymph node hypertrophy, and an increase in mitotic activity in the spleen. Probably it was a simple immunological response to antigens in the extract
Thyroid	Melatonin or pineal extracts inhibit thyroid function, possibly through regulation of TSH release from the adenohypophysis

under constant dark conditions. Plasma melatonin rhythm is a consequence of a circadian rhythm in NAT activity (Fig. 6-6). This enzymatic rhythm is controlled by neural signals from the suprachiasmatic nucleus (SCN) of the hypothalamus. Information on photoperiod detected by the retina is responsible for entraining the SCN to light/dark cycles. The SCN probably controls a number of circadian rhythms in mammals. Blood levels of melatonin are greatest during the dark portion of the day. Nighttime melatonin secretion occurs in three basic patterns showing differences in latency of the response after light disappears or in the temporal relationship to when light returns (Fig. 6-7).

Most studies of pineal activity and light have been done in rodents, which show greater sensitivity to light than do large mammals such as humans and sheep. Nevertheless, similar mechanisms appear to be operating in most species. Light stimulates the retina of the eye, which sends impulses via two pathways to alter pineal secretion (Fig. 6-8). The retinohypothalamic pathway innervates the SCN, which in turn operates through the brain stem and spinal cord to reduce the activity of sympathetic fibers traveling from the **superior cervical ganglion (SCG)** to the pinealocytes of the pineal gland. Normally, these postganglionic fibers release norepinephrine (NE), which increases cAMP in the pinealocytes. Elevated cAMP is associated with increased activity of NAT and subsequent melatonin synthesis. Light shining on the retina reduces cAMP levels, NAT activity, and melatonin synthesis via this pathway. Activity of HIOMT also is reduced slightly by exposure of the animal to light.

Lesions in the retinohypothalamic pathway or the SCN do not necessarily abolish pineal secretory rhythms. A second pathway has been discovered between the retina and the brain stem that travels via the inferior accessory optic tract (Fig. 6-8). There is as yet no adequate explanation for this redundancy.

2. Pineal Secretions and Reproduction

The major effects of the pineal gland in mammals relate to reproduction and are most pronounced in species that breed only during spring or fall. In 1941, Fiske reported that keeping rats under conditions of constant light increased the frequency of estrus (a time of enhanced female receptivity to the male). Several years later, Wurtman discovered that pinealectomy also increased the frequency of estrus in rats maintained under normal photoperiods, and a surge of investigation was launched into possible roles of photoperiod, the pineal gland, and melatonin in controlling sexual maturity and reproductive cycles in mammals. A mass of data soon appeared to suggest that melatonin released from the pineal gland acted through either the blood or cerebrospinal fluid on the hypothalamus or directly on the pituitary to lower circulating LH levels. As mentioned above, light inhibits sympathetic input to the pineal, resulting in decreased melatonin synthesis followed by increased levels of LH leading to estrus.

Considerable evidence has accumulated from clinical studies that implicates the pineal gland in controlling the onset of puberty in humans, although little experimental work on humans is available. Circulating melatonin decreases by 75% between the ages of 7 and 12 years, when LH levels are observed to rise. Furthermore, many cases of precocious puberty, especially in males, are associated with nonsecreting pituitary tumors that presumably allow for an early release of luteinizing hormone (LH) and consequent gonadal stimulation (for additional information on puberty and reproduction, see Chapter 11).

Many studies of the relation between the pineal and gonadal function have been conducted with the golden hamster, *Mesocricetus auratus*. This animal exhibits marked gonadal collapse when subjected to short photoperiods (less than 12 hr of light per day) when the daily period of elevated plasma melatonin is longest. Pinealectomized golden hamsters

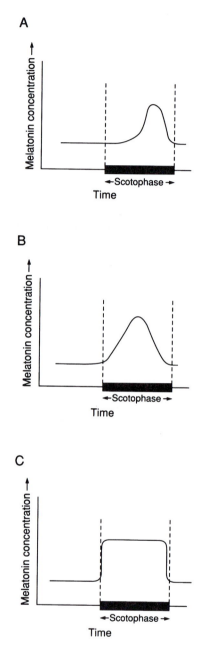

Figure 6-7. Patterns of melatonin secretion. Three distinct secretory patterns for nocturnal secretion of melatonin are known. (A) Increase observed only during second half of the scotophase (house mouse, Syrian hamster, Mongolian gerbil); (B) most common pattern with secretion beginning soon after onset of darkness, peaking at midscotophase, and decreasing prior to resumption of the photophase; (C) reaches highest level as soon as lights go off and remains high until just before light returns (Djungarian or Siberian hamster; domestic sheep). [Modified from Reiter, R. J. Pineal melatonin: Cell biology of its synthesis and its physiological interactions. *Endocr. Rev.* **12**, 151–180, 1991. © The Endocrine Society.]

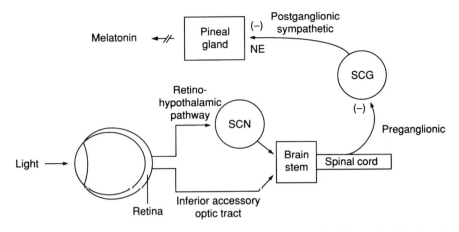

Figure 6-8. Pathway whereby light inhibits pineal secretion in mammals. Light received by the retina inhibits pineal function via two central pathways, both of which reach the pineal through the superior cervical ganglion (SCG) of the sympathetic system. Secretion of less norepinephrine (NE) by postganglionic sympathetic neurons reduces cAMP formation and causes a marked reduction in activity of *N*-acetyltransferase (NAT), the rate-limiting enzyme for melatonin synthesis.

do not exhibit gonadal collapse when subjected to short photoperiods, and subdermal melatonin implants (administered in Silastic capsules) cause testicular atrophy in hamsters maintained on long photoperiods.

Some studies have provided a different explanation for the effects of light on estrus. Pinealectomy or injection of massive doses of melatonin never produces marked effects on rat reproduction, and some workers have not found any effect of melatonin on rat reproduction. One study, in fact, reported stimulation of rat gonads by melatonin treatments. If rats are made anosmic (olfaction blocked either mechanically or surgically) and are blinded, more marked gonadal atrophy occurs than was seen following pinealectomy.

For many mammals, longer photoperiods stimulate gonadal development only at certain times and at other times the animal does not respond; that is, it is **photorefractory.** Many animals exhibit **photorefractoriness** after breeding and cannot be induced to reenter estrus. Support for a stimulatory gonadal role for melatonin has been reported by Thorpe and Herbert for the ferret, in which melatonin may be responsible for bringing the animal out of photorefractoriness. Melatonin treatment restores the gonadal growth response to long photoperiods in photorefractory (postbreeding) ferrets maintained on artificially long photoperiods. In Djungarian hamsters, photorefractoriness appears in a different form. Short photoperiods cause gonadal collapse due to inhibitory effects of elevated melatonin secretion on the reproductive endocrine axis. However, after a period of continued short-day exposure, the animals exhibit photorefractoriness to the short photoperiod, and the gonads undergo recrudescence (regrowth).

Ependymal cells of fetal human and rat pineals synthesize **arginine vasotocin (AVT),** which also has been found in adult pineals of some species. If AVT is administered to neonatal mice during the period when the brain is undergoing sexual differentiation, increased growth of reproductive organs on entering adulthood is observed. In contrast, if AVT is administered after the brain has undergone sexual differentiation, the growth of accessory organs and in some cases growth of the gonads themselves is inhibited.

The hypertrophy of the remaining ovary after unilateral ovariectomy of mice is a response to increased gonadotropin levels caused by an effective reduction in circulating estrogens (a feedback effect). Arginine vasotocin administered intraperitoneally or directly into the third ventricle of the brain prevents this **compensatory ovarian hypertrophy (COH).** Much less AVT is required if it is administered through the third ventricle than if it is given intraperitoneally. Several related basic nonapeptides, including arginine vasopressin (AVP), lysine vasopressin (LVP), and [4]Leu-AVT (Leu replaces Gln at position 4), also inhibits COH. Oxytocin, a neutral nonapeptide, is ineffective. All of the active compounds have an identical ring structure and a basic amino acid at position 8. Treatment of these active molecules with 2-mercaptoethanol disrupts the disulfide bridges necessary for maintenance of the ring structure. Such reduced nonapeptides no longer prevent COH; in fact, they enhance it.

Arginine vasotocin exhibits effects on 5α-reductase and MAO activities similar to those reported for melatonin. In fact, AVT may be the physiological regulator instead of melatonin. An antigonadotropic peptide (which is not AVT) has been isolated from bovine pineals, and it prevents LH release from the pituitary. Release of LH or FSH from rat adenohypophysial cells in culture, however, is not influenced by AVT over a concentration range of 10^{-18} to 10^{-7} mol/liter of culture medium, supporting a role for AVT at the level of the hypothalamus.

The presence of other hypothalamic peptides including oxytocin, AVP, thyrotropin-releasing hormone (TRH), and somatostatin has been demonstrated in human pineals. In addition to sympathetic innervation coming indirectly from the SCN through the superior cervical ganglion, numerous peptidergic fibers are found in the pineal gland that originate in other brain regions (e.g., paraventricular and habenular nuclei). Immunoreactive vasoactive intestinal peptide (VIP), neuropeptide Y (NPY), AVP, and oxytocin have been demonstrated in pineal nerve endings. Receptors for VIP are present on pinealocytes and binding of VIP activates a cAMP-mediated increase in NAT function. This mechanism enhances the stimulation of cAMP and NAT by NE.

3. The Pineal Gland and Other Tropic Hormones

In addition to the inhibition of gonadotropin secretion, melatonin may have important influences on secretion of other tropic hormones as well (Fig. 6-9). Evidence for effects of melatonin on thyrotropin (TSH), prolactin, and ACTH release is less convincing than for gonadotropin release and more studies are needed.

Thyroid function in at least some mammals is strongly affected by photoperiod, which appears to be acting through the control of melatonin secretion. Melatonin treatment reduces thyroid function, presumably by limiting hypothalamic secretion of TRH and not by a direct action at the thyrotropic cells of the adenohypophysis.

The rise in prolactin release observed in rats at the onset of the photophase has been linked to a reduction in melatonin release. Long photoperiods are correlated with increased prolactin secretion in ruminant ungulates (sheep, cattle, goats), and melatonin treatment decreases prolactin secretions in both sheep and goats but not in cattle. Although pinealectomy or denervation of the pineal sometimes produces increased prolactin secretion in goats and sheep, pinealectomy has no effect in cattle. Similarly, pineal activity is associated with seasonal breeding in sheep and goats but is not associated with reproduction in cattle. This difference among ungulates may be due to artificial selection as a result of selective breeding of cattle over many centuries.

Evidence for an effect of melatonin on ACTH release relies mainly on a few observa-

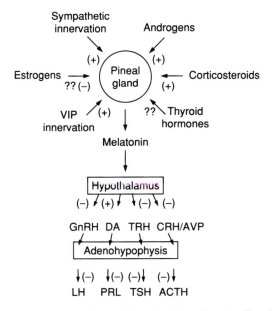

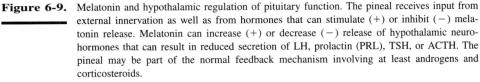

Figure 6-9. Melatonin and hypothalamic regulation of pituitary function. The pineal receives input from external innervation as well as from hormones that can stimulate (+) or inhibit (−) melatonin release. Melatonin can increase (+) or decrease (−) release of hypothalamic neurohormones that can result in reduced secretion of LH, prolactin (PRL), TSH, or ACTH. The pineal may be part of the normal feedback mechanism involving at least androgens and corticosteroids.

tions. Exogenous corticosteroids stimulate melatonin release, which in turn might be involved in reducing ACTH release as part of a negative feedback loop.

4. The Pineal Gland and Aging

Biological aging is a complex phenomenon that is often complicated by attendant pathologies. There are many theories of the causes and progression of aging and no single explanation seems adequate. One popular theory of aging involves the formation and accumulation of **free radicals,** compounds that can interact with and damage in particular proteins, phospholipids, nucleic acids, and sugars. One of the most dangerous free radicals is produced during the breakdown of hydrogen peroxide. The presence of melatonin in certain *in vitro* systems reduces free radicals whereas 5-HT, another pineal secretion, elevates free radicals. In a number of pathological disorders (including Parkinson's disease, atherosclerosis, muscular dystrophy, multiple sclerosis, and rheumatoid arthritis), free radicals are responsible for cell damage. As a mammal ages, the SCN, which sends important regulatory messages to the pineal as well as to cells controlling pituitary function, becomes dysfunctional and the pineal gland reduces production of melatonin and elevates production of 5-HT.

   Another influence of the pineal on aging may be related to effects of melatonin on the immune response system. Melatonin appears to enhance the functioning of the immune response system and has been claimed to increase immune surveillance and decrease the incidence of cancer. In contrast, 5-HT of pineal origin may impair immune functions. Since decreased immune functions in general are associated with aging, melatonin may have a dual retarding effect on the aging process (e.g., by reducing free radicals and enhancing immune surveillance).

C. Other Factors Affecting the Pineal Gland

In addition to the well-known actions of light on the pineal gland, hypophysectomy, stress, and gonadal steroids all influence pineal function. Androgens (e.g., testosterone, dihydro-testosterone) inhibit MAO activity in the pineal, which in turn allows for increased mela-tonin synthesis. Increased melatonin can reduce LH release and in turn decrease androgen levels. Estrogens increase MAO activity and decrease pineal secretion of melatonin, bring-ing about enhancement of gonadotropin release. Hypophysectomy or administration of his-tamine also reduces melatonin synthesis by reducing activity of HIOMT.

Acute stress stimulates pineal activity, presumably through enhanced sympathetic stimu-lation. Chronic stress (for example, starvation) is associated with elevated corticosteroids, which reduce pineal MAO activity and allow for increased melatonin production. Thus, stress working through these two pathways can depress reproductive function via the pineal gland.

D. Extrapineal Sources of Melatonin

The **harderian gland,** or Harder's lacrimal gland, was described by Harder in 1694 in the red deer. It is located directly behind and around the eye in all vertebrates that possess nictitating membranes (reptiles, birds, and most mammals). In humans, the harderian gland is rudimentary. Reddish porphyrin pigments present in the harderian gland undergo fluc-tuations correlated with lighting conditions. Prior to 12 days of age, the rat harderian gland contains little porphyrin pigment. Blinded 12-day-old rats exhibit an increase in pineal 5-HT as well as HIOMT activity during the scotophase (dark portion of the photoperiod cycle), but this rhythm is abolished if the harderian glands also are removed. Melatonin has been demonstrated in the rat harderian gland, and continuous illumination causes enlarge-ment of the rat harderian gland and an increase in HIOMT activity. Harderian HIOMT differs somewhat from the HIOMT found in the pineal gland and from that found in the retina of the eye. In contrast, continuous illumination decreases pineal weight and pineal HIOMT activity. The importance of these observations to overall involvement of pineal indoles or harderian indoles to observations on reproduction or other pineal-influenced processes remains to be determined.

The retina of the eye may be another viable source for melatonin, but the nocturnal increase in circulating melatonin is clearly of pineal origin. Melatonin also has been found in the colon of rats, but the significance of this observation is unknown.

E. Comparative Aspects of the Epiphysial Complex

Pineal function has been studied extensively in fishes and nonmammalian tetrapods, espe-cially with respect to photoperiodic stimuli and daily and seasonal biological rhythms. Examination of these groups of vertebrates reveals a progressive decrease in the reliance of the pineal as a photoreceptor and an increasing dependence on the lateral eyes and the SCN–pineal pathway to control photoperiodically linked events such as seasonal repro-

Table 6-5. Effects of Melatonin Treatment in Vertebrates

Effect	Vertebrate group					
	Agnatha	Osteichthyes	Amphibia	Reptilia	Aves	Mammalia
Concentration of melanosomes or inhibition of melanin synthesis	Yes	Yes	Yes	Yes		Yes
Preferred temperature		Increase	Decrease	Decrease		
Thermogenic effect on body temperature				Decrease	Decrease	Increase
Gonad function						
Adults		Stimulate/ inhibit	Inhibit	Inhibit	Inhibit	Stimulate/ inhibit
Juvenile					Stimulate	
Thyroid function			Inhibit			Inhibit

duction via the secretion of melatonin. A summary of some of the actions of melatonin in vertebrates is provided in Table 6-5.

1. Agnathan Fishes: Cyclostomes

The epiphysial complex of cyclostomes consists of a pineal organ and a parapineal organ (Fig. 6-5), although the latter may be lacking in some species. Both organs possess variably shaped end vesicles that project dorsally against the roof of the brain case. The end vesicles contain sensory and ependymal cells structurally organized to suggest that they are photoreceptors. Efferent neural fibers from the pineal organ end at the posterior commissure, whereas those from the parapineal terminate at the habenular commissure. These organs probably relay photoperiodic information to other regions of the brain.

Nocturnal blanching has been observed in the ammocetes larvae of some lampreys, and epiphysial levels of HIOMT are correlated with nocturnal lightening in *Geotria australis.* Removal of the epiphysial complex causes persistent expansion of melanophores in *Lampetra planeri* and *G. australis* but not in *Mordax mordacia,* which lacks a parapineal organ and does not exhibit nocturnal lightening. Thus, melanophore expansion may be mediated via the parapineal. Hypophysectomy of cyclostomes also results in blanching due to the absence of melanotropin (*α*-MSH), which normally stimulates melanophore expansion. The role of the parapineal may be to inhibit release of melanotropin, but melatonin could be the factor responsible for the observed melanophore contraction.

There is evidence for a possible antithyroid effect of the pineal organ. Pinealectomy inhibits metamorphosis of the ammocetes in all of the above-mentioned lamprey species, but additional data are needed to establish the site of action for pineal secretions.

2. Chondrichthyean Fishes

The shark pineal organ contains photosensory and supportive cells. No experimental data, however, have been reported, and the functional importance of the pineal is not known for these fishes. A parapineal organ has not been described for any species in this group. This vertebrate taxon demands further investigation.

3. Bony Fishes: Teleosts

The epiphysial complex of teleosts consists of a pineal organ (Fig. 6-4) that is extremely variable in both size and degree of development. There is usually a prominent lumen, which in some cases is open to the third ventricle. Photosensory cells and ependymal cells are present in the pineals of several species. A reduced parapineal has been described in some teleosts.

Pigmentation, responses to light, and thyroid changes are influenced by the pineal organ in teleosts. Melanophore changes have been reported in several species of teleost, including rainbow trout, following injection of pharmacological amounts of melatonin. Some of these species also respond in a similar manner to epinephrine treatment. Other species respond only to epinephrine with no reaction to melatonin. Circulatory melatonin levels exhibit no correlations to adaptation by rainbow trout to different backgrounds, indicating that melatonin may not be a factor influencing normal pigmentary responses in this species. Furthermore, levels of HIOMT in rainbow trout pineals are not altered by either continuous light or darkness.

William Gern and colleagues were the first to demonstrate that direct control of pineal secretion is determined by the presence or absence of light in teleosts and not by some endogenous mechanism or rhythm. Rainbow trout pineal glands *in vitro* show no endogenous secretory rhythm in melatonin secretion and release melatonin only when in darkness regardless of the photic regimen (i.e., long photophase, short photophase, or continuous light or darkness). Similar observations later were reported for goldfish pineals by Japanese workers. It appears that the pineal glands of some teleostean fishes respond directly to light and may not be controlled through the lateral eyes and the SCN pathway as shown for mammals. However, it would be premature to extend these observations since the pineals of some teleosts do exhibit an endogenous, light-independent secretory pattern when examined *in vitro.*

Numerous studies by Martin Kavaliers and co-investigators have focused on the responses of the white sucker, *Catostomus commersoni,* to light intensity and/or thermal gradients. This sensitivity is mediated through the pineal. Shielding this sucker's pineal from light causes the fish to choose the warmer portion of horizontal temperature gradients or the better illuminated portions of chambers maintained at a constant temperature. When the shield is removed, the fish returns to its original preference. The pineal may provide information on light intensity for use in mediating thermal behavior. When there is no temperature gradient, retinal receptors alone determine the location of the fish.

Structural correlations have been described between the pineal organ and phototactic responses by fishes. Species with a translucent covering over the pineal organ (a definitive pineal spot) exhibit predominantly positive phototaxis, whereas species with a pigmented, opaque skeletal covering do not show phototaxis. Species that have pigment cells located so that dispersal and concentration of pigment granules could regulate the intensity of light reaching the pineal organ exhibit responses varying from positive phototaxis to no response. The pineal organ might be involved in some other functions for the species showing no phototaxis. The presence of a pineal spot is more common in deep-sea fishes than in freshwater or shallow-water marine species and may relate to the influence of light on vertical migrations performed by deep-sea fishes.

Pinealectomy of *Poecilia reticulata* (guppy) causes pituitary enlargement and hyperplasia of the thyroid. This effect also occurs in *Fundulus heteroclitus* if pinealectomy is performed during the winter months (December through March). Pinealectomy between February and June produces no effect on the thyroid, however. Pituitary and thyroid of the characin, *Astyanax mexicanus,* are not affected by pinealectomy, but both stimulation and inhibition of the goldfish thyroid have been described. A possible influence of the pineal on gonadal development has been reported for *F. heteroclitus* and *F. similis,* but no relationship was found in goldfish or in *A. mexicanus.* Additional studies performed on a seasonal basis involving a large number of species and paying close attention to photoperiodic regimens are needed before any definitive statements can be made with respect to the pineal and reproductive or thyroid functions.

Melatonin has been measured in the retina of rainbow trout, and HIOMT is present in the retinas of several teleostean species. Levels of melatonin in trout retina exceed levels reported in the pineal. Melatonin may cause concentration of pigment in retinal melanophores to increase sensitivity of retinal cells in dim light (see Section II,F).

4. Amphibians

The epiphysial complex of amphibians consists of a pineal organ, dorsal sac, choroid plexus, SCO, paraphysis, and often a parapineal organ (Fig. 6-4). The proximal or basal portion of the amphibian epiphysial complex as well as the retina of the eye contain HIOMT activity and melatonin. Pinealectomy reduces circulating melatonin to daytime levels, suggesting that the retina may be responsible for basal levels of melatonin in the blood.

As mentioned earlier, the role of melatonin in the melanophores of larval amphibians was first suggested by the observations of McCord and Allen. Since that time, it has been shown that melatonin is the pineal agent responsible for the blanching of tadpoles or larval salamanders when held in the dark. As little as 0.0001 μg/ml medium causes aggregation of melanin granules in melanophores of *Xenopus laevis* tadpoles.

Attempts to relate pineal function or melatonin with thyroid function have been equivocal in amphibians. Pinealectomy of tadpoles of the midwife toad *Alytes obstetricans* accelerates metamorphosis, but a similar operation in larvae of the newt *Taricha torosa* is without effect. Earlier observations in *Bufo americanus,* however, indicate that feeding mammalian pineal to tadpoles accelerated metamorphosis. Pinealectomized larval tiger salamanders, *Ambystoma tigrinum,* exhibit decreased thyroidal uptake of injected radioiodide, but neither purified melatonin nor commercial bovine pineal powder influences iodide uptake of intact larvae. Certainly, additional studies of the relationship of pineal factors to thyroid function would help to resolve some of these apparent contradictions.

Reproduction may be under the inhibitory influence of the pineal organ, at least in anurans. Accelerated gonadal development follows pinealectomy of *A. obstetricans* and *Hyla cinerea.* Gonadotropin-induced ovulation from *R. pipiens* ovaries *in vitro* is inhibited by addition of melatonin to the culture medium. Bovine pineal extract similarly inhibited human chorionic gonadotropin-induced spermiation in male *R. esculenta,* but purified melatonin had no effect. This dichotomy of the influence of melatonin in male and female anurans warrants further investigation. The possible influence of pineal principles on reproduction in urodeles has not been studied.

Unlike the other jawed vertebrates, some anurans have retained a well-developed parapineal end vesicle known also as the frontal organ or *stirnorgan* (Fig. 6-4). Because of the presence of photosensory cells, the parapineal is often referred to as the parietal eye in these species. Photic information collected by the parietal eye is conducted directly to the pineal.

5. Reptiles

Reptiles can be separated into several groups on the basis of the anatomy of the epiphysial complex. Melatonin has been localized in the blood, pineal gland, and retinas of snakes, lizards, and turtles. Modern crocodilians have no pineal, parapineal, or parietal structures. Although a pineal is absent in alligators, melatonin, presumably of retinal origin, is present in the blood. Lizards possess an elaborate saclike, pigmented pineal organ containing both photosensory and ependymal cells (Fig. 6-4). The lumen of the lizard pineal lies close to the third ventricle but does not join with it. A parapineal organ penetrates the skull, forming a parietal spot on the surface. As in amphibians, the parapineal is often termed the parietal eye. Turtles and snakes have retained only the basal (glandular) portion of the pineal organ and have lost the end vesicle and stalk of the pineal organ as well as the complete parapineal organ. Nevertheless, the turtle pineal organ is the largest and the best-developed epiphysial structure among reptiles (Fig. 6-4).

The parietal eye of lizards has been examined with respect to thyroid function, thermoregulation, and reproduction. The lizard parietal eye contains HIOMT activity, suggesting that it synthesizes melatonin, and removal of the parietal eye stimulates thyroid hyperplasia and oxygen consumption. These data suggest that melatonin or some other pineal principle influences pituitary function although direct effects on the thyroid are not ruled out. The reptilian parietal eye may be a photo/thermal radiation dosimeter that monitors solar radiation and, in turn, regulates activity patterns of lizards. Indeed, excision of the pineal or parietal eye alters thermal responses of numerous lizard species, and it may well be the major functional vote for the epiphysial complex in lizards.

Definitive effects of the epiphysial complex on reproduction also have been reported in lizards. For example, excision of the parietal eye of the lizard *Anolis carolinensis* stimulates ovarian development in reproductively quiescent animals. This effect is blocked by administration of melatonin. The onset of gonadal recrudescence in this lizard is induced by long photoperiod and warm temperatures. The parietal eye appears to be the transducer through which photoperiod influences reproduction.

6. Birds

The pineal organ of birds has been reduced to the glandular basal portion, the pineal gland. No parapineal or remnant thereof is present, and distinct photoreceptors are absent in the pineal. Structurally, the avian pineal exhibits considerable diversity, and it is composed of several cellular types. The avian pineal is innervated by sympathetic fibers as reported for mammals.

Avian pineals are biochemically like their mammalian counterpart. Variations have been reported in HIOMT and NAT activities with respect to lighting conditions, but the most dramatic effects involve NAT. Activity of this enzyme exhibits a marked increase with onset of the scotophase, a peak about the middle of the scotophase, and a decrease rapidly following the onset of the photophase. Brief exposure to light at the peak of NAT activity causes a rapid reduction to photophase levels. Melatonin rhythms that correlate with rhythms in pineal NAT activity have been reported for brain, pineal, retina, and serum of birds. The brain and especially the hypothalamus in birds may be the primary site of action for melatonin and may explain effects of melatonin on gonadal function, thermoregulation, and locomotor activity.

Unlike mammals and other vertebrates, the role of the pineal in the reproductive biology of birds may be progonadal. Pinealectomy inhibits androgen synthesis, whereas administration of melatonin stimulates androgen synthesis, presumably by altering gonadotropin release from the adenohypophysis. Pinealectomy of quail delays ovarian development, an observation that also supports a progonadal role. Melatonin injections can decrease gonadal weight, suggesting that the progonadal agent might be a peptide that normally overpowers effects of endogenous melatonin on reproduction. Marked species differences may occur, however. For example, no effects of melatonin treatment on parameters of gonadal development and their relationship to photoperiod could be demonstrated in either white-throated sparrows or border canaries.

Neural control of melatonin secretion in birds also differs markedly in contrast to mammals. Increased activity of sympathetic fibers from the superior cervical ganglion decreases melatonin secretion in birds, and NE turnover is greatest during the photophase. This inhibitory action of NE appears to involve G_i-proteins in pinealocytes and subsequent reduction in cAMP formation, NAT activity, and melatonin synthesis and release (see Chapter 3). Pinealectomy abolishes endogenous body temperature rhythms as well as free-running locomotor activity rhythms in house sparrows, *Passer domesticus*. Effects on locomotor activity appear to involve two pathways, one of which can bypass the pineal organ. Pinealectomized birds that have lost free-running locomotor activity in the dark still exhibit entrainment to light–dark cycles, supporting the presence of a bypass system. Both systems can be entrained by light, but the pineal has control only over the bypass system (Fig. 6-10).

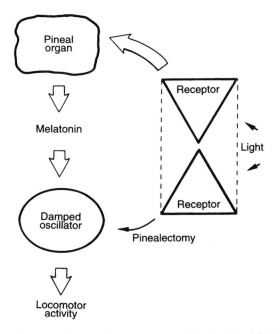

Figure 6-10. Role of the pineal in controlling locomotor activity in birds. Diurnal fluctuations in pineal activity entrain a damped oscillator that drives locomotor activity. Following pinealectomy, a secondary endogenous pathway (also entrained by photoperiod) takes over control of the rhythm. [Based on Menaker, M., and Zimmerman, N. (1976). Role of the pineal in the circadian system of birds. *Am. Zool.* **16**, 45–55.]

F. Evolution of the Functions of Melatonin

An intriguing hypothesis, proposed by William Gern, for the original function of melatonin and the evolution of other functions, is based on the presence of melatonin-synthesizing systems in retinas, parietal eyes, and pineals and on the observations that pineals of more primitive vertebrate groups are photoreceptive. Evidence suggests that both retinal and pineal melatonin exhibit nighttime (scotophasic) peaks of synthesis. This hypothesis proposes that melatonin was initially a local hormone for regulating the distribution of melanosomes in the retina. During the day, melanosomes are dispersed in retinal pigment cells which protect the photoreceptors from intense light. At night, elevated melatonin causes concentration of melanosomes and allows dim light to stimulate the photoreceptors maximally. This sensitivity of melanosomes to melatonin is retained in the melanophores of skin and brain ependymas of both modern fishes and amphibians. Similar mechanisms presumedly operate in the photoreceptive outer portion of the pineal, in the parietal eye, and in the amphibian frontal organ. The increase in melatonin synthesis during the scotophase causes a greater proportion of melatonin to appear in the blood. Consequently, the scotophasic elevation in melatonin is a reliable internal cue for obtaining information about seasonal photoperiods. Information concerning length of the scotophase is reflected in circulating melatonin levels. Thus, according to Gern's hypothesis, the diurnal rhythm in melatonin has been coopted as the blood-borne signal entraining a number of other internal events during the evolution of vertebrates.

III. Summary

A. Pars Nervosa

There are typically two nonapeptide types produced in the hypothalamus and stored in the vertebrate pars nervosa. One is a basic nonapeptide (AVP, LVP, PVP, AVT), which has a predominantly pressor/antidiuretic activity, and the other is a neutral nonapeptide (oxytocin, MST, etc.) that stimulates contraction of reproductive smooth muscles (i.e., uterotonic action). The most primitive nonapeptide, AVT, unlike other basic peptides, has rather good uterotonic activity related to its ring structure which is like that of oxytocin. Jawless fishes (agnathans) have only AVT but all other vertebrates studied exhibit at least one basic and one neutral nonapeptide.

B. Epiphysial Complex

The major secretory product of the pineal gland is melatonin, although peptides such as AVT may play important roles as well. Pineal secretion is inhibited by light, and the pineal gland plays a major role in mediating seasonal and daily endocrine activity primarily through effects on the hypothalamus. Melatonin generally acts as an antigonadal (except in birds), antithyroid, and anti-prolactin-releasing factor. Roles in thermoregulation (teleosts, reptiles) and in aging (mammals) are also important.

The pineal receives innervation from several sources. In mammals, the most important pathway is the sympathetic innervation of pinealocytes which is controlled by two retinal pathways both passing through the superior cervical ganglion. One of these pathways involves the hypothalamic SCN, a major determinant of biological rhythms in mammals. Details of pineal regulation have not been examined thoroughly in nonmammals.

Suggested Reading

Books

Binkley, S. A. (1990). "The Clockwork Sparrow: Time, Clocks, and Calendars in Biological Organisms." Prentice Hall, Englewood Cliffs, NJ.

Pierpaoli, W., Regelson, W., and Fabris, N. (1994). The aging clock: The pineal gland and other pacemakers in the progression of aging and carcinogenesis. The New York Academy of Sciences, New York.

Reiter, R, J. (1994). "The Pineal Gland and Melatonin." CRC Press, Boca Raton, FL.

Saito, T., Kurokawa, K., and Yoshida, S. (1995). "Neurohypophysis: Recent Progress of Vasopressin and Oxytocin." Elsevier, Amsterdam.

Articles

Acher, R. (1995). Evolution of neurohypophysial control of water homeostasis: Integrative biology of molecular, cellular and organismal aspects. *In* "Neurohypophysis: Recent Progress of Vasopressin and Oxytocin" (T. Saito, K. Kurokawa, and S. Yoshida, eds.), pp. 39–54. Elsevier, Amsterdam.

Armstrong, D. L., and White, R. E. (1994). Natriuretic peptides and receptors. *Sci. Am. Sci. Med.* **Mar/Apr,** 34–43.

Baker, B. I. (1994). Melanin-concentrating hormone updated. *Trends Endocrinol. Metab.* **5,** 120–126.

Crowley, W. R., and Armstrong, W. E. (1992). Neurochemical regulation of oxytocin secretion in lactation. *Endocr. Rev.* **13,** 33–65.

Gern, W. A., Nervina, J. M., and Greenhouse, S. S. (1987). Pineal involvement in seasonality of reproduction. *In* "Hormones and Reproduction in Fishes, Amphibians, and Reptiles" (D. O. Norris and R. E. Jones, eds.), pp. 433–460. Plenum, New York.

Kloas, W. (1994). Localization of binding sites for atrial natriuretic peptide and angiotensin II in gill and kidney of various fish species. *In* "Fish Ecotoxicology and Ecophysiology" (T. Braunbeck, W. Hanke, and H. Segner, eds.), pp. 367–384. VCH, Weinheim.

Reinecke, M., Betzler, D., Aoki, A., and Forssmann, W.-G. (1993). Atrial natriuretic peptides (ANP) in fish heart. *In* "Fish Ecotoxicology and Ecophysiology" (T. Braunbeck, W. Hanke, and H. Segner, eds.), pp. 385–404. VCH, Weinheim.

Reiter, R. J. (1991). Pineal gland: Interface between the photoperiodic environment and the endocrine system. *Trends Endocrinol. Metab.* **2,** 13–19.

Reiter, R. J. (1991). Pineal melatonin: Cell biology of its synthesis and its physiological interactions. *Endocr. Rev.* **12,** 151–180.

Saarela, S., and Reiter, R. J. (1994). Function of melatonin in thermoregulatory processes. *Life Sci.* **54,** 295–311.

Samson, W. K. (1992). Natriuretic peptides: A family of hormones. *Trends Endocrinol. Metab.* **3,** 86–90.

Watkins, W. B., and Choy, V. J. (1988). Identification of neurohypophysial peptides in the ovaries of several mammalian and nonmammalian species. *Peptide* **9,** 927–932.

Urano, A., Kubokawa, K., and Hiraoka, S. (1994). Expression of the vasotocin and isotocin gene family in fish. *In* "Fish Physiology, Vol. XIII, Molecular Endocrinology of Fish" (N. M. Sherwood and C. L. Hew, eds), pp. 101–132. Academic Press, San Diego.

7

The Hypothalamo–Hypophysial–Thyroid Axis of Mammals

THE THYROID gland (Fig. 7-1) is unique among vertebrate endocrine glands in that it stores its secretory products (thyroid hormones) extracellularly. It is possibly the most highly vascularized endocrine gland in mammals and appears to be one of the oldest endocrine glands phylogenetically (see Chapter 8). Two separate hormones are synthesized from the amino acid tyrosine, which first is iodinated, and then two iodinated tyrosines are linked together to form **triiodothyronine (T₃)** and **tetraiodothyronine** or **thyroxine (T₄).** The structures of these compounds are found in Chapter 2.

Thyroid hormones influence many aspects of reproduction, growth, differentiation, and metabolism. Many of these actions occur cooperatively with other hormones, and the thyroid hormones enhance their effectiveness. This cooperative role for thyroid hormones is referred to as a **permissive action,** whereby thyroid hormones produce changes in target tissues that "allow" these tissues to be more responsive to another hormone, to neural stimulation, or possibly to certain environmental stimuli such as light. Thyroid hormones may maintain maximal sensitivity to other regulating agents in many types of tissues. The importance of thyroid hormones is reflected in the observation that the incidence of thyroid disease in humans is exceeded only by the incidence of diabetes mellitus (see Chapter 14). Although rarely lethal, thyroid disorders have widespread effects due to their many actions with other hormones.

I. Some Historical Aspects

Either deficient or excessive production of thyroid hormones may lead to serious pathological states with overt symptoms (Table 7-1). The first description of thyroid disease was of abnormal enlargement of the thyroid in humans, recognized by Chinese physicians about 3000 B.C. As a remedy, they recommended ingestion of seaweed and burned sponge or desiccated deer thyroids. The first two substances contained therapeutic quantities of iodide and the last sufficient thyroid hormones to alleviate the pathological symptoms in most cases. Hypothyroid deficiencies of this sort were not recognized in Western culture as clinical disorders until many centuries later. In 1526, the **cretinism** syndrome (see Section VI,B) was described clinically in Europe. Cretinism is manifest very early in life as a consequence of severe thyroid deficiency. This syndrome is characterized by dwarfism and a number of other physical abnormalities in addition to severe mental retardation, slow mental and physical activity, bradycardia (slowing of heart beat), and hypothermia. In 1880–1890, another classic clinical disorder in adults, **myxedema** (see Section VI,B), was linked to hypothyroid function. Myxedematous symptoms in adults are related to abnormal accumulation of water and protein throughout the body as well as to other disturbances in general metabolism. These accumulations of protein and fluid alter facial features, causing the patient to appear expressionless. In later stages of the disorder, the sufferer becomes less interested in both self and environment, and if untreated would eventually enter a coma and die. Juvenile myxedema is similar to cretinism except that early growth and development are normal but become severely retarded in later childhood. All of these different clinical syndromes have the same basic cause: hypofunction of the thyroid gland.

Bauman discovered in 1896 that an organic iodine-containing compound could be extracted from thyroid glands. Subsequently, it was demonstrated that this "thyroidin" substance could reverse the adverse effects of iodide deficiency. In the early 1900s, the thyroid gland and its hormones were implicated in elevating basal metabolic rate, primarily through effects on certain tissues, for example, liver, kidney, and muscle. This observation has strongly influenced the direction of thyroid research in mammals as well as in many

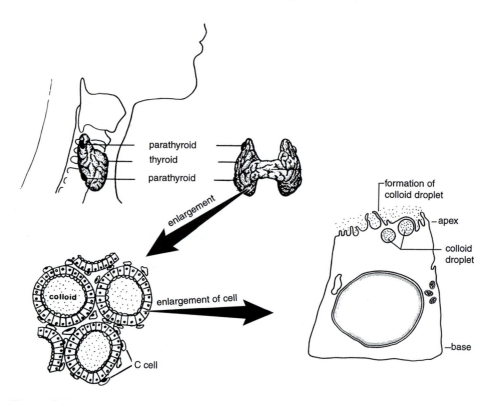

Figure 7-1. The mammalian thyroid. The thyroid gland is located in the neck region. It consists of many hollow follicles. The lumen of each follicle is filled with a proteinaceous fluid called *colloid*. The follicular cells secrete the colloid as well as thyroid hormones. The C cells or parafollicular cells are of ultimobranchial origin and secrete the calcium-regulating hormone, calcitonin (see Chapter 15). [Modified from McNabb, F. M. A. (1993). "Thyroid Hormones." Prentice-Hall, Englewood Cliffs, New Jersey; and Bolander, F. F. (1989). "Molecular Endocrinology." Academic Press, San Diego.]

nonmammalian vertebrates. The action of thyroid hormones on metabolism is reflected in clinical thyroid states (Table 7-1).

The iodine-containing hormone, T_4, was isolated, crystallized, and reported in a publi-

Table 7-1. Symptoms of Thyroid Deficiency and Hyperactivity in Humans

Type of symptom	Hypothyroid	Hyperthyroid
Appearance	Myxedema; deficient growth	Exophthalmus
Behavioral	Mental retardation; mentally and physically sluggish; somnolent; sensitive to cold	Often quick mentally; restless, irritable, anxious, hyperkinetic; wakeful; sensitive to heat
Metabolism	Hypophagia; low basal metabolic rate; reduced $QO_2{}^a$ of liver, kidney, and muscle *in vitro*; decrease in oxidative enzymes; constipation	Hyperphagia; high basal metabolic rate; increased QO_2 of liver, kidney, and muscle *in vitro*; increased oxidative enzymes; diarrhea
Muscle function	Weakness; hypotonia	Weakness; fibrillary twitchings, tremors

[a] QO_2, Respiratory quotient.

cation by Edward C. Kendall in 1915. This event marked a significant milestone, not only in thyroid research but in endocrinology as a whole, for T_4 was the first hormone to be isolated in pure form. It was not until 1952, however, that a second thyroid hormone, T_3, was identified by J. Gross and R. Pitt-Rivers. The importance of this discovery will become evident as the mechanisms of synthesis and action for thyroid hormones are discussed. It was the discovery of the so-called antithyroid drugs in the early 1940s as well as the ready availability of radioactive isotopes of iodide (radioiodide) following developments in nuclear physics that made it possible to elucidate the details of thyroid hormone synthesis, metabolism, and mechanisms of action.

II. Development and Organization
of the Mammalian Thyroid Gland

The mammalian thyroid gland consists of many follicles encapsulated by a connective tissue sheath. The gland is highly vascularized with a dense capillary network surrounding each follicle. The thyroid receives sympathetic innervation, and it appears that both the vasculature and the follicle cells may be innervated.

Development of the thyroid gland begins by formation of a ventral bud in the floor of the embryonic pharynx (endoderm) between the first and second pharyngeal pouches. The gland initially differentiates as cellular cords that later separate into clusters of cells destined to become thyroid follicles. The cells of a cluster secrete a proteinaceous fluid termed **colloid** that accumulates extracellularly in the center of the cluster. This secretory activity eventually leads to formation of a colloid-filled space, the lumen of the follicle, surrounded by a single layer of epithelial cells, the epithelium of the follicle (Fig. 7-1). The portion of the follicular cell that borders on the lumen of the follicle is known as the apical part. The nucleus is generally found in the basal portion of the cell that is farthest from the lumen and closest to the capillaries.

In addition to capillaries and follicles, **parafollicular** or **C cells** occur in the regions between or adjacent to the follicles. Parafollicular cells may occur within the follicular epithelium or may even form separate follicular structures in some species. These cells are derived from another pharyngeal derivative, the **ultimobranchial body,** and secrete a hormone, **calcitonin,** that influences calcium metabolism (see Chapter 15). A comparison of parafollicular cells and follicular cells (Table 7-2) emphasizes their different structural and functional features. In some mammals the parathyroid glands may be embedded within the mass of the thyroid (Fig. 7-2). The parathyroids, like the parafollicular cells, have their origin nearby, from the embryonic pharynx, and in some species become embedded in the mass of thyroid follicles during development. The parathyroid glands are also discussed in Chapter 15.

III. Biochemistry of Thyroid Hormones

The events related to the ability of thyroid follicles to synthesize and release thyroid hormones are discussed separately for simplicity, but it is important to keep in mind that many of these events may be occurring simultaneously (see Fig. 7-3). The processes discussed in this section include the following:

Table 7-2. Comparison of Characteristics of Thyroid Follicular and Parafollicular Cells

Thyroid follicular cell	Thyroid parafollicular cell
Absence of secretion granules	Large number of eosinophilic granules, 0.2-μm diameter; stain with silver nitrate
Endoplasmic reticulum cisternae of larger diameter, containing flocculent precipitate like that found in albumin-secreting cells	Many mitochondria and high level of the mitochondrial enzyme α-glycerophosphate dehydrogenase
Carbohydrate added at Golgi apparatus, which is rather inconspicuous in these cells	No lumenal surface present
	Nucleus more irregular in outline than those of follicular cells
Enlargement of Golgi apparatus from TSH treatment	Golgi apparatus prominent
Binds antibody to thyroglobulin but not to calcitonin	Binds antibody to calcitonin
Cytology not altered by high blood calcium level	Degranulation due to high blood calcium level
Readily accumulates iodide	

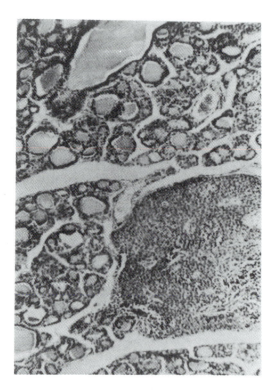

Figure 7-2. Thyroid and parathyroid glands. This photograph of a slice of thyroid gland shows the follicular structure that characterizes the gland. An embedded parathyroid gland also is shown (PT).

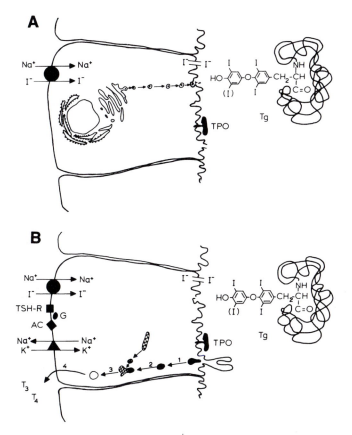

Figure 7-3. Thyroid hormone biosynthesis. (A) Na^+/I^- cotransport occurs at the basal surface of a follicular cell. The enzyme thyroid peroxidase (TPO) located at the apical surface is responsible for activating I^-, for iodinating thyroglobulin (Tg), and for coupling iodinated tyrosines to form thyronines (T_3 and T_4). (B) Release of thyroid hormones requires engulfing colloid (endocytosis) to form intracellular endosomes that merge with lysosomes. This causes degradation of Tg and liberation of T_3 and T_4 from the basal surface of the cell into the blood. MIT and DIT are deiodinated and the products recycled within the follicular cell. TSH-R, TSH receptor; G, G_s protein; AC, adenylate cyclose. [Reprinted from *Biochem. Biophys. Acta* **1154**, Carrasco, N. Iodide transport in the thyroid gland. 65–82. © 1993 with kind permission of Elsevier Science-NL, Sara Burgerhartstraat 25, 1055 KV Amsterdam, The Netherlands.]

1. Accumulation of inorganic iodide by follicular cells
2. Synthesis of **thyroglobulin,** a glycoprotein that contains numerous tyrosine residues for hormone synthesis
3. Binding of inorganic iodide to tyrosine residues in thyroglobulin
4. Synthesis of T_3 and T_4 from iodinated tyrosines
5. Storage of thyroglobulin containing the thyroid hormones in the lumen of the follicle
6. Engulfing of colloid by follicular cells and hydrolysis of thyroglobulin to release thyroid hormones

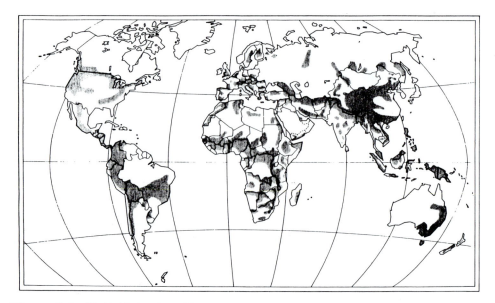

Figure 7-4. Worldwide location of iodide-poor regions. The shaded portions indicate the iodide-poor regions. World Health Organization.

7. Diffusion of T_3 and T_4 into the general circulation and their transport to targets
8. Conversion of T_4 to T_3 in the liver
9. Mechanisms of action for thyroid hormones
10. Metabolism and excretion of thyroid hormones

A. Dietary Iodide and Iodide Uptake

The principal source for inorganic iodide is dietary. In certain portions of the world, environmental iodide is in short supply; for example, the Great Lakes and Rocky Mountain regions of the United States, and northeastern Europe (Fig. 7-4). Consequently, in these regions, unsupplemented human diets are low in naturally occurring iodide, and hypothyroid states commonly are encountered unless an iodide supplement is used. At one time, hypothyroid **goiters** or enlarged thyroids were common in people who inhabited these low-iodide regions or "goiter belts," but the use of iodized salt has almost eliminated this condition in developed countries. Unfortunately, hypothyroidism due to low iodide is rampant in developing countries (Table 7-3), where it has devastating effects on the well-being of millions of people. (The term *goiter* or *goitre* originally meant any tumor or abnormal glandular enlargement in the neck but has come to mean an enlarged thyroid.)

Inorganic iodide is readily absorbed from the intestine into the blood, from which it is selectively accumulated by thyroid follicular cells. There is an energy-dependent, active transport mechanism, or "iodide pump," in the follicular cell basal membrane that is specific for iodide. Iodide is cotransported with Na^+ at the basal membrane and passively diffuses across the apical membrane into the colloid. Thus, uptake of iodide is enhanced by translocation of inorganic iodide at the apical surface and its conversion to organically bound forms (that is, iodinated tyrosines). Thyroidal iodide accumulation is dependent on

Table 7-3. Percentages of School Children Exhibiting Goiter[a]

Developing countries	Percent	Developed countries	Percent
Syria	73	Turkey	36
Central African Republic	63	Bulgaria	20
Zambia	51	Italy	20
Nepal	44	Portugal	15
Albania	41	Germany	10
Peru	36	Greece	10
Iran	30	Denmark	5
Malaysia	20	France	5

[a] Based on World Health Organization figures for 1993.

an Na^+/K^+-ATPase-related mechanism and is not affected by other halide anions, including Cl^-, Br^-, and Fl^-. The resulting concentration of iodide normally exceeds plasma levels by 20 to 40 times and under some conditions may be even greater.

The events of iodide uptake, accumulation, and binding to tyrosine have been examined using radioiodide. For many years, the radioisotope employed most frequently for iodide uptake studies was ^{131}I, a strong β- and γ-emitting isotope with a short radiation half-life (about 8 days). This particular isotope can be detected in blood or tissues with relatively unsophisticated detection equipment because of the high-energy γ radiation it emits. Another isotope, ^{125}I, now is employed in most thyroid studies, making use of the lower energy radiation produced by this isotope and its longer radiation half-life (about 60 days). This isotope emits β and low-energy γ radiation and is suitable for high-resolution autoradiography at both the light and electron microscope level. Because of its lower energy emission and longer radiation half-life, ^{125}I is more suited for metabolic studies, for it is much less destructive to cells than ^{131}I and safer to use. Uptake of radioiodide, like that of the normal isotope (^{127}I), is stimulated by thyrotropin (TSH) from the adenohypophysis, and there is no discrimination among the various isotopes in the formation of organically bound iodide associated with thyroid hormone synthesis.

Calculation of a rate for radioiodide accumulation following administration of a given dose provides a quantitative estimate of the degree of TSH stimulation and a reflection of pituitary TSH release. Hence, measurement of radioiodide uptake and accumulation provides a simple and rapid method for estimating endogenous activities of the hypothalamo–thyroid axis as well as a means to assess responsiveness of thyroid follicular cells to exogenous TSH. Usually, radioiodide uptake is expressed as the percent uptake of the injected dose at some predetermined time, such as 24 hr following administration of radioiodide (Table 7-4).

The ability to accumulate and bind iodide organically is not a feature unique to thyroid follicular cells, and most cells will accumulate some iodide. Cells such as melanophores (melanocytes), pigmented retinal cells, and the epithelial cells found in sweat glands, salivary glands, lactating mammary glands, and kidney tubules readily accumulate radioiodide following injection of either radioactive isotope and may add radioiodide to their secretions. Furthermore, oocytes of many oviparous (egg-laying) vertebrates readily accumulate large amounts of radioiodide. This ovarian accumulation is associated with the normal process of ensuring a source of iodide in the egg that can be used by the young animal for early synthesis of thyroid hormones until an adequate dietary source becomes available. In

Table 7-4. Comparison of Thyroid Function in a Monotreme (Echidna), Marsupial (Bandicoot), and Placental Mammal (Rabbit)[a]

	Euthyroid parameters			Effect of thyroidectomy	
	Iodide uptake (% injected dose)	Plasma T_4 (nmol/liter)	Plasma T_3 (nmol/liter)	BMR[b]	Body temperature
Echidna, *Tachyglossus aculeatus*	6.4	15.7	0.7	No effect	No effect
Bandicoot, *Perameles nasuta*	13.7	22.0	1.5	Decrease	No effect
Rabbit, *Oryctolagus cuniculus*	22.9	57.9	6.9	Decrease	No effect

[a] Data from Hurlbert and Augee (1982). *Physiol. Zool.* **55**, 220–228.
[b] BMR, Basal metabolic rate.

addition, thyroid hormones, presumably of maternal origin, have been found in the oocytes of several teleosts and at least one amphibian. Marsupials, placental mammals, and other live-bearing vertebrates are known to transfer iodide and thyroid hormones to the developing young from the maternal blood or via the milk to suckling newborns.

Thyroid hormones may not be released into the circulation in proportion to the uptake and binding of radioiodide, however. Uptake, binding, and release of thyroid hormones are separate events independently influenced by a variety of factors, as evidenced in the following discussions. Nevertheless, measurement of radioiodide uptake or uptake of related ions is a rapid and convenient method to assay for TSH activity with respect to endogenous levels or exogenous treatment, and it is widely employed in thyroid diagnosis and research.

B. Biosynthesis of Thyroid Hormones

As described in Chapter 2, the synthesis of thyroid hormones in the follicular cells involves the synthesis of thyroglobulin and the binding of accumulated inorganic iodide to the tyrosines. The final step is the linking together (coupling) of two iodinated tyrosines contained within thyroglobulin to form the iodinated hormones T_3 and T_4. The use of radioiodide, polarized monolayer cultures of porcine thyroid cells, and a clone of normal rat thyroid cells called FRTL-5 have been instrumental in the elucidation of the following events.

1. Synthesis of Thyroglobulin

Thyroglobulin may occur in more than one molecular form. Thyroglobulins occur in several sizes, ranging from 12S to 27S (S stands for Svedberg sedimentation coefficient, which is related to size, shape, and, to a lesser degree, electrostatic charge of a molecule. These parameters determine the migration and final position of molecules in a molecular gradient following high-speed centrifugation). The basic monomer is the 12S form. Comparative analysis of thyroglobulins from different vertebrates indicates different proportions of oligomers of thyroglobulins (Table 7-5). Most mammalian thyroglobulin preparations

Table 7-5. Sedimentation Coefficients for Mammalian
Iodoproteins (Thyroglobulins)

Species	Percent		
	12S	19S	27S
Cavia porcellus (guinea pig)	14	83	3
Rattus rattus (rat)	Trace	93	7
Mus musculus (mouse)	—	94	6
Oryctolagus cuniculus (rabbit)	Trace	98	2
Canis familiaris (dog)	—	94	6
Felis catus (cat)	—	92	8
Bos taurus (ox)	—	91	9
Bulbalus bulalis (brahma)	—	93	7
Capra hircus (goat)	—	90	10
Ovis aries (sheep)	—	88	12
Sus scrofa (pig)	—	88	12
Equus caballus (horse)	—	100	—
Equus asinus (donkey)	—	100	—
E. caballus × *E. asinus* (mule)	—	100	—
Homo sapiens (human)	—	92	8
Macaca mulatta (rhesus monkey)	—	92	8

exhibit a predominant 19S component (87–100% of the total iodinated protein in thyroid preparations) with a small proportion of larger 27S and occasionally a small quantity of 12S thyroglobulin (rabbit, rat, and especially the guinea pig). The 19S form consists of two 12S subunits, and the 27S form is composed of four subunits. Regardless of its actual form, thyroglobulin will be treated in the following discussions as though it were a single molecular species.

Thyroglobulin synthesis occurs at the rough endoplasmic reticulum, and it is packaged into membrane-bound secretion granules in the Golgi apparatus. It appears that noniodinated tyrosines are incorporated into thyroglobulins first. There are no transfer RNAs for iodinated tyrosines in follicular cells, and studies show that iodination occurs at the cell–colloid interface. However, relatively few of the tyrosine residues are iodinated, and the number of thyroid hormone molecules incorporated into a thyroglobulin molecule is estimated at between four and eight.

2. Iodination of Tyrosine Residues in Thyroglobulin

Organic binding of iodine begins with conversion of inorganic iodide to **active iodide,** a form of inorganic iodide that is readily incorporated into the phenolic ring of tyrosine. Although the exact chemical nature of active iodide has never been determined with certainty, it is apparently formed in the colloid compartment through the action of an enzymatic **thyroid peroxidase (TPO)** system located on the extracellular side of the apical membrane of the follicular cell. TPO catalyzes glucose oxidation and reduction of pyridine nucleotides to form hydrogen peroxide, H_2O_2. Inorganic iodide reacts with H_2O_2 to form active iodide, which in turn is converted immediately to iodine attached to tyrosine residues in thyroglobulin. Treatment of thyroid hormone-synthesizing systems with the enzyme catalase specifically hydrolyzes H_2O_2 to water and oxygen. This blocks iodination, supporting the role of peroxides in formation of organically bound iodine.

The binding of one iodine to tyrosine at position 3 on the phenolic ring yields **3-mono-**

iodotyrosine or **MIT.** A second iodine may attach at position 5 of the same tyrosine residue, resulting in conversion of MIT to **3,5-diiodotyrosine** or **DIT.** In reality, there is no difference between the 3 position and the 5 position, due to the symmetry of the phenolic ring, and the numbering is arbitrary. Hence, by convention, the first one iodinated always yields a 3-monoiodotyrosine but never a 5-monoiodotyrosine.

The proportion of MIT to DIT formed will be determined by the amount of active iodide available, which in turn depends on the size of the inorganic iodide pool. The structures of MIT and DIT and their coupling to form T_3 and T_4 are illustrated in Chapter 2 (Fig. 2-36).

3. Coupling of Iodinated Tyrosines

The exact way in which thyroid hormones are formed from iodinated tyrosines is not known. Coupling appears to be an enzymatically controlled process that involves two iodinated tyrosines, usually two DIT molecules but in some cases one DIT plus one MIT. The alanine side chain on one of the iodinated tyrosines is cleaved off, and the remaining iodinated phenolic ring is joined to the other iodinated tyrosine through formation of an ether $(-O-)$ linkage. The resultant structure is known as an iodinated **thyronine.** Appropriate coupling of MIT and DIT yields 3,5,3'-triiodothyronine or T_3. Similarly, coupling of two DIT molecules forms 3,5,3',5'-tetraiodothyronine or T_4. The proportion of MIT and DIT available for coupling will influence the proportions of T_3 and T_4 formed. Normally, much more T_4 than T_3 is synthesized, but the relative proportion of T_3 may increase markedly if iodide is in short supply.

Coupling of iodinated tyrosines follows hydrolysis of some peptide bonds to release smaller peptides of 15 to 20 kDa from thyroglobulin. Combination of adjacent residues in the folded, globular thyroglobulin molecule and the peptide fragment accomplishes the coupling (Fig. 7-5). Only a fraction of the iodinated tyrosines are actually coupled, and iodinated thyronines are formed from only about 10% of the iodinated tyrosine residues present in thyroglobulin. Most of the extractable organic iodide is still in the form of MIT and DIT.

The specificity of the coupling reaction implies an enzymatic conversion of iodinated tyrosines to thyronines. Only one form of triiodothyronine usually is produced in the thyroid, although it would seem possible to produce equal amounts of the alternative form, 3,3',5'-triiodothyronine or **reverse T_3 (rT_3).** As we shall see, formation of rT_3 is performed primarily by the liver through partial deiodination of T_4 (see Section III,C).

In the synthesis of T_3 or T_4 by thyroid cells, there is a requirement that DIT be positioned "on the right" in the peptide fragment, whereas either MIT or DIT may occur to the "left" in thyroglobulin for the coupling reaction to proceed. In other words, the tyrosine residue in the thyroglobulin is thought to donate its phenolic ring and is converted to a variant of alanine called **dehydroalanine** (missing one hydrogen). The general stereospecificity for DIT is further supported by the failure of thyroid systems to synthesize, even under conditions of severe iodide deficiency, any diiodothyronines (T_2) utilizing two MIT molecules or little if any of rT_3.

4. Hormone Release: Hydrolysis of Thyroglobulin

Release of thyroid hormones following administration of TSH is not linked directly to iodide uptake and iodothyronine synthesis. Thyrotropin independently stimulates engulfment of colloid by the follicular cell and its intracellular hydrolysis to amino acids, MIT, DIT, T_3, and T_4 (Fig. 7-3). Autoradiographic studies indicate that the first event observed

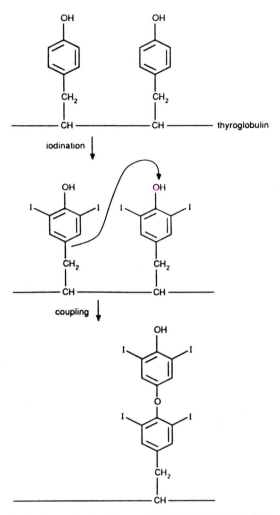

Figure 7-5. Synthesis of a thyronine. In this example, the iodinated ring of one molecule of DIT is re-moved and linked to the oxygen on a second DIT nearby. This coupling is a result of the activity of the thyroid peroxidase system (TPO). The resultant molecule is called a thyronine. This particular thyronine would be thyroxine. [From Bolander, F. F. (1989). "Molecular Endocrinology." Academic Press, San Diego.]

following TSH administration is the engulfment of colloid through a process of endocytosis. These colloid droplets or **endosomes** migrate from the apical portion of the cells toward the basal portion and become associated with lysosomes, which contain a number of hydrolytic enzymes. Fusion of endosomes with lysosomes results in formation of "fusion droplets" or **endolysosomes.** As the endolysosomes migrate toward the basal portion of the cell, they become progressively degranulated, presumably because of hydrolysis of thyroglobulin and diffusion of the hydrolysis products into the cytosol.

Thyroglobulin hydrolysis within the endolysosome releases MIT, DIT, T_3, T_4, and amino acids. Essentially it is only T_3 and T_4 that diffuse from the cell into the vast capillary network surrounding the follicles because a special cytoplasmic **deiodinase** hydrolyzes MIT and DIT to tyrosine and inorganic iodide. Thyroidal deiodinase may also deiodinate a very small proportion of T_4 as well. Hence, this enzyme is sometimes called iodotyrosine dehalogenase (iodine is one of a group of elements called halogens). Deiodination is apparently a conservation mechanism to reuse inorganic iodide incorporated in free cytoplasmic MIT and DIT for later iodination. These iodinated tyrosines cannot be used in thyroglobulin synthesis, as stated previously, and they must be either deiodinated or allowed to diffuse from the cell. Approximately 85 to 90% of the iodide released through deiodination of MIT and DIT enters a "second iodide pool" within the follicular cell, which is then available for diffusion into the colloid and iodination of newly synthesized thyroglobulin.

C. Transport of Thyroid Hormones in the Blood

Almost all of the circulating thyroid hormones are bound reversibly to serum proteins. Serum binding and transporting of thyroid hormones are essential, because T_3 and T_4 are somewhat hydrophobic and are not very soluble in blood. Serum-bound hormones provide a ready reservoir to quickly replenish the levels of free hormones in the blood. Being somewhat hydrophobic, free thyroid hormones appear to move across cell membranes readily and are rapidly removed from the blood and metabolized, especially by the liver and kidneys. Several different serum proteins are capable of binding and transporting thyroid hormones. About 75 to 85% of the bound hormones are linked to the α_2-globulins called **thyroid-binding globulin (TBG).** Only a very small fraction (<0.1%) are transported free in the blood, and the remainder is bound to **prealbumin (TBPA)** and **albumin (TBA).**

Under normal conditions, circulating T_4 levels are much greater than T_3 levels. Only about one-seventh to one-half of the circulating T_3, however, is of thyroid origin (depending on the species and/or physiological parameters), and the remainder is produced through peripheral deiodination of T_4. This deiodination is accomplished primarily in the liver, with some additional deiodination occurring in kidneys and skeletal muscle. Free T_3 or T_4 in the serum is in equilibrium with bound hormones so that, as free hormones are metabolized or enter target cells, there is a proportionate dissociation of bound hormone to replace the lost free hormone.

Thyroxine is more tightly bound than T_3 to the serum proteins, and consequently T_3 is more rapidly eliminated from the blood. The biological half-life for T_3 in humans is about 24 hr, whereas T_4 has a much longer biological half-life (about 7 days), attributable to the greater affinity of T_4 for the serum-binding proteins. The reduced serum protein binding and more rapid clearance of T_3 from the circulation and binding to target cell receptors are responsible for the observed greater effectiveness of T_3 over T_4 when administered to humans or rats. The nuclear thyroid hormone receptors also have a greater affinity for T_3 than T_4. Many consider T_4 to be only a storage reservoir (a prohormone) from which T_3 is synthesized peripherally.

Low levels of thyroglobulin have been reported in the circulation of several mammals under normal conditions. Levels of about 4 ng of thyroglobulin per milliliter have been found in humans. Elevated plasma thyroglobulin occurs in the presence of thyroid carcinomas.

D. Peripheral Metabolism of Thyroid Hormones

Thyroxine has several metabolic fates after being released from the thyroid gland. Most T_4 is deiodinated to either T_3 or rT_3. Thyroid hormone metabolites are illustrated in Chap-

Table 7-6. Types of Deiodinase Activity

Characteristic	Type of deiodinase		
	I	II	III
Location	Liver, kidney, thyroid	Brain, pituitary, placenta, brown adipose tissue (BAT)	Brain, skin, placenta
Substrate preference	Inner or outer ring	Outer ring	Inner ring
Effect of PTU[a]	Inhibition	No effect	No effect
Reactions catalyzed	$T_4 \rightarrow T_3$	$T_4 \rightarrow T_3$	$T_3 \rightarrow T_2$
	$T_4 \rightarrow rT_3$	$rT_3 \rightarrow T_2$	$T_4 \rightarrow rT_3$
	$rT_3 \rightarrow T_2$		
	$T_3 \rightarrow T_2$		

[a] PTU, Propylthiouracil; rT_3, reverse T_3.

ter 2 (Fig. 2-37). Approximately one-half of the T_4 entering the liver is converted to rT_3 by deiodination. However, as a result of this rapid clearance by the kidney, rT_3 levels in the blood are rather low. In most mammals, the majority of circulating T_3 has its origin from deiodination of T_4 in the liver. Increases or decreases in circulating T_3 levels are always accompanied by reciprocal changes in rT_3 levels.

A second pathway for T_4 metabolism is conversion to **tetraiodothyroacetate (TETRAC),** which has no significant physiological activity and is excreted rapidly in urine or bile. Some T_3 similarly is converted to **triiodothyroacetate (TRIAC),** but the majority of nondeiodinated T_3 is conjugated to sulfate for excretion in humans. Some T_3 and T_4 are eliminated intact via the feces as a result of hormone loss through the hepatic–intestinal route.

Three types of deiodinase (Table 7-6) are recognized in mammals. **Type I deiodinase** is found primarily in liver, kidney, and thyroid. This enzyme will deiodinate either the outer or inner ring of iodinated thyronines and is strongly inhibited by the antithyroid drug, propylthiouracil (see IV,B,3). Type I deiodinase has a strong preference for rT_3 and, second, for T_4. Some rT_3 is rapidly deiodinated to **diiodothyronine (T_2).** Peripheral conversion of T_4 to both rT_3 and T_3 also occurs in the liver. **Type II deiodinase** occurs primarily in brain, pituitary, brown adipose tissue (rodents), and the placenta. It prefers T_4 over rT_3 as a substrate, deiodinates only the outer ring, and is responsible for intracellular levels of T_3 in these target tissues. **Type III deiodinase** also is found in the brain and placenta as well as in the intestine and fetal skin. This enzyme attacks only the inner ring, preferring T_3 as a substrate, although it can convert T_4 into rT_3 as well. Neither type II nor type III deiodinase is affected by propylthiouracil.

The dynamics of thyroid hormone metabolism involve not only the interactions between free and bound hormones but also their distributions in the body. Deiodination rates differ markedly in different tissues, and complicated models have been proposed to evaluate thyroid hormone kinetics (see McNabb, 1993, in Suggested Reading section for a discussion and introduction to the literature).

E. Mechanism of Action of Thyroid Hormones

Thyroid hormones, because of their somewhat hydrophobic nature, readily enter target cells, where T_4 is converted to T_3 by a cytoplasmic deiodinase. T_3 diffuses into the nucleus and binds to nuclear receptor proteins (see Chapter 2). Occupied receptors influence the synthesis of new proteins by regulating gene transcription. Nuclear receptors for thyroid hormones have greater affinity for T_3 than for T_4, supporting the hypothesis that conversion of T_4 to T_3 is a requisite for thyroid hormone action. Mitochondrial receptor proteins for thyroid hormones also have been demonstrated, and these mitochondrial receptors may be associated with observed effects on mitochondrial protein synthesis and oxidative metabolism. Unoccupied receptors for thyroid hormones have not been demonstrated in the cytosol.

Some permissive actions of thyroid hormones may be a consequence of effects of thyroid hormones on nuclear-directed synthesis of adenylate cyclase or on availability of ATP through their actions on mitochondria or on both. The levels of adenylate cyclase and ATP would influence the effects of hormones that normally produce their actions through some cAMP-dependent mechanism (see Chapter 2). Another mechanism for their permissive actions could be related to thyroid hormone-induced synthesis of receptor proteins for other regulators.

IV. Factors That Influence Thyroid Function in Mammals

The hypothalamus and pituitary exert direct control over thyroid gland functions but in turn are influenced by environmental factors working through the nervous system. Many other factors, such as diet, influence the thyroid state in mammals, which may influence processes controlled by other hormones.

A. Endocrine Factors Affecting Thyroid Gland Function

The hypothalamus exerts regulatory control over release of TSH from the pars distalis via secretion of thyrotropin-releasing hormone (TRH; see Chapter 4). Thyroid hormones themselves play important roles through negative feedback on the thyroid axis. In addition, the pineal gland may have a negative influence on thyroid function under certain conditions.

1. Thyrotropin-Releasing Hormone and Thyrotropin

The hypothalamus is the source of TRH that is released from the median eminence and travels to the pituitary thyrotrope via the hypothalamo–hypophysial portal system. Thyrotropin release is stimulated by TRH, which in turn produces an increase in circulating thyroid hormones. Both synthesis and release of TSH are regulated through an inositol 1,4,5-trisphosphate (IP_3) second-messenger system (see Chapter 2). Calcium ions also are involved in release of TSH from the thyrotropic cell.

Thyrotropin enhances uptake of radioiodide and the synthesis of thyroglobulin and thyroid hormones. In addition, TSH induces endocytosis and hydrolysis of thyroglobulin, causing thyroid hormones to be released into the blood. Continued stimulation by TSH causes structural changes in the follicular cells that are related to thyroid hormone synthesis and release. The follicular cells in an inactive or unstimulated follicle are usually flat or squamous cells. Thyrotropin can cause such flat cells to assume a cuboidal or even columnar shape, resulting in visible thickening of the follicular epithelium. Much of this enlargement of the follicular cells is due to an increase in rough endoplasmic reticulum and Golgi apparatus for thyroglobulin synthesis. This increase in cellular size due to increased cellular growth is referred to as hypertrophy. Chronically stimulated thyroid glands may exhibit hyperplasia as well, which is an increase in cellular numbers due to mitotic divisions by the stimulated cells. Hypertrophy or hyperplasia or both can lead to formation of a goiter.

The increase in the cellular portion of the follicle due to hypertrophy and the concomitant reduction in colloid are reflected in a change in the diameter of the follicle with respect to the thickness of the epithelium or to the volume of the lumen. The ratio of follicle diameter to thickness of the epithelium or diameter of the follicular lumen changes predictably with TSH levels and is frequently used as a measure of the degree of stimulation by TSH. Generally, a "stimulated" histology is indicative of thyroid hormone deficiencies and enhanced TSH secretion to compensate for these deficiencies. Other factors, such as cold stress, may be operating at the hypothalamus, however, to elevate TSH secretion above that normally maintained through negative feedback by the thyroid hormones. This effect also may cause follicular cell hypertrophy.

One of the first cellular events that occurs in follicular cells following administration of TSH is activation of adenylate cyclase and resultant increase in the intracellular levels of cAMP. It is not clear which of the succeeding cellular events are mediated by cAMP (that is, iodide uptake, thyroglobulin synthesis, formation of organic iodide, or engulfment and hydrolysis of colloid). Coincident with increased iodide uptake, formation of organic iodide, and endocytosis of colloid is an increase in glucose oxidation that may be caused by cAMP. Glucose oxidation is thought to be the "driving force" for both endocytosis and iodination, the latter involving oxidation of pyridine nucleotides and formation of H_2O_2. By controlling reactions such as glucose oxidation, cAMP could mediate several different cellular events associated with the action of TSH on the follicular cells.

2. Triiodothyronine and Thyroxine Feedback Effects

The release of TSH is regulated by negative feedback produced by thyroid hormones, and the administration of exogenous thyroid hormones decreases circulating TSH and associated thyroid gland activities. The major site for negative feedback is on the thyrotropic cells directly and not the hypothalamic thyrotropic center responsible for TRH production. Thyrotropes contain receptors that bind T_3 more effectively than T_4. Occupied thyroid hormone receptors interfere with the cAMP-dependent releasing mechanism by stimulating synthesis of an inhibitory protein or peptide. Type II deiodinase present in thyrotropes converts most of the T_4 to T_3, which enhances feedback. Thyrotropin levels seem to be maintained by direct negative feedback, and the role of TRH may be to override the system during times of increased demand for thyroid hormones. In other words, thyroid hormones determine the level of TSH secretion that regulates daily thyroid gland activities whereas the hypothalamus adjusts that level through TRH secretion as dictated by other neural factors and/or by environmental cues.

Evidence has been reported for a stimulatory role by thyroid hormones on hypothalamic TRH release. These experimental observations have not established this as a major regulatory pathway in mammals, but they do provide the basis for further investigation into the mechanism whereby hypothalamic control can override the adenohypophysial set point under conditions of increased demand for thyroid hormones.

3. The Epiphysial Complex

The pineal gland of the epiphysial complex in mammals has been implicated as a factor regulating thyroid function (see also Chapter 6). This action on thyroid function is probably mediated through inhibitory effects on hypothalamic TRH release. Numerous studies have demonstrated an inhibitory action on the thyroid gland by melatonin, one of the principles synthesized and released from the pineal gland. Photoperiodic influences on thyroid activity also may be mediated through inhibition or stimulation of the pineal gland.

B. Nonendocrine Factors

Thyroid activity is affected by a number of chemicals, including the levels of iodide and other chemicals in the diet that block one or another biochemical step in thyroid secretion. In addition, a number of environmental factors, such as photoperiod and temperature, can influence the activity of this system.

1. Diet

Low iodide availability reduces the synthesis of thyroid hormones and leads to development of hypothyroid states. Most seriously affected by the reduction in iodide is the synthesis of DIT, which in turn reduces the proportion of T_4 that can be synthesized.

Conversely, an excessive level of blood iodide inhibits uptake and accumulation of iodide by the follicular cells, presumably by inhibiting the iodide-pumping mechanism. In nature, it would be most unusual for a mammal to be subjected to an excess of iodide, but several instances of excessive iodide are known for humans. Ingestion by Japanese fisherman of large quantities of seaweed, which is rich in natural iodide, induces a hypothyroid state. In the early 1980s, occurrence of high levels of iodide in milk and fast foods due to artificial additives or contaminants raised medical concerns of potential induction of hypothyroidism, especially in children whose nervous systems would be especially sensitive to insufficient thyroid hormone. However, the low levels of iodide added to commercial salt preparations do not pose a threat but represent an important deterrent to prevent hypothyroidism from insufficient iodide availability.

Reduced caloric intake or fasting depresses circulating T_3 with (in rats) or without (in humans) depression of T_4 levels. This is an adaptive response that limits growth or metabolic rate when energy sources are low. Studies in rats suggest that overfeeding (especially of carbohydrates) elevates T_3 levels.

2. Inhibitors of Iodide Uptake

Certain anions are effective in blocking accumulation of iodide by the follicular cells through competitive inhibition of iodide transport. **Thiocyanate (SCN⁻), perchlorate ions (ClO_4^-), and pertechnate (TcO_4^-)** are particularly effective at blocking iodide uptake. These agents can be used to block thyroid function and particularly to block iodide uptake mechanisms. Because of its ability to compete with iodide for uptake by thyroid cells, radioactive pertechnate may be used in place of radioiodide for determining thyroid activity. Pertechnate may be useful as a blocking agent since it is less toxic than some other agents, especially SCN⁻. Labeled pertechnate is also less active than radioiodide and presents a lower radiation exposure to patient as well as to medical technicians and researchers.

3. Inhibitors of Iodination

Compounds that interfere with thyroid hormone synthesis by inhibiting iodination of tyrosines are often termed **goitrogens.** The resultant reduction in circulating hormones causes increased TSH secretion as a consequence of reduced negative feedback. Continuous stimulation of the thyroid gland by TSH results in enlargement of the thyroid and production of a goiter. Goitrogens, in addition to blocking formation of active iodide, also induce goiter formation. Many of these compounds secondarily inhibit iodide uptake by increasing the size of the intracellular inorganic iodide pool. Because agents that selectively inhibit iodide uptake secondarily block thyroid hormone synthesis and can lead to goiter formation, these inhibitory anions, such as $HClO_3^-$, are often termed goitrogens, too.

Several synthetic drugs are capable of blocking formation of active iodide, including the classically used drugs, **thiourea (TU)** and **propylthiouracil (PTU),** as well as newer drugs such as **methimazole** (Fig. 7-6). These drugs interfere with the peroxidase system responsible for generation of H_2O_2. Such drugs can be used to "chemically thyroidectomize" an animal reversibly. Certain reduced compounds such as ascorbic acid, reduced glutathione, and reduced pyrimidines remove H_2O_2 from the system and also block formation of active iodide.

Many flowering plants of the family Brassicaceae (cabbage, Brussels sprouts, rutabaga, and turnips) naturally contain a compound known as **progoitrin** that can be enzymatically converted to a goitrogenic compound called **goitrin** (Fig. 7-6). If sufficient quantities of goitrin are absorbed from the intestine into the general circulation, the synthesis of thyroid hormones is impaired and a hypothyroid state ensues. People or animals that consume large quantities of these plants are at risk for hypothyroidism, which may be accentuated if coupled with an iodide-poor diet. Cooking the plants normally destroys the enzyme that converts progoitrin to goitrin, but progoitrin itself is not affected by the quantity of heat applied in cooking the vegetables. Bacteria in the human intestinal flora are capable of converting all ingested progoitrin into goitrin, reversing the effect of cooking.

Some vascular plants such as cauliflower contain a glycoside of thiocyanate that can be converted to free thiocyanate in the body. One would have to ingest about 10 kg of cauliflower per day to produce any serious effects on thyroid function unless dietary iodide was extremely low.

4. Inhibitors of Deiodination and of Receptor Binding

Deiodination of T_4 to T_3 can be prevented by treatment with drugs such as **iopanoate** or **ipodate.** These drugs were injected initially as radioopaque agents that were removed from the blood by the liver and secreted into bile, where they aided visual examination of the gallbladder in which they accumulated. They later were discovered to be very potent blockers of deiodinating enzymes. It is also possible to block thyroid hormone action with the drug **amiodarone,** which binds to thyroid receptors and blocks binding of thyroid hormones. This drug was first used to treat cardiac arrhythmia. We now know that amiodarone also is an inhibitor of deiodinase activity.

5. Environmental Factors

Environmental factors such as photoperiod and temperature may influence thyroid hormone secretion rates through neurocrine or endocrine agents. Such factors may influence synthesis and release of hypothalamic and hypophysial hormones or may alter thyroid function through sympathetic innervation of the thyroid gland itself.

Internal biological clocks may be related to the actions of environmental factors in regulating thyroid cycles. Cyclical variations have been reported for thyroid hormones on both a diurnal and seasonal basis. Internal secretory rhythms of hypothalamic regulators might

A

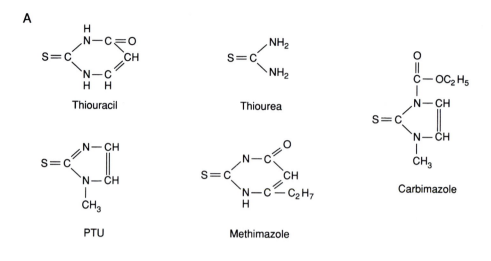

Thiouracil

Thiourea

Carbimazole

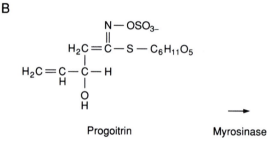

PTU

Methimazole

B

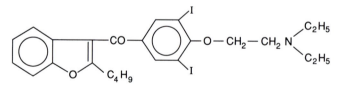

Progoitrin

Myrosinase

Goitrin (5-Vinyl-2-thiooxazolidone)

C

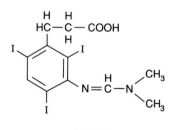

Amiodarone

Ipodate

be influenced by environmental factors or might regulate the sensitivity of other effectors to external factors. Feedback by thyroid hormones may alter sensitivity as well.

V. Biological Actions of Thyroid Hormones in Mammals

Thyroid hormones affect many diverse tissues and influence major processes such as metabolism, growth, differentiation, and reproduction. They are responsible for maintaining a general state of well-being for many cells so that they are capable of maximal responses to other stimuli. Although several effects of thyroid hormones are discussed below, this "permissive action" of thyroid hormones may be their single most important role.

A. Metabolic Actions

The effects produced by thyroid hormones on mammalian metabolism include a calorigenic or **thermogenic action** (heat generating) and specific effects related to carbohydrate, lipid, and protein metabolism. These actions of thyroid hormones become more meaningful when considered together with the actions of other hormones on metabolism (see Chapter 14). Many of these metabolic actions are possibly permissive actions occurring in cooperation with other hormones such as epinephrine and growth hormone.

Thyroid hormones cause thermogenic actions in certain tissues and may be involved in certain physiological responses to cold stress. They can accelerate the rate at which glucose is oxidized and thus increase the amount of metabolic heat produced in a given time. This excess heat production could be used to warm the body. Increased glucose oxidation is reflected in an increased basal metabolic rate (BMR) as measured by the change in rate of oxygen consumption. Decreased nutrient intake operates through neural mechanisms to reduce thyroid hormone secretion and lower metabolic rate. In addition, thyroid hormones have been claimed to "uncouple" oxidative phosphorylation, which would decrease the efficiency of ATP synthesis in the mitochondria and increase the quantity of heat released per mole of glucose oxidized. Although such uncoupling may occur in a hyperthyroid pathological state known as thyroid storm (see Section VI,A), its postulated role in chronic cold stress is probably not so important as the ability of thyroid hormones to increase the total rate of glucose oxidation.

Thyroid hormones induce increased synthesis of several mitochondrial respiratory proteins, especially cytochrome c, cytochrome oxidase, and succinoxidase. This mitochondrial action to augment oxidative metabolism would be advantageous in adapting to chronic cold stress. In general, thyroid activity in mammals is greater during prolonged periods of cold stress (winter) than during warmer periods. Acute cold responses probably are mediated by epinephrine from the adrenal medulla (Chapter 9) rather than by thyroid hormones.

In many nonhibernating mammals such as beaver and muskrat, thyroid activity is depressed during the winter months. Hypothyroidism has been described for hibernating ground squirrels and badgers, but there does not appear to be a causal relationship between

Figure 7-6. Thyroid inhibitors. (A) Thiourea, thiouracil, propylthiouracil (PTU), carbimazole, and methimazole are all goitrogens that block iodide uptake and/or the iodination and coupling reactions (Fig. 7-5). (B) Goitrin is a naturally occurring goitrogen that is made from a precursor called progoitrin by the enzyme myrosinase. (C) Ipodate and amiodarone block liver deiodinases. PTU also blocks type-1 deiodinase activity (see Table 7-6).

reduced thyroid function and the onset of hibernation. Additional field studies employing sophisticated methods for assessing thyroid functions are needed before the endocrine factors related to either onset or termination of hibernation will be established.

In addition to increasing glucose oxidation, thyroid hormones cause hyperglycemia and may secondarily stimulate lipolysis (hydrolysis of fats). These actions may in part be associated with potentiation of the hyperglycemic and lipolytic actions of epinephrine (see Chapters 9 and 14). Thyroid hormones alter nitrogen balance and are either protein anabolic or catabolic, depending on the tissue being examined and under what experimental conditions it is examined. These actions are probably related to enhancement of effects normally regulated by other hormones.

B. Growth and Differentiation

Thyroid hormones are essential for normal growth and differentiation in mammals as evidenced in cretinism and juvenile myxedema in humans (see Section VI,B). These growth-promoting actions of thyroid hormones are closely related to the role of pituitary growth hormone (GH), and they probably represent a permissive action on GH-sensitive target cells. Thyroid hormones also may stimulate insulin-like growth factor (IGF) production and hence augment the action of GH.

The major tissue affected by the lack of thyroid hormones during differentiation is the nervous system. Normal development of the nervous system as well as attainment of normal mental capacities are strongly influenced by thyroid hormones. Hypothyroidism during gestation and up to about 2 years of age in humans, as for comparable time periods in other mammals, seriously impairs differentiation and functioning of the nervous system. A reduction in mental capacity can occur in hypothyroid adults, too.

Replacement of hair in adult mammals is stimulated by thyroid hormones. The postnuptial molt cycle in harbor seals, *Phoca vitulina,* involves thyroid hormones and cortisol from the adrenal cortex. Hair loss is correlated with low thyroid function and high cortisol levels, whereas resumption of hair growth is correlated with increased T_4 and return of cortisol to basal levels. Thyroid activity also is related to molting of hair in other mammals, including red fox and mink.

C. Reproduction

Another cooperative role for thyroid hormones occurs with respect to gonadal development and function (see Chapter 11). In general, sexual maturation is delayed in hypothyroid mammals. In hypothyroid males, spermatogenesis may occur, but androgen synthesis is low. Ovarian weight is reduced and ovarian cycles become irregular in hypothyroid females. These correlations to hypothyroidism have been attributed to reduced gonadotropin levels and can be alleviated by treatment with thyroid hormones. Experimental studies support the notion that thyroid hormones influence gonadotropin release through an effect at the level of the hypothalamus. Reduction of gonadotropin, however, was not observed in hypothyroid female rats.

VI. Clinical Aspects of Thyroid Function

Although thyroid disorders are among the most common human endocrine problems (second only to diabetes mellitus), these disorders are mostly underactive thyroid conditions and include inherited disorders as well as environmentally induced disorders. Hyperactive thyroid disorders are less common, and thyroid carcinomas are rare. Either hypothyroid or hyperthyroid conditions may be associated with goiter formation. Generalized symptoms of thyroid disorders are listed in Table 7-1.

A. Thyrotoxicosis and Hyperthyroidism

Thyrotoxicosis is a general term referring to an excess of thyroid hormone. If this condition results from thyroid hypersecretion, it is known as **hyperthyroidism.** Primary hyperthyroidism may be due to toxic multinodular goiters consisting of multiple aggregates of small, hyperactive follicles **(Marine–Lenhart syndrome),** or several large TSH-dependent hyperactive follicles **(Plummer's disease).** Follicular adenomas are sometimes autonomously hyperactive as well. Circulating TSH levels typically are low when autonomously hyperactive multinodular goiter or adenomas are present.

Hyperthyroidism may be of a secondary nature caused by a rare pituitary adenoma of TSH-secreting cells. Certain cancerous tumors such as choriocarcinomas may elaborate TRH-like or TSH-like molecules that stimulate thyroid activity. Secretory tumors of these types typically are insensitive to any feedback by thyroid hormones.

Graves' disease is a secondary hyperthyroid state that may be mediated by an immunoglobulin known as **LATS (long-acting thyroid stimulator).** The basis for production of LATS is not understood. Apparently, LATS binds to TSH receptors and activates thyroid cell functions even in the absence of circulating TSH. These individuals present a hyperthyroid state with very low circulating TSH. Graves' disease is much more common in women than men. Reduction of thyroid tissue by radioiodide treatment is the most common form of treatment for Graves' disease, although surgical removal of a portion of the thyroid is sometimes employed.

Rarely, hyperthyroidism is due to a nonthyroid source of thyroid hormones. For example, ovarian dermoid tumors can synthesize sufficient thyroid hormones to bring about hyperthyroidism.

Juvenile thyrotoxicosis is characterized by nervousness, tremor, accelerated heart rate, and goiter. This syndrome is manifest in children beyond age 10 years (80% of the cases). Exophthalmus (protrusion of the eyeballs) occurs in about half of these children, and weight gain usually is retarded.

A somewhat rare but dramatic condition is thyrotoxic crisis or **thyroid storm.** This disorder involves a sudden increase in thyroid secretion, severe hypermetabolism, fever, and other, more variable symptoms. It occurs only in hyperthyroid patients and may be precipitated following incomplete thyroidectomy, interruption of antithyroid therapy, or even as a reaction to an infection or tooth extraction. It also may be induced by periods of excessive summer heat. The actual cause of thyroid storm is not certain and the condition could encompass a variety of different disorders.

B. Myxedema and Hypothyroidism

Myxedema is the extreme condition in which no thyroid hormones are secreted. In these patients swelling of the skin and subcutaneous tissues is caused by the extracellular accumulation of a high-protein fluid. Hypothyroidism refers to conditions of insufficient thyroid hormones of primary (at the thyroid) or secondary (hypothalamus or pituitary) origins. It is especially serious in children because of marked effects on both general and neural development. The term **juvenile hypothyroidism** refers to cases of hypothyroidism in children that do not lead to severe retardation in somatic and intellectual development. When development is markedly retarded, it is called **cretinism.** This syndrome is remarkably common and 1 in every 8500 births exhibits cretinism. However, if hypothyroidism is detected at birth by using a simple thyroid test, it can be alleviated with thyroid hormone therapy so that growth and development are normal. However, because of poor medical care and widespread iodide deficiency, cretinism is a serious, crippling disorder in many developing countries.

There are a number of symptoms characteristic of hypothyroidism, including rough and dry skin, yellow pallor, coarse scalp hair, hoarse voice, and slow thought and action (Table 7-1). However, sometimes the hypothyroid person exhibits none or only a few of these symptoms. Obesity is often listed as a characteristic but it does not always accompany hypothyroidism and occasionally occurs in patients presenting hyperthyroidism. Conversely, persons suffering from secondary hypothyroidism are often thin, as is frequently the case for hyperthyroid patients. Exophthalmus, usually correlated with hyperthyroid states, may occur occasionally in primary myxedema.

C. Goiters

Any enlarged thyroid is referred to as a goiter regardless of the cause or nature of the enlargement. Actually, there are four kinds of clinical goiters. The first kind is a hypothyroid goiter caused by failing hormone production resulting in a shortage of T_3 and T_4. Circulating TSH levels are elevated because of reduced negative feedback, and this increased TSH causes enlargement of the thyroid gland and formation of a goiter.

The second and third types are hyperthyroid goiters: the **hyperfunctioning goiter** and the **hyperfunctioning goiter of pregnancy.** They are not as common as hypothyroid goiters. In the first case, circulating thyroid hormones are high and TSH levels are low. Diffuse thyrotoxic goiter is often termed Graves' disease, whereas toxic nodular goiters are associated with Plummer's disease or Marine–Lenhart syndrome. The second case of hyperthyroid goiter occurs during normal pregnancy. There is an increase in thyroxine-binding globulins and a consequent decrease in free T_4 and T_3 in maternal plasma. This results in elevated TSH through decreased negative feedback and in a slight thyroid enlargement to maintain normal functional levels of thyroid hormones during pregnancy. This condition usually returns to normal in women after birth.

The fourth kind of goiter develops in people with otherwise normal thyroid function. Such enlargements have many different causes, including infiltration of the gland with tuberculosis bacteria, syphilitic bacteria, or parasites as well as by the presence of adenomas or carcinomas. Inflammation due to autoimmune disease (thyroiditis; see below) can also cause enlargement of the thyroid.

D. Thyroiditis

Thyroiditis is a general term often applied to a collection of autoimmune disorders involving production of antibodies that attack thyroid proteins, especially peroxidase and thyroglobulin. Thyroid cells also may be attacked by lymphocytes, which in some cases are attracted by the actions of antibodies with thyroid antigens. Thyroiditis is more prevalent in women than in men. It is sometimes the cause of hypothyroidism that follows pregnancy because of a rebound in immune activity following suppression during pregnancy.

The most common types are **Hashimoto's thyroiditis** (struma lymphomatosa) and **Reidel's thyroiditis** (struma fibrosa). Hashimoto's disease is characterized by the presence of numerous lymphoid follicles, diffuse infiltration by lymphocytes, extensive increase in the connective tissue components of the gland, some changes in follicular structure, and a marked reduction in production of thyroid hormones. One of the most common causes for goiter among adolescents is Hashimoto's thyroiditis. In Reidel's thyroiditis, the gland is progressively replaced by fibrous connective tissue. In either type of thyroiditis, a gradual reduction in circulating thyroid hormones brings about an elevation of TSH secretion (negative feedback) that causes goiter formation. Antibodies produced against thyroid iodoproteins and peroxidase are demonstrable in the blood of these patients.

E. Inherited Disorders

More than 20 heritable defects in thyroid gland metabolism are known and may be classified according to the type of causative metabolic defect. For example, there are several defects related to unresponsiveness of thyroid cells to TSH, including insufficient numbers of receptors or abnormal receptors. There also have been numerous mutations discovered in the gene responsible for synthesis of a particular thyroid hormone receptor. TRβ-1 (see Chapter 2). These mutations are linked to generalized resistance to thyroid hormones. In a number of these cases, resistance to thyroid hormones has been correlated with **attention-deficit hyperactivity disorder (ADD).**

Inherited defects also occur in iodide transport mechanisms, iodination, and coupling of iodinated tyrosines. Several defects have been identified in relation to synthesis of abnormal thyroglobulins. Deficiencies in thyroid deiodinase and total body deiodinase as well as the presence of abnormal plasma transport proteins sometimes occur as heritable errors.

F. Euthyroid Sick Syndrome

Many nonthyroid conditions can alter thyroid function, so that even though the person is actually euthyroid, he/she will appear to be hypothyroid. For example, fasting, anorexia nervosa, protein–calorie malnutrition, and untreated diabetes mellitus are all metabolic disturbances that bring about marked decreases in circulating T_3 and increases in rT_3. Thyroid hormones also are altered in a variety of liver and renal diseases as a consequence of numerous infections and following myocardial infarctions. Some conditions may alter levels of T_4, TSH, T_3, and/or rT_3. These alterations in thyroid parameters by peripheral disorders constitute the so-called **euthyroid sick syndrome,** reflecting normal responses of the euthyroid person to the generalized disease state that present superficially as abnormal thyroid function.

VII. Summary

The functional unit of the vertebrate thyroid gland is the thyroid follicle, which is unique for storing thyroid hormones (T_3 and its prohormone T_4) extracellularly in the colloid of the follicular lumen. Normal thyroid gland functioning depends on a constant supply of iodide in the diet as well as on regulatory stimuli from the hypothalamus (TRH), and the adenohypophysis (TSH). Iodide is accumulated by the follicular cells and at the apical cell surface is incorporated as iodine into tyrosine residues of large glycoproteins collectively referred to as thyroglobulin. Some of the iodinated tyrosines (MIT and DIT) are coupled to form iodinated thyronines (T_3 and T_4), and the iodinated thyroglobulins are stored in the follicular lumen. Endocytosis of colloid droplets followed by intracellular hydrolysis of thyroglobulin in endolysosomes releases iodinated thyronines and allows them to diffuse into the circulation. Intracellular MIT and DIT are enzymatically degraded (deiodinase) to tyrosine and inorganic iodide for new synthesis of thyroglobulin and subsequent iodination of its tyrosine residues. Both synthesis and release of thyroid hormones are stimulated by TSH. Antithyroid drugs, including certain anions (SCN^-, CIO_4^-, and TcO_4^-) and the "traditional" goitrogens (PTU, TU, and methimazole) interfere with iodide uptake or the iodination process or with both, causing thyroid deficiencies.

Thyroid hormones are transported in the circulation bound to plasma proteins (TBG, TBPA, and TBA), and only a small proportion of free hormones is present. Thyroxine is more tightly bound to plasma proteins than T_3, and this greater affinity of T_4 for plasma proteins is partly responsible for its much longer biological half-life. Free T_3 apparently enters cells more rapidly than does T_4. Metabolism of most of the T_4 to T_3 and rT_3 occurs in the liver by the action of a type I deiodinase. Since most of the circulating T_3 arises peripherally from T_4, it has been suggested that T_4 acts as a reservoir for T_3, which is the "active form" of the hormone. Mitochondrial and nuclear receptors have been prepared from target tissues. These receptors have greater affinity for T_3 than for T_4, which supports the hypothesis that deiodination of T_4 to T_3 may be the important first step in its mechanism of action.

It is difficult to characterize the actions of thyroid hormones with respect to their functions. In addition to their direct actions on metabolism, thyroid hormones play permissive roles. They maintain "responsiveness" in many cells that allows these cells to become more sensitive to other endocrine or neural stimuli. Thyroid hormones also cooperate with other hormones to enhance their actions. Processes that are affected by permissive or cooperative actions include growth, metabolism, and reproduction.

The action of thyroid hormones on peripheral tissues following specific binding of the hormone to nuclear receptors is followed by activation or inhibition of mRNA transcription and eventual enzyme syntheses. The permissive actions of thyroid hormones may be related to such events as the stimulation of the synthesis of components of second-messenger systems, up-regulation of receptors for another regulator, effects on structural components, etc. Binding of thyroid hormones to mitochondrial receptors or indirect actions on mitochondria, especially in liver and kidney, may be important in the thermogenic and oxidative actions of thyroid hormones on the BMR.

Suggested Reading

Books

Braverman, L. E., and Utiger, R. D. (1996). "Werner and Ingbar's The Thyroid: A Fundamental and Clinical Text," 7th Ed. Lippincott-Raven, Hagerstown, MD.
Greer, M. A. (1990). "The Thyroid Gland. Comprehensive Endocrinology." Rev. Ser. Raven, New York.
McNabb, F. M. A. (1993). "Thyroid Hormones." Prentice-Hall, Englewood Cliffs, NJ.

General Articles

Benvenga, S., and Robbins, J. (1993). Lipoprotein–thyroid hormone interactions. *Trends Endocrinol. Metab.* **4,** 194–198.
Carrasco, N. (1993). Iodide transport in the thyroid gland. *Biochim. Biophys. Acta* **1154,** 65–82.
Christophe, D., and Vassart, G. (1990). The thyroglobulin gene: Evolutionary and regulatory issues. *Trends Endocrinol. Metab.* **1,** 356–362.

Danforth, E., Jr., and Burger, A. G. (1989). The impact of nutrition on thyroid hormone physiology and action. *Annu. Rev. Nutr.* **9,** 201–227.

Dumont, J. E., Maenhaut, C., and Lamy, F. (1992). Control of thyroid cell proliferation and goitrogenesis. *Trends Endocrinol. Metab.* **3,** 12–17.

Franklyn, J. A., and Sheppard, M. C. (1994). Amiodarone and thyroid function. *Trends Endocrinol. Metab.* **5,** 128–131.

Hafner, R. P. (1987). Thyroid hormone uptake into the cell and its subsequent localisation to the mitochondria. *FEBS Lett.* **224,** 251–256.

Maayan, M. L., Volpert, E. M., and Debons, A. F. (1987). Neurotransmitter regulation of thyroid activity. *Endocr. Rev.* **13,** 199–214.

Rodriguez-Arnao, J., Miell, J. P., and Rosa, R. J. M. (1993). Influence of thyroid hormones on the GH–IGF-l axis. *Trends Endocrinol. Metab.* **4,** 169–173.

St. Germain, D. L. (1994). Iodothyronine deiodinases. *Trends Endocrinol. Metab.* **5,** 36–42.

Silva, J. E. (1993). Hormonal control of thermogenesis and energy dissipation. *Trends Endocrinol. Metab.* **4,** 25–32.

Studer, H., and Gerber, H. (1991). Intrathyroidal iodine: Heterogeneity of iodocompounds and kinetic compartmentalization. *Trends Endocrinol. Metab.* **2,** 29–34.

Clinical Articles

Edwins, D. L., and McGregor, A. M. (1990). Pregnancy and autoimmune thyroid disease. *Trends Endocrinol. Metab.* **1,** 296–300.

Fisfalen, M.-E., and DeGroot, L. J. (1995). Graves' disease and autoimmune thyroiditis. *In* "Molecular Endocrinology: Basic Concepts and Clinical Implications" (B. D. Weintraub, ed.), pp. 319–370. Raven, New York.

Gaitau, E., and Dunn, J. T. (1992). Epidemiology of iodine deficiency. *Trends Endocrinol. Metab.* **3,** 170–175.

Hauser, P., Zametkin, A. J., Martinez, P., Vitiello, B., Matochik, J. A., Mixson, A. J., and Weintraub, B. D. (1993). Attention-deficit-hyperactivity disorder in people with generalized resistance to thyroid hormone. *N. Engl. J. Med.* **328,** 997–1040.

O'Connor, G., and Davies, T. F. (1990). Human autoimmune thyroid disease: A mechanistic update. *Trends Endocrinol. Metab.* **1,** 266–274.

Usala, S. J., and Weintraub, B. D. (1991). Thyroid hormone resistance syndromes. *Trends Endocrinol. Metab.* **2,** 140–144.

Wong, T. K., and Hershman, J. M. (1992). Changes in thyroid function in nonthyroid illness. *Trends Endocrinol. Metab.* **3,** 8–12.

8

Comparative Aspects of Vertebrate Thyroids

IN THIS chapter, we discuss the functions of the thyroid gland in nonmammalian vertebrates, with some emphasis on the evolution of the thyroid and thyroid functions in vertebrates. The reader first should examine Chapter 7 for an understanding of mammalian thyroid functions at the cellular and organismic levels before attempting this chapter, since all basic definitions and explanations of thyroid function are provided in Chapter 7 and are not repeated here.

I. Evolution of the Thyroid Gland and Its Functions

The thyroid gland structurally is a conservative endocrine gland in vertebrates and is one of the oldest vertebrate endocrine organs. It is generally found as one or two masses of highly vascularized follicles surrounded by a connective tissue capsule or as scattered follicles throughout the pharyngeal region, as is the case in most fishes (Figs. 8-1 and 8-2). Regardless of any gross morphological differences, follicular structure and function are mammalian-like in all the gnathostomes with respect to iodide metabolism (Table 8-1), hormone synthesis, production and storage of **thyroglobulin** (Table 8-2), hormone release, and responsiveness to **thyrotropin (TSH).** Biochemical differences are quantitative and not qualitative, and the same thyroid hormones, **triiodothyronine (T_3) and thyroxine (T_4),** are present in all vertebrates (Table 8-3) and are found in some invertebrates as well (Table 8-4).

The most primitive thyroid condition is considered to be that found in the adult cyclostomes, in which only scattered follicles may be present. Extracellular storage of thyroid hormones does not occur in the cyclostome follicle, and iodoproteins are retained in the follicular cells of these primitive fishes. The follicles of all other vertebrates are organized and appear to function like those of mammals (see Chapter 7).

A. The Origin of the Thyroid Gland

Although many biochemical aspects of the synthesis of thyroid hormones are well known in vertebrates, the evolutionary origin of the vertebrate thyroid and acquisition of functional significance for these iodinated compounds are uncertain. The first evidence in chordates of formation of thyroid hormones comes from studies of the **endostyle** of protochordates, such as the tunicates and especially the cephalochordate, amphioxus. The endostyle is a mucus-secreting gland located in the pharyngeal region. This mucus is involved in the feeding mechanisms to trap food particles brought into the pharynx by water currents caused by movements of pharyngeal cilia. These cilia then move the mucus with trapped food into the digestive tract. Specialized cells near the opening of the endostyle of protochordates accumulate iodide from seawater and use it to iodinate tyrosines in glycoproteins secreted by the endostyle cells. Furthermore, some of those tyrosines are coupled to form the iodinated **thyronines,** T_3 and T_4.

However, acceptance of the simple notion that the thyroid cells evolved from specialized cells of the endostyle of protochordates leaves some basic questions unanswered. For example, why were these iodinated compounds first synthesized? What was their primitive role? What selective forces were responsible for adoption of these iodinated compounds as metabolic regulators? Certainly the ability of iodide to be incorporated into tyrosine residues in a protein and even formation of **3-monoiodotyrosine (MIT), 3,5-diiodotyrosine (DIT),** and iodinated thyronines occur repeatedly among the invertebrate phyla (Table 8-4). Relatively few studies have examined these iodinated molecules and their functions among invertebrates. In the primitive cnidarians, thyroid hormones are associated with a dramatic developmental change known as **metamorphosis.** This process, which occurs in vertebrates as well, not only includes specific gene activation resulting in biochemical, physiological, and behavioral changes but also a major revision of the ecological role of the organism results.

It has been hypothesized that these iodine-containing compounds were obtained originally

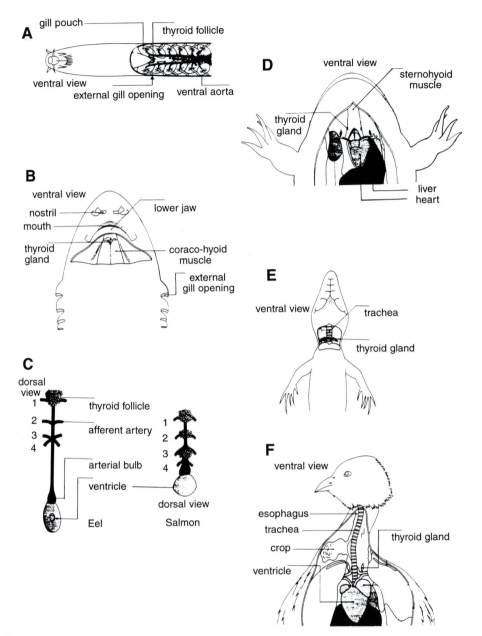

Figure 8-1. Location of thyroid tissue in nonmammalian vertebrates. (A) Scattered thyroid follicles in a hagfish (*Eptatretus burgeri*); (B) discrete thyroid gland of the shark, *Triakis scyllium*; (C) diffuse thyroids of the eel, *Anguilla japonica* (left), and the Pacific salmon, *Oncorhynchus masou* (right); (D) paired thyroids in the bullfrog, *Rana catesbeiana*; (E) medial thyroid gland in neck of the lizard, *Takydromas tachydromoides*; (F) paired thyroid glands in a bird, the Japanese quail, *Coturnix coturnix japonicus*. [Dissections of thyroid regions modified from Matsumoto, A., and Ishii, S. (1989). "Atlas of Endocrine Organs." Springer-Verlag, Berlin.]

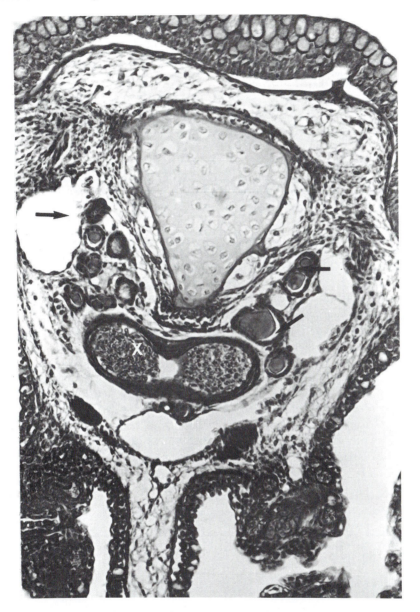

Figure 8-2. Salmon thyroid follicles. Cross-section through the lower jaw of a fingerling chinook salmon, *Oncorhynchus tshawytscha*, through the second arotic arch (x). Thyroid follicles (arrows) appear dorsal to the arch on either side.

from feeding activities of protochordates. These dietary compounds supposedly were first utilized opportunistically as regulatory substances of some sort. Absorption of T_4 liberated during digestion and its internal deiodination to T_3 might have been important prerequisites for development of a regulatory role. Later, the ability to synthesize these specific compounds was acquired, culminating in the concentration of such a mechanism in certain cells

Table 8-1. Some Effects of Thyrotropin and Thyroid
Inhibitors on [131]I Uptake[a] by Thyroids of Larval Salamanders,
Ambystoma tigrinum

N	Mean body weight (g ± SEM)	Daily injections; pretreatment for 7 days	Mean thyroid uptake (% injected dose ± SEM)
6	11 ± 0.5	None	4.6 ± 1.6
6	8 ± 0.6	0.25 μg TSH[b]	12.4 ± 2.5
6	11 ± 0.6	2.5 μg TSH	29.2 ± 4.7
6	17 ± 1.6	25 μg TSH	38.3 ± 1.0
6	9 ± 0.6	2.5 μg TSH + 10 μg PTU[c]	10.2 ± 2.9
6	7 ± 1.0	2.5 μg TSH + 0.5 mg NaSCN[d]	3.9 ± 0.7

[a] Radioiodide uptake was determined 24 hr after intraperitoneal injection of 5 μCi of [131]I.
[b] Ovine TSH (NIH-TSH-S6).
[c] PTU, Propylthiouracil.
[d] Sodium thiocyanate.

of the endostyle (Figure 8-3). These specialized cells later became the follicular cells of the thyroid gland as evidenced from studies in lampreys. Iodide-concentrating cells of the endostyle in the larval ammocete are reformed into follicles of the thyroid gland during metamorphosis into an adult lamprey. The production of extracellular colloid and its resultant accumulation in the follicular lumen do not occur in modern cyclostomes and presumably were acquired at a later time. Another hypothesis for the origin of these primitive iodoproteins is that they first served some enzymatic or structural function and that the basic mechanism became modified into a hormonal synthetic pathway.

There is a basic similarity among thyroid iodoproteins in all vertebrates. Thyroglobulins from birds are similar to those of rats, with about 94% as 19S, about 5% as 27S, and a trace of 12S thyroglobulin (Table 8-2). Turtles, teleostean fishes, and selachians exhibit lower

Table 8-2. Sedimentation Coefficients for Chordate
Iodoproteins (Thyroglobulins)

Species	Percent iodoproteins found			
	< 12S	12S	19S	27S
Urochordata				
Ciona intestinalis (tunicate)	100	—	—	—
Vertebrata				
Agnatha				
Lampetra fluviatilis (river lamprey)	32	68	—	—
Chondrichthyes				
Scyliorhinus stellaris (dogfish shark)	—	16	80	4
Osteichthyes				
Conger conger (Congo eel; teleost)	—	6	73	6
Reptilia				
Thalassochelis caretta (sea turtle)	—	11	84	5
Testudo hermanni (land turtle)	—	12	84	4
Aves				
Anas platyrhynchos (domestic duck)	—	Trace	94	6
Gallus gallus (domestic chicken)	—	Trace	94	5

Table 8-3. Levels of Thyroid Hormones for Selected Vertebrate Species

Class/Order	Species	Conditions	T_3	T_4
Agnatha/Cyclostomata	Petromyzon marinus (sea lamprey)	Mature females	0.5–1.77 nM	69.5–139 nM
		Ammocetes	—	4.9–18.5 µg/dl
		Adults	—	0.46 µg/dl
	Eptatretus stouti (Pacific hagfish)			2.2–10.5 µg/dl
Osteichthyes/Teleostei	Salvelinus fontinalis (brook trout)			0.15–3.36 µg/dl
	Paralichthys olivaceus (flounder)	Eggs	6–7 ng/g	1 ng/g
		At metamorphic climax	1–1.5 ng/g	10–15 ng/g
	Marone saxatilis (striped bass)	Eggs	4.5 ng/g	5.3 ng/g
		6 weeks postlarvae	6.5 ng/g	7.0 ng/g
Amphibia/Anura	Bufo marinus (cane toad)	Fertilized eggs	3.1 ng/g	0.44 ng/g
		At metamorphic climax	5.3 ng/g	8.4 ng/g
Amphibia/Caudata	Ambystoma tigrinum (tiger salamander)	Sexually mature larvae	0.07–0.73 ng/ml	0.25–9.0 ng/ml
Reptilia/Squamata	Naja naja (cobra)		—	1.25–1.55 µg/dl
	Trachydosaurus ragosus (shingleback lizard)	At preferred temperature	0.28 M	2.6 nM
	Calotes versicolor (garden lizard)	Prior to hibernation	1.2 ng/ml	4.3 ng/ml
		During hibernation	0.6 ng/ml	1 ng/ml
Reptilia/Crocodilia	Crocodylus johnstoni (freshwater crocodile)	At preferred temperature	0.51 nM	3.24 nM
Reptilia/Chelonia	Chelodina longicollis (long-necked tortoise)	At preferred temperature	0.28 nM	0.55 nM
Aves/Galliformes	Gallus gallus (domestic chicks, 1 day to 6 weeks old)		—	1.45–3.0 µg/dl
Mammalia	Axis axis (axis deer)	March	115 ng/dl	12.1 µg/dl
		April	70 ng/dl	4.5 µg/dl

Table 8-4. Occurrence of Thyroid Hormones and Their
Precursors in Animals and Algae

Organism	MIT	DIT	T_3	T_4
Algae	+	+		
Nereis (Annelida)	+	+	+	+
Sponge (Porifera)	+	+		
Coral (Cnidaria)	+	+		
Periplaneta (Arthropoda: cockroach)	+	+		+
Musca (Arthropoda: fly)	+	+		+
Planorbis (Mollusca)	+	+	+	+
Amphioxus (Cephalochordata)	+	+	+	+
Hagfish (Agnatha: Cyclostome)	+	+	+	+
All other vertebrates	+	+	+	+

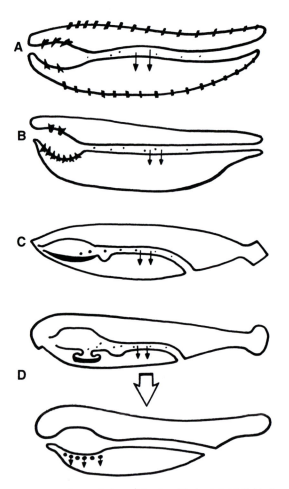

Figure 8-3. Generalized pattern for evolution of the thyroid gland. Initially, iodinated mucoproteins were distributed over the body surface and in the anterior digestive tract (A) and later became restricted more to the mouth region. Iodinated mucoproteins released from the mouth region entered the gut where they were digested, liberating iodinated tyrosines and thyronines (B). Amphioxus, the cephalochordate (C), has iodinated protein production confined to the endostyle in the mouth and pharynx. The metamorphosis of the endostyle to a thyroid gland occurs during the transformation of the ammocetes larva to the adult lamprey (D).

Table 8-5. Plasma Levels of Thyroglobulin in Bullfrogs and Chickens[a]

Species: stage/strain	Thyroglobulin (ng/ml)
Rana catesbeiana	
Premetamorphic (stage XX)	56
Prometamorphic (stage XV)	179
Metamorphic (stages XVIII–XXV)	347–477
Froglet	154
Adults	267
Chickens	
Newly hatched (three normal strains)	44–73
Newly hatched, obese strain[b]	140–186

[a] Chicks of the obese strain are highly susceptible to autoimmune thyroiditis.
[b] Data from Suzuki, S., and Fujikura, K. (1994). Circulating thyroglobulin levels in tadpoles and adult frogs of *Rana catesbeiana*. *Gen. Comp. Endocrinol.* **94,** 72–77; Sanker, A. J., Sundick, R. S., and Brown, T. R. (1983). Analysis of the serum concentration and antigenic determinants of thyroglobulin in chickens susceptible to autoimmune thyroiditis. *J. Immunol.* **131,** 1252–1256.

amounts of 19S (73–84%), but a significant quantity of 12S thyroglobulin is present (6–16%). Cyclostomes are the most primitive living vertebrates, and they exhibit what may be considered the most primitive condition. Lampreys produce largely a 12S component (68%), and the remainder of iodinated protein is in the form of a much smaller component (5.4S). Iodoproteins of about 7.6S have been isolated from urochordates and yield MIT, DIT, T_3, and T_4 on hydrolysis. These data suggest that thyroglobulin exists as 12S monomers (or an even smaller unit) that aggregate to form 19S and 27S complexes under certain conditions. These data also may be interpreted as reflecting degrees of disruption from different vertebrates of the intact thyroglobulin during the extraction procedures. Thyroglobulin has been reported in the circulation of chickens and frogs as it has been seen in mammals on occasion. Increases in circulating thyroglobulin occur in obese chicks and during metamorphosis of bullfrogs (see Table 8-5).

B. Regulation of Vertebrate Thyroids

In Chapters 4 and 5, we described the regulation of thyroid function by the hypothalamus (thyrotropin-releasing hormone, TRH) and the adenohypophysis (TSH). The focus for feedback regulation is the hypothalamus and pituitary. This system maintains a sufficient level of free T_4, which can be converted peripherally by type I deiodinase to T_3 in the liver and kidney or by type II deiodinase in target cells (see Chapter 7).

C. The Origin of Thyroid Function

There appears to be no consistency in the actions of thyroid hormones when the entire subphylum Vertebrata is surveyed, although in part the inconsistency may be due to lack of thorough investigation in lower vertebrates, especially the fishes and amphibians. It was proposed by Martin Sage in 1973 that thyroid function evolved hand in hand with endocrine control of reproduction and that the basic function for thyroid hormones is associated primitively with gonadal maturation. This hypothesis is supported by a variety of observations, including the close parallel of thyroid activity and reproductive cycles in elasmo-branch and bony fishes.

Pituitary thyrotropes and TSH are apparently absent in agnathans, and these cells may have evolved later from gonadotropes. This proposed origin for thyrotropes is supported by their cytological similarity to gonadotropes, their location in the adenohypophysis of elasmobranchs (pars ventralis) and teleosts (proximal pars distalis), and by the biochemical similarity of TSH to the gonadotropins (see Chapter 4). Exogenous thyroid hormones and gonadal steroids both have inhibitory actions on thyrotropes as well as on gonadotropes in teleosts. Similar influences of thyroid hormones on gonadal function and, reciprocally, go-nadal steroids on thyroid function have been reported in amphibians, reptiles, and birds. The observations that mammalian luteinizing hormone (LH) stimulates thyroid function in fishes may be interpreted as supportive of this relationship or it may be due simply to a chance similarity in the structures of mammalian LH and teleostean TSH. However, these same gonadotropins are ineffective when tested on mammalian thyroids. The effects of thyroid hormones that modify behavior through actions on the central nervous system pre-sumably evolved from the general feedback effects on the hypothalamus originally associ-ated with the gonadal axis. According to Martin Sage, the actions of thyroid hormones related to growth, metabolism, development, and the integument would be later evolution-ary events. Location of the major feedback of thyroid hormones on TSH release in the adenohypophysis rather than at the hypothalamus might be a consequence of this evolu-tionary sequence.

D. Prolactin and Its Interactions with the Thyroid Axis

A general antagonism has been reported between **prolactin (PRL)** and the thyroid axis of nonmammalian vertebrates. Some studies have demonstrated what appears to be a goitro-genic action of PRL directly on thyroids of teleosts, amphibians, lizards, and birds, but the peripheral antagonism between thyroxine and PRL in certain tissues of larval amphibians is best known. Isolated reports of enhancement of thyroid secretion following treatment with mammalian PRL require further investigation to verify the general applicability of this effect to vertebrates. The significance of an antagonistic interaction between thyroid hormones and PRL in fish, reptiles, birds, and mammals is not clear. In amphibians, the administration of anti-PRL agents (PRL antibodies or ergot derivatives) enhances re-sponses of larval amphibians to endogenous and exogenous thyroid hormones. These ob-servations support an endogenous role for PRL in preventing premature metamorphosis (see Section II,F,3).

Mammalian synthetic **thyrotropin-releasing hormone (TRH)** causes PRL release from pituitaries of bullfrogs, turtles, birds, and mammals but not from the pituitaries of red-spotted newts (see Chapter 5). These observations further complicate the picture and raise some important questions concerning the biological significance, if any, of demonstrations that PRL produces antithyroid effects. It also suggests the role of the TRH tripeptide, which is present abundantly in the amphibian brain, is that of a PRL-releasing agent, especially in light of many failed attempts to demonstrate a role for TRH in the activation of the am-phibian thyroid axis.

E. Surgical and Chemical Thyroidectomy and Radiothyroidectomy

A basic approach employed in thyroid studies involves hypophysectomy or thyroidectomy or a combination of the two followed by classical replacement therapy. Sometimes, however, it is desirable to make an animal only slightly hypothyroid or reversibly hypothyroid or both. Chemical thyroidectomy involves administration of a chemical goitrogen (see Chapter 7) such as **propylthiouracil (PTU)** or **thiourea (TU)** at a predetermined dose for a given period. Withdrawal of the goitrogen or addition of thyroid hormones in the continued presence of the goitrogen may then allow the animal to return to a euthyroid condition for comparisons. With this approach, changes in thyroid function before, during, and after treatment may be examined in each individual. Caution must be exercised in that some goitrogens (e.g., TU) have been shown to produce effects on other tissues (especially the liver) that do not appear following surgical thyroidectomy. Such "nonspecific" actions of chemical inhibitors are well known to investigators, and, when possible, other approaches should be used to corroborate actions of goitrogens.

Large doses of radioiodide often are employed as therapeutic agents in humans to destroy excessive amounts of thyroid tissue in certain hyperthyroid conditions. Accumulated radioiodide destroys cells because of the destructive effects of radiation on the cell that has incorporated it. Very large doses can be used to destroy thyroid tissue completely, especially in most species of teleostean fishes from which it is impossible to remove it surgically. Radiothyroidectomy, however, must be interpreted with caution since radioiodide accumulation may occur in other tissues (see Table 8-6 and Fig. 8-4) and result in destructive changes that may not be thyroid related. The general effects of whole-body radiation from this powerful γ emitter should be a concern as well.

II. Comparative Thyroid Physiology

The following account and Table 8-7 emphasize the biological actions of thyroid hormones on vertebrate target tissues. The secretion of TSH and the effects of various hypothalamic hormones on TSH secretion are discussed in Chapter 5; the focus here is on the functions of T_3 and T_4.

A. Agnathan Fishes: Cyclostomes

No function for thyroid hormones has been verified in cyclostome fishes, although in lampreys the binding of iodide by the larval endostyle increases following administration of T_4. Circulating levels of T_4 decrease markedly as free-living, filter-feeding **ammocetes larvae** of the sea lamprey, *Petromyzon marinus,* undergo metamorphosis to become ectoparasitic adult lampreys.

Table 8-6. Accumulation of Radioiodide by Thyroids and Gonads of Sexually Mature Vertebrates

Class	Species	Sex	Radioiodide uptake Gonad/thyroid
Osteichthyes[a]	*Micropterus dolomieu*	M	0.012
	(smallmouth bass)	F	3.780
Amphibia[a]	*Ambystoma tigrinum*	M	0.045
	(tiger salamander, neotene)	F	0.512
Aves	*Coturnix coturnix japonica*	F	4 to 10
	(Japanese quail)		

[a] Norris, unpublished data (1966).

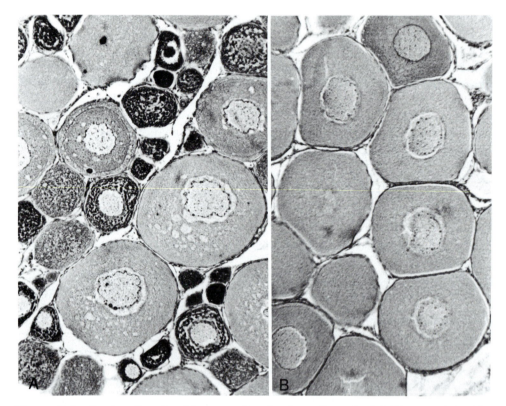

Figure 8-4. Oocytes of juvenile rainbow trout, *Oncorhynchus mykiss*, and radioiodide accumulation. The ovaries of nonmammalian vertebrates readily concentrate iodide as efficiently as the thyroid and store it in the oocytes. The control ovary at left exhibits a range of oocytes. The ovary at right accumulated radioiodide shortly after hatching. Although the radioactivity quickly decays to undetectable levels, the ovary at the right contains only the largest class of oocytes when examined 1 year later.

Hormone synthesis in agnathans differs from that of other vertebrates in that organic binding of iodide and storage of the iodinated proteins occur intracellularly. Hypophysectomy does not alter thyroid function in adult lampreys or in hagfishes, and TSH is absent

Table 8-7. Some Major Actions of Thyroid Hormones in Vertebrates[a]

Vertebrate group	Development	Molting	Growth	Metabolism	Reproduction	Neural behavior	Neural differentiation
Sharks					+	+	
Bony fishes	+		+		+	+	
Amphibians	+	+[b]	−[c]		−	+	+
Reptiles		+	+	?	+		
Birds	+	+	+	+	+/−	+	
Mammals	+	+	+	+	+	+	+

[a] +, Stimulatory; −, inhibitory. The stimulatory action of thyroid hormones on metabolism seems to be associated with the evolution of homeothermy.

[b] Urodeles only. Molting in anurans is controlled by corticosteroids.

[c] Prevents growth in larval amphibians by inducing metamorphosis.

in agnathans. These observations suggest that evolution of thyrotropes and TSH occurred first in jawed fishes, possibly in the placoderms, and has persisted in all other extant vertebrates. T_4 is present in serum from mature female sea lampreys, comparable to levels reported for ammocetes of this species. Much lower levels of T_4 are reported for mature male sea lampreys. T_3 occurs in the serum of both immature and mature sea lampreys, but the levels are higher in the mature individuals. These data suggest a relationship to reproductive maturation, but definitive data are needed before a firm relationship can be accepted.

B. Chondrichthyean Fishes

Only limited data are available with respect to thyroid function in sharks, in which the pituitary–thyroid axis appears to be well established. Most studies on sharks are related to reproduction or oxygen consumption, and the thyroid systems of other chondrichthyean fishes are virtually unknown.

1. Thyroid and Reproduction in Selachians

Thyroid cycles are positively correlated with reproductive cycles in sharks. However, increased thyroid activity observed in at least one species is correlated with migratory behavior related to reproduction rather than with reproduction or gonadal maturation per se. This area requires further investigation.

2. Thyroid and Oxygen Consumption in Selachians

Late embryos of *Squalus suckleyi* exhibit a transient increase in oxygen consumption following treatment with T_3 or T_4, but this response cannot be maintained by continued treatment with thyroid hormones. Treatment with PTU has no effect on oxygen consumption in this species. These observations do not provide strong support of a role for thyroid hormones in oxidative metabolism.

3. Thyroid and Differentiation in Selachians

Differentiation of hypothalamic neurosecretory centers is accelerated in the embryos of the oviparous shark, *S. suckleyi,* following treatment with T_3 or T_4. This effect of thyroid hormones is manifest in both the preoptico–hypophysial fiber tracts and in the neurohypophysis. These data are suggestive of a role for thyroid hormones in nervous tissue differentiation and maturation of the hypothalamo–hypophysial system.

C. Bony Fishes: Chondrosteans

Some limited data for sturgeons indicate that peak thyroid activity coincides with spawning behavior. Furthermore, thyroid treatments can reverse the degenerative changes that occur in gonads of captive sturgeon, suggesting a direct relationship between thyroid hormones and reproduction. Additional studies are needed in chondrostean fishes to examine other possible roles of thyroid hormones and to confirm this suggestive relationship with reproduction.

D. Bony Fishes: Teleosts

Thyroid function has been studied intensively in teleostean fishes and considerable information is available with respect to iodide uptake, hormone synthesis, secretion rates, clearance rates, and metabolism of thyroid hormones. Most of this knowledge about teleost thyroids stems from the work of J. G. Eales and colleagues, working principally with the rainbow trout, *Oncorhynchus mykiss*.

Several features of teleostean thyroid function differ markedly from what we know about mammals. Probably the most striking difference is that neuroglandular hypothalamic control in all species studied is inhibitory, with both TRH and somatostatin capable of preventing TSH release. However, both TRH and catecholamines are stimulatory in salmonids. Teleosts readily accumulate iodide from the surrounding water through their gills, and some species maintain high circulating levels of protein-bound inorganic iodide. Consequently, these fish rarely show iodide deficiencies in nature, and their thyroid iodide metabolism is resistant to excessive iodide inhibition. There is negligible enterohepatic circulation of thyroid hormones or their metabolites in teleosts, and thyroid hormones are readily lost in urine, too. The active form of thyroid hormone is T_3, but its level is apparently controlled through the activity of deiodinases. Administration of T_4 generally has no effect on circulating levels of T_3. Although some species have been shown to metabolize T_4 to **reverse T_3 (rT$_3$)**, in rainbow trout the production of rT_3 is curiously absent (Fig. 8-5) unless the fish is treated first with T_3. This is an interesting observation since regulation of the type III deiodinase that causes this conversion is implicated in the occurrence of T_3 surges in other vertebrates.

Thyroid hormones are implicated in reproduction and aspects of behavior in teleostean fishes. Their role with respect to oxygen consumption is not so clear, however. Thyroid hormones are essential for normal development and growth and may be involved in meta-

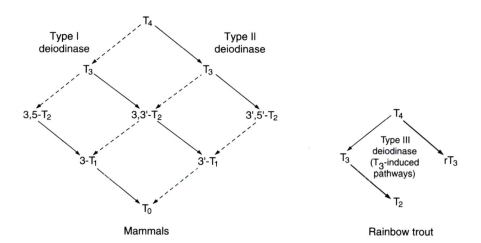

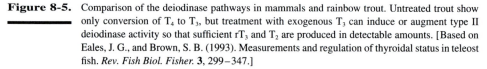

Figure 8-5. Comparison of the deiodinase pathways in mammals and rainbow trout. Untreated trout show only conversion of T_4 to T_3, but treatment with exogenous T_3 can induce or augment type II deiodinase activity so that sufficient rT_3 and T_2 are produced in detectable amounts. [Based on Eales, J. G., and Brown, S. B. (1993). Measurements and regulation of thyroidal status in teleost fish. *Rev. Fish Biol. Fisher.* **3**, 299–347.]

morphosis of young fishes and in the parr–smolt transformation of salmonid fishes, another metamorphosis-like change. Although thyroid hormones have been implicated in osmo-regulation, carbohydrate metabolism, nitrogen metabolism, and growth, other hormones are known to play more direct roles. In fact, some effects of thyroid hormones may be permissive actions similar to those described for mammals (Chapter 7).

1. Heterotopic Thyroid Tissue in Teleosts

The teleostean thyroid consists of scattered thyroid follicles throughout the pharyngeal region. The diffuse nature of the teleostean thyroid gland, with only a few exceptions (for example, the tuna and the Bermuda parrot fish), makes assessment of thyroid function difficult and renders surgical thyroidectomy impossible. Most of the thyroid tissue is located in the pharyngeal area, where follicles are found between the second and fourth aortic arches (Fig. 8-6). Because of the absence of a covering connective tissue sheath that holds the follicles into one or more masses, thyroid follicles are frequently found outside this pharyngeal region. These extrapharyngeal thyroid follicles are termed accessory or **heterotopic thyroid follicles** because of their location outside the normal site. Heterotopic thyroid follicles occur commonly in species of some of the more recent teleostean families (Fig. 8-6). Relatively large numbers of thyroid follicles may be found embedded within the head kidney of some species and occasionally in other locations such as the pericardium or ovary. In such species, it may be important to examine the

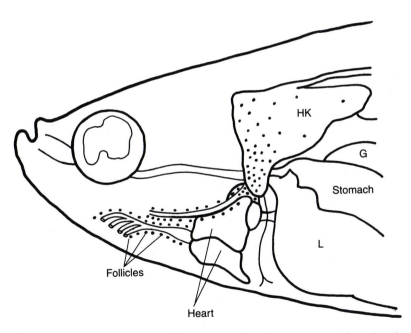

Figure 8-6. Heterotopic thyroid tissue. This teleostean fish, *Xiphophorus maculatus*, has thyroid follicles (black dots) located not only throughout the pharyngeal region but in the heart, head kidney (HK), and sometimes even in the gonads (G) and liver (L).

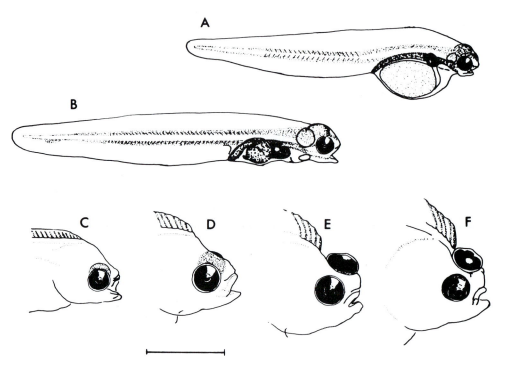

Figure 8-7. Metamorphosis of the flounder, *Pleuronectes platessa.* During metamorphosis (A–F), which is controlled by thyroid hormones, one eye migrates from one side of the body to the other. Levels of plasma T$_4$ rise during this process. [From Blaxter, J. H. S. (1988). Pattern and variety in development. *In* "Fish Physiology," (W. S. Hoar and D. J. Randall, eds.), Vol. 11A, pp. 1–58. Academic Press, San Diego.]

activity of heterotopic thyroid as well as of pharyngeal thyroid when using techniques such as radioiodide uptake.

2. Thyroid and Development in Teleosts

Thyroid hormones are responsible for a posthatching metamorphosis in many species of teleosts. Probably the most dramatic metamorphosis is that of flounders and other flatfishes; in these species metamorphosis involves the migration of the eye and attendant neural structures from one side of the head to the other (Fig. 8-7), while the mouth and associated structures migrate to the opposite side. The adult animal then behaves with the left or right side (depending on the species) acting as the ventral surface, with the mouth located there for bottom feeding and the other side having both eyes and acting as the dorsal surface. Levels of plasma T$_4$ rise dramatically during flounder metamorphosis, as does cortisol, whereas T$_3$ levels remain low. Measurements of T$_3$ turnover are lacking, however.

The **parr–smolt transformation** or **smoltification** occurs prior to seaward migration in many salmonids (Fig. 8-8). This metamorphosis from a sedentary, cryptically marked fish, the **parr,** with a freshwater physiology to an active, silvery **smolt** that is preadapted for osmoregulation in salt water is accompanied by increased cortisol secretion (see Chapter 10), although surges in T$_4$, insulin, and PRL also have been reported (Fig. 8-9).

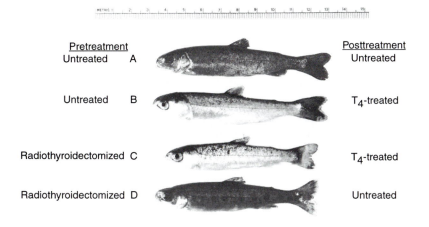

Figure 8-8. Smolting in rainbow trout. Thyroid hormone stimulates deposition of guanine in the scales. (A) Normal fingerling; (B) normal fingerling immersed for 4 weeks in fresh water containing 10^{-8} M T_4; (C) radiothyroidectomized trout treated with T_4 as in (B); (D) Untreated radiothyroidectomized trout.

3. Thyroid and Reproduction in Teleosts

There is a strong positive correlation in many teleostean species between thyroid state and reproductive cycles. Thyroxine consistently stimulates precocial gonadal maturation, whereas radiothyroidectomy or treatment with goitrogens inhibits or retards gonadal development. In several species, thyroid activity is greatest at spawning. Furthermore, increased thyroid activity is coincident with migrations and spawning in Pacific salmon and may represent a response to increased metabolic demands. Similarly, this increased

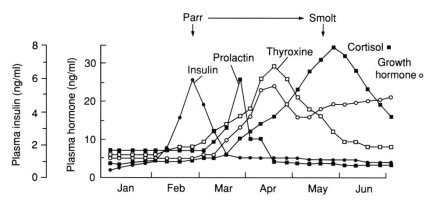

Figure 8-9. Plasma hormone levels during smoltification of coho salmon. Prolactin, growth hormone, thyroxine, and cortisol all peak during smoltification. Insulin peaks in the parr and declines as smoltification gets under way. [From Fish and amphibian models for developmental endocrinology. Dickhoff, W., Brown, C. L., Sullivan, C. V., and Bern, H. A. *J. Exp. Zool.* Copyright © 1990 John Wiley & Sons. Reprinted by permission of John Wiley & Sons, Inc.]

thyroid activity appears to be more closely correlated with the spawning migration than with reproductive events.

4. Thyroid and Oxygen Consumption in Teleosts

Observations have yielded opposing results with respect to the importance of thyroid hormones in oxygen consumption by teleostean fishes. Most studies suggest that neither thyroid hormones nor goitrogens have any effects on oxygen consumption. A positive correlation has been reported for cyclic seasonal variations in thyroid activity and oxygen consumption by *Heteropneustes fossilis*. Studies of oxygen consumption are difficult to interpret because temperature, changes in behavior, and potential side effects of treatments used to render the fish hypothyroid (goitrogens, radiothyroidectomy) may have direct effects on metabolism.

5. Thyroid and Osmoregulation in Teleosts

The endocrine regulation of osmoregulation is primarily under the control of prolactin and cortisol with respect to maintaining Na^+ balance (see Chapter 10) and with arginine vasotocin (AVT) responsible for water balance (see Chapter 6). Calcium regulation is probably controlled by hormones from the corpuscles of Stannius embedded in the kidney and the ultimobranchial body found in the pharyngeal region (see Chapter 14). Less direct roles in ionic and osmotic regulation are attributable to catecholamines, thyroid hormones, and factors from the pineal gland (Chapter 6) as well as the caudal neurosecretory system (Chapter 10).

Thyroid hormones may enhance seawater adaptation and may influence migratory behavior, especially in species that migrate between salt and fresh water. These proposed actions may be related to a permissive type of action rather than to a causative role for thyroid hormones or may be only correlated events and not causally related.

6. Behavioral Actions of Thyroid Hormones in Teleosts

Thyroid hormones are elevated in relation to spawning, premigratory, and migratory behaviors. Increased thyroid function has been reported for a variety of migrating fishes, and premigratory restless behavior has been correlated with increased thyroid function. Thyroid activity is greater in young salmon smolts migrating to the ocean and in adults during their upstream return to fresh water for spawning. Again, the action of thyroid hormones may only enhance the sensitivities of neural components to environmental stimuli.

E. Bony Fishes: Sarcopterygians

The function of thyroid hormones in lungfishes has received only a cursory examination. Thyroid hormones have been linked to a particularly fascinating aspect of the life history of African lungfishes: the ability to survive periods of drought while encased in a "cocoon." Awakening of estivating lungfishes *Protopterus annectens* from the cocoon of dried mud when moistened may involve a neuroendocrine mechanism associated with the thyroid axis. One hypothesis, based on limited experimental data, suggests that increasing humidity activates the hypothalamic apparatus controlling TSH release and stimulates thyroid hormone secretion. Thyroid hormones in turn would increase the sensitivity of olfactory centers that evoke normal feeding behavior as well as other behaviors associated with wakening. Additional studies of thyroid function in these fishes are needed to substantiate this interesting hypothesis.

F. Thyroid Functions in Amphibians

Thyroid hormones influence many processes in both anurans and urodeles, including reproduction, metamorphosis, metabolism, growth, and molting. The most dramatic and best studied event among amphibians is the endocrine induction of metamorphosis from an aquatic larva to a terrestrial or semiterrestrial form.

1. Thyroid and Reproduction in Amphibians

Experimental reduction of thyroid function, such as by thyroidectomy or administration of goitrogens, accelerates gonadal development in several anurans. In one urodele, *Ambystoma tigrinum,* circulating levels of T_4 are inversely correlated with seasonal gonadal development, although such correlations do not substantiate cause–effect relationships. Some conflicting results with respect to thyroid hormones and seasonal maturation have been reported for some adult anurans, but the general relationship in amphibians appears to be an antagonistic one. All of the data reported to date are circumstantial and an absolute antagonism by endogenous thyroid hormones in natural populations is yet to be demonstrated.

2. Thyroid and Oxygen Consumption in Amphibians

The relationship between thyroid hormones and metabolism is not clear. Some studies have shown a positive correlation between thyroid state and oxygen consumption, whereas most studies show no effects. Oxygen consumption is not elevated during either spontaneous metamorphosis caused by elevated endogenous thyroid hormones or during induced metamorphosis caused by exogenous thyroid hormones.

Liver slices prepared from T_4-treated adult frogs, *Rana pipiens,* exhibit significantly greater oxygen consumption *in vitro* than appropriate controls at 25°C. Oxygen consumption of treated slices is the same as for controls when observed at 15°C. These data suggest that the respiratory response of amphibian tissues to thyroid hormones may be temperature dependent and that the response occurs only at higher temperatures. Most studies reporting no effect for thyroid hormones on oxygen consumption were performed below 20°C. What the biological importance is for enhanced oxygen consumption at higher temperatures is open to speculation.

3. Thyroid and Metamorphosis in Amphibians

Thyroid hormones induce metamorphosis, a marked biochemical, physiological, morphological, and behavioral transformation from an aquatic larva to a terrestrial or semiterrestrial form. These events occur in the life history of every amphibian, although the developmental sequence and duration of the various stages may be modified extensively in those species that incubate their eggs on land or are live bearing.

Table 8-8. Genes Affected by Thyroid Hormones during
Amphibian Metamorphosis

Gene	Tissue	Regulation
Early Response Genes		
Tail 1 (zinc finger region of SP1)	Tail	Up
TRβ (thyroid hormone receptor)	Tail and intestine	Up
Tail 8/9 (bZIP^F of E4BP4)	Tail and intestine	Up
Tail 14 (stromelysin 3)	Tail and intestine	Up
Tail 15 (type I 5′-deiodinase)	Tail	Up
Late Response Genes		
Carbamyl-phosphate synthetase I	Liver	Up
Argininosuccinate synthetase	Liver	Up
Arginase	Liver	Up
Albumin	Liver	Up
Myosin heavy chain	Limb	Up
Keratin	Epidermis	Up
Trypsin	Pancreas	Down
Intestinal fatty acid-binding protein	Intestine	Down

Amphibian metamorphosis has been studied at all levels of organization and provides an excellent illustration of hormonal interactions of thyroid hormones with several additional hormones (e.g., prolactin, corticosteroids) and a wide variety of tissues. Furthermore, this process is closely linked to the interaction of complex environmental factors that together with endocrine-related events constitute the metamorphosis phenomenon.

Metamorphosis in amphibians involves regulation of specific genes, often initiated by an external factor, that orchestrate a programmed sequence of biochemical (see Table 8-8), morphological (e.g., tail and gill resorption; see Fig. 8-10 and Table 8-9), physiological (e.g., nitrogen excretion; see Fig. 8-11), and behavioral events. Three decades ago, William Etkin proposed a model to explain the then-known stimulatory role of the thyroid axis and observed inhibitory actions of prolactin on metamorphosis (see Fig. 8-12). The **Etkin hypothesis** has necessarily undergone considerable revision since the discoveries that corticosteroids also are involved, that **corticotropin-releasing factor (CRH)** may release TSH as well as corticotropin (ACTH) from the pituitary, and that TRH may be responsible for the surge of PRL that also accompanies metamorphosis (Fig. 8-13). In fact, CRH may be the endogenous hypothalamic stimulator of the thyroid axis and the TRH peptide may be a PRL releaser. Dopamine appears to be the PRL release-inhibiting hormone of amphibians.

Environmental factors such as photoperiod may operate through catecholaminergic mechanisms in the brain to influence neurohormone release from hypothalamic nuclei (Fig. 8-14). The pineal gland may play an important role in mediating the observed stimulatory actions of light on thyroid function and metamorphosis, at least in urodeles. Detailed studies are needed to identify these pathways.

An important mechanism for induction of metamorphosis involves changes in deiodinase activities. In premetamorphic larvae, type III deiodinase is present, which keeps T_3 levels low by converting T_4 preferentially to rT_3 and hastening deiodination of T_3. During

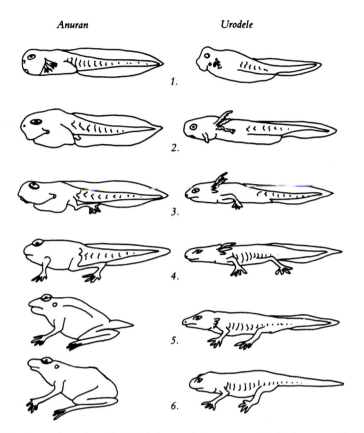

Figure 8-10. Comparison of morphological changes during metamorphosis of an anuran (left) and a urodele (right) amphibian. Urodeles quickly reach stage 4 and remain in this stage most of their larval lives. Anurans may spend up to 2 years at stage 2 before limbs emerge and the tail quickly is resorbed. Note that when urodeles undergo metamorphosis, the external gills are resorbed, as is the tail fin.

the activation of metamorphosis, there is a measurable increase in type II deiodinase that enhances formation of T_3 from T_4 as well as a decrease in type III deiodinase activity. Patterns of deiodinating activity in liver and target tissues (Fig. 8-15) affect the availability

Table 8-9. Life History of *Rana pipiens* at 23°C

Period	Events	Duration
Embryonic	Embryogenesis	8 days
Premetamorphic	Growth	5–6 weeks
Prometamorphic	Accelerated growth of hind limbs; skin changes occur	3 weeks
Metamorphic climax	Forelimbs emerge; tail resorbs; gills resorb; head and gut reconstruction; etc.	1 week
Juvenile	Growth	—
Adult	Reproductive maturation and breeding	—

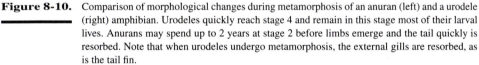

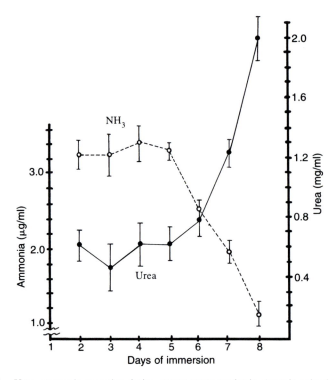

Figure 8-11. Urea–ammonia excretion during anuran metamorphosis. Aquatic animals excrete ammonia as their principal nitrogenous wastes whereas terrestrial amphibians produce urea. Immersion of tadpoles in water containing thyroxine induces a switch from ammonia excretion to urea excretion, similar to that observed during normal metamorphosis.

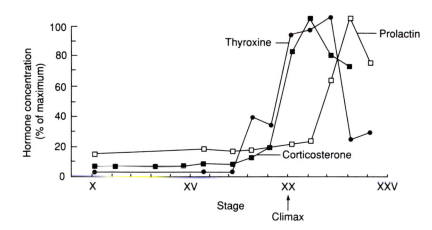

Figure 8-12. Hormonal changes during amphibian metamorphosis. Thyroxine, prolactin, and corticosteroids all peak in the final stages of anuran metamorphosis (climax). These three hormones appear to be involved in urodele metamorphosis as well. Compare to the pattern of hormone secretion during smoltification in salmonid fishes (Fig. 8-9). [From Fish and amphibian models for developmental endocrinology. Dickhoff, W., Brown, C. L., Sullivan, C. V., and Bern, H. A. *J. Exp. Zool.* Copyright © 1990 John Wiley & Sons. Reprinted by permission of John Wiley & Sons, Inc.]

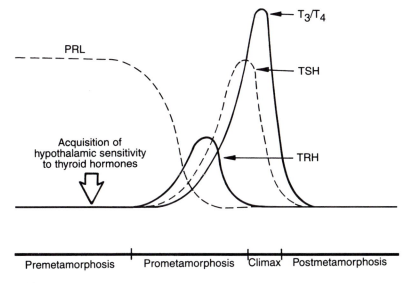

Figure 8-13. The Etkin hypothesis. William Etkin proposed that low levels of thyroid hormones secreted autonomously by the thyroid gland mature the hypothalamus to produce a "TRH" that results in TSH secretion and eventually in accelerated thyroid hormone production, which causes metamorphosis. Etkin also proposed that prolactin (PRL) levels would be high prior to metamorphosis and then drop off. Later studies have modified various aspects of the original hypothesis with the discovery of involvement of other hormones as well as the role of CRH and deiodinase activity. See text for more explanation.

of active T_3 and facilitate metamorphosis. Marked changes in tissue thyroid receptor levels also occur during metamorphosis. Changes in both α- and β-type thyroid receptors similar to the changes in mammalian receptors have been demonstrated in different amphibian tissues, but their roles in the metamorphosis process are uncertain at this time.

Thyroid hormones influence gene activity in target cells and initiate metamorphic changes (Table 8-8). For example, in the hind limbs of anurans, thyroid hormones have been shown to activate 14 genes while 5 others are down-regulated, resulting in rapid hind-limb growth characteristic of the later stages of metamorphosis. In tadpole tail fins, researchers have identified 15 genes activated during thyroid hormone-induced tail resorption and 4 others that are down-regulated. Twenty-two genes are activated in the intestine of metamorphosing anurans but only one is down-regulated. Many of these genes are called direct- or **early-response genes** and are activated directly by thyroid hormones whereas **late-response genes** (e.g., urea cycle genes in liver; keratin genes in skin) are not activated until about 2 days after application of thyroid hormones. This delay implies that synthesis of other molecules including transcription factors is dependent on protein synthesis activated by the original stimulus.

Programmed cell death or **apoptosis** is another component of tail regression that is influenced by thyroid hormones. Induction of **ubiquitin,** a biochemical marker for apoptosis, is caused by T_4 treatment of isolated tadpole tail tips.

Thyroid hormones are also involved in a second metamorphic event, the so-called **water drive** associated with reproduction in newts. Water-drive behavior is a well-known

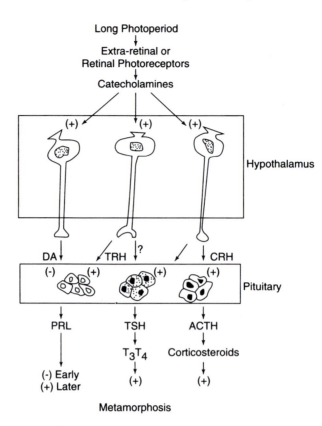

Figure 8-14. The role of long photoperiod in urodele metamorphosis. Light working through pineal or retinal photoreceptors activates neural centers that initially prevent prolactin (PRL) release and activate TSH and ACTH release (via CRH), causing surges in thyroid hormones and corticosteroids, respectively. Later, a surge of PRL occurs, probably due to TRH or VIP release. See also Fig. 8-12.

bioassay for PRL activity (see Chapter 5). Small quantities of thyroid hormones facilitate the preinduced water drive and associated morphological changes in the integument. Larger amounts of thyroid hormones inhibit water drive, and thyroid hormones appear to be responsible for land-drive behavior of recently metamorphosed amphibians (see Chapter 5).

4. Thyroid and Growth of Amphibians

The involvement of thyroid hormones in growth of amphibians, like the relationship to reproduction, deviates from the general vertebrate pattern. The surge of thyroid activity that initiates metamorphosis in larval amphibians arrests growth. Treatment of anuran tadpoles with goitrogens or mammalian PRL accelerates larval growth and blocks metamorphosis. Growth in natural populations is arrested during metamorphosis. The complicated metamorphic process itself involves many drastic physiological changes and tissue rearrangements that may be responsible for arrested growth. These observations associated with larval growth do not eliminate participation of low levels of thyroid hormones in growth of

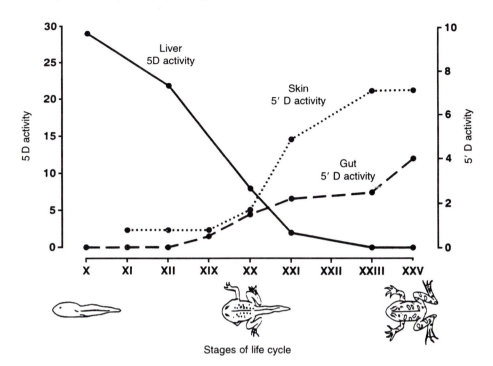

Figure 8-15. Changes in deiodinase (D) activities during metamorphosis. The reduction of type III (5D) deiodinase activity in liver leads to elevated circulating T_3. 5'D, Type II deiodinase. [Reprinted by permission of the publisher from Galton, V. A. The role of thyroid hormone in amphibian metamorphosis. *Trends Endocrinol. Metab.* **3**, 96–100. Copyright 1992 by Elsevier Science Inc.]

the subadult or postmetamorphic juvenile forms. Such participation for thyroid hormones in the growth of metamorphosed amphibians has not been reported.

5. Molting and Other Skin Effects of Thyroid Hormones in Amphibians

Molting or shedding of skin (ecdysis) in larval and adult urodeles is under the direct stimulatory control by thyroid hormones, and frequent molting accompanies and follows metamorphosis in these animals. Molting in adult anurans, however, does not appear to be influenced by thyroid hormones. Here molting is stimulated by corticosteroids or indirectly by ACTH operating through its action of increasing corticosteroid secretion. The reasons for this marked difference in hormonal control of molting between urodeles and anurans are not known.

Thyroid hormones cause a number of skin changes related to invasion of the terrestrial environment. Both a thickening of the epidermis and keratinization are induced by thyroid hormones in urodeles and anurans. In the newts, thyroid hormones antagonize the effects of PRL on the skin of the eft prior to its return to water. Prolactin reduces the keratinization and helps return the skin to a smooth, moist condition characteristic of aquatic newts.

The large and conspicuous **Leydig skin cells,** which are characteristic of urodele larvae, disappear during metamorphosis, presumably a consequence of thyroid hormone actions. The role of these cells in the larvae is unknown. Histochemical observations suggest a contribution of their contents to formation of a keratinized layer in the skin that probably aids in reducing desiccation.

G. Thyroid Functions in Reptiles

Changes in thyroid state in reptiles are correlated with reproduction, environmental temperature, and activity level, although cause–effect relationships have not been established clearly. Exogenous thyroid hormones also stimulate oxygen consumption, growth, and molting under certain conditions.

The presence of iodinated tyrosines in the blood is a situation unique to reptiles. Blood of the striped racer *Elaphe taeniura* and the cobra *Naja naja* contains MIT and DIT. Thyrotropin stimulates *in vitro* release of MIT, DIT, and T_4 from chunks of thyroid tissue from a turtle (*Geochemys reevsii*), a lizard (*Gekko gecko*), and a snake (*Elaphe rachiata*). These studies suggest that reptilian thyroid glands may lack any deiodinase to convert MIT and DIT to inorganic iodide and tyrosine. These data, however, are open to other interpretations, and additional research is needed. Evidence for a type I deiodinase (PTU sensitive) has been reported for the lizard *Sceloporus occidentalis,* and conversion of T_4 to T_3 is blocked in another lizard, *Calotes versicolor,* following treatment with the deiodinase inhibitor, **iopodate.**

1. Thyroid and Reproduction in Reptiles

Seasonal changes in thyroid function have been correlated positively with a number of reproductive events in lizards, snakes, and turtles. An active thyroid, usually assessed histologically, is associated with spermatogenesis, ovulation, and mating in a number of lizards. Similar observations have been reported in snakes and in at least one turtle.

Among live-bearing reptiles, there is no evidence for increased thyroid function during gestation. However, thyroidectomy of pregnant lizards *Lacerta vivipara* 6 weeks prior to term caused premature discharge of most eggs, and those eggs that were retained failed to hatch.

2. Environmental Temperature and Reptilian Thyroid

In general, thyroid function in reptiles varies proportionally with changes in temperature, although the cause–effect relationships are still obscure. Behaviors such as basking in lizards make it difficult to estimate body temperature when assessing thyroid state. Thyroid activity in temperate lizards is highest during the warmer seasons. Lowering the environmental temperature artificially during the summer months causes reduction in thyroid activity, and lizards maintained in the laboratory at high temperatures ($35°C$) exhibit greater thyroid function than when maintained at $15°C$. Thyroid function in snakes is greater during warm periods and lowest during hibernation. The greatest level of thyroid activity, however, correlates with reproductive events in some lizards and snakes and not with temperature. Similar data have been reported for turtles.

An exception to the relationship of thyroid activity and temperature described above is the situation found in several species of lizard inhabiting warmer climates. These lizards do not exhibit depressed thyroid function at lower temperatures. In fact, thyroids of such lizards tend to be more active during cool periods. There appears to be a positive correlation also between thyroid state and activity in lizards and snakes. Increased humidity and activity are correlated with increased thyroid function in the lizard *Agama agama savattieri,* suggesting that a more complex relationship exists among thyroid state, activity level of the animal, and environmental factors.

3. Thyroid and Oxygen Consumption in Reptiles

The relationship between thyroid state and oxygen consumption is temperature dependent. Thyroid hormones, TSH, and thyroidectomy have little effect on lizards maintained at 20°C, but a positive relationship appears between oxygen consumption and thyroid state at 30°C. Heart, brain, liver, muscle, and lung tissues from lizards incubated *in vitro* at 30°C respond to T_4 with increased consumption of oxygen. Homogenates of both liver and skeletal muscles from T_4-treated snakes *Natrix piscator* exhibit higher oxygen consumption at 30°C than do tissues from untreated snakes. Thyroidectomy causes a decrease in oxygen consumption in these animals that is restored to normal by treatment with T_4. Metabolic rate and cytochrome oxidase activity also are stimulated by T_4 at 25°C. These observations suggest that homeothermy in birds and mammals (as well as proposed for certain large, extinct reptiles) is an adaptation to maximize oxidative metabolism and its stimulation by thyroid hormone.

4. Thyroid and Molting in Reptiles

Shedding of the skin is stimulated in lizards by thyroid hormones and is retarded by thyroidectomy. Implants of thyroid tissue into muscles of thyroidectomized *Lacerta* spp. restore molting that has been interrupted by thyroidectomy. Apparently either mammalian TSH or PRL is capable of restoring molting in some hypophysectomized lizards, and PRL enhances the effect of T_4 on intact animals. Such a relationship between PRL and the thyroid axis in lizards is unlike the role for PRL in premetamorphic amphibians, but fits with the synergistic roles of these hormones observed during late stages of metamorphosis in both urodele and anuran amphibians.

In marked contrast to lizards, thyroidectomy in snakes increases molting frequency, and cessation of molting follows administration of thyroid hormones. No explanation for this contradiction between snakes and lizards has been offered. This situation in snakes is different from that found in amphibians, in which thyroid hormones and corticosteroids are stimulators of molting in urodeles and anurans, respectively.

5. Thyroid and Growth of Reptiles

Although detailed studies of the relationship of thyroid state to growth are lacking for reptiles, a variety of studies employing embryonic, juvenile, and adult reptiles suggest a relationship between thyroid hormones and growth that is similar to the mammalian pattern. Thyroidectomy of the lizard *Sceloporus undulatus* retards growth. Other studies in reptiles are needed to substantiate this relationship.

H. Thyroid Functions in Birds

The structural and functional features of avian thyroid glands are similar to those of mammals. Thyroid hormones have been reported to affect reproduction, growth, metabolism, temperature regulation, molting, and various behaviors. Most studies on avian species have concentrated heavily on certain domestic birds (chicken, duck, pigeon, Japanese quail), and few data are available for wild bird species. Nevertheless, with respect to most of these processes, the relationships are similar for wild and domestic species.

1. Development of Thyroid Function in Birds

The thyroid axis develops early in some birds, being functional in the chicken by day 13 of development. The embryonic chick thyroid actually responds to TRH and TSH administration on the sixth day of incubation, indicating that although the secretory capacity of the thyroid gland has matured, the endogenous activity of the hypothalamic–pituitary connection is still immature. Plasma levels of T_4 increase steadily after day 13 to maximal levels by day 20. In contrast, levels of T_3 remain low until day 19, when there is a surge of T_3 production during pipping and hatching. This surge corresponds to a rise in type I deiodinase activity in the liver and a marked decrease in type III deiodinase similar to events described during amphibian metamorphosis. Thus, conversion of T_4 to T_3 increases and degradation of T_3 decreases. This relationship continues after hatching. Hypophysectomy of hatchlings, however, causes an increase in type III deiodinase activity that can be prevented by treatment with growth hormone (GH). A similar effect of GH is seen in adult chickens as well, suggesting a function beyond the traditional synergistic actions between thyroid hormones and GH in growth (see Chapters 4 and 5). Type I deiodinase is not affected by GH. This dramatic change in thyroid activity has been documented in chicken and Japanese quail, which are **precocial birds,** but not in **altricial birds** such as ring doves, which, unlike precocial species, are totally dependent on their parents after hatching. In ring doves, thyroid and deiodinase activity change much more gradually as the birds develop greater independence from their parents.

2. Thyroid and Reproduction in Birds

Domestic birds require thyroid hormones for normal gonadal development. For example, T_4 stimulates testicular growth, whereas thyroidectomy or goitrogen administration impairs testicular function or induces gonadal regression. T_4 can stimulate testicular growth out of season in some wild birds as well. However, exogenous T_4 under some circumstances may exert suppressive actions on gonads. In several wild species, thyroidectomy prior to the time for normal gonadal growth results in precocious gonadal growth and maintenance at maximal condition. Treatment with T_4 results in gonadal regression in thyroidectomized birds at any time of year. In other wild species, thyroidectomy simply prolongs the active gonadal phase and shortens the time during which the gonads are regressed. Obviously, one cannot generalize about the relationships between thyroid hormones and reproductive events in birds since each species may prove to be a special case.

3. Thyroid, Thermogenesis, and Oxygen Consumption in Birds

It appears that thyroid hormones are directly involved in cold adaptation. **Thermogenesis** (heat production) in birds, as in mammals, is closely linked to oxygen consumption. Some wild birds in temperate regions exhibit heightened thyroid activity in late autumn and early winter as evidenced by histological examination and changes in thyroid gland weight. Thyroidectomy of adult birds depresses their ability to produce heat, and treatment with thyroid hormones increases oxygen consumption.

4. Thyroid, Carbohydrate Metabolism, and Growth in Birds

Thyroid hormones reduce glycogen stores in liver, increase free fatty acid levels, and induce mild hyperglycemia. Thyroidectomy causes a decrease in blood glucose and liver fatty acid levels but an increase in blood cholesterol levels. These effects may not be of primary importance since a number of other hormones, such as glucagon and epinephrine, are known to be more important than thyroid hormones with respect to controlling carbohydrate metabolism in birds. Thyroid hormones may produce a permissive effect with respect to the actions of these other "glucohormones," possibly through an effect on adenylate cyclase levels in the target tissues. These effects on carbohydrate metabolism may be related to the thermogenic action of thyroid hormones.

There is evidence that thyroid hormones act cooperatively with GH. Thyroidectomy causes depression or retardation of growth in birds and seasonal changes in circulating GH and T_4 are correlated. Release of GH is reduced by elevated thyroid hormones as if this were a normal negative feedback loop for controlling GH release.

5. Thyroid and Molting in Birds

Thyroid hormones produce stimulatory effects on the skin and feathers that are usually associated with the molting process. It is generally accepted that the gonadal axis provides the factors that directly regulate the molting process in birds. Unlike the condition for urodele amphibians and lizards, the role for thyroid hormones in the molting process of birds may be permissive rather than causative.

6. Thyroid and Migration in Birds

Migratory birds have been shown to possess active thyroid glands as compared to non-migrating or postmigrating individuals of the same species, suggesting some interaction in the migratory process. Although some of the older literature indicates that thyroid hormones may directly influence migratory behavior, the nature of this influence has not been confirmed. Thyroid hormones may alter metabolic patterns associated with energy requirements during migration. Another suggestion is that thyroid hormones "tune" the nervous system so that it is more sensitive to environmental or other endocrine cues or both. In the Canada goose (*Branta canadensis interior*), thyroid histology appeared activated and plasma T_4 levels increased in association with spring premigratory restlessness or *Zugunruhe* as compared to postmigrating birds, but the plasma T_3/T_4 ratio decreased dramatically (Table 8-10). However, in relation to the fall migration, there was a marked increase in T_3 in postmigrants and the T_3/T_4 ratio increased. This latter observation may reflect differences in thermal conditions and the role of thyroid hormones in thermogenesis during the cooler fall.

III. Summary

The thyroid axis of vertebrates shows many parallels with respect to thyroid structure and to the synthesis and metabolism of thyroid hormones as well as to their actions (see Table 8-7). T_3 appears to be the more active form of thyroid hormones in all vertebrates. Thyroid hormones appear to interact with a variety of other endocrine-regulated systems, where

Table 8-10. Plasma levels of T_3 and T_4 in Migrating and Postmigrating Canada Goose, *Branta canadensis interior*[a]

Sampling time	Plasma T_4 (ng/ml)	Plasma T_3 (ng/ml)	T_3/T_4
Spring premigration	11.8	3.5	0.28
Spring postmigration	17.0	2.0	0.10
Fall premigration	13.3	1.0	0.05
Fall postmigration	11.8	2.0	0.18

[a] Values estimated from Fig. 27 in George, J. C., and John, T. M., (1990). Canada goose thyroid: Seasonal ultrastructural changes in relation to circulating levels of its hormones. *Cytobios* **61**, 97–115.

they play permissive or possibly synergistic roles especially in development, growth, and reproduction. The systems of deiodinases seem to be responsible for allowing surges in T_3 to occur at critical times. Direct actions of thyroid hormones on development occur in all vertebrates, but their participation in lipid metabolism and thermogenesis is correlated with homeothermy (Fig. 8-16).

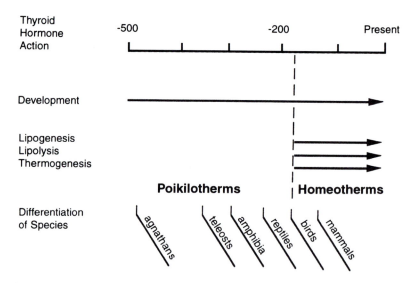

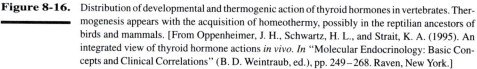

Figure 8-16.   Distribution of developmental and thermogenic action of thyroid hormones in vertebrates. Thermogenesis appears with the acquisition of homeothermy, possibly in the reptilian ancestors of birds and mammals. [From Oppenheimer, J. H., Schwartz, H. L., and Strait, K. A. (1995). An integrated view of thyroid hormone actions *in vivo*. *In* "Molecular Endocrinology: Basic Concepts and Clinical Correlations" (B. D. Weintraub, ed.), pp. 249–268. Raven, New York.]

Some notable phylogenetic differences stand out with respect to hypothalamic regulation in teleostean fishes and amphibians. TRH appears to exert a negative control in many fishes over TSH release, and in amphibians evidence is mounting to identify the CRH peptide as the endogenous TSH-releasing hormone.

Considerable work remains to be done on thyroid systems in nonmammalian vertebrates, especially in selachian fishes, a greater variety of teleosts, and reptiles. Once this has been accomplished, a clearer pattern of vertebrate thyroid function should appear.

Suggested Reading

Books

Armstrong, J. B., and Malacinski, G. M. (1989). "Developmental Biology of the Axolotl." Oxford Univ. Press, New York.

Balls, M., and Bownes, M. (1985). "Metamorphosis." Oxford Univ. Press, Oxford.

Gilbert, L. I., and Freiden, E. (1981). "Metamorphosis: A Problem in Developmental Biology," 2nd Ed. Plenum, New York.

Gilbert, L. I., Tata, J. R., and Atkinson, B. G. (1996). "Metamorphosis: Postembryonic Reprogramming of Gene Expression in Amphibian and Insect Cells." Academic Press, San Diego.

Matsuda, R. (1987). "Animal Evolution in Changing Environments with Special Reference to Abnormal Metamorphosis." Wiley, New York.

Articles

Blaxter, J. H. S. (1988). Pattern and variety in development. In "Fish Physiology, The Physiology of Developing Fish" (W. S. Hoar and D. J. Randall, eds.), Vol. 11A, pp. 1–58. Academic Press, San Diego.

Decuypere, E., Kuhn, E. R., and Chadwick, A. (1985). Rhythms in circulating prolactin and thyroid hormones in the early postnatal life of the domestic fowl: Influence of fasting and feeding on thyroid rhythmicity. In "The Endocrine System and the Environment" (B. K. Follett, S. Ishii, and A. Chandola, eds.), pp. 189–200. Japan Sci. Soc./Springer-Verlag, Berlin.

Dent, J. N. (1986). The thyroid gland. In "Vertebrate Endocrinology: Fundamentals and Biomedical Implications" (P. K. T. Pang and M. P. Schreibman, eds.), Vol. 1, pp. 175–206. Academic Press, San Diego.

Dickhoff, W. W., and Darling, D. S. (1983). Evolution of thyroid function and its control in lower vertebrates. Am. Zool. 23, 697–708.

Eales, J. G., and Brown, S. B. (1993). Measurements and regulation of thyroidal status in teleost fish. Rev. Fish Biol. Fisher. 3, 299–347.

Fritzsch, B. (1990). The evolution of metamorphosis in amphibians. J. Neurobiol. 21, 1011–1021.

Galton, V. A. (1992). Thyroid hormone receptors and iodothyronine deiodinases in the developing Mexican axolotl. Ambystoma mexicanum. Gen. Comp. Endocrinol. 85, 62–70.

Galton, V. A. (1992). The role of thyroid hormone in amphibian metamorphosis. Trends Endocrinol. Metab. 3, 96–100.

Hulbert, A. J. (1985). A comparative study of thyroid function in reptiles and mammals. In "The Endocrine System and the Environment" (B. K. Follett, S. Ishii, and A. Chandola, eds.), pp. 105–115. Japan Sci. Soc./Springer-Verlag, Berlin.

Kikuyama, S., Kawamura, K., Tanaka, S., and Yamamoto, K. (1993). Aspects of amphibians metamorphosis: Hormonal control. Int. Rev. Cytol. 145, 105–148.

Kuhn, E. R., Mol, K. A., and Darras, V. M. (1993). Control strategies of thyroid hormone monodeiodination in vertebrates. Zool. Sci. 10, 873–885.

Leatherland, J. F. (1988). Endocrine factors affecting thyroid economy of teleost fish. Am. Zool. 28, 319–328.

Leloup, J., and de Luze, A. (1985). Environmental effects of temperature and salinity on thyroid function in teleost fishes. In "The Endocrine System and the Environment" (B. K. Follett, S. Ishii, and A. Chandola, eds.), pp. 23–32. Japan Sci. Soc./Springer-Verlag, Berlin.

May, J. D. (1989). The role of the thyroid in avian species. Crit. Rev. Poultry Biol. 2, 171–186.

Norris, D. O., and Dent, J. N. (1989). Neuroendocrine aspects of amphibian metamorphosis. *In* "Development, Maturation, and Senescence of Neuroendocrine Systems" (M. P. Schreibman and C. G. Scanes, eds.), pp. 63–90. Academic Press, San Diego.

Sharp, P. J., and Klandorf, H. (1985). Environmental and physiological factors controlling thyroid function in galliformes. *In* "The Endocrine System and the Environment" (B. K. Follett, S. Ishii, and A. Chandola, eds.), pp. 175–188. Japan Sci. Soc./Springer-Verlag.

Shi, Y.-B. (1994). Molecular biology of amphibian metamorphosis: A new approach to an old problem. *Trends Endocrinol. Metab.* **5,** 14–20.

Wakahara, M., Miyashita, N., Sakamoto, A., and Arai, T. (1994). Several biochemical alterations from larval to adult types are independent of morphological metamorphosis in a salamander, *Hynobius retardatus. Zool. Sci.* **11,** 583–588.

9

The Mammalian Adrenal Glands: Cortical and Chromaffin Cells

THE RESPONSES an animal makes to any stressful stimulus lead to a stress response, which includes the release of hormones from the adrenal glands. The response of the animal is defined as **stress.** Adrenal hormones induce changes in metabolism and/or ionic regulation that work to combat physiological and psychological symptoms of stress and eventually to eliminate the stressful stimulus. Knowledge of this system contributes to our understanding of how animals adapt physiologically to physical and psychological traumas.

Mammals typically possess two adrenal glands, one located superior to each kidney (*ad renal* or *suprarenal;* Fig. 9-1). This anatomical arrangement is responsible for their present name, adrenals, and for an alternative name that is no longer in common usage, the suprarenal glands.

Each adrenal gland actually consists internally of two almost separate endocrine glands. The outer portion, or **adrenal cortex,** is composed largely of lipid-containing, steroidogenic **adrenocortical cells,** and the cortex surrounds an inner mass of **chromaffin cells** called the **adrenal medulla.** Chromaffin cells are named for their staining affinity for certain chromium compounds (see Section V).

The adrenocortical cells are derived from the coelomic epithelium in the pronephric re-

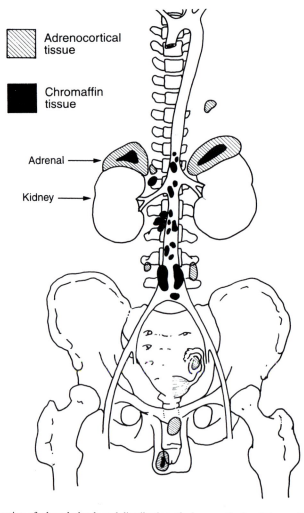

Figure 9-1. Location of adrenal glands and distribution of adrenocortical and chromaffin tissues in humans. Note the heterotopic locations where either tissue may be found in either sex.

gion of the embryo adjacent to the genital ridge that gives rise to the gonads. These cells produce the **glucocorticoids, mineralocorticoids,** and weak androgens such as **dehydro-epiandrosterone (DHEA).** Secretion of glucocorticoids and probably DHEA under the direct stimulatory influence of **corticotropin (ACTH)** from the pituitary gland. Mineralocorticoid secretion is regulated mainly by the **renin–angiotensin system** (Sections II,A and V,B,2). Glucocorticoids are named for their influences on glucose metabolism and mineralocorticoids for their effects on Na^+ and K^+ balance. The major glucocorticoids in mammals are **cortisol, corticosterone,** and **11-deoxycortisol.** The major mineralocorticoids are **aldosterone** and **deoxycorticosterone.** Representative blood levels for glucocorticoids and mineralocorticoids in mammals are provided in Table 9-1. Relative glucocorticoid and mineralocorticoid activities of some common natural and synthetic corticosteroids can be found in Table 9-2.

The chromaffin cells are of neural crest origin, and the medulla functions essentially like a modified sympathetic ganglion. The adrenal medulla is under direct neural control (preganglionic cholinergic sympathetic neurons) and releases **norepinephrine** and **epinephrine** into the blood. These secretions are usually termed "hormones" but, considering the embryonic origin of the medulla, they could be called "neurohormones," too.

Table 9-1. Circulating Levels of Corticoids in Selected Mammalian Species (Prototheria, Metatheria, and Eutheria)

Species	Mean steroid levels ± SD in peripheral blood				
	Aldosterone (ng/dl)	Corticosterone (μg/dl)	Cortisol (μg/dl)	Deoxycorticosterone (ng/dl)	11-Deoxycortisol (μg/dl)
Prototheria					
Echidna[a]	1.5	0.35	0.18	—	—
Echidna[b]	—	0.14 ± 0.07	0.07 ± 0.03	—	—
Echidna + ACTH[b]	—	1.06 ± 0.56	0.42 ± 0.23	—	—
Metatheria					
Black-tailed wallaby	8.2 ± 4.3	0.12 ± 0.02	1.1 ± 0.2	3.3 ± 0.6	0.13 ± 0.08
Common wombat	0.9 ± 1.4	0.06 ± 0.02	0.04 ± 0.04	2.0 ± 1.3	0.14 ± 0.08
Dingo	6.7 ± 4.3	0.04 ± 0.21	1.90 ± 1.90	23.6 ± 23.7	0.19 ± 0.09
Koala	1.6 ± 2.2	0.20 ± 0.05	*[c]	6.0 ± 8.3	0.06 ± 0.11
Eutheria					
Sheep	2.1 ± 1.7	0.09 ± 0.04	0.52 ± 0.50	2.5 ± 2.5	0.05 ± 0.03
Dog	2.1 ± 3.6	0.20 ± 0.13	0.85 ± 0.39	11.3 ± 6.0	0.07 ± 0.05
Fox	13.9 ± 3.2	0.59 ± 0.17	2.30 ± 1.2	37.4 ± 22.7	0.17 ± 0.03
BALB/cfC3H mice[d]	—	12.00 ± 2.16	—	—	—
Sand rat[e]	2.8 ± 13.3	0.20 ± 0.8	10.7 ± 24.9	—	—
Human, male	12.4	0.42	14.4	6.6	0.05

[a] Oddie, G. J., Blaine, E. H., Bradshaw, J. P., Coghlan, J. P., Denton, D. A., Nelson, J. F., and Scoggins, B. A. (1976). Blood corticosteroids in Australian marsupial and placental mammals and one monotreme. *J. Endocrinol.* **69,** 341–348.

[b] Sernia, C., and McDonald, I. R. (1977). Adrenocortical function in a protetherian mammal. *J. Endocrinol.* **72,** 41–52.

[c] *, Undetectable.

[d] Hawkins, E. F., Young, P. N., Hawkins, A. M. C., and Bern, H. A. (1975). Adrenocortical function: Corticosterone levels in female BALB/C and C34 mice under various conditions. *J. Exp. Zool.* **194,** 479–484.

[e] Amirat, Z., Khammar, F., and Brudieux, R. (1980). Seasonal changes in plasma and adrenal concentrations of cortisol, corticosterone, aldosterone, and electrolytes in the adult male sand rat (*Psammomys obesus*). *Gen. Comp. Endocrinol.* **40,** 36–43.

Table 9-2. Activities of Corticosteroids

Steroid	Relative activity in assay				
	MR[a]	Na^{+b}	GLY[c]	GR[d]	AI[e]
Aldosterone	1	1	0.15	1	—
DOC[f]	0.8	0.03	0.02	1	—
Corticosterone	0.2	0.004	0.36	2	0.3
Cortisol	0.1	0.001	1	1	1
18-OH-DOC[f]	0.015	0.004	—	0.02	—
Dexamethasone	0.05	0.001	17–250	10	25–169

[a]MR, Mineralocorticoid; MR receptor assay is based on competition for [³H]aldosterone-binding sites in rat kidney.

[b]Na$^+$ bioassay, normally based on change in urinary Na$^+$/K$^+$ in the adrenalectomized rat.

[c]GLY, Glycogen; GLY bioassay is based on glycogen deposition in the adrenalectomized rat liver.

[d]GR, Glucocorticoid receptor; GR assay is based on competition with [3H]dexamethasone-binding sites in rat kidney.

[e]AI, Antiinflammatory; AI bioassay is based on antiinflammatory activity.

[f]DOC, Deoxycorticosterone; 18-OH-DOC, 18-hydroxy-DOC.

At first glance, there seems to be no functional significance for the close anatomical relationship of these two distinctly different tissues (adrenocortical and chromaffin). Although there is participation of both systems with respect to adaptations to stressful stimuli, the factors controlling release of their secretions are not obviously related, nor do their biological actions overlap. The anatomical closeness of the cortex and medulla in mammals as well as the relationships of their homologous tissues in nonmammals may simply be a function of the physical closeness of their embryological sites of origin, although the progressive evolution of this arrangement is obvious when other vertebrate systems are examined (see Chapter 10). We shall see in the following discussion that there are some actions of ACTH and glucocorticoids on catecholamine synthesis by chromaffin cells and that these systems are not entirely independent.

I. The Mammalian Adrenal Cortex

The adrenal cortex of adult mammals may be subdivided on the basis of histological criteria into three well-defined regions: the **zona glomerulosa, zona fasciculata,** and **zona reticularis** (Fig. 9-2). These regions are arranged as concentric shells surrounding the adrenal medulla. In addition, there are inner zones described between the outer zones and the medulla in some mammals. These regions may perform unique functions and may be transitory.

A. Zonation of the Adrenal Cortex

The cells of the outermost region of the adrenal cortex, the zona glomerulosa, are smaller, more rounded, and contain less lipid than those of the more central zona fasciculata. The zona glomerulosa is responsible for synthesis of **aldosterone** as well as some other corticosteroids. There are few cytological changes in the zona glomerulosa following hypophysectomy or administration of ACTH, suggesting that the secretion of aldosterone is independent of pituitary control. Although ACTH is not necessary for steroidogenesis and

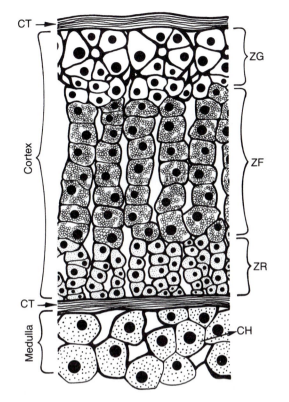

Figure 9-2. Zonation of the adrenal gland. The cortex consists of an outermost layer of connective tissue (CT), the zona glomerulosa (ZG) producing aldosterone, the zona fasciculata (ZF) secreting most of the glucocorticoids, and the inner zona reticularis (ZR) specializing in adrenal androgen production. The adrenal medulla is separated from the cortex by another layer of CT and consists primarily of chromaffin cells (CH) that secrete catecholamines and endogenous opiates.

release of aldosterone, the responsiveness of the glomerulosal cells to agents that normally elicit these events is reduced in hypophysectomized mammals and is enhanced with ACTH treatment.

The zona fasciculata is located between the zona glomerulosa and the zona reticularis and is histologically distinct from both. It consists of polyhedral (many-sided) cells that are sources of the glucocorticoids. The cells of the zona fasciculata are arranged in narrow columns, or *cords,* surrounding blood sinusoids that allow the cells to be bathed directly with blood. The proportions of cortisol and corticosterone secreted differ markedly, from secretion of primarily cortisol (human), through mixtures of both (cat), to primarily corticosterone (rat). Thickness of the zona fasciculata is most sensitive to the activity of the hypothalamo–hypophysial release of ACTH. It exhibits hypertrophy and hyperplasia in response to prolonged elevation of ACTH secretion caused by stress or treatment with the drug **metyrapone,** which blocks the 11-hydroxylating step necessary for glucocorticoid synthesis. Conversely, the zona fasciculata atrophies markedly following hypophysectomy or prolonged glucocorticoid therapy.

The zona reticularis typically borders the adrenal medulla, and it contains numerous reticular fibers (hence its name). It is a primary source of adrenal androgens but some glucocorticoids are synthesized as well. The zona reticularis also hypertrophies in response to ACTH and atrophies following hypophysectomy, but not so dramatically as the zona fasciculata.

It should be noted that the "typical" anatomical pattern described here within the adrenal cortex and the anatomical relationship of cortex to medulla vary considerably within mammals as a group. Furthermore, ectopic nodules of functional cortical tissue are not uncommon (Fig. 9-1), and this accessory adrenocortical tissue may become an important source for corticosteroids following surgical adrenalectomy.

B. Additional Zonation

Several unique adrenocortical zones are known only for certain species, but the adrenals of most mammalian species have not been examined in detail. These unique zones may be conspicuous only at certain times in the life of an animal or in only one sex.

1. The Fetal Zone

In primates, a very conspicuous zone occupies the bulk of the adrenal gland prior to birth. This region is called the **fetal zone** and is responsible for the relatively large size of the adrenal at birth (Fig. 9-3). In humans, the neonate adrenal may be as large as the adrenal gland of a 10 to 12 year old. During gestation, the fetal zone synthesizes and releases relatively large quantities of DHEA and lesser amounts of its sulfated derivative, **DHEA-S.** These adrenal androgens serve as precursors for the synthesis of estrogens by the placenta. Failure of the fetal zone to produce adequate amounts of DHEA/DHEA-S results in pre-

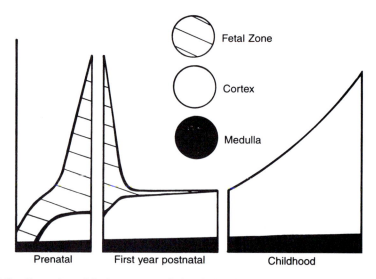

Figure 9-3. Comparison of fetal zone to remainder of adrenal gland in humans. Prior to birth, the bulk of the adrenal consists of the androgen-secreting fetal zone, which regresses rapidly after birth. Whereas the fetal zone is gone by 1 year of age, the medulla and cortex continue to grow until puberty.

mature termination of gestation. Formerly, it was believed that chorionic gonadotropin from the placenta (see Chapters 4 and 11) was responsible for stimulating fetal adrenal androgen production, but recent studies suggest adrenal androgen secretion is influenced by ACTH and prolactin.

Following birth, the fetal zone ceases to function and degenerates rapidly. The zona reticularis continues to synthesize DHEA-S/DHEA until about age 20 years, when its production declines. It has been suggested that DHEA may be an antitumor substance and/or a precursor in the synthesis of other androgens or estrogens, especially in postmenopausal women.

2. The Mouse X-Zone

The cortex of the mouse adrenal contains a unique **X-zone** between the zona reticularis and the medulla. The X-zone degenerates in males at puberty and in females of most mouse strains during the first pregnancy. Degeneration in the males is linked to production of gonadal androgens. The function of this region is not known.

3. The "Special Zone"

In at least one marsupial, the brush-tailed possum (*Trichosurus vulpecula*), there is a large, inner special zone that appears only in the adult female. Its function has not been elucidated, and it is not known whether it is comparable to the X-zone of mice.

II. Biosynthesis and Transport of Corticosteroids

Glucocorticoids and mineralocorticoids differ not only in their actions but also in their synthesis and transport in the blood. As outlined in Chapter 2, the corticosteroids are derived from conversion of **cholesterol** to **pregnenolone** in the mitochondria by the action of the side-chain cleavage enzyme, P-450_{scc}. Pregnenolone is converted into progesterone and then further modified (see Chapter 2, Figs. 2-30 to 2-32). Cholesterol is readily obtained from circulating **low-density lipoprotein (LDL) droplets** or from stored deposits in the adrenal cells. Synthesis of cholesterol from acetyl groups also can occur in adrenal cells, as described in Chapter 2.

A. Synthesis of Corticosteroids

At the membranes of the smooth endoplasmic reticulum, pregnenolone can be metabolized along either the **17α-hydroxysteroid pathway** or the **17-deoxysteroid pathway.** The former pathway leads to **11-deoxycortisol** which is accomplished by the enzyme **21-hydroxylase.** 11-Deoxycortisol may be secreted or may reenter the mitochondria, where the **11β-hydroxylase enzyme (P-$450_{11\beta}$)** converts it to cortisol. Conversion of pregnenolone to progesterone in the zona fasciculata or glomerulosa, via the 17-deoxysteroid pathway, yields **11-deoxycorticosterone,** which is converted to corticosterone by employing the same enzymes as in the 17α-hydroxysteroid pathway. Cortisol and/or corticosterone are the major end products of corticosteroidogenesis in the zona fasciculata and to a lesser extent in the zona reticularis.

In the zona glomerulosa, the 17-deoxysteroid pathway is favored and corticosterone is further modified to **18-hydroxycorticosterone** and then to aldosterone. Another mitochondrial enzyme, called **aldosterone synthase** or $P\text{-}450_{C18}$, is responsible for aldosterone synthesis in rodents and humans. In the zona glomerulosa of cows and pigs, this step is accompanied by $P\text{-}450_{11\beta}$. However, $P\text{-}450_{11\beta}$ does not produce aldosterone in any other region of the cortex. Both of these steroids have mineralocorticoid activity, but aldosterone is more potent and is typically the dominant secretory product of the zona glomerulosa *in vivo*. However, deoxycorticosterone may be the major secretory product in some species (see ` Table 9-1).

Adrenal androgens are synthesized primarily in the zona reticularis as well as by the fetal zone of primates. Both the Δ^4- and Δ^5- pathways may be utilized (see Chapter 2) but the Δ^5- pathway is more common, resulting in DHEA, which is then sulfated to form DHEA-S. The primate fetal zone makes little or no **androstenedione,** which is more commonly found in the adult adrenal. DHEA is a weak androgen and the sulfated form does not penetrate readily into potential target cells, reducing the threat of masculinization to a female fetus or the mother. DHEA-S is converted by the placenta into **16α-hydroxy-DHEA** and then by the **aromatase** enzyme to estriol, which is essential for maintaining pregnancy. The placenta cannot synthesize androgens and depends entirely on the fetal adrenal to provide the androgen precursor for estrogen synthesis.

B. Release of Corticosteroids

Circulating corticosteroids reflect synthesis rates since little hormone is stored in the adrenal, and corticosteroids are released as they are made. In adult humans, daily episodic release of cortisol is maximal between 0600 and 0900 hr in the morning, with lower values occurring in the afternoon and minimal levels at night (see Chapter 3, Fig. 3-2). This daily rhythm can be disrupted by eating lunch but is not affected by eating dinner. In nocturnal animals, such as the rat, the rhythmic pattern of secretion is reversed.

Aldosterone secretion shows a rhythm similar to that of glucocorticoids in humans, with peak levels observed between 0600 and 0900 hr. Unlike the glucocorticoid rhythm, the aldosterone rhythm is independent of ACTH levels and is not influenced by circulating sodium or potassium levels. The drug **propranolol,** a β-adrenergic blocker, abolishes the aldosterone rhythm in rats, suggesting that catecholamines are responsible for the rhythm.

In humans, as is generally true for other mammals, adrenal androgen synthesis peaks or reaches **adrenarche** at puberty in both males and females (Fig. 9-4) and continues at this level until the early 20s, when it begins to decline. This decline in adrenal androgen production has been termed **adrenopause.**

Because of the episodic nature of corticosteroid release and the strong diurnal aspect of secretion, measurements to determine appropriate levels of corticosteroids are complicated by the need for repeated sampling of blood or saliva. (Saliva contains corticosteroids in proportion to their concentrations in the blood, and their measurement can be used as an index of blood levels.) Consequently, analysis of steroid metabolites in 24-hr urine samples provides an index of integrated secretion over that time period. Quantitative analysis of urinary steroids can be accomplished rapidly using such techniques as gas chromatography or fast atom bombardment mass spectrometry.

C. Transport of Corticosteroids in the Blood

Probably, more than 90% of the glucocorticoids are bound to a specific plasma transport protein called **corticosteroid-binding globulin (CBG),** or transcortin. Some glucocorti-

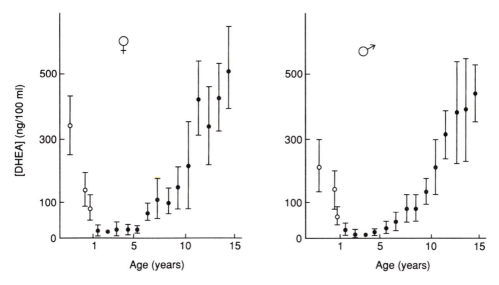

Figure 9-4. Adrenarche in humans. After birth, DHEA continues to be secreted by the regressing fetal zone (open circles). Although the fetal zone has regressed completely by age 1 year, production of the adrenal androgen DHEA continues in the zona reticularis (solid circles). DHEA secretion begins to increase at about age 5 years and peaks at puberty. [Modified from de Peretti, E., and Forest, M. G. (1976). Unconjugated dehydroepiandrosterone plasma levels in normal subjects from birth to adolescence in human: The use of a sensitive radioimmunoassay. *J. Clin. Endocrinol. Metab.* **43**, 982–991.]

coid also binds nonspecifically to albumin as was described for thyroid hormones in Chapter 7. Although there is a bias for assuming free hormone is most important for entering target cells and producing effects, there also is evidence to support a role for interaction of occupied CBG with target cell membranes that facilitates dissociation of glucocorticoids from CBG and their entrance into the target cell.

Aldosterone does not bind significantly to CBG or other plasma proteins. Consequently, aldosterone is present at much lower concentrations in the blood than are the glucocorticoids because aldosterone is cleared from the blood much more rapidly.

D. Metabolism of Corticosteroids

The liver is the major site for metabolizing corticosteroids although other tissues, such as the kidneys and intestines, also can perform this task. Most corticosteroids are modified to increase their solubility in water and are excreted via the urine, bile, or feces. In humans, the majority of corticosteroid metabolites are excreted via the urine whereas in the rat the biliary and fecal routes are more popular.

One way in which corticosteroids are metabolized is by reduction of the A-ring and addition of hydroxyl groups at C-3 or C-20, followed by side-chain cleavage to yield C_{19} steroids. Hydroxylation by cytochrome *P*-450 enzymes also occurs at C-6 and C-16. In humans, up to a third of the cortisol may be oxidized to corotic acids. Many steroids are conjugated as sulfates or glucuronides. Cortisol, aldosterone, and most adrenal androgens (except DHEA) are conjugated as glucuronides, but DHEA is excreted as DHEA-S. Some selected metabolites of corticosteroids are depicted in Chapter 2 (Fig. 2-35).

III. Secretion and Actions of Glucocorticoids

Secretion of glucocorticoids from the zona fasciculata and zona reticularis is under direct control of the hypothalamo–hypophysial axis involving corticotropin-releasing hormone and ACTH, respectively (see Chapter 4). Circulating ACTH levels are depressed by elevated levels of glucocorticoids and are increased following adrenalectomy, establishing the existence of direct negative feedback of corticosteroids on ACTH release. **Arginine vasopressin (AVP)** potentiates the action of **corticotropin-releasing hormone (CRH)** on the corticotropic cells of the pars distalis during responses to chronic stress. Corticotropin also may be released by direct neural input to the hypothalamus as a result of such stressful events as physical trauma (injury, surgery), and this response may occur even in the presence of elevated glucocorticoid levels. Thus, stressful stimuli can bring about and maintain a severalfold increase in glucocorticoid levels. The ability of traumatic stimuli to override the normal negative feedback mechanism verifies the importance of neural control over ACTH release.

A. Actions of Glucocorticoids

Glucocorticoids produce marked effects on energy metabolism at physiological doses as a result of changes in transport of materials into cells and induction of new enzyme syntheses. Their major effect is to supplant and conserve the energy normally derived from circulating glucose by (1) inhibiting glucose utilization by peripheral tissues (especially skeletal muscle), (2) stimulating entry of amino acids into nonneural cells and their conversion to glucose and storage as glycogen, and (3) enhancing the mobilization of fat stores in nonneural tissues. By blocking glucose utilization, glucocorticoids spare blood glucose so that the brain has a preferential source of glucose.

1. Glucocorticoids and Metabolism

The metabolic conversion of amino acids into glucose is termed **gluconeogenesis,** the production of glucose from a noncarbohydrate precursor; in this case, amino acids. Glucocorticoids also enhance the actions of growth hormone on lipolysis in adipose tissue. This causes release of fatty acids and glycerol into the circulation and their conversion by liver cells into glucose, another form of gluconeogenesis.

2. Glucocorticoids and Immunity

It is common to observe an increased incidence of disease in stressed animals. In fact, the immune response system is inhibited by the hypothalamo–pituitary–adrenal axis at several levels. For example, elevated levels of glucocorticoids inhibit the inflammatory response as well as production of antibodies. Furthermore, CRH not only stimulates ACTH release but also causes release of somatostatin (SS) and dopamine, which reduce secretion of growth hormone (GH) and prolactin (PRL), respectively. Either GH or PRL can restore immune function that is normally depressed following hypophysectomy. The observation that GH and PRL secretion occurs at night, when ACTH and glucocorticoid levels normally are lowest, supports these relationships since renewal of immune function also occurs at night.

   Evidence suggests that cytokines influence the adrenal axis and this may be part of the mechanisms for homeostatic regulation of the immune response system. A variety of cytokine effects on brain function have been described, including effects of interleukins on the blood–brain barrier allowing access of other elements into the nervous system. Cytokines may enter the brain through special structures called **circumventricular organs** such as the organum vasculosum lateralis terminae (OVLT) of the preoptic area. Circumventricular organs contain fenestrated capillaries that allow access of blood-borne materials to the brain. Other evidence suggests they can enter at many sites via carrier-mediated transport. Regardless of how they enter, cytokines such as interleukins and interferons released during increased immune activity can activate CRH release from the hypothalamus, which ultimately causes elevated glucocorticoids and suppression of the immune response. Studies of cytokine effects on brain function suggest that at least some of their effects are mediated via sensory fibers of the vagus nerve. In rats, vagotomy blocks several responses classically attributed to cytokine effects on the brain including fever, hypothalamic depletion of norepinephrine, and elevated plasma corticosterone. These latter observations suggest there is still a great deal to learn about the nature of regulatory pathways in vertebrates.

3. Glucocorticoids and Stress

The responses characteristic of adaptations to chronic stress are mediated by glucocorticoids. These responses have been incorporated by Selye into a general theory of adaptation to stress termed the **general adaptation syndrome.** According to Selye, there are three stages of adaptation by an organism to stressful stimuli. Earlier, stress was defined as an all-encompassing term to include the responses to all stimuli (stressors) that are harmful or potentially harmful to the organism. The first phase of the general adaptation syndrome is the **alarm reaction,** which includes a generalized increase in sympathetic stimulation involving the adrenal medulla as well as increased secretion of glucocorticoids. Because of the activation of glucocorticoid secretion, the alarm reaction is not identical to the "emergency" or "fight-or-flight" response of Cannon (see Section VI,B,2). This phase is frequently marked by enlargement of the adrenal glands, primarily due to hypertrophy of the zona fasciculata and to some extent the zona reticularis of the cortex. Under the influence of glucocorticoids the organism adapts to the continued presence of the stressful stimuli. This phase of adaptation is termed the **stage of resistance** and is characterized by prolonged, increased secretion of glucocorticoids. (It can be argued whether this is truly adaptation or the lack of adaptation.) Finally, under continuation of extremely stressful conditions, the ability of the organism to function normally is impaired. The continuous presence of the stressful stimuli causes the organism to enter the final **stage of exhaustion** that leads to death. Although some aspects of Selye's theory are still being debated, the general pattern of the stress response is heavily documented.

The relative roles of the adrenal cortex and the adrenal medulla are not fixed but can vary according to the circumstances. For example, humans were examined while performing the same task under self-paced conditions or externally paced conditions that challenged their ability to keep up. Plasma epinephrine levels were similar in both groups, suggesting an equal intensity of effort, but cortisol levels were much higher in the externally challenged subjects. Social position, especially in species exhibiting a clear dominance hierarchy, also can influence the physiological response to a potentially stressful stimulus. Following establishment of a new laboratory colony, mice becoming subordinate were found to have higher levels of plasma corticosterone whereas mice emerging as dominants had higher levels of catecholamines as evidenced by increased tyrosine hydroxylase activity.

Stress of experimental animals is important when attempting to interpret data on corticosteroid levels or nutrient levels (amino acids, glucose, etc.). Laboratory conditions alone can influence the adrenal axis. Stress can also influence the levels of other hormones. For example, the order in which groups of rats were removed from a common holding facility and killed at a remote site influenced the mean levels of PRL that were measured for each group. Since it is not possible to undertake experimental work on animals free from stress, detailed knowledge of how animals were maintained, and of all procedures involved in an experiment, is essential when interpreting results.

4. Permissive Actions of Glucocorticoids

It has been suggested that, in addition to their roles during stress and participation in immune suppression, glucocorticoids may act as "permissive agents" as described earlier for thyroid hormones (see Chapter 7). Through changes in membrane permeabilities to important metabolites and by stimulating the synthesis of new enzymes or receptors, glucocorticoids may provide the appropriate cellular environment in which other hormones operate.

5. Pharmacological Actions of Glucocorticoids

The glucocorticoids possibly are better known for their pharmacological actions and therapeutic effects than for their biological actions. The tremendous potential for glucocorticoids in the treatment of the rare hypoadrenocorticism described by Thomas Addison in 1885 (Addison's disease) was not recognized for many years. However, the discovery of the **antiinflammatory effects** of glucocorticoids and their use for treatment of rheumatoid arthritis spurred a tremendous explosion in therapeutic applications of glucocorticoids. The debilitating symptoms of rheumatoid arthritis are the consequence of inflammation associated with an autoimmune response in which the patient produces antibody against his/her own connective tissue. Glucocorticoid therapy alleviates painful inflammation occurring as a result of the immune reaction but does nothing to correct the causative factors.

One of the most recent therapeutic applications of glucocorticoids is suppression of the entire immune response to allow tissue and organ transplantations. Although normal therapeutic antiinflammatory doses of glucocorticoids do not interfere with normal antigen–antibody interactions, very high doses depress new antibody synthesis.

Glucocorticoids interfere with the elaboration of histamine or with its actions in mediating the inflammatory response, which includes local hyperemia and resultant edema, or with both. One postulated mechanism for this interference is the glucocorticoid-induced inhibition of the **kallikreins,** enzymes that catalyze formation of **kinins** from a plasma precursor protein. Kinins induce inflammation by causing release of histamine normally observed following the combination of antigen and antibody. Another suggestion for glucocorticoid antiinflammatory activity stems from observations of their effects on lysosomes. Glucocorticoids stabilize lysosomal membranes, thereby reducing release of hydrolytic enzymes following cell injury and hence reducing the spread of the inflammatory reaction. Glucocorticoids also may inhibit the synthesis of leukotrienes and cytokine agents (e.g., interleukins) that mediate inflammation and cell-mediated immune responses.

6. Mechanism of Glucocorticoid Cellular Action

Glucocorticoid effects on target cells are described in Chapter 2, and only a brief summary is provided here. The initial requirement for glucocorticoid action on liver cells is the binding of glucocorticoid to type II glucocorticoid receptors, translocation of occupied receptors into the nucleus, and eventual stimulation of nuclear RNA synthesis (both messenger RNA and ribosomal synthesis). Approximately 2 to 4 hr following application of glucocorticoids, there is an increase in new enzymes that bring about the changes in cellular metabolism characteristic of glucocorticoid action. In liver cells, these new enzymes include those associated with the conversion of amino acids into glucose and the polymerization of glucose to form glycogen. Amino acid transport into the liver cell also is stimulated. In contrast, glucocorticoids inhibit the uptake of amino acids and the metabolism of glucose in peripheral tissues such as skin, skeletal muscle, and adipose cells.

In the brain, glucocorticoids may bind to plasma membrane receptors and bring about rapid effects in target cells. Membrane-bound receptors have been identified in nonmammals, where corticosterone is responsible for inducing reproductive behaviors. Presumably, these membrane receptors operate through known second-messenger systems to produce their rapid effects.

Excessive doses of glucocorticoids inhibit protein synthesis in certain tissues (for example, skeletal muscle, bone, and lymphoid tissue), which has been related to some of the adverse side effects of glucocorticoid therapy. Curiously, these protein-catabolic effects are not manifest in liver cells even at very high doses. The basis for this difference between liver and other organs is not known.

Early feedback effects on pituitary corticotropes during a stress response reduce the availability of intracellular calcium ions that normally are increased by CRH or AVP working through second messengers. Consequently, ACTH release is blocked. Later actions of glucocorticoids involve blocking ACTH synthesis and down-regulation of CRH receptors in corticotropes.

IV. Aldosterone: The Principal Mammalian "Mineralocorticoid"

The zona glomerulosa secretes aldosterone independent of direct pituitary control, although, as mentioned earlier, ACTH appears to play a permissive role in maintaining the responsiveness of these cells to other controlling factors. The major action of aldosterone is to maintain the normal sodium–potassium balance in body fluids, and its secondary action is to regulate extracellular fluid volume. Aldosterone stimulates sodium reabsorption into the blood and potassium secretion into the urine by the nephrons in the kidney. The mechanism controlling secretion of aldosterone involves a most complex and seemingly round-about series of events involving both liver and kidney: the renin–angiotensin system.

A. The Renin–Angiotensin System and Aldosterone Secretion

Renin is an enzyme produced in the kidney by the **juxtaglomerular body,** a modified group of cells in the afferent arteriole carrying blood to the glomerulus (Fig. 9-5). This enzyme is a glycoprotein (about 40,000 Da) possibly secreted as a larger, inactive form or **prorenin** (MW 63,000). Conversion of prorenin to renin may be accomplished by the activity of kidney kallikrein. Development of renin activity from prorenin also occurs following mild acidification of the plasma. Renin has a plasma biological half-life of about 15 min and is rapidly degraded.

The juxtaglomerular body is intimately associated with a modified region of the distal

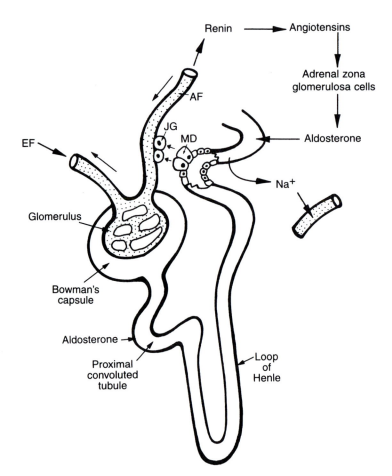

Figure 9-5. Schematic representation of the juxtaglomerular apparatus. The stippled structures represent the
afferent arteriole (AF) bringing blood to the glomerulus housed in Bowman's capsule of the
nephron and exiting via the efferent arteriole (EF) prior to entering the peritubular capillaries.
The juxtaglomerular apparatus consists of the renin-secreting juxtaglomerular cells (JG) embedded in the walls of the afferent arteriole and the macula densa (MD), which consists of modified
cells in the distal convoluted portion of the nephron.

convoluted portion of the nephron known as the **macula densa.** Together these two structures comprise the **juxtaglomerular apparatus.**

Blood volume (as reflected by pressure within the renal arterioles) and sodium concentration in the glomerular filtrate as it enters the proximal convoluted tubule control renin
release. Intrarenal arteriolar pressure is monitored by stretch receptors in the juxtaglomerular body. Renin is released in response to a decrease in this pressure. Sodium concentration in the tubular lumen is monitored by cells of the macula densa, and low sodium levels
somehow trigger communication between the macula densa and the juxtaglomerular cells,
resulting in renin release. Changes in either or both parameters influence renin secretion.

Once renin enters the blood, it comes in contact with a plasma protein termed **renin
substrate (angiotensinogen)** that was synthesized by the liver. Mammalian renin sub-

strates are large glycoproteins (MW 58,000–110,000) that behave like α_1-globulins, α_2-globulins, or albumin in humans, herbivores, and rodents, respectively. Renin causes the enzymatic release of a small peptide (decapeptide) known as **angiotensin I** from renin substrate. Although angiotensin I can facilitate release of norepinephrine from the adrenal medulla and can produce direct and indirect pressor effects on the cardiovascular system, it is generally believed to be of little physiological importance because **angiotensin-converting enzyme (ACE)** hydrolyzes most of the angiotensin I to an octapeptide, **angiotensin II.** Angiotensin II has a biological half-life of about 2 min. One product of its hydrolysis is a heptapeptide known as **angiotensin III,** which may also stimulate aldosterone release. Angiotensin III is as potent as angiotensin II in releasing aldosterone but is a less effective pressor agent. It is probably an important regulator in rats, where it accounts for 58% of the angiotensin activity. However, in humans and dogs, only about 12% of the angiotensin activity is attributable to angiotensin III. The actions of the angiotensins are summarized in Figs. 9-6 and 9-7.

Angiotensin II is a potent vasoconstricting agent and helps restore blood pressure to normal by causing contraction of vascular smooth muscle, resulting in a decrease in arteriole diameter. Angiotensin II is about 40 times more potent than norepinephrine at inducing vasoconstriction. ACE is found in a number of capillary beds (especially in the lungs), suggesting that angiotensin II produces vascular effects in a number of tissues. **Captopril** is a potent synthetic inhibitor of converting enzyme and can be used to block formation of angiotensin II.

Another important action of angiotensin II is to stimulate synthesis and release of aldosterone from cells of the zona glomerulosa. Aldosterone release appears to be a result of activation of inositol triphosphate and diacylglycerol second messengers and their effects on calcium channels in zona glomerulosa cells.

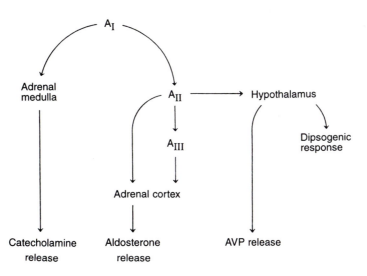

Figure 9-6. Actions of angiotensins. The major endocrine actions of angiotensins are included but the vasopressor action is not. In humans, the major angiotensin is angiotensin II (A_{II}) but in rats angiotensin III (A_{III}) may be equally important.

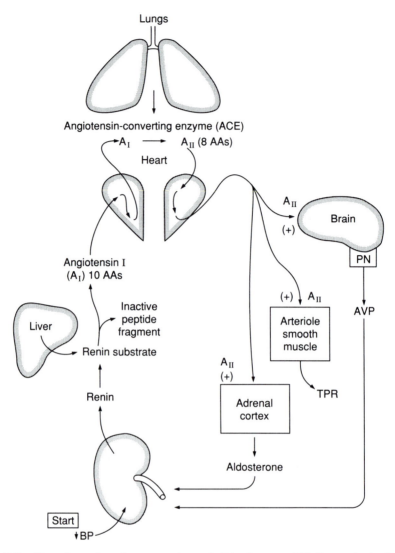

Figure 9-7. The renin–angiotensin system and control of blood pressure (BP). See text for details. Aldosterone also increases Na^+ reabsorption and K^+ secretion by the kidney and renin release can be stimulated by low sodium via the macula densa. TPR, Total peripheral resistance; TPR of the blood vascular system is due primarily to the diameter of arterioles. PN, Pars nervosa.

Aldosterone travels through the blood to the kidney and stimulates increased reabsorption of sodium and some increased excretion of potassium. This increased sodium reabsorption aids in water retention, reduces urine volume, and helps to restore normal fluid volume.

The renin–angiotensin–aldosterone mechanism is closely linked functionally to the role of vasopressins (arginine and lysine vasopressin) in maintaining normal blood osmotic conditions. Angiotensin II also stimulates drinking (dipsogenic response) through actions on the subfornical organ in the brain and causes release of vasopressins through actions on the hypothalamus. These two effects contribute to the maintenance of normal fluid balance.

Angiotensin II can induce hypertrophy of cardiac myocytes as well as mitosis in mesangial cells. Gonadal tumor cells proliferate following treatment with angiotenin II. The hypertrophic response in cardiac cells is mediated by transcription and synthesis of **transforming growth factor β (TGF-β)**. Angiotensin II also causes release of norepinephrine from atrial sympathetic nerve endings.

1. Independent Renin–Angiotensin Systems

Since the discovery of the renin angiotensin system, which involves the cooperative efforts of several organs to regulate blood pressure and sodium/potassium balance, complete renin–angiotensin mechanisms have been demonstrated within the brain, pituitary, gonads, and the adrenal cortex itself. The presence of a complete renin–angiotensin system within the adrenal cortex suggests the possibility of an adrenal paracrine regulatory system for controlling basal secretion of aldosterone. Evidence suggests a more active brain renin–angiotensin system may be present in hypertensive rats. In the pituitary, angiotensin II may stimulate proliferation of tropic hormone-secreting cells. There is evidence to support a role for angiotensin II in ovulation. In the cow ovary, the number of angiotensin II receptors increases with size of the follicle. The amount of prorenin in seminal fluid is proportional to the sperm count, but neither the origin nor the significance of seminal prorenin is known.

B. Additional Factors Controlling Aldosterone Secretion

Aldosterone secretion is influenced by several factors in addition to the renin–angiotensin system. For example, blood potassium levels can alter aldosterone release directly. Release of aldosterone also can be inhibited by a natriuretic peptide secreted by the heart (see Section IV,B,2).

1. Potassium

High levels of potassium in extracellular fluids directly stimulate aldosterone secretion from cells of the zona glomerulosa, which in turn promotes renal potassium loss. Potassium may increase the sensitivity of cells in the zona glomerulosa to angiotensin II or may directly cause aldosterone release. In contrast, extracellular sodium variations do not directly influence aldosterone secretion unless unusually large variations are produced.

2. Natriuretic Peptides

Two similar peptides are known to inhibit release of aldosterone through direct actions on cells of the zona glomerulosa. The first such peptide was found in the atria of the heart and was named **atrial natriuretic peptide (ANP)** for its ability to increase sodium excretion via the urine. The second peptide was discovered first in the brain and became known as **brain natriuretic peptide (BNP)**. However, both forms are found in the heart and a third form (**CNP**) has been found in the brain. These natriuretic peptides also cause relaxation of smooth muscle cells.

In humans, ANP is the predominant circulating form and consists of 28 amino acids (Fig. 9-8). BNP occurs in two forms of 26 and 32 amino acids, respectively, both of which

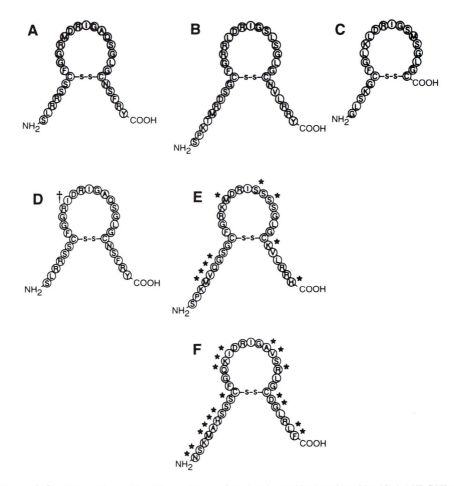

Figure 9-8. Natriuretic peptides. Three subtypes of natriuretic peptides have been identified: ANP, BNP, and CNP. Asterisks denote variability in amino acid sites within a class. Amino acid designations are presented in Appendix C. (A) Porcine/human α-ANP; (B) porcine BNP; (C) porcine CNP; (D) rat ANP; (E) human BNP; (F) rat BNP. Human and pig ANP differ from rat in only one amino acid (dagger). [Reprinted by permission of the publisher from Samson, W. K. Natriuretic peptides. A family of hormones. *Trends Endocrinol. Metab.* **3**, 86–90. Copyright 1992 Elsevier Science Inc.]

are structurally similar to ANP. Peripheral receptors appear to bind ANP and BNP equally well whereas brain receptors preferentially bind BNP and CNP over ANP.

Chronically elevated blood volume and/or pressure stretches the atria and causes release of ANP, which not only inhibits aldosterone release, but also inhibits vasopressin release and directly promotes sodium excretion and water loss in the kidney. All of these effects will reduce blood volume and blood pressure. Release of ANP also is stimulated directly by elevated plasma sodium.

3. Neurotransmitters

The adrenal cortex is well supplied with neurons employing a variety of neurotransmitters. Serotonin, norepinephrine, acetylcholine, vasoactive intestinal peptide (VIP), vasopressin, and prostaglandins are found in adrenal cortex and can stimulate aldosterone release. Somatostatin may be produced locally and is known to inhibit angiotensin-II-induced release of aldosterone.

C. Mechanism of Aldosterone Action

Aldosterone stimulates sodium reabsorption and potassium excretion in the distal convoluted tubule of the kidney and possibly enhances some sodium reabsorption in the proximal convoluted portion as well as in the intestinal mucosa, salivary glands, and sweat glands. Aldosterone produces the typical steroidal pattern of action on target cells. Increased nuclear RNA synthesis results in production of a specific protein, **aldosterone-induced protein,** which somehow mediates the movement of sodium across the target cells, somewhat reminiscent of the mechanism for movement of calcium ions across cells of the duodenal mucosa (see Chapter 15).

V. Pathologies of the Adrenal Axis

Glucocorticoid pathologies generally are distinct from those for aldosterone. However, high levels of glucocorticoids can bind to and activate aldosterone receptors. Primary and secondary disorders as well as hypersecretion and hyposecretion pathologies have been described.

A. Glucocorticoid Hypersecretion

The major disorders of glucocorticoid hypersecretion or **hyperadrenocorticism** are **Cushing's disease** and **Cushing's syndrome.** The latter term encompasses any adrenal disorder accompanied by elevated glucocorticoids whereas the former refers to the specific condition of hyperadrenocorticism caused by hypersecretion of ACTH by a corticotropic adenoma.

1. Cushing's Disease

This condition was described by Cushing in 1932. Hypersecretion of ACTH by a basophilic pituitary adenoma is the cause of this condition. Such adenomas are not sensitive to feedback by glucocorticoids. In this secondary adrenal disorder, the adrenal cortices are hypertrophied and plasma cortisol levels also are elevated (hypercortisolism). Hyperpigmentation (excessive darkening of the skin) may occur owing to the melanotropin (α-MSH)-like actions of ACTH (see Chapter 4). Excessive cortisol levels have adverse effects on the metabolism of many tissues, including the brain, muscles, skin, vascular tissue, kidney, liver, and skeleton. Hypophysectomy alleviates the symptoms (see Table 9-3) but necessitates extensive replacement therapies since all of the tropic hormones are removed by this procedure. Adrenalectomy requires only corticosteroid therapy.

2. Cushing's Syndrome

The symptoms of this disorder are the same as those for Cushing's disease, but the cause is not due to excessive secretion of ACTH by pituitary cells, but rather may be due to elevated cortisol from adrenal adenomas or adrenal carcinomas or possibly due to ACTH secreted

Table 9-3. Some Symptoms Commonly
Presented with Cushing's Disorders

Symptom	Percentage of cases
Obesity or weight gain	80
Thin skin	80
Hypertension	75
Purple skin striae	65
Hirsutism	65
Amenorrhea (women only)	60
Abnormal glucose tolerance test	55
Acne	45
Osteoporosis	40
Hyperpigmentation	20

by nonpituitary tumors. Such corticosteroid-secreting tumors suppress CRH and ACTH, which brings about atrophy of normal adrenal cortical cells. Consequently, hyperpigmentation usually does not occur in this form of hyperadrenocorticism.

B. Glucocorticoid Hyposecretion

The best known examples of **hypoadrenocorticism** are the primary disorders called **Addison's disease** and **congenital adrenal hyperplasia (CAH).** Secondary hypoadrenocorticism is not so common.

1. Addison's Disease

Addison's disease is characterized by a shortage or absence of cortisol, resulting in hypersecretion of ACTH. Hyperpigmentation occurs in most cases. Aldosterone and adrenal androgen production are depressed.

A person with Addison's disease is usually hypoglycemic owing to lack of cortisol. Absence of aldosterone causes other symptoms including muscle weakness, water losses, hypotension, and salt-craving. The loss of adrenal androgens becomes a problem only for postmenopausal or castrate women (see Chapter 11), who rely on the adrenal as their sole androgen source.

Although Addison's disease once was commonly associated with tuberculosis infections, today the most common cause of Addison's disease is bilateral atrophy of the adrenals resulting from an autoimmune attack. It also may occur as a result of drug-induced or congenital deficiencies in steroidogenetic enzymes or as a complication in acquired immunodeficiency syndrome (AIDS).

2. Congenital Adrenal Hyperplasia

Genetic defects in one or more steroidogenic enzymes necessary for corticosteroid synthesis can cause excessive ACTH secretion due to the absence of normal feedback. Hypersecretion of ACTH, in turn, causes hypertrophy and hyperplasia of the adrenal cortex, principally the zona fasciculata and zona reticularis. This condition is known as **congenital adrenal hyperplasia (CAH).** Depending on what enzyme is affected, there will be a buildup of inappropriate secretory products.

The most common form of CAH is caused by the absence of 21-hydroxylase, which blocks synthesis of cortisol, corticosterone, and aldosterone. This disorder is responsible for 90 to 95% of the cases of CAH, with the absence of enzymes such as 11β-hydroxylase, P-450$_{scc}$, and others being responsible for the remaining cases. Failure to convert progesterone to deoxycorticosterone in the absence of a functional C_{21}-hydroxylase causes a buildup of 17α-hydroxypregnenolone and 17α-hydroxyprogesterone, which are then converted into adrenal androgens, particularly DHEA and androstenedione. These androgens can be converted by peripheral tissues into testosterone, causing virilization. In extreme cases of CAH, a newborn female may be diagnosed as a male because of masculinized external genitalia. Both sexes suffering from CAH exhibit rapid somatic growth, early cessation of long bone growth, and short stature. The penis and clitoris show continuous growth. Aldosterone synthesis is impaired in about two-thirds of these cases, resulting in loss of sodium and elevation of plasma potassium (see Section IV,C,2). These subjects also are said to suffer from salt-losing or **salt-wasting disease.**

There is a much milder form of CAH in which excessive androgen production does not occur until puberty. Subjects with this cryptic form of C_{21}-hydroxylase deficiency can be identified by a simple test that shows elevated 17α-hydroxyprogesterone levels in the blood of all C_{21}-hydroxylase-deficient people following a 60-min challenge with exogenous ACTH.

3. Secondary Hypoadrenocorticism

Hypothalamic or pituitary lesions that block production of CRH or ACTH can produce the symptoms of hypoadrenocorticism. People with secondary hypoadrenocorticism lack the hyperpigmentation that usually accompanies Addison's disease but have the other symptoms. This condition can develop as a result of hypophysectomy, autoimmune disorders, viral illnesses, prolonged morphine administration, and other causes. The most common origin of the disorder is a consequence of prolonged therapy with cortisol or related steroids employed as antiinflammatory agents or immune suppressants.

C. Disorders of Aldosterone Secretion

Hyperaldosteronism is characterized by low blood potassium, high blood sodium, and muscle weakness. Elevated plasma sodium levels bring about water retention and may produce hypertension. Hyperaldosteronism may result from an adenoma or carcinoma in the zona glomerulosa that autonomously secretes excessive amounts of aldosterone. These symptoms also develop when ACTH or ACTH-like peptides are elevated chronically. Treatment with aldosterone inhibitors can reduce these symptoms until the source of the excessive aldosterone is removed.

Loss of the zona glomerulosa through adrenalectomy or Addison's disease, congenital or drug-induced depression of aldosterone production, and defects in the renin–angiotensin system can induce **hypoaldosteronism.** Potassium excretion is reduced, sodium is lost in the urine, and water retention is impaired. Imbalances in sodium/potassium ratios alter muscle and nerve function. These conditions can be alleviated by treatment with aldosterone.

D. Adrenal Excesses in Androgen Production

Adrenal androgen production may be elevated in several hyperadrenocorticoid conditions such as Cushing's syndrome or with hypoadrenocorticism as in the case of CAH. Adrenal tumors may secrete excessive quantities of adrenal androgens. Symptoms of excessive production of adrenal androgens in women include hirsutism, acne, seborrhea, irregular menses, reduced fertility, lowered voice pitch, atrophy of the breasts, possible thinning of hair and recession of the scalp in the temporal region, clitoral enlargement, and hypertrophy of skeletal muscles. These same symptoms may be present in males but are not as noticeable, for obvious reasons. Young adult women that exhibit anorexia nervosa (a disorder characterized by greatly reduced food intake) and an elevated adrenal axis may develop facial hair (hirsutism) as a result of ACTH-induced secretion of adrenal androgens.

E. Side Effects of Corticosteroid Therapy

Corticosteroids typically are administered in high doses to achieve their therapeutic effects. This is especially true of glucocorticoid use in the treatment of inflammations, varying in intensity from skin rashes to muscle injuries or arthritis. Adverse side effects that occur as a consequence of prolonged therapy are predictable.

1. Adverse Effects of Glucocorticoid Therapy

The beneficial therapeutic effects of glucocorticoids are manifest only when applied at doses two to three times physiological levels. Consequently a number of adverse side effects occur with prolonged administration of glucocorticoids, including mild diabetes mellitus as a consequence of a prolonged hyperglycemic antagonism induced by glucocorticoids (see Chapter 13). Subjects also experience muscle weakness due to extensive protein catabolism (an effect that does not occur with physiological doses), osteoporosis due to destruction of bone substance, reduced activity of the immunological response system, and mental depression. Glucocorticoids also produce mineralocorticoid-like effects on Na^+ and K^+ balance owing to high levels that through saturation overcome the weak binding of glucocorticoids to mineralocorticoid receptors.

2. Adverse Effects of Aldosterone Therapy

Excessive doses of mineralocorticoids (as well as glucocorticoids) cause sodium retention and consequent accumulation of fluids (edema), as evidenced by rapid weight gain following their administration. However, an "escape" phenomenon (due to unknown factors) occurs, and the condition is alleviated before serious complications arise and irreparable damage has occurred. In cases of cardiac failure, the escape mechanism also fails.

Aldosterone therapy results in decreased blood potassium and increased urine potassium. Excessive aldosterone causes severe potassium losses that can induce muscle cramps and muscle weakness. These events occur primarily because of adverse effects on cell membrane characteristics and resultant alterations of normal muscle cell physiology.

VI. The Mammalian Adrenal Medulla

The medullary portion of the mammalian adrenal consists of sympathetic preganglionic neuronal endings (cholinergic) and modified cells derived from the neural crest and homologous to postganglionic sympathetic neurons (adrenergic). These adrenal cells secrete either norepinephrine or epinephrine directly into the blood. In other words, the adrenal medulla is a modified sympathetic ganglion. Both epinephrine and norepinephrine (as well

Table 9-4. Percentage of Catecholamines
in Adult Adrenal Medulla Represented
by Norepinephrine

Vertebrate group/species	Percent NE
Whale	83[a]
Ungulates	15–50[a]
Carnivores	27–60[a]
Rodents	2–50[a]
Lagomorphs	0–12[a]
Rabbit (*Oryctolagus cuniculus*)	8–13[b]
Primates	0–20[a]

[a] Data from Gorhman, A., and Bern, H. A. (1962). "A
Textbook of Comparative Endocrinology." Wiley, New
York.
[b] Data from Coupland, R. E. (1953). On the morphology
and adrenaline-noradrenaline content of chromaffin tissue. *J. Endocrinol.* **9**, 194–203.

as small quantities of dopamine) can be extracted from the adrenal medulla, but the ratio
in adult mammals strongly favors epinephrine (Table 9-4). This proportion of norepinephrine to epinephrine varies throughout life, however. Fetal and neonatal adrenals secrete
predominantly norepinephrine followed by a gradual increase in the proportion of epineph-
rines, so that eventually epinephrine dominates in adult mammals. Whales are an apparent
exception in that the adult adrenal consists of about 83% norepinephrine.

Treatment of adrenal medullary cells with potassium dichromate or chromic acid results
in formation of a yellowish or brown oxidation product, the **chromaffin reaction.** Cells
that exhibit a positive chromaffin reaction are termed chromaffin cells. The catecholamine-
secreting cells of the adrenal medulla show a positive chromaffin reaction, but so do other
cells in the body (for example, certain cells in the intestinal epithelium and skin). Cells
containing the tryptophan derivative serotonin also exhibit a positive chromaffin reaction.
Norepinephrine-secreting cells can be distinguished from epinephrine-secreting cells by
the formaldehyde treatment devised by Hillarp and Falck. Formaldehyde combines chemically with norepinephrine storage granules and the resulting complex will fluoresce. Today,
we can readily separate epinephrine- and norepinephrine-secreting cells using immunohistochemical techniques to localize either the specific catecholamine or the enzymes responsible for its synthesis.

A. Synthesis and Metabolism of Adrenal Catecholamines

Two distinct cellular types in the mammalian adrenal medulla are related to production of
norepinephrine and epinephrine, respectively. Both epinephrine and norepinephrine are
synthesized from the amino acid tyrosine and employ the same biochemical pathway (see
Chapter 2, Fig. 2-3). However, only one cellular type possesses the critical enzyme **phenylethanolamine *N*-methyltransferase (PNMT)** necessary for converting norepinephrine to
epinephrine through addition of a methyl group donated by *S*-adenosylmethionine.

Norepinephrine and epinephrine may be found circulating free in the plasma or as conjugates with sulfate or glucuronide. Most of the circulating epinephrine is bound to plasma
proteins, especially albumin. Norepinephrine binds to plasma proteins to a much lesser
degree than does epinephrine. Small amounts of dopamine also may be released.

Circulating catecholamines have a short biological half-life and are rapidly excreted via the urine in either free or conjugated forms. The biological half-life for epinephrine is about 5 min. The most common metabolic pathway for inactivation of catecholamines involves liver monoamine oxidase (MAO) followed by aldehyde oxidase to produce inactive metabolites that appear in the urine. The most common urinary metabolites are **3-methoxy-4-hydroxymandelic acid, metanephrine,** and **normetanephrine** (see Chapter 2, Fig. 2-4).

B. Regulation of Catecholamine Secretion

Primary control of secretion by the adrenal medulla is by the sympathetic nervous system. Environmental stimuli operate through the sympathetic system under emergency conditions or stress. In addition, some control of medullary secretion is exerted directly by ACTH and by glucocorticoids.

1. The Central Nervous Pathway

Stimulation of the cholinergic sympathetic fibers innervating the medulla causes local release of acetylcholine (ACh), which in turn stimulates release of norepinephrine, epinephrine, or both from the chromaffin cells. This action of ACh involves Ca^{2+} uptake by the chromaffin cells. Apparently separate control centers for release of epinephrine and norepinephrine are located in the anterior and medial hypothalamus, and selective release of these medullary hormones occurs through neural pathways. Norepinephrine and epinephrine may be released separately under differing physiological conditions, and these two catecholamines have independent physiological roles in homeostasis.

2. Environmental Factors and Catecholamine Release

Walter B. Cannon first formulated an **emergency reaction** hypothesis involving secretions of the adrenal medulla and activity of other portions of the sympathetic nervous system. These emergency responses include increased heart rate, vasodilation of arterioles in skeletal muscle, general venoconstriction, relaxation of bronchiolar muscles, pupillary dilatation, piloerection (elevation of hair), and mobilization of liver glycogen and free fatty acids. All of these responses contribute to increased efficiency of operation so that the organism can best respond to whatever emergency has arisen. This type of response to short-term stress may be distinguished from the response to chronic stress associated with the glucocorticoids and the general adaptation syndrome of Selye described earlier. However, the emergency reaction of Cannon may be thought of as being similar to the medullary component of the alarm reaction of the Selye hypothesis.

The major physiological actions of adrenal medullary hormones are their effects on metabolism in response to emotional stress (anxiety, apprehension), physical stress (injury, exercise), or what has been distinguished as physiological stress (temperature, pH, oxygen availability, hypotension, and hypoglycemia). The actions of adrenal catecholamines on cardiovascular events other than the acceleration of heart rate (which is due to metabolic effects of epinephrine on cardiac muscle) are probably secondary to the effects of norepinephrine released from postganglionic sympathetic fibers or, in the case of the skeletal muscle arterioles, the release of ACh from postganglionic sympathetic fibers. This secondary role for adrenal catecholamines is further supported by the relatively low percentages of norepinephrine in the adrenals of most adult mammals. Sympathetic control mechanisms are not well developed in fetal and neonatal mammals, and this observation could be linked to the high proportion of norepinephrine in their adrenals.

Epinephrine stimulates hydrolysis of liver glycogen to glucose and production of lactate from muscle glycogen stores. Circulating norepinephrine produces a similar effect on liver glycogen, but muscle glycogen stores are not affected by norepinephrine. This would explain the ineffectiveness of exogenous norepinephrine as a cardioacceleratory drug, whereas epinephrine is very potent. The mobilization of lipids and release of free fatty acids from adipose tissue is under neural sympathetic control and is not regulated by adrenal catecholamines.

Emotional and severe physical stress increase circulatory levels of catecholamines via hypothalamus–adrenal neural pathways. Response to emotional stress such as is often induced by written or oral examinations involves an increase only in epinephrine, whereas the adrenal response to anticipation involves primarily norepinephrine. Exercise causes an increase in norepinephrine levels, presumably from both adrenal and neural sources. Epinephrine secretion is not influenced by moderate exercise, but it is markedly increased during long-distance running. Several physiological factors such as cold and heat stress, alkalosis or acidosis, and hypotension do not appear to involve primary actions of adrenal medullary hormones. However, responses to asphyxia or anoxia and hypoglycemia are major factors influencing epinephrine release from the adrenal medulla in adult mammals. Asphyxia causes an increase in epinephrine release, probably through direct actions of oxygen deprivation on the nervous system. In fetal or neonatal animals, asphyxia directly evokes catecholamine release from the adrenal.

Insulin-induced hypoglycemia results in cardiac acceleration through increased epinephrine release. Hypoglycemia induces epinephrine release primarily through direct effects on glucose-sensitive centers in the hypothalamus. Epinephrine also retards the insulin-induced decrease in blood sugar through its antagonistic actions on liver glycogen.

3. Corticotropin and Glucocorticoid Effects on Catecholamine Secretion

Development of a close anatomical association between adrenocortical and chromaffin tissues during vertebrate evolution has suggested a concomitant development of a functional relationship as well. Extensive studies by Wurtman and coworkers have demonstrated that ACTH exerts a stimulatory effect on epinephrine secretion through the action of the former on circulating glucocorticoid levels. Hypophysectomy reduces adrenal epinephrine levels, and treatment with either ACTH or glucocorticoids restores adrenal levels of epinephrine to normal. Furthermore, glucocorticoids increase the activity of adrenal medullary PNMT, the enzyme responsible for conversion (methylation) of norepinephrine to epinephrine. Some studies indicate that ACTH may have a direct action on the medulla as well. Levels of both **tyrosine hydroxylase** (tyrosine → Dopa) and **dopamine β-hydroxylase** (dopamine → norepinephrine) but not PNMT are increased by ACTH treatment. These observations indicate that chronic stress may influence epinephrine secretion not only during the alarm reaction but also in the later stages of the response. It has been reported that epinephrine can cause release of ACTH through actions at either the hypothalamic or adenohypophysial level, but the physiological significance of these observations is not clear.

C. Mechanism of Action for Adrenal Catecholamines

The presence of specific receptors for epinephrine was first postulated in 1906 by Sir Henry Dale, who showed that ergot alkaloids (drugs such as ergocornine and ergocryptine) blocked some of the actions of epinephrine. Later, studies suggested there are two kinds of adrenergic receptors in target cells that are capable of binding adrenal catecholamines:

Table 9-5. Some Catecholamine Agonists and Antagonists and the Receptor Type to Which They Preferentially Bind

Adrenal catecholamine	Receptor type			
	α_1	α_2	β_1	β_2
Agonists				
Clonidine		×		
Isoproterenol			×	×
Phenylephrine	×			
Ritodrine				×
Antagonists				
Butoxamine				×
Metoprolol			×	
Propranolol			×	×
Phentolamine	+	×		
Yohimbine	×	+		

[a]Epinephrine binds to all types, but best to β-receptors. Norepinephrine binds best to α-receptors, but will bind β_1-receptors (although not as well as epinephrine). ×, Binding; +, weak binding compared to other receptor types.

α- and β-receptors. These receptors also respond to a number of epinephrine-like drugs that have been termed *sympathomimetic drugs* since they mimic actions of sympathetic catecholamines (see Table 9-5). Two common sympathomimetic drugs are **isoproterenol** and **phenylephrine.** Norepinephrine binds mainly to α-receptors, whereas epinephrine binds to both. When both α- and β-receptors are present on a target cell that binds epinephrine, the α-effect predominates unless epinephrine is administered with an α-blocking agent (for example, **phentolamine**).

Detailed studies of the mechanism of adrenal catecholamine actions on target cells have concentrated on the effects of epinephrine in cardiac muscle and liver cells. In fact, it was studies in cardiac cells that led Earl Sutherland and coworkers to the discovery of the second-messenger role for cyclic adenosine 3′,5′ monophosphate (cAMP) and an eventual Nobel Prize in Physiology or Medicine (see Chapter 2). Epinephrine stimulates the breakdown of glycogen to glucose in both liver and muscle cells by first stimulating an increase in intracellular cAMP. The glucose released from liver glycogen tends to enter the general circulation, whereas the glucose liberated from muscle glycogen is utilized for rapid ATP synthesis and production of lactate. For a further explanation of this difference, see Chapter 14.

VII. Summary

The mammalian adrenal gland consists of an outer region (cortex) of adrenocortical cells (steroidogenic) and an inner region (medulla) of chromaffin cells (adrenergic). The cortex consists of a zona glomerulosa that secretes primarily the mineralocorticoid aldosterone and two inner zones, the zona fasciculata and zona reticularis, that secrete primarily glucocorticoids (typically cortisol or corticosterone). Adrenal androgens (DHEA and DHEA-S) are produced primarily in the zona reticularis. The medulla contains two chromaffin cellular types that secrete the two catecholamine hormones, norepinephrine and epinephrine, respectively.

The synthesis and release of aldosterone is controlled by the renin–angiotensin system. Renin is released from the juxtaglomerular apparatus in the kidney in response to reduction in sodium levels of extracellular fluids or reduction in blood pressure. Renin acts on a protein substrate (renin substrate or angiotensinogen) in the blood to release angiotensin I. Angiotensin-converting enzyme (ACE) in lung capillaries transforms angiotensin I to angiotensin II, which in turn stimulates aldosterone release. Angiotensin II may be converted to angiotensin III, which is also active in causing aldosterone release. ACTH plays only a permissive role in aldosterone secretion. Aldosterone regulates sodium levels by increasing sodium reabsorption by the kidney. This response of the cells of certain regions in the nephron to aldosterone involves the synthesis of a specific protein that is responsible for sodium reabsorption. Aldosterone plays a secondary role in regulating volume of the extracellular fluids through its action on sodium reabsorption. It also regulates potassium excretion. Angiotensins may stimulate hypertension, drinking, and vasopressin release to aid in fluid volume regulation. The renin–angiotensin system may have evolved to regulate blood pressure. Natriuretic peptides (ANP, BNP, CNP) released by high blood pressure inhibit release and actions of aldosterone.

Secretion of cortisol or corticosterone is under direct pituitary control through ACTH. The main physiological actions of these glucocorticoids are related to their effects on transport of materials into cells and the induction of new cellular enzymes. Glucocorticoids inhibit glucose utilization by peripheral tissues, stimulate amino acid uptake and conversion to glucose and storage as glycogen, and stimulate mobilization of fat stores. They are also antiinflammatory and may produce immunosuppression. Cytokines from immune responses may also activate adrenal activity. Another important contribution may be the permissive action whereby glucocorticoids create an intracellular environment favorable to the actions of many other hormones. Glucocorticoids certainly are important in the adaptive mechanisms whereby an organism combats chronic stress (general adaptation syndrome of Selye). Some important pharmacological actions of glucocorticoids include antiinflammatory properties, a diabetogenic action, and excessive catabolism of body proteins.

Release of epinephrine and norepinephrine in mammals is directed by separate centers in the hypothalamus that innervate the medulla through preganglionic sympathetic neurons. Glucocorticoids and ACTH may also influence the ability of the medulla to secrete catecholamines by stimulating the synthesis of key enzymes. Epinephrine is primarily a metabolic hormone. For example, it stimulates hydrolysis of glycogen to provide glucose for combating hypoglycemia (liver) or for use as an immediate energy source (in skeletal muscle). Epinephrine is responsible for the emergency response of Cannon and is also involved in the alarm reaction of the general adaptation syndrome of Selye. Administration of epinephrine causes increased glucose metabolism and ATP availability within cardiac cells, leading to cardiac acceleration. Epinephrine binds to either α- or β-receptors in target cell membranes. Norepinephrine binds significantly only to α-receptors, and its major physiological action is venoconstriction. Circulating norepinephrine is not a cardiac stimulator since cardiac muscle cells possess predominantly β-receptors on their exposed surfaces. In most adult mammals, circulating norepinephrine appears to be of secondary importance to the sympathetic postganglionic neurons for control of vascular tone.

Suggested Reading

Books

Brown, M. R., Koob, G. F., and Rivier, C. (1990). "Stress: Neurobiology and Neuroendocrinology." Dekker, New York.

Mulrow, P. J. (1986). "The Adrenal Gland." Elsevier, New York.

Vinson, G. P., Whitehouse, B., and Hinson, J. (1993). "The Adrenal Cortex." Prentice-Hall, Englewood Cliffs, NJ.

General Articles

Black, P. H. (1995). Psychoneuroimmunology: Brain and immunity. *Sci. Med.* **2,** 16–25.

Dallman, M. F. (1993). Stress update: Adaptation of the hypothalamic–pituitary–adrenal axis to chronic stress. *Trends Endocrincl. Metab.* **4,** 62–69.

Fuller, R. W. (1992). The involvement of serotonin in regulation of pituitary–adrenocortical function. *Front. Neuroendocrinol.* **13,** 250–270.

Funder, J. W. (1993). Aldosterone action. *Annu. Rev. Physiol.* **55,** 115–130.

Gaillard, R. C. (1994). Neuroendocrine–immune system interactions: The immune–hypothalamo–pituitary–adrenal axis. *Trends Endocrinol. Metab.* **5,** 303–309.

Henry, J. P. (1993). Biological basis of the stress response. *NIPS* **8,** 69–73.

Jacobson, L., and Sapolsky, R. (1991). The role of the hippocampus in feedback regulation of the hypothalmic–pituitary–adrenocortical axis. *Endocr. Rev.* **12,** 118–134.

Shipston, M. J. (1995). Mechanism(s) of early glucocorticoid inhibition of adrenocorticotropin secretion from anterior pituitary corticotropes. *Trends Endocrinol. Metab.* **6,** 261–266.

Vinson, G. P., Pudney, J., and Whitehouse, B. J. (1985). The mammalian adrenal circulation and the relationship between adrenal blood flow and steroidogenesis. *J. Endocrinol.* **105,** 285–294.

Watkins, L. R., Maier, S., and Goehler, L. E. (1995). Cytokine-to-brain communication: A review and analysis of alternative mechanisms. *Life Sci.* **57,** 1011–1026.

Wong, P. C. (1992). Angiotensin II receptor antagonists and receptor subtypes. *Trends Endocrinol. Metab.* **3,** 211–217.

Clinical Articles

Kemppainen, R. J., and Peterson, M. E. (1994). Animal models of Cushing's syndrome. *Trends Endocrinol. Metab.* **5,** 21–28.

Magiakou, M.-A., and Chrousos, G. P. (1995). Diagnosis and treatment of Cushing's disease. *In* "The Pituitary Gland" (H. Imura, ed.), 2nd Ed., pp. 491–508. Raven, New York.

Malchoff, C. D., and Malchoff, D. M. (1995). Glucocorticoid resistance in humans. *Trends Endocrinol. Metab.* **6,** 89–94.

Pang, S., and Clark, A. (1990). Newborn screening, prenatal diagnosis, and prenatal treatment of congenital hyperplasia due to 21-hydroxylase deficiency. *Trends Endocrinol. Metab.* **1,** 300–307.

Trainer, P. J., and Besser, M. (1990). Cushing's syndrome. Difficulties in diagnosis. *Trends Endocrinol. Metab.* **1,** 292–295.

White, P. C., and Pascoe, L. (1992). Disorders of steroid 11β-hydroxylase isozymes. *Trends Endocrinol. Metab.* **3,** 229–234.

10

Comparative Aspects
of Vertebrate Adrenals

THE ANATOMICAL organization of adrenocortical homologs and chromaffin cells in nonmammalian vertebrates differs markedly, with the only obvious uniformity being a tendency in amniotes for combining both cellular types into one organ, the adrenal gland (Fig. 10-1). Although the term **chromaffin** may be used to designate the catecholamine-secreting cells responsible for elaborating **epinephrine** and **norepinephrine** in all vertebrates, several terminologies have been proposed in attempting to deal with the diverse character of the adrenocortical homologs, including **interrenal,** corticosteroidogenic, and

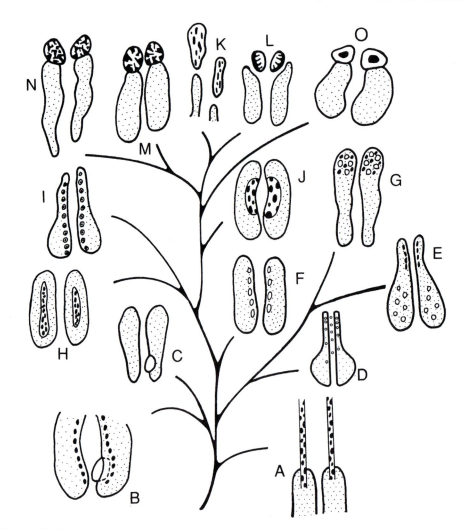

Figure 10-1. Comparative anatomy of adrenal tissues. Comparison of distribution of chromaffin (black) and adrenocortical tissue (clear) associated with the kidneys (stippled) in the vertebrates. Chromaffin tissue is not shown for the ratfish (C) and sturgeon (D). (A) Cyclostomes; (B) selachians; (C) holocephalans; (D) chondrosteans; (E) holosteans; (F) dipnoans; (G) teleosts; (H) anurans; (I) urodeles; (J) chelonians; (K) snakes; (L) lizards; (M) crocodilians; (N) birds; (O) mammals.

adrenocortical. Although the adrenals of nonmammals lack the anatomical cortex–medulla relationship of mammals (and in many cases no distinctly separate adrenal gland is present), the term "adrenocortical" will be used here to designate these cells because it denotes the functional and evolutionary relationship of this cellular type to those of the mammalian adrenal cortex.

The structures of the more common corticosteroids and details about their synthesis and metabolism are provided in Chapter 9. The mechanism of corticosteroid action on target cells is described in Chapter 2.

I. Comparative Aspects of Adrenocortical Tissue

Cytologically, the adrenocortical cells of nonmammals resemble the steroidogenic cells of the mammalian zona fasciculata. Adrenocortical cellular types of cyclostomes, teleosts, and nonmammalian tetrapods possess well-developed smooth endoplasmic reticula, mitochondria with tubular cristae, and numerous osmiophilic (lipoidal) inclusions. Following stimulation with corticotropin, pituitary extracts, or appropriate environmental stimuli, these cells exhibit increased basophilia, increased activity of 3β-hydroxy-Δ^5-steroid dehydrogenase (3β-HSD), and decreased lipid content. Nuclear and cellular enlargement and hyperplasia occur as a consequence of chronically elevated **corticotropin (ACTH)** or treatment with the drug **metyrapone (SU4885)**, which blocks the 11β-hydroxylation step, thus preventing glucocorticoid synthesis and elevating endogenous ACTH. As in mammals, atrophy of the adrenocortical cells follows hypophysectomy.

You will recall that there are three distinct zones in the mammalian adrenal cortex, and that each zone consists of specialized cells that secrete different steroids. Zonation of adrenocortical cells is suggested cytologically in some anurans, reptiles, and birds, and two separate cellular types have been claimed for the bullfrog *Rana catesbeiana* and for birds. However, little work has been done to establish firmly the existence of more than one functional type of adrenocortical cell in fishes and most other nonmammals.

Adrenocortical cells of fishes differ most from the general mammalian pattern of corticosteroidogenesis with respect to some of the hormones produced. However, the general sequences for corticosteroidogenesis are similar in all the vertebrates with respect to precursor–product relationships, with many of the same enzymes being involved (Chapter 2). Among the nonmammalian tetrapods, the nature of corticosteroid secretion is nearly identical, and the pattern of secretion is very similar to that described for cells of the mammalian zona fasciculata.

Daily rhythms and seasonal variations in corticosteroid secretory patterns occur in nonmammals as well as in mammals. Peak seasonal adrenocortical activity is roughly correlated to periods of reproductive activity, although a cause–effect relationship cannot be categorically applied. Some studies suggest that "stress" may be the critical factor and that stressors associated with reproduction may be only one component involved in stimulating adrenocortical function (albeit a major one). The responses of nonmammals to stressors such as surgery, forced exercise, and handling are very much like those described for mammals.

A. Agnathan Fishes: Cyclostomes

In lampreys, presumed adrenocortical cells have been identified as islands of cells above the pronephric funnels in the kidney as well as in the walls of the large dorsal blood vessels (postcardinal veins) in this same region. However, *in vitro* studies employing radioactively labeled steroidal precursors (pregnenolone, progesterone, or both) have not demonstrated the ability of these presumed adrenocortical cells in lampreys or hagfishes to produce corticosteroids. Nevertheless, cortisol, cortisone, corticosterone, and 11-deoxycortisol have been isolated from hagfish and lamprey plasma and are believed to be of adrenal origin (Table 10-1). Steroidogenic enzymes, such as 3β-HSD, have not been demonstrated in cyclostome adrenocortical cells, however. The site for corticosteroidogenesis has not been established in cyclostomes, and more studies of this group are needed.

Table 10-1. Plasma Levels[a] of Corticosteroids[b] in Fishes

Source	Cortisol	Corticosterone	DOC	11-DOC	1α-OHB	Aldosterone
Agnathans						
Myxine glutinosa	70	20				
Eptatretus stouti		8		27		
Petromyzon marinus	5	2				
Selachians						
Raja laevis		160	420	30	940	
Squalus acanthias		2500		160	2800	
Chondrosteans						
Acipenser oxyrhynchus	1.8	0.7	0.8	0.7		
Holosteans						
Amia calva	7.6					
Teleosteans						
Salmo trutta	24	0.22				
Carassius auratus	440	72	8			1.1
Clupea harengus	750	0.6		0.3		
Dipnoans						
Lepidosiren paradoxa	60	1.6		0.3		5.8
Neoceratodus fosteri	2.8	8.8				0.3

[a] In nanograms per milliliter.
[b] DOC, Deoxycorticosterone; 11-DOC, 11-deoxycorticosterone; 1α-OHB, 1α-hydroxycorticosterone.

No evidence of either renin or a juxtaglomerular apparatus has been found in cyclostomes. This kidney specialization may not have evolved until later, in the chondrichthyean and bony fishes. Extracts prepared from atria and ventricles of the hagfish, *Myxine glutinosa* produce relaxation of the precontracted rabbit aorta, a standard bioassay for **atrial natriuretic peptide (ANP)**. Bioassay data also suggest ANP activity is present in the brain of *Myxine* as well. In addition, ANP-binding sites have been identified in gill and kidney of this species, suggesting a mammalian-like role for ANP-like peptides. No effect of ANP on corticosteroid levels has been reported to date.

B. Chondrichthyean Fishes

Some selachians and holocephalans have one large unpaired adrenal gland. In others, the adrenocortical tissue may occur as paired strands along the medial border of the posterior kidney. Since this gland consists exclusively of adrenocortical cells and is located between the posterior ends of the kidneys, it is truly interrenal. In addition, small islands or islets of adrenocortical cells may be found on the surface of the kidneys extending anteriorly. Chromaffin tissue occurs as small masses along the medial border of each kidney.

In 1934, Grollman and coworkers used extracts of the interrenal glands from three skates (*Raja* species) to maintain adrenalectomized rats, demonstrating the presence of corticosteroids in these glands. Soon, selachians were shown to produce a unique corticosteroid, **1α-hydroxycorticosterone (1α-OHB).** The enzyme necessary for synthesizing this unique steroid, **1α-hydroxylase,** is found only in the elasmobranch interrenal gland and nowhere else. Even holocephalans, such as the ratfish *Hydrolagus colliei,* lack 1α-hydroxylase and secrete primarily cortisol.

Shark interrenals (*Scyliorhinus canicula*) incubated *in vitro* with exogenous pregnenolone as a substrate synthesize primarily corticosterone and 11-deoxycorticosterone, with lesser amounts of 1α-OHB. When endogenous precursors only are involved, the primary product *in vitro* becomes the anticipated 1α-OHB. There are no 18- or 17α-hydroxylases present in the shark interrenal, and consequently aldosterone, cortisol, cortisone, and 11-deoxycorticosterone are not synthesized. Data from plasma analyses, however, indicate that not only is 1α-OHB secreted in the dogfish shark, *Squalus acanthias,* but also cortisol, corticosterone, and 11-deoxycortisol (Table 10-1). Rays and skates produce 11-deoxycorticosterone as well. Such discrepancies between *in vivo* plasma levels and *in vitro* synthesis occur in other vertebrate classes as well, and one should be extremely cautious in extrapolating from *in vitro* capabilities (in which various intermediates and products accumulate) to *in vivo* situations, in which the final products are removed (secreted into the blood). Accumulation of intermediates and products in the vicinity of the secretory cells under *in vitro* conditions may upset chemical equilibria so that unusual ratios of steroids are observed.

1. Renin–Angiotensin System in Chondrichthyean Fishes

Although it has been reported repeatedly that the renin–angiotensin system is absent in chondrichthyean fishes and that attempts to demonstrate the presence of renin have been unsuccessful, careful cytological and histological studies have shown otherwise. Examination of four selachian species (two sharks, one ray, and one skate) reveals that definite modified smooth muscle cells containing renin-like secretion granules are associated with the afferent arterioles and form juxtaglomerular-like structures in the kidneys. Furthermore, a distinct macula densa-like modification of the distal tubule where it passes between the afferent and efferent arterioles is present. **Angiotensin-converting enzyme-like action (ACELA)** is present in gills and spleen as well as in lesser amounts in brain and kidney of *S. canicula*. Angiotensin II stimulates production of 1α-OHB by shark interrenals. Additional physiological studies are still needed, but it seems likely that a mammalian-like renin–angiotensin system is present in these primitive fishes.

2. Atrial Natriuretic Peptide in Chondrichthyean Fishes

Immunoreactive ANP cells are found in the hearts of three selachians (*S. acanthias, S. canicula,* and *Raja clavata*) and in at least one holocephalan (*Chimaera monstrosa*), with many more cells being present in the atria than in the ventricles. More ANP-like activity is present in atrial than in ventricular cardiac extracts. Although ANP-like activity could not be demonstrated in the brains of *S. canicula* or *C. monstrosa,* weak activity was present in the brain of *S. acanthias*. Radiolabeled ANP binds to the secondary lamellae of gill filaments and to kidney glomeruli in *S. canicula*.

C. Ray-Finned Osteichthyean Fishes: Actinopterygians

The anatomical arrangement of adrenocortical cells in the actinopterygian fishes differs markedly from that described for all other fish groups and could be described anatomically as "intrarenal." In the sturgeons and polypterine fishes (chondrosteans) and in the ganoids (holosteans), the adrenocortical cells are scattered in small clumps throughout the kidney. The identification of these cells is hampered in the ganoids (*Amia, Lepisosteus*) by the presence in the kidney of large numbers of **corpuscles of Stannius,** which although not steroidogenic, do resemble cytologically the adrenocortical cells. The corpuscles of Stan-

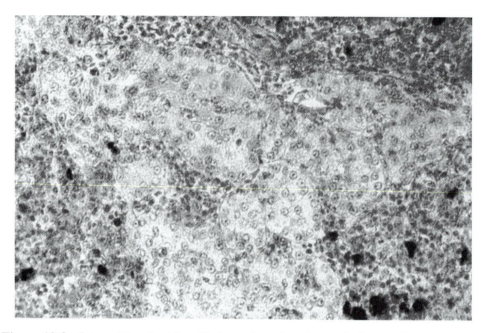

Figure 10-2. Interrenal tissue in a teleost. The larger, clear cells are interrenal cells located in clumps among the much smaller lymphoid cells of the head kidney of a brown trout.

nius may play a role in calcium metabolism (see Chapter 15). The teleostean adrenocortical cells are embedded in the most anterior portion of the kidney, known as the **head kidney** (Fig. 10-2). Frequently, these cells are associated with the dorsal posterior cardinal veins as described for cyclostomes. The head kidney has lost its renal function and consists mostly of lymphoid tissue, nonfunctional pronephric tubules, and small islands of adrenocortical cells. In a few species, all of the adrenocortical cells surround the posterior cardinal veins, and none are associated with kidney elements (Fig. 10-1). Because of the diffuse nature of the adrenocortical tissue in teleosts, it is not possible to remove these cells surgically, and one must resort to the use of selective inhibitors of corticosteroidal synthesis such as metyrapone, which blocks the synthesis of corticosteroids (see Chapter 9).

The principal circulating steroid in the chondrostean, holostean, and teleostean fishes is cortisol, with corticosterone, aldosterone, and some others present in lesser quantities in the teleosts (Table 10-1). Bony fishes lack the 1α-hydroxylase of elasmobranchs and consequently do not synthesize 1α-OHB. *In vitro* studies with teleostean adrenocortical cells indicate that they convert pregnenolone preferentially to cortisol. When progesterone is supplied as a precursor, the principal product is corticosterone. *In vivo,* cortisol primarily is produced.

Teleosts respond to ACTH and to acute or chronic stress with markedly elevated corticosteroids (Table 10-2). Treatment with metyrapone elevates ACTH secretion, causing adrenocortical hypertrophy and hyperplasia. The adrenocortical cells respond to treatment with ACTH by secreting more cortisol, and ACTH levels as well as cortisol levels are

Table 10-2. Effects of Stress on Plasma Cortisol Levels in Salmonid Fishes[a]

Species	Fish size	Treatment	Unstressed (ng of cortisol/ml)	Stressed (ng of cortisol/ml)
Salmo trutta (brown trout)	326 g	Mild confinement, 1 hr	20	75
		Severe confinement, 1 hr	—	150
Salmo salar (Atlantic salmon)	600 g	Mild confinement, 30 min	—	200 (50–600)
Salvelinus namaycush (lake trout)	23.1 cm	Acute handling and examined 1 hr later	20	270
Oncorhynchus mykiss (rainbow trout)	22.4 cm	Acute handling and examined 1 hr later	2–25	300

[a] Most of these data are estimated from published figures.

elevated under conditions of stress. Release of ACTH occurs directly though innervation by neurons containing **corticotropin-releasing hormone (CRH)** that enter the adenohypophysis, where the corticotropes reside. Immunoreactive CRH cells in the preoptic nucleus (PON) of the common white sucker also contain **arginine vasotocin (AVT),** but those in the nucleus lateralis tuberis (NLT) that send fibers into the adenohypophysis contain only CRH. Studies in the goldfish suggest that the PON regulates release of ACTH whereas the NLT controls synthesis of ACTH. The peptides sauvagine and urotensin I are chemically similar to CRH and also are effective ACTH releasers. It remains to be proven that they play an endogenous role in ACTH release, but the CRH-like activity in the NLT may be due to either urotensin I or a urotensin-like peptide.

1. Actinopterygian Renin–Angiotensin System

Renin activity is present in all of the spiny-rayed fish groups and juxtaglomerular-like cells have been identified in several teleosts (Fig. 10-3 and Table 10-3). Histological and cytological identification of renal cells exhibiting renin granules has not been verified in any of the nonteleostean actinopterygian fishes, however. No macula densa-like structure has been reported for any bony fish. Angiotensin II has been shown to stimulate secretion of cortisol, which is the principal salt-regulating corticosteroid in these fishes.

2. Atrial Natiuretic Peptide in Actinopterygians

Bioassayable ANP-like activity has been demonstrated in the hearts of teleosts (Table 10-4), with atrial extracts being 10 to 50 times more potent than ventricular extracts. Although ANP-like activity was not detectable in the brain of rainbow trout, *Oncorhynchus mykiss,* it was present in the brains of two marine species. Binding of ANP to secondary lamellae of gills occurs in both chondrosteans and teleosts but kidney binding was either nonexistent or barely detectable. Release of cortisol by ANP has been demonstrated in the Japanese eel, *Anguilla japonica,* both *in vivo* and *in vitro.* This is not what one might predict from

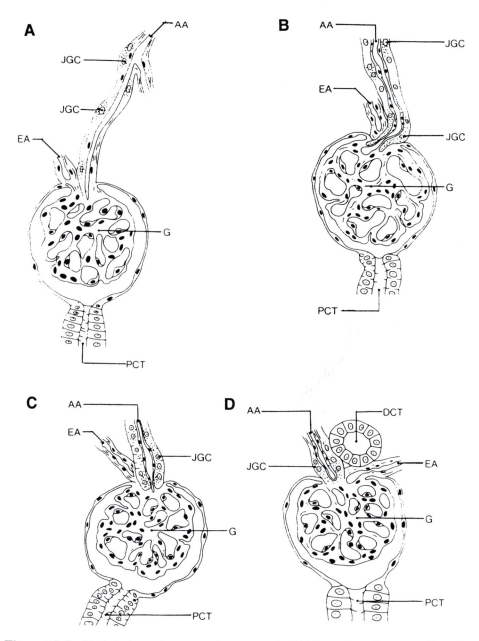

Figure 10-3.   The juxtaglomerular apparatus in nonmammals. (A) Teleost, *Carassius auratus*; (B) bullfrog, *Rana catesbeiana*; (C) snake, *Elaphe quadrivirgata*; (D) domestic chicken. JGC, Juxtaglomerular cells; G, glomerulus; PCT, proximal convoluted tubule; DCT, distal convoluted tubule; AA, afferent arteriole; EA, efferent arteriole. [Modified from Matsumoto, A., and Ishii, S. (1989). "Atlas of Endocrine Organs." Springer-Verlag, Berlin.]

the known role of ANP in mammals and suggests the inhibitory role of ANP, as evidenced in most birds and all mammals examined to date, is a later evolutionary development. Unfortunately, data on ANP actions in amphibians and reptiles are not available.

Table 10-3. Distribution of Renin Activity and Presence of Renal Cells Containing Renin Granules in Fishes

Classification	Species	Renin activity	Renin granules
Chondrichthyes			
Selachid	*Squalus acanthias*	—	+
	Mustelus canis		+
	Raja erinacea		+
Osteichthyes			
Actinopterygii			
Polypteri	*Polypterus senegalis*	+	—
	Erpetoichthys calabaricus		—
Chondrostei	*Acipenser breviorostris*	+ ?	—
Holostei	*Amia calva*	+	—
	Lepisosteus osseus	+	—
Teleostei	*Anguilla rostrata*	+	
	Opsanus tau	+	+
Sarcopterygii			
Crossopterygii	*Latimeria chalumnae*	+	+
Dipnoi	*Lepidosiren paradoxa*	+	+
	Protopterus aethiopicus	+	+

D. Lobe-Finned Osteichthyean Fishes: Sarcopterygians

The lungfishes (Dipnoi) have been of special interest to comparative endocrinologists seeking to understand the evolution of corticosteroids since they represent close relatives to both the fish and tetrapod lines. In dipnoan fishes, the adrenocortical cells are found as small cords located between renal and perirenal tissues adjacent to branches of the post-

Table 10-4. Distribution of Atrial Natriuretic Peptide-Like Immunoreactivity in Heart and Plasma of Vertebrates[a]

Species	Atrium (ng/mg)	Ventricle (ng/mg)	Plasma (pg/ml)
Teleosts			
Anguilla japonica	0.08	ND[b]	NR[b]
Zano platypus	0.21	0.47	NR
Pelteobagrus fulvidraco	0.16	0.21	NR
Amphibians			
Rana nigromaculata	27.5	1.0	NR
Reptiles			
Amyda japonica	0.8–0.9	0.08	ND
Birds			
Chicken	0.62–0.71	0.051	NR
Mammals			
Rat	237–344	NR	135
Mouse	28–46	0.6	NR
Rabbit	12–13	NR	72.5

[a] Based on Kim, S. H., Cho, K. W., Koh, G. Y., Seul, K. H., So, J. N., and Ryu, H. (1989). Phylogenetic study on the immunoreactive atrial natriuretic peptide in the heart. *Gen. Comp. Endocrinol.* **74,** 127–135.
[b] ND, Not detectable; NR, not reported.

cardinal veins. Adrenocortical cells from estivating *Protopterus* synthesize corticosterone *in vitro* from progesterone. However, only cortisol was identified in the plasma of the aquatic phase, suggesting that a tetrapod-like secretion (corticosterone) occurs during its moist, air-breathing phase and that a teleostean-like secretion (cortisol) occurs during its aquatic phase. Although it is tempting to speculate on the evolutionary significance of these data, it would be premature to do so without further investigation.

Aldosterone, cortisol, corticosterone, and a trace of 11-deoxycortisol (Table 10-1) circulate in the blood of the predominantly aquatic South American lungfish, *Lepidosiren paradoxa*. Studies show that the obligate aquatic Australian lungfish, *Neoceratodus fosteri*, secretes aldosterone, corticosterone, and cortisol (Table 10-1). Cortisterone levels are three to four times greater than cortisol levels. These data suggest that the lungfishes are intermediate between the tetrapod condition (aldosterone and corticosterone) and the actinopterygian fishes (cortisol) with respect to which corticosteroids are prominent, but closer to tetrapods than these fishes. It would be interesting to know which corticosteroids are secreted by adrenocortical cells of the crossopterygian coelacanth, *Latimeria chalumnae*.

1. Sarcopterygian Renin–Angiotensin System and Atrial Natriuretic Peptide

Renin activity and the presence of renal cells containing renin granules have been observed in the coelacanth and in two genera of lungfishes (Table 10-2). Angiotensin II causes secretion of aldosterone in *N. fosteri*. ANP is present in myocardial tissue of the African lungfish, *Protopterus aethiopicus*, and may play a role in controlling corticosteroid release.

E. Amphibians

The adrenocortical cells of amphibians are extrarenal and extremely variable with respect to their location. The anatomical pattern of anurans generally differs markedly from that of apodans and urodeles. However, the specific adrenal secretions correlate more to habitat than to anatomy or phylogeny. Adrenal corticosteroid secretion is stimulated by ACTH, which in turn is controlled by CRH. Evidence suggests that CRH also causes release of thyrotropin (TSH) from the adenohypophysis of larval amphibians (see Chapter 8).

1. Anatomical Features of Amphibian Adrenocortical Tissue

In anurans, adrenocortical tissue is found in irregular nodules organized loosely into a pair of interrenal glands on the ventral surface of the kidneys (Fig. 10-1). In most anurans, some chromaffin cells are associated with the interrenal glands, and in one anuran, *Rana hexadactyla*, there are more chromaffin cells than adrenocortical cells in the interrenal glands. In addition to the adrenocortical cells and chromaffin cells, a third cellular type, the summer or **Stilling cell**, has been found in the interrenal glands of ranid frogs (Fig. 10-4). These Stilling cells appear in summer and regress in winter frogs. They are eosinophilic and resemble mast cells (histamine-producing cells). The functional significance of the Stilling cell is unknown.

Adrenocortical cells of both apodans and urodeles occur in scattered islands on the ventral surface of the kidney (Fig. 10-1). This anatomical arrangement in part explains the virtual absence of studies employing apodan or urodele adrenocortical cells *in vitro*. Curiously, in one anuran, *Xenopus laevis,* the adrenocortical tissue also is organized as small islets on the ventral surface of the kidney. Each of these adrenocortical islets contains two or three chromaffin cells as well.

2. Amphibian Adrenocortical Secretions

Studies with adult anuran adrenocortical tissue *in vitro* have shown that the major corticosteroids synthesized are aldosterone and corticosterone, and both hormones have been identified in adult amphibian plasma (Table 10-5). In addition, *in vitro* syntheses result in production of a large quantity of 18-hydroxycorticosterone, which can function as a precursor for the synthesis of aldosterone. The ratio of aldosterone to 18-hydroxycorticosterone to corticosterone *in vitro* is $6:3:1$. However, corticosterone is the predominant circulating corticosteroid, and the high levels of this aldosterone precursor and aldosterone may simply be an artifact of *in vitro* conditions, as reported for mammals (see Chapter 9). Ovarian production of significant quantities of 11-deoxycorticosterone has been reported, and this may be an important source for circulating corticosteroids in sexually mature females. It is not clear if these ovarian corticosteroids play a reproductive role in ovulation as suggested in certain teleosts.

Although corticosterone is the dominant corticosteroid reported for terrestrial amphibians, cortisol has been reported to be the major corticosteroid in metamorphosing ranid tadpoles, in the permanently aquatic frog *X. laevis,* and in some permanently aquatic urodeles. Aquatic-phase, adult red-spotted newts, *Notophthalmus viridescens,* produce substantial amounts of cortisol whereas the terrestrial efts produce primarily corticosterone. These observations would support the hypothesis that cortisol is important for maintaining sodium balance in freshwater amphibians, as reported for fishes, and that corticosterone becomes more important following metamorphosis to a terrestrial-phase amphibian. Corticosterone is also the major corticosteroid in reptiles, birds, and many mammals. It may be that there is a transition from cortisol to corticosterone secretion that takes place during or immediately following metamorphosis (see Chapter 8).

3. Direct Regulation by Arginine Vasotocin and Atrial Natriuretic Peptide in Amphibians

AVT directly stimulates secretion of corticosterone (*X. laevis*) and aldosterone (*X. laevis, Rana catesbeiana*) by adrenocortical cells. Recall that in mammals, AVT also has a positive influence on the adrenal axis but does so by increasing ACTH release.

Amphibian ANP-like peptide has been isolated, and activation of ANP receptors in *X. laevis* adrenocortical cells causes a decrease in aldosterone release but has no effect on corticosterone secretion. Note that this response differs from mammals, in which ANP has no effect by itself on the adrenal cortex but influences the responsiveness of cells in the zona glomerulosa to ACTH, angiotensin II, and K^+ (see Chapter 9).

4. Amphibian Renin–Angiotensin System

Renin activity is present in amphibian renal tissue, although the secretory renin-containing granules differ morphologically from those of mammals. Several studies report no macula

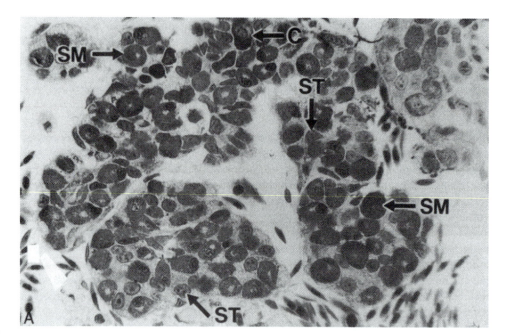

Figure 10-4. Amphibian adrenal tissue. (A) Light micrograph from the bullfrog, *Rana catesbeiana*, showing adrenocortical cells (ST), chromaffin (C), and summer or Stilling cells (SM). (B) Electron micrograph from the bullfrog, *R. catesbeiana*. The cells with the large secretory granules are Stilling cells (SM). The steroidogenic cells contain lipid droplets (LD) and mitochondria (M) with tubular cristae. N, Nucleus. (C) Adrenocortical cells in a salamander are located close to kidney tubules (T). [(A) and (B) from Matsumoto, A., and Ishii, S. (1989). "Atlas of Endocrine Organs." Spring-Verlag, Berlin; (C) from THE ADRENAL CORTEX by Vinson/Whitehouse/Hinson, © 1993. Adapted by permission of Prentice-Hall, Inc., Upper Saddle River, NJ.]

densa in amphibians (Fig. 10-3), but a macula densa-like structure has been reported for one toad. This report is of considerable interest since a macula densa has not been reported for any reptile.

F. Reptiles

Chelonians, crocodilians, and most snakes have paired, suprarenally positioned adrenal glands similar to those of mammals. There is a variable degree of intermingling of chromaffin cell cords within a mass of adrenocortical cells, which may show evidence of zonation (Fig. 10-5). In lizards and some snakes, the adrenocortical cells are partially encapsulated by chromaffin cells, resulting in a "cortex" homologous to the mammalian medulla. Some chromaffin cells also are found within the central mass of adrenocortical cells. The chromaffin cells of *Sphenodon* (the primitive rhynchocephalian reptile) surround the dorsal aspect of the gland as well as form islets within the mass of adrenocortical cells. For the first time in vertebrates, the interrenals of reptiles have attained their own vascular supply and venous drainage, no longer relying on the kidney and a renal portal system for distribution of their secretory products. The metanephric kidney appears first in reptiles and has its own blood supply.

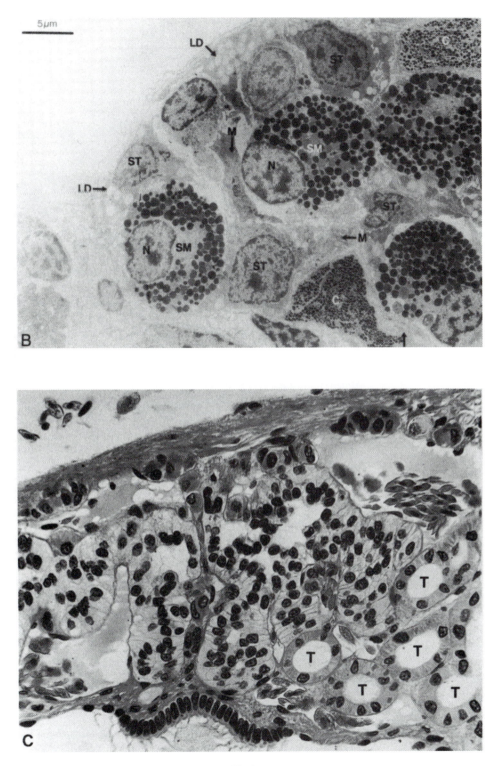

Figure 10-4 (*continued*).

Table 10-5. Circulating Corticosteroids [a] in
Amphibian Species (Order Anura)

Species	Corticosterone	Aldosterone
Rana catesbeiana		
Control	8.8	1.8–50.2
ACTH treated	18.0	0.2
Freshly collected	0.2–1.6	
Hypophysectomized	1.3	0.15
Rana esculenta	—	0.82
Bufo marinus		
In distilled water	380.6	54.7
Saline adapted	66.2	15.2
Bufo americanus		
October animals	1.4	—

[a] In micrograms per deciliter.

Although very few species have been studied, all of those examined under *in vitro* conditions (including turtles, lizards, snakes, and the American alligator) synthesize aldosterone and corticosterone as the major corticosteroids and what appears to be 18-

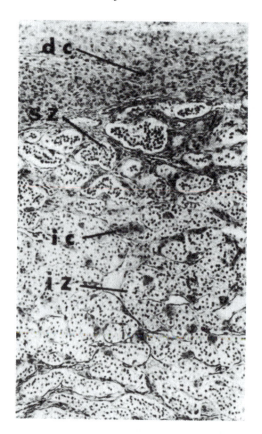

Figure 10-5. Interrenal of the cobra, *Naja naja.* The dorsal band of chromaffin tissue is prominent. The steroidogenic tissue appears as two distinct zones. dc, Chromaffin tissue; sz, outer subcapsular zone; ic, intercortical chromaffin islets; iz, inner cortical zone. [From Lofts, B., Phillips, J. G., and Tam, W. H. (1971). Seasonal changes in the histology of the adrenal gland of the cobra, *Naja naja. Gen Comp. Endocrinol.* **16**, 121–131.]

Table 10-6. Plasma Corticosteroids in Reptiles

Species	Corticosterone (μg/dl)	Aldosterone (ng/dl)
Caiman crocodilus	2.2 ± 0.47	—
Sceloporus cyanogenys		
Control	5.96 ± 1.13	
ACTH treated	12.92 ± 1.17	
Varamus gouldi		
Hydrated		42 ± 7
Salt loaded		18.3 ± 6

hydroxycorticosterone. Corticosterone synthesis predominates *in vitro,* with the amounts of 18-hydroxycorticosterone exceeding the levels of aldosterone.

Adrenocortical cells from turtles (*Chrysemys picta*) secrete corticosterone *in vitro* when mACTH or crude extracts of avian, chelonian, or anuran pituitaries are added to the culture medium. Adrenocorticoid synthesis in the cobra, *Naja naja,* also is stimulated by mACTH. Corticosterone levels *in vivo* (Table 10-6) and *in vitro* are elevated following administration of ACTH to *Caiman crocodilus* or *Caiman sclerops.* ACTH also elevates corticosterone levels in lizards but does not influence aldosterone levels. Hypophysectomy causes a reduction in adrenal weight of reptiles whereas treatment with metyrapone elevates ACTH levels and causes adrenal hypertrophy and hyperplasia.

One extraadrenocortical source of corticosteroids has been reported for reptiles. As in fishes and amphibians, isolated ovaries from the night lizard, *Xantusia vigilis,* synthesize 11-deoxycorticosterone, but the importance of this observation either to corticosteroid physiology or to reproductive function in reptiles is uncertain.

1. Reptilian Renin–Angiotensin System

Renin activity has been reported for kidneys of turtles, lizards, and snakes, but no macula densa has been described (Fig. 10-3). This is curious considering the possibility of a macula-like structure in amphibians and the evidence for a distinct macula in selachians, birds, and mammals. Angiotensin II stimulates corticosterone release in both turtles and lizards. No data are available on other reptilian groups.

2. Reptilian Atrial Natriuretic Peptide

Immunoreactive ANP granules are present in the atria of a snake (*Python reticulata*), lizards (*Anolis carolinensis, Lacerta viridis*), and a turtle (*Amyda japonica*) although they are scarce or absent in ventricles. However, ANP has been extracted from the atria and ventricles of the heart of the freshwater turtle, *Pseudemys scripta,* suggesting its presence in both atria and ventricles.

G. Birds

The adrenal glands of birds are organized in the same manner as described for turtles, crocodilians, and most snakes. The relative quantities of chromaffin with respect to adrenocortical cells vary, however. There appears to be some zonation of cellular

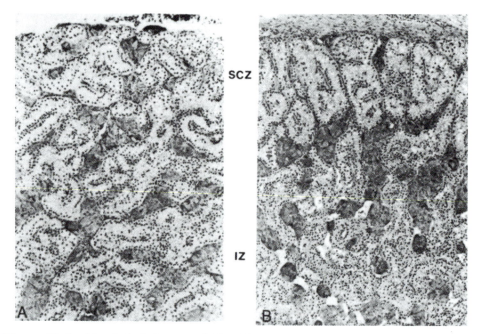

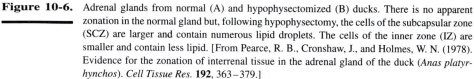

Figure 10-6.   Adrenal glands from normal (A) and hypophysectomized (B) ducks. There is no apparent
                   zonation in the normal gland but, following hypophysectomy, the cells of the subcapsular zone
                   (SCZ) are larger and contain numerous lipid droplets. The cells of the inner zone (IZ) are
                   smaller and contain less lipid. [From Pearce, R. B., Cronshaw, J., and Holmes, W. N. (1978).
                   Evidence for the zonation of interrenal tissue in the adrenal gland of the duck (*Anas platyr-
                   hynchos*). *Cell Tissue Res.* **192**, 363–379.]

types on histochemical and cytological bases similar to that seen in mammals (Figs. 10-6
and 10-7).

   The major corticosteroids synthesized by adrenocortical cells taken from domestic species
and incubated *in vitro* are (predictably) corticosterone, aldosterone, and 18-hydroxycorti-
costerone. This sequence of steroidogenesis (corticosterone→18-hydroxycorticoster-
one→aldosterone) is characteristic of birds as well as anuran amphibians and reptiles and
is essentially the same pattern found in cells of the mammalian zona glomerulosa.

   Studies with duck adrenal slices *in vitro* suggest that corticosterone is the immediate
precursor for 18-hydroxycorticosterone and that the latter is converted to aldosterone. However, only corticosterone and aldosterone have been found in avian plasma (Table 10-7). The
absence of 18-hydroxycorticosterone in plasma further supports the conclusion that it is
only a precursor for aldosterone synthesis, and its accumulation *in vitro* is an artifact. In
addition to corticosterone and aldosterone, 11-deoxycorticosterone has been reported in
the herring gull. The biological half-lives for corticosteroids in avian plasma are provided
in Table 10-8.

   Although mACTH stimulates adrenocorticoid secretion (corticosterone) in chickens, hypophysectomy does not cause complete cessation of corticosteroidogenesis (Table 10-7).
Thus, there may be a basal level of secretion that does not require ACTH. It also has been
proposed that melanophore-stimulating hormone (α-MSH) produced and released by the
hypothalamus maintains corticosteroidogenesis in hypophysectomized chickens. Appar-

ently, bird adrenals also secrete corticosteroids in response to prolactin (PRL), human growth hormone (hGH), serotonin, or parathyroid hormone (PTH).

1. Atrial Natriuretic Peptide in Birds

Secretion of aldosterone is reduced on a high Na^+ diet and elevated on a low Na^+ diet. Treatment with ANP also lowers aldosterone levels and it is likely that ANP is the intermediary whereby high Na^+ reduces aldosterone secretion, as it does in mammals.

2. Avian Renin–Angiotensin System

Renin activity is present in birds, and both the juxtaglomerular apparatus and a macula densa have been described. Cells of the avian macula densa are similar to those of mammals, with only some minor differences.

II. Physiological Roles for Corticosteroids in Nonmammalian Vertebrates

Many of the studies of corticosteroid function in nonmammals have focused on their effects on salt transport, particularly sodium, that is, mineralocorticoid activity. Unlike the mammalian condition, aldosterone, cortisol, or corticosterone may possess strong mineralocorticoid activity when tested in nonmammals. It has been hypothesized that the renin–angiotensin system evolved as a mechanism for regulating blood pressure. Control over mineralocorticoid synthesis and regulation of sodium/potassium balance presumably was acquired later.

The effects of glucocorticoids on metabolic activities have not been studied extensively but considerable effort has been made to examine the response of glucocorticoid secretion to various natural and artificial stressors. In this regard, one must distinguish between natural events (reproduction, migration, feeding, etc.) and unpredictable events (storms, droughts, reduction in food supply) that can result in elevated corticosteroids. Whereas the natural events may result in elevated corticosteroids, typically there is no disruption of normal reproductive activities. The unpredictable events would result in a greater secretion of corticosteroids and may even shut down reproductive function. Such severe stressor effects often are observed in captive populations, but these responses may be abnormal in the sense that placing the animal in the laboratory has severely limited the repertoire of activities in which it might engage to eliminate the stressor.

A. Agnathan Fishes: Cyclostomes

The blood of myxinoids (hagfishes) is isosmotic to sea water, but there are some minor differences in concentrations of specific ions. Therefore, although osmotic balance per se is no problem, the differential distribution of certain ions must be maintained actively. Injections of aldosterone or deoxycorticosterone acetate alter electrolyte composition of the body fluids with respect to sodium ions, but cortisone has no effect.

Lampreys are either freshwater organisms or migrate between fresh water and the sea. While lampreys are in fresh water, their body fluids are hyperosmotic to their surroundings,

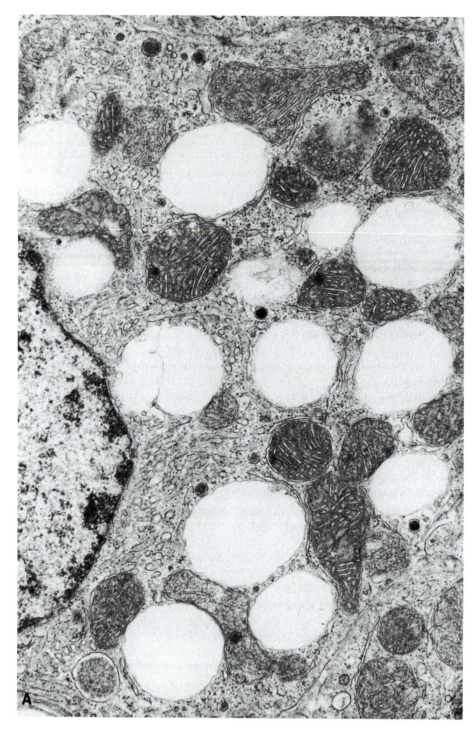

Figure 10-7. Ultrastructural comparison of adrenocortical cells from the subcapsular (A) and inner zones (B) of the duck interrenal. Subcapsular zone cells have mitochondria with platelike cristae and many free ribosomes in the cytoplasm. Cells of the inner zone have mitochondria with tubular cristae and occasional paracrystalline inclusions. [From Pearce, R. B., Cronshaw, J., and Holmes, W. N. (1978). Evidence for the zonation of interrenal tissue in the adrenal gland of the duck (*Anas platyrhynchos*). *Cell Tissue Res.* **192**, 363–379.]

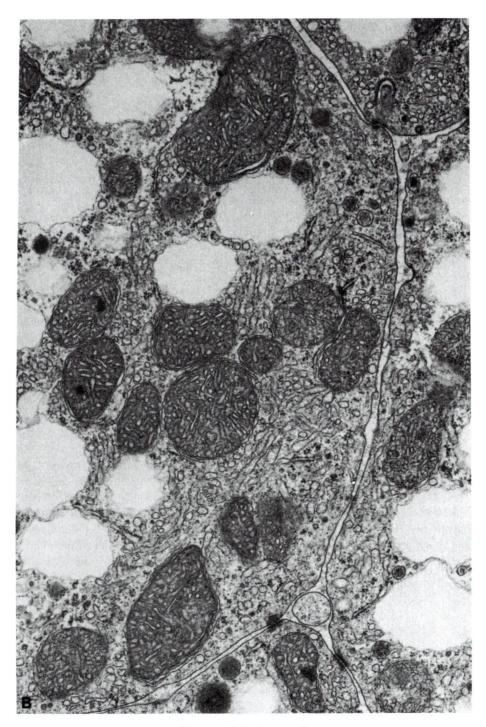

Figure 10-7 (*continued*).

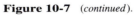

Table 10-7. Plasma Corticosteroids[a] in Birds
Following Certain Experimental Manipulations

Species	Treatment	Corticosterone	Aldosterone
Duck	Control	19.88 ± 1.66	—
	Hypox	9.92 ± 2.62	—
Chicken	Control	5.1 ± 0.49	—
	Hypox	1.06 ± 0.20	0.06 ± 0.01
	ACTH	15.16 ± 3.81	0.11 ± 0.02
	Surgical stress	6.58 ± 1.47	0.21 ± 0.06
Pigeon	Control	9.50 ± 0.63	—
	Hypox	5.13 ± 0.55	—
	Hypox with pituitary autotransplant	7.32 ± 0.79	—
Quail	Control	10.51 ± 1.35	—
	Hypox	4.68 ± 0.47	—
	Hypox with pituitary autotransplant	9.39 ± 1.01	—

[a] In micrograms per deciliter.

and they produce a large quantity of dilute urine. Sea lampreys (*Petromyzon marinus*) do not secrete corticosteroids when in sea water, but when they are in fresh water corticosteroids can be identified in the circulation. Aldosterone treatment can reduce renal and extrarenal sodium losses from lampreys held in fresh water. The significance of this latter observation to osmoregulatory control in lampreys must await demonstration of aldosterone in lampreys.

B. Chondrichthyean Fishes

Selachians have plasma that is hyperosmotic to sea water and hence do not have the osmoregulatory problems exhibited by most marine fishes. This hyperosmotic condition is achieved by maintaining high circulating levels of urea and trimethylamine oxide. There is no clearly defined role for any corticosteroids, including 1α-hydroxycorticosterone, in selachians. However, 1α-hydroxycorticosterone does bind to cells in the gills, rectal gland, and nephrons of the kidney. Although corticosteroids stimulate salt excretion by the rectal gland, this salt-secreting gland reportedly plays only a minor role in ionoosmotic

Table 10-8. Biological Half-Life of Adrenocorticosteroids in Avian Species[a]

Species	Corticosterone half-life (min ± SE)	Aldosterone half-life (min ± SE)
Duck	7.5 ± 0.6	6.2 ± 0.6
Pigeon	18.4 ± 0.1	12.8 ± 1.6

[a] Modified from Holmes, W. N., and Phillips, J. G. (1976). The adrenal cortex of birds. *In* "General, Comparative and Clinical Endocrinology of the Adrenal Cortex" (I. Chester Jones and I. W. Henderson, eds.), pp. 293–420. Academic Press, New York.

homeostasis. Interrenalectomy of the skate, *Raja radiata*, had no effect on plasma osmolarity or on the concentrations of Na^+, Cl^-, Mg^{2+}, K^+, or urea. Exogenous angiotensin stimulates release of 1α-hydroxycorticosterone, and suggests a mammalian-like regulatory pattern for corticosteroid secretion. However, hANP also causes release of 1α-hydroxycorticosterone, which is not consistent with the inhibitory action of ANP on corticoid release in other vertebrates. A possible role for corticosteroids in carbohydrate metabolism of selachians has been suggested. There are no reports of corticosteroid actions in holocephalans.

C. Osteichthyean Fishes

Among the bony fishes, corticosteroids have been investigated with respect to function only in the teleosts. In general, corticosteroids (especially cortisol) stimulate sodium transport across gills (both influx and efflux), across the mucosa of the gut, and in the kidney of freshwater fish. Cortisol appears to be the major corticosteroid in bony fishes (Table 10-1), and it appears to regulate sodium fluxes. The most complete studies have been conducted on freshwater- and seawater-adapted eels (*Anguilla* spp.). Seawater-adapted eels exhibit a marked turnover of Na^+ (50–60%/hr), but ion flux is very low in freshwater-adapted eels (<1%/hr). Eels in sea water are faced with an influx of Na^+ that they must eliminate whereas freshwater eels must conserve body Na^+, which readily can be lost to their Na^+-poor surroundings. Cortisol treatment increases the activity of **Na^+/K^+-dependent adenosine triphosphatase (Na^+/K^+-ATPase)** in gills, gut epithelial cells, and kidneys. Circulating levels of cortisol are similar in freshwater-adapted and seawater-adapted eels, suggesting cortisol is not important. However, other studies show that elevated cortisol occurs during the initial process of adapting to fresh water and later drops back to a basal level. Neither cortisol nor ACTH is completely effective in maintaining normal Na^+ balance in hypophysectomized freshwater eels, and PRL, another osmoregulatory hormone in fishes, may be important as well.

Cortisol also has been reported to maintain gluconeogenesis and a balance between lipid, carbohydrate, and protein metabolism in several species. A comprehensive study of rainbow trout with chronically elevated cortisol, however, does not support a gluconeogenic role for cortisol in that species. It may be premature to draw any conclusions about possible actions of corticosteroids on the metabolism of bony fishes until more species are examined in a similar manner but it appears that cortisol is generally gluconeogenic.

Stressors produce marked elevations of corticosteroids in fishes (Table 10-2). Comparisons of absolute corticosteroid levels cannot be made among different species, or in some cases even within the same species, since it is not possible to assess the role of stressors in most cases. Capture and confinement of wild trout cause a marked elevation in cortisol over time (Fig. 10-8), and bringing wild trout into the laboratory can induce sustained increases in levels of plasma cortisol. Exposure of brown trout, *Salmo trutta*, in streams to cadmium and zinc pollution is correlated with stimulation of adrenocortical cells. Elevated corticosteroids in migrating juvenile and adult Pacific salmon (genus *Oncorhynchus*) probably reflect a response of the adrenal axis to chronic stress. This condition leads to pathological changes similar to those described for Cushing's disease in humans (see Chapter 9) and what appears to be adrenal exhaustion after spawning that contributes to the death of the spawned fish (Table 10-9).

Cortisol or 11-deoxycorticosterone is elevated during spawning in a number of species, but it is not clear whether this is a stress response or part of the reproductive sce-

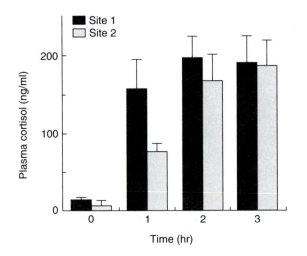

Figure 10-8. Stress in trout. Effect of acute confinement stress on plasma cortisol levels in brown trout, *Salmo trutta,* at two sites in Eagle River, Colorado. Site 2 is contaminated with heavy metals (cadmium and zinc) and these fish show a delayed cortisol secretory response compared to fish living 3 miles upstream in an uncontaminated area (site 1). [From D. O. Norris, and J. Woodling, unpublished data, 1995.]

nario. Extensive protein catabolism observed in spawning Pacific salmon may be due to high circulating levels of cortisol. In the killifish, cortisol peaks along with estradiol at ovulation. Secretion of ovarian 11-deoxycorticosterone in the Indian catfish, *Heteropneus-*

Table 10-9. Comparison of Cushing's Syndrome with Experimental Hyperadrenocorticism and with Natural Hyperadrenocorticism of Spawning Salmon

Tissue	Spawning salmon	Cushing's syndrome	Experimental hyperadrenocorticism
Adrenal	Hyperplasia and degeneration	Hyperplasia and tumors	Hyperplasia
Pituitary	Degeneration	Hyaline change in basophils	Minimal degeneration
Spleen	Depletion of lymphocytes, fibrosis	No reported change	Depletion of lymphocytes
Thymus	Involution, depletion of thymocytes	Involution, occasional tumor	Depletion of thymocytes
Liver	Degeneration	Occasional fatty degeneration	No change
Kidney	Degeneration	Degeneration	Degeneration
Pancreas	Hypertrophy of islets	Hypertrophy of islets; variable	Hypertrophy of islets
Stomach	Atrophy and generation	Occasional ulcers	Atrophy of epithelium and occasional ulcers
Thyroid	Atrophy and generation	Atrophy of follicular epithelium	Atrophy of follicular epithelium
Gonads	Degeneration of testes	Atrophy	Degeneration
Muscle	Degeneration of masseter	Atrophy	No change
Cardiovascular system	Degeneration and beginning arteriosclerosis	Arteriosclerosis	Arteriosclerosis
Skin		Hypertrophy	Atrophy

[a] Adapted from Robertson, O. H., and Wexler, B. C. (1960). Histological changes in the organs and tissues of migrating and spawning Pacific salmon (genus *Oncorhynchus*). *Endocrinology* **66,** 222–239.

tes fossilis, may be responsible for induction of ovulation. In spawning Pacific salmon, however, there appears to be a decline in corticosteroid levels, possibly related to adrenal exhaustion.

The renin–angiotensin system appears to regulate osmotic and ionic adaptation in teleosts. Freshwater-adapted eels exhibit greater renin activity in sea water than in fresh water. Furthermore, there are gradual changes in plasma renin activity during adaptation of eels to either fresh water or sea water. Administration of either ACTH or renin causes elevation of circulating cortisol levels in eels. Treatment of freshwater-adapted eels with captopril, an inhibitor of the angiotensin-converting enzyme, ACE (see Chapter 9), lowers the secretory response of corticosterone following a seawater challenge. Infusion of isosmotic sodium chloride solution into the lungfish *Neoceratodus fosteri* causes reduction in plasma renin activity, suggesting the involvement of a renin mechanism in osmotic adaptation.

Seasonal and daily rhythms in circulating levels of cortisol have been reported for several species. Maximal levels of corticosteroids in rainbow trout maintained on a long photoperiod are observed at night (2400 hr). In the spring, plasma corticosteroids rise markedly in juvenile salmonid fishes prior to their migration from fresh water to the ocean or while adapting to sea water. Part of the osmoregulatory stress prior to migration may be due to decreased PRL secretion, a hormone considered necessary for freshwater adaptation. Adult salmonids exhibit adrenocortical cell hyperplasia and elevated corticosteroid levels during their spawning migration from the ocean to fresh water. This increase in corticosteroid levels is accompanied by accelerated catabolism of protein, suggesting a major metabolic role for corticosteroids. However, during spawning, there is degeneration of interrenal tissue and a decrease in the ability to secrete cortisol.

D. Amphibians

Hypophysectomy or exposure of ranid frogs to sea water induces atrophy of the adrenocortical tissue, presumably because of reduced requirements for corticosteroids. Adrenalectomy of these frogs causes a decrease in plasma sodium and an increase in plasma potassium similar to that observed in mammals. Winter ranid frogs usually survive adrenalectomy, but summer frogs die: possibly, death is related to the Stilling cells, whose role is unknown in summer frogs.

Aldosterone seems to be the important salt-regulating hormone (mineralocorticoid) acting on the skin and urinary bladder, increasing sodium influx and retention, respectively. The action of aldosterone on the urinary bladder involves synthesis of new protein and would appear to be analogous to the production of aldosterone-induced protein in the mammalian distal convoluted tubule (see Chapter 9). ACTH elevates aldosterone levels in *Rana esculenta,* implying a direct action of ACTH on aldosterone, a condition very different from that described for mammals. Salt-depleted frogs exhibit elevated kidney levels of renin, suggesting the presence of both mechanisms for regulating aldosterone production, although plasma renin levels do not appear to differ between distilled water-adapted toads and saline-adapted toads. Angiotensin II stimulates corticosterone and aldosterone synthesis *in vitro* by adrenocortical fragments from *Rana ridibunda.*

Aldosterone treatment reverses the depression of plasma sodium caused by aminoglutethamide in the tiger salamander. Higher doses of corticosterone than aldosterone were needed to bring plasma sodium back to normal, supporting a physiological role for aldosterone principally as a mineralocorticoid in urodele amphibians. Both aldosterone and corticosterone induce marked hyperglycemia as well as some increase in liver and muscle glycogen levels, but it is not certain that both molecules function as physiological glucocorticoids.

Seasonal and daily rhythms of cortisol levels have been reported in plasma samples from *R. esculenta* and *Bufo americanus*. The time of the daily maximum varied in *R. esculenta* from 2400 hr in May to 1900 hr in July and 0800 hr in November. Greatest secretion of corticosteroids coincided with reproduction in the spring. Two peaks of corticosteroid secretion were observed in captive and free *B. americanus*. One peak coincided with reproduction in the spring but the second peak occurred in the fall at the time of prehibernating migrations. Daily maxima corresponded to increased locomotor activity.

Stressors cause elevation in plasma corticosteroids in amphibians. Placing freshly collected bullfrogs in sacks for up to 24 hr doubles their corticosterone levels. Stress also can alter secretion of other hormones. The secretion of PRL, gonadotropin, and androgen all decrease in some species of stressed amphibians following capture. The failure of certain species to breed in captivity and the stimulus for some larvae to undergo metamorphosis (see Chapter 8) following capture and laboratory confinement may be species-specific consequences of stress. Other species, however, breed readily in captivity and should be examined more closely in this regard.

Extensive studies in male rough-skinned newts (*Taricha granulosa*) by Frank Moore and associates at Oregon State University have implicated CRH and corticosterone in the inhibition of amphibian reproductive behavior. Corticosterone operating through neural membrane receptors blocks clasping behavior by male newts (see Chapter 12).

E. Reptiles

The roles of corticosteroids in ionic regulation by reptiles have not been studied extensively, but effects on Na^+ balance have been shown. Like the situation in the nonmammalian vertebrate groups discussed previously, there does not seem to be any clear distinction between mineralocorticoids and glucocorticoids, with both aldosterone and corticosterone producing similar effects. Reptiles have nasal (orbital) salt-excreting glands similar to those found in aquatic birds. Corticosterone increases Na^+ secretion by these glands, and aldosterone decreases Na^+ excretion while stimulating Na^+ reabsorption in the kidneys and urinary bladder. Injections of concentrated sodium chloride solutions (salt-loading) depress plasma aldosterone levels in several species of lizard. Similar observations are reported for the tortoise, *Testudo hermanni*. Salt-loaded lizards also have elevated levels of AVT, which is consistent with its antidiuretic role (see Chapter 6).

In the lizard, *Dipsosaurus dorsalis,* it appears that corticosterone does not play much of a role in lactate utilization by skeletal muscles during recovery from exercise. Epinephrine, however, produces a marked stimulation of lactate incorporation into glycogen under similar conditions. Both hormones are elevated in the blood after 5 min of exhaustive exercise in the laboratory. Following aggressive bouts in the American chameleon, *Anolis carolinensis,* winners had much higher levels of epinephrine and norepinephrine than did losers, which may relate to this metabolic action. In comparing dominant to subordinate *A. carolinensis,* corticosterone plasma levels were greater in the subordinate animal.

Angiotensin II stimulates corticosterone and aldosterone secretion in lizards but only corticosterone secretion in turtles. However, sodium depletion apparently does not affect renin activity in turtles. Whether or not there is a role for the renin–angiotensin system *in vivo* is not clear.

F. Birds

Corticosteroids cause the nasal salt-excreting glands of certain aquatic birds to secrete a hypertonic NaCl solution. This salt secretion is enhanced by treatment with ACTH or corticosterone, and corticosterone uptake by nasal salt glands is followed by an increase in salt excretion. Adrenalectomized ducks cannot excrete a salt load, but corticosteroid therapy restores this salt-excreting ability. Metyrapone treatment blocks nasal gland secretion. Aldosterone also stimulates nasal salt excretion, unlike the situation in reptiles, where aldosterone diminishes excretion as it normally does in kidney. However, due to the powerful action of aldosterone on Na^+ reabsorption in the kidney, there is a net positive effect on Na^+ retention in spite of what occurs locally at the salt glands. Both aldosterone and corticosterone promote Na^+ retention in the lower intestine (copradeum).

Increases in environmental salinity are correlated with increases in Na^+/K^+-ATPase in salt gland cells, and this event is possibly related to increased protein synthesis caused by corticosteroids. Marine birds have larger adrenal glands than do freshwater or terrestrial birds, which supports the role for corticosterone in nasal salt gland regulation. Predictably, birds inhabiting brackish water have intermediate-sized adrenals.

In addition to mineralocorticoid effects, corticosterone causes metabolic effects in birds including weight loss, which probably is a reflection of changes in protein catabolism. In addition, chronic treatment with corticosterone elevates blood glucose and liver glycogen, probably through gluconeogenesis.

Renin activity and angiotensin activity in blood plasma of ducks and pigeons increase following hemorrhage, establishing a physiological role for the renin–angiotensin system that is similar to its role in mammals. Furthermore, elevated dietary Na^+ causes a reduction in circulating aldosterone without affecting corticosterone levels. Conversely, a reduction in dietary Na^+ raises aldosterone levels. Angiotensin II stimulation of aldosterone secretion is reduced by administration of ANP. However, in turkeys, ANP stimulates aldosterone release as it does in teleosts.

Circadian rhythms for corticosteroids have been described in some birds, usually peaking early in the morning and exhibiting lowest levels at dusk. The time of the daily corticosteroid peak is believed to determine the behavioral response following injection of PRL into migratory white-crowned sparrows. These observations have led to hypotheses concerning the control of migratory behavior by the phase relationships between rhythms of corticosterone and PRL secretion. North American migratory birds exposed to a day–night cycle with a long photophase release PRL, which induces premigratory fattening. Seasonal PRL and corticosteroid rhythms differ markedly throughout the year, and fattening can be altered simply by changing the timing of the corticosteroid peak with respect to the pattern in timing of the daily PRL peak.

Stress activates the hypothalamo–adrenal axis in birds. This involves the ACTH–corticosterone axis as well as release of epinephrine from the adrenal medulla. Activation of the endocrine stress mechanism also influences other endocrine glands. In 10- to 14-day-old chicken embryos, there is a surge of catecholamines (dopamine and epinephrine) in the face of short-term stress. Profound effects of stress are noted in adult birds, too. For example, chasing (the stressor) of male zebra finches for 15 min results in a significant depression in plasma androgens measured 2 hr later. Isolation of males in small cages for 12 hr virtually obliterates testosterone in the circulation.

III. Evolution of Chromaffin Tissue and Adrenal Medullary Hormones

Morphologically, there has been a general trend to develop a close anatomical relationship between chromaffin tissues homologous to the mammalian adrenal medullary cells and the adrenocortical cells. Chromaffin tissue in agnathans is found in association with the posterior cardinal veins, as are the separate clusters of adrenocortical cells. In the cartilaginous fishes the chromaffin tissue is intrarenal and is more widely separated from the "interrenal" adrenocortical cells than in any other vertebrate group. Although some chromaffin cells have been described in association with the cardinal veins of lungfishes, chromaffin cells are found mostly in the heart. These cardiac chromaffin cells are claimed to represent the homolog of the adrenal medulla in lungfishes. There is considerable anatomical variation among the adrenals of teleosts, but no species are known to exhibit cardiac chromaffin tissue. The chromaffin cells may form clumps entirely separate from the adrenocortical cell clusters in the head kidney or may be intermingled with adrenocortical cells. In some species, both conditions may be found (i.e., separate and mixed clusters of cell types). Clusters of adrenocortical cells (islets) in amphibians typically contain only a few chromaffin cells, and the latter usually are separated from adrenocortical cells. The anatomical distribution of chromaffin cells in reptiles is varied. The adrenal glands of squamates and the tuatara *Sphenodon punctatus* have peripherally located norepinephrine-secreting chromaffin cells and central islets of epinephrine-secreting cells intermingled among the adrenocortical cells. The avian adrenal glands consist of such a mixture of cortical and chromaffin cells that no distinction of "cortex" and "medulla" is possible.

Early studies on the proportion of adrenal catecholamines in nonmammals suggested that the mammalian fetal pattern of high norepinephrine production with increasing production of epinephrine following birth (see Chapter 9) is an example of the old principle that ontogeny recapitulates phylogeny. Analysis of extractable catecholamines from the dogfish shark *S. acanthias* (which has entirely separate chromaffin tissue), the frog *Rana temporaria* (which exhibits some mixing of chromaffin and adrenocortical cells), and the rabbit *Oryctolagus cuniculus* suggests an evolutionary change from reliance on norepinephrine to the progressive reliance on epinephrine (see Table 10-10). Although it seems logical to assume that the methylation step (norepinephrine→epinephrine) was a later evolutionary event, the picture is even more complex. Histochemical and chromatographic procedures indicate that norepinephrine predominates in some bird species whereas others produce mainly epinephrine. Still other groups show only a slight preference for producing one catecholamine over the other. Although norepinephrine predominates in extracts of adrenals prepared from chickens, turkeys, and pigeons, the major circulating catecholamine in these species is epinephrine. These data point out the dangers of extrapolating too broadly from gland content to secretory activities, especially with respect to substrate–production relationships.

Few comparative studies of adrenal catecholamines on carbohydrate metabolism have been reported. The hyperglycemic action of epinephrine seems to be a primitive active in nonmammalian vertebrates although the mechanisms involved are not clear. In chinook salmon, for example, both epinephrine and norepinephrine stimulate glycogen utilization in the liver and release of glucose. Incorporation of lactate into glycogen also is stimulated by epinephrine in carp liver, but this effect is overshadowed by a general breakdown of glycogen that occurs concomitantly. Epinephrine produces glycogen breakdown in the liver

Table 10-10. Proportion of Norepinephrine of
Total Catecholamines in Adult Adrenal Extracts

Species	Percent norepinephrine
Elasmobranchs	66–73[a]
Dogfish shark *(Squalus acanthias)*	100[b]
Amphibians	40–60[a]
Frog *(Rana temporaria)*	55–69[b]
Reptiles	60[a]
Snake *(Xenodon merremii)*	
Peripheral chromaffin cells	97[c]
Central chromaffin islets	15[c]
Birds	55–80[a]
Some passerine birds	0[a]

[a]Data from Gorbman, A., and Bern, H. A. (1962). "A Textbook of Comparative Endocrinology." Wiley, New York.
[b]Data from Coupland, R. E. (1953). On the morphology of adrenaline-nonadrenaline content of chromaffin tissue. *J. Endocrinol.* **9,** 194–203.
[c]Data from Gabe, M. (1970). The adrenal. *In* "The Biology of the Reptilia" (C. Gans, ed.), Vol. 3, pp. 263–318. Academic Press, New York.

of all birds examined. Norepinephrine, on the other hand, produces hypoglycemia in the ratfish and in 6- to 9-week-old Japanese quail. The bases for these unusual observations on norepinephrine action are unknown.

Studies of catecholamine effects on lipid metabolism are less common than studies of catecholamine effects on carbohydrate metabolism in lower vertebrates, although numerous studies of effects by other hormones have been reported (e.g., GH, PRL, thyroxine). One of the more thorough studies shows that while acting through β-adrenergic receptors, norepinephrine stimulates, in a dose-dependent manner, a triglyceride lipase that causes lipolysis in liver slices from coho salmon. Epinephrine has no lipolytic action in this system although it has been shown to have lipolytic effects in other fishes. More work in this area clearly is needed.

IV. Summary

Nonmammals do not exhibit "cortex" to "medulla" relationships similar to those described for mammals in Chapter 9. However, there is a general trend throughout the vertebrates to establish a closer anatomical relationship between the steroidogenic adrenocortical cells and the chromaffin tissue, which secretes catecholamines. In the more primitive fishes (cyclostomes and elasmobranchs), the chromaffin and adrenocortical cells are entirely separate. In teleosts, both the adrenocortical and chromaffin cells occur as scattered independent groups within the head kidney. Chromaffin and adrenocortical tissues are more or less scattered over the ventral surfaces of the kidneys in amphibians, but in reptiles they become consolidated into distinct adrenal glands. Bird adrenals are organized along the reptilian lines as well, but they exhibit some zonation of adrenocortical cells like that of mammals.

The major salt-regulating corticosteroid in fishes is cortisol, which also may produce glucocorticoid-like activities. Larval amphibians and permanently aquatic amphibians exhibit a preponderance of cortisol over corticosterone. In terrestrial tetrapods (including terrestrial amphibians), primarily corticosterone and aldosterone are secreted by the adrenocortical cells. Corticosterone is primarily a mineralocorticoid in these groups but may be a glucocorticoid as well. Aldosterone has distinct mineralocorticoid roles in tetrapods but its importance in fishes awaits the interpretation of further research.

Norepinephrine predominates in the adrenals of nonmammals, as it does in fetal and neonatal mammals. Tetrapods exhibit more reliance on epinephrine than norepinephrine, although the picture is variable. The most primitive function for epinephrine appears to be its ability to elevate blood sugar levels. The role of norepinephrine from the chromaffin cells in nonmammals is not understood.

Elements of the renin–angiotensin system are present in fishes but the juxtaglomerular cells have been identified with certainty only in the bony fishes. Although all tetrapods have juxtaglomerular cells, the macula densa is present only in birds and mammals. Angiotensin II stimulates corticosteroid release throughout the vertebrates, and these corticosteroids produce mineralocorticoid-like effects.

Natriuretic peptides apparently are present in all vertebrates, implying that they perform a primitive functional role. However, that role in fishes appears to be stimulation of corticosteroid secretion, which is the opposite of its role in amphibians, most birds, and mammals. No information on functions of natriuretic peptides is available for reptiles.

Suggested Reading

Books

Chester-Jones, I., and Henderson, I. W. (1976–1980). "General, Comparative, and Clinical Endocrinology of the Adrenal Cortex." Academic Press, New York.

Delrio, G., and Brachet, J. (1984). "Steroids and Their Mechanism of Action in Nonmammalian Vertebrates." Raven, New York.

Pickering, A. D. (1981). "Stress and Fish." Academic Press, New York.

Vinson, G. P. (1993). "The Adrenal Cortex." Prentice-Hall, Englewood Cliffs, NJ.

General Articles

Chester Jones, I. (1987). Structure of the adrenal and interrenal glands. In "Fundamentals of Comparative Vertebrate Endocrinology" (I. Chester Jones, P. M. Ingleton, and J. G. Phillips, eds.), pp. 95–121. Plenum, New York.

Chester-Jones, I., and Phillips, J. G. (1986). The adrenal and interrenal glands. In "Vertebrate Endocrinology: Vol. I, Fundamentals and Biomedical Implications" (P. K. T. Pang and M. P. Schreibman, eds.), pp. 319–349. Academic Press, San Diego.

Greenberg, N., and Wingfield, J. C. (1987). Stress and reproduction: Reciprocal relationships. In "Reproduction in Fishes, Amphibians, and Reptiles" (D. O. Norris and R. E. Jones, eds.), pp. 461–503. Plenum, New York.

Henderson, I. W., and Kime, D. E. (1987). The adrenal cortical steroids. In "Vertebrate Endocrinology: Fundamental and Biomedical Implications" (P. K. T. Pang and M. P. Schreibman, eds.), Vol 2, pp. 121–142. Academic Press, San Diego.

Fishes

Andersen, D. E., Reid, S. D., Moon, T. W., and Perry, S. F. (1991). Metabolic effects associated with chronically elevated cortisol in rainbow trout (*Oncorhynchus mykiss*). *Can. J. Aquat. Sci.* **48,** 1811–1817.

Balm, P. H. M., and Pottinger, T. G. (1995). Corticotrope and melanotrope POMC-derived peptides in relation to interrenal function during stress in rainbow trout (*Oncorhynchus mykiss*). *Gen. Comp. Endocrinol.* **98,** 279–288.

Felvolden, S. E., and Roed, K. H. (1993). Cortisol and immune characteristics of rainbow trout (*Oncorhynchus mykiss*) selected for high or low tolerance to stress. *J. Fish Biol.* **43,** 919–930.

Foo, J. T. W., and Lam, T. J. (1993). Serum cortisol response to handling stress and the effect of cortisol implantation on testosterone level in the tilapia, *Oreochromis mossambicus*. *Aquaculture* **115,** 145–158.

Hanke, W., Hegab, S. A., Assem, H., Berkowsky, B., Gerhard, A., Gupta, O., and Reiter, S. (1993). Mechanisms of hormonal action on osmotic adaptation in teleost fish. *In* "Fish Ecotoxicity and Ecophysiology" (T. Braunbeck, W. Hanke, and H. Segner, eds.), pp. 315–326. VCH, Weinheim.

Holloway, A. C., Reddy, P. K., Sheridan, M. A., and Leatherland, J. F. (1994). Diurnal rhythms of plasma growth hormone, somatostatin, thyroid hormones, cortisol and glucose concentrations in rainbow trout *Oncorhynchus mykiss,* during progressive food deprivation. *Biol. Rhythm Res.* **25,** 415–432.

Hontela, A., Rasmussen, J. B., Audet, C., and Chevalier, G. (1992). Impaired cortisol stress response in fish from environments polluted by PAHs, PCBs, and mercury. *Arch. Environ. Contam. Toxicol.* **22,** 278–283.

Iger, Y., Balm, P. H. M., and Wendelaar Bonga, S. E. (1994). Cellular responses of the skin and changes in plasma cortisol levels of trout (*Oncorhynchus mykiss*) exposed to acidified water. *Cell Tissue Res.* **278,** 535–542.

McCormick, S. D. (1995). Hormonal control of gill Na^+,K^+-ATPase and chloride cell function. *In* "Fish Physiology, Vol. 14, Cellular and Molecular Approaches to Fish Ionic Regulation" (C. M. Wood and T. J. Suttleworth, eds.), pp. 285–315. Academic Press, San Diego.

Wedemeyer, G. A., Barton, B. A., and McLeay, D. J. (1990). Stress and acclimation. *In* "Methods in Fish Biology," (C. B. Schreck and P. B. Moyle, eds.), pp. 451–489. Amer. Fish Soc., Bethesda, MD.

Amphibians

Ceballos, N. R., Shackleton, C. H., Harnik, M., Cozza, E. N., Gros, E. G., and Lantos, C. P. (1993). Corticosteroidogenesis in the toad *Bufo arenarum* H. Evidence for a precursor role for an aldosterone 3β-hydroxy-5-ene gene analogue (3β,11β,21-trihydroxy-20-oxo-5-pregnen-18-al). *Biochem. J.* **292,** 143–147.

Guardabassi, A., Muccioli, G., Andreoletti, G. E., Pattono, P., and Usai, P. (1991). Prolactin and interrenal hormone balance in *Xenopus laevis* adult specimens adapted to brackish water. *Atti. Acad. Sci. Torino* **125,** 55–69.

Iwamuro, S., Hayashi, H., Yamashita, M., and Kikuyama, S. (1991). Arginine vasotocin (AVT) and AVT-related peptide are major aldosterone-releasing factors in the bullfrog intermediate lobe. *Gen. Comp. Endocrinol.* **84,** 412–418.

LaForgia, V., and Capaldo, A. (1992). The interrenal gland of *Triturus cristatus* after insulin administration during the annual cycle. *J. Morphol.* **211,** 87–93.

Lihrmann, I., Netchitailo, P., Feuilloley, M., and Cantin, M., Delarue, C., Leboulenger, F., De Lean, A., and Vaudry, H. (1988). Effect of atrial natriuretic factor on corticosteroid production by perifused frog interrenal slices. *Gen. Comp. Endocrinol.* **71,** 55–62.

Zerani, M., and Gobetti, A. (1991). Effects of β-endorphin and naloxone on corticosterone and cortisol release in the newt (*Triturus carnifex*): Studies *in vivo* and *in vitro*. *J. Endocrinol.* **131,** 295–302.

Reptiles

Dauphin-Villement, C., and Xavier, F. (1987). Nychthemeral variations of plasma corticosteroids in captive female *Lacerta vivipara* Jacquin: Influence of stress and reproductive state. *Gen. Comp. Endocrinol.* **67,** 292–302.

Gabe, M. (1970). The adrenal. *In* "Biology of the Reptilia, Vol. 3, Morphology C" (C. Gans, ed.), pp. 263–318. Academic Press, New York.

Gleeson, T. T. (1993). Plasma catecholamine and corticosterone and their *in vitro* effects on lizard skeletal muscle lactate metabolism. *Am. J. Physiol.* **265,** R632–R639.

Mahapatra, M. S., Mahata, S. K., and Maiti, B. R. (1987). Influence of age on diurnal rhythms of adrenal norepi-
    nephrine, epinephrine, and corticosterone levels in soft-shelled turtles (*Lyssemys punctata punctata*). *Gen.
    Comp. Endocrinol.* **67,** 279–281.
Reinhart, G. A., and Zehr, J. E. (1994). Atrial natriuretic factor in the freshwater turtle *Pseudemys scripta:* A
    partial characterization, *Gen. Comp. Endocrinol.* **96,** 259–269.

Birds

Gray, D. A., Schutz, H., and Gerstberger, R. (1991). Interaction of atrial natriuretic factor and osmoregulatory
    hormone in the Peking duck. *Gen. Comp. Endocrinol.* **81,** 246–255.
Kocsis, J. F., McIlroy, P. J., and Carsia, R. V. (1995). Atrial natriuretic peptide stimulates aldosterone production
    by turkey (*Meleagris gallopavo*) adrenal steroidogenic cells. *Gen. Comp. Endocrinol.* **99,** 364–372.

11

The Endocrinology of Mammalian Reproduction

REPRODUCTION IS a process of evolutionary adaptation that extends far beyond the inclusion of what is usually termed the reproductive system. This process in vertebrates includes sexual determination, development and birth, sexual maturation or puberty, development of gametes, physiological and behavioral aspects of mating, fusion of gametes, and development of the zygote, which begins the process anew. In addition, a period of complex parental care is intercalated between birth and sexual maturation and possibly extends longer, such as in the case of humans. Every step in this complicated reproductive process is controlled internally or is modified by chemical regulators. Furthermore, interactions between individuals profoundly may affect internal processes as well.

The "reproductive system" includes the complex hypothalamo–hypophysial–gonadal axis as well as their associated target structures. Environmental factors (chemical, visual, photic, thermal, and tactile stimuli) operating through effects on neural and endocrine factors frequently determine the timing of many reproductive events. Other endocrine factors such as the thyroid and adrenal axes may have important effects on reproductive events, too.

The major difficulty that arises when generalizations are made about reproduction and reproductive mechanisms stems from the central importance of these processes with respect to past and future evolutionary events. The reproductive system is a central focus for selective agents since reproductive success is the major determinant of evolutionary success. Consequently, the reproductive system has been highly responsive to selective forces throughout the long evolutionary history of vertebrates. Owing to the many different selective forces that may act on a species at a given time and the variety of ways that a species may evolve in response to a selective factor, it should not come as a surprise that it is difficult to generalize about reproductive systems. In fact, it is remarkable that we can generalize at all.

Many features of extant species may be viewed as examples of specific adaptations to solve common environmental problems rather than as representatives of progressive "improvements" in the basic system that "culminated" in the placental mammals. A case in point would be the achievement of viviparity in all but two extant vertebrate classes, the Agnatha and Aves, as specific adaptations that, when coupled with varying degrees of parental care, result in greater percentage survival of a small number of offspring. Viviparity represents only one solution, however, to similar selective pressures that confront all species. In spite of the problems of environmental adaptations that tend to confuse evolutionary relationships, there remain numerous conservative features in regulatory mechanisms of reproductive biology, and it is these that are emphasized in this chapter.

I. General Features of Mammalian Reproduction

Mammals can be separated into three, distinct taxonomic groups: the Prototheria (monotremes), the Metatheria (marsupials), and the Eutheria (placentals). All possess **mammary glands,** specialized skin glands that are employed in secretion of milk to feed their young. The egg-laying monotremes comprise the most primitive group, of which only a few extant species remain. The marsupials or pouched mammals (Gr. **marsupi,** pouch) are confined mostly to Australia, with a few species in North, Central, and South America. There are about 230 extant species of marsupials. The placental mammals are the dominant mammalian group in number of species, distribution, and abundance. The **placenta** is a specialized structure that develops through interactions of zygote-derived extraembryonic tissues and maternal uterine tissues. It provides nutritional, respiratory, and excretory support for the offspring developing within the uterus.

Reproduction is closely regulated primarily through the hypothalamo–hypophysial–gonadal axis, which coordinates specific gonadal events through regulation of circulating gonadotropins. This gonadal axis is modified by other systems, especially the thyroid (Chapter 7) and the adrenal axes (Chapter 9). Factors influencing gonadotropin release and hence affecting gonadal functions are discussed in Chapter 4 and are summarized only briefly here.

Reproductive events in mammals are controlled through the hypothalamo–hypophysial–gonadal axis. The release of **luteinizing hormone (LH)** and **follicle-stimulating hormone (FSH)** from the adenohypophysis appears to be under the control of a single hypothalamic hormone, the **gonadotropin-releasing hormone, GnRH** (Table 11-1). Pulsatile release is an innate feature of GnRH neurons, which synthesize GnRH in two separate centers in the hypothalamus (see Chapter 4). The **tonic center,** occurring in both males and females, maintains a relatively constant pattern of pulsatile release of GnRH and produces rather static circulating levels of both LH and FSH. The **surge center** is responsible for the mid-cycle **LH surge** observed in mature females. In some cases, these centers are distinctly separated and in others there is no obvious anatomical separation.

Pulsatile secretion of GnRH is under stimulatory control of catecholaminergic neurons (norepinephrine, NE; dopamine, DA), and the NE-secreting neurons are inhibited by endogenous opioid peptides (EOPs, such as endorphins, dynorphins, and enkephalins). Other neurotransmitters, such as γ-aminobutyric acid (GABA), also may inhibit GnRH release. One hypothesis suggests that combined increases in adrenergic activity and an inhibition of opioid-secreting neurons allow the midcycle ovulatory LH surge to occur.

Surges in both LH and FSH occur in response to elevated GnRH prior to ovulation, but the magnitude of the LH surge greatly exceeds the FSH surge. LH release is enhanced selectively by the neuropeptide **galanin,** which is coreleased with GnRH just prior to the midcycle LH surge. Galanin has no effect on FSH release. Under the influence of FSH, the ovaries secrete a peptide called **inhibin** (see Section IV,C) that selectively blocks FSH release from the pituitary and contributes to the reduced FSH surge. In polyestrous species, the importance of the FSH surge may be related to initiation of follicle development in the next cycle.

Gonadotropins stimulate gamete maturation in both males and females as well as steroidogenesis and release of **estrogens, androgens,** and **progestogens** into the general circulation. Gametogenesis (oogenesis and spermatogenesis) is controlled primarily by FSH, whereas LH is mainly responsible for controlling steroidogenesis as well as release of gametes in both sexes. Induction of ovulation and formation of corpora lutea in the ovaries is due to LH. Steroidogenesis also can be influenced by FSH in both males and females. In some species, **prolactin (PRL)** may play a role in regulating ovarian steroidogenesis. The details of these events are discussed below. Gonadotropins may be responsible for synthesis of a variety of paracrine or autocrine factors by the gonads. These factors may play important roles in steroidogenesis or gametogenesis (see Section IV,C).

The gonadal steroids, released through the action of gonadotropins on special cells in ovaries and testes, control differentiation and maintenance of many **primary sexual characters** (such as the uterus or vas deferens) and **secondary sexual characters** (such as muscle development and beard growth in men). These gonadal actions were recognized in the twelfth century and probably much earlier by the Chinese, who used gonadal (and

Table 11-1. Summary of Generalized Hormone Actions in Mammalian Reproduction (Eutheria)

Hormone	Action in: Females	Action in: Males
GnRH	Stimulates FSH and LH secretion	Stimulates FSH and LH secretion
FSH	Initiates follicle growth; conversion of androgen to estrogen; synthesis of inhibin	Initiates spermatogenesis; secretion of androgen-binding protein and inhibin by Sertoli cells; conversion of androgen to estrogen by Sertoli cell
LH	Androgen synthesis; ovulation; formation of corpus luteum from granulosa; secretion of progesterone initiated in corpus luteum	Androgen secretion by interstitial cell (Leydig)
Prolactin	Synthesis of milk	Stimulates certain sex accessory structures (with androgen)
Oxytocin	Contraction of uterine smooth muscle: menstrual sloughing, birth, orgasm, milk ejection from mammary	Ejaculation of sperm; orgasm
Androgens	Precursors for estrogen synthesis; stimulates sexual behavior	Completes FSH-initiated spermatogenesis; stimulates prostate gland, other sex accessory structures; stimulates secondary sexual characters, such as beard growth in man
Estrogens	Stimulates proliferation of endometrium; induces LH surge; sensitizes uterus to oxytocin; negative feedback on pituitary release; may be primate luteolytic factor (estrone); may induce PRL surge	Converted from androgens; induces male hypothalamus; stimulates sexual behavior
Progesterone	Maintains secretory phase of uterus; inhibits release of gonadotropins from adenohypophysis; maintains pregnancy	
Prostaglandins	Causes corpus luteum to degenerate at end of luteal phase (not in primate); may be involved in birth initiation (induction of labor)	Ejaculation
Relaxin	Softens pelvic ligaments and cervix; possible role in lactation (?)	
Chorionic gonadotropin	Stimulates corpus luteum to produce progesterone	
Chorionic somatomammotropin	Stimulates mammary to synthesize milk during late pregnancy; growth hormone-like (somatotropin) actions on metabolism	
Inhibin (Sertoli cell factor, folliculostatin)	Inhibits FSH secretion from pituitary	Inhibits FSH secretion from pituitary

placental) preparations routinely to treat conditions ranging from impotence in men to the inability of a woman to bear sons.

In addition to their roles in development of male characteristics, androgens also are important in controlling sexual behavior in both male and female mammals. Generally speaking, the male mammal is larger and more aggressive than the female, usually playing a dominant social role with respect to the female as well as to other males. In many species, males accumulate harems of females and defend them against other males. A notable exception is the spotted hyena, *Crocuta crocuta,* in which the females are masculinized prior to birth. During development, female hyenas have higher levels of androgens than males. As adults, females have higher **androstenedione** levels than males, are larger and more aggressive, and have genitalia that closely resemble that of males. The clitoris is as large and erectile as the penis, and the urogenital canal traverses the entire length of the clitoris. Both mating and birth are accomplished through the clitoris. In addition, the vaginal labia of the females are fused to form a scrotum.

First births usually involve only one pup, and maternal mortality is high. Surviving fe males typically produce a litter of two in future breedings. Even at birth, the female spotted hyena hits the ground showing extremely aggressive behavior to its sibling. This aggressive behavior is associated with the higher circulating levels of androstenedione. In the wild, female–male twins are much more common then are female–female twins, probably because the males readily exhibit submissive behavior that enhances their survival, whereas only one female survives in female–female cases. Among triplets, aggression is directed against the last born and those female–female pairs that appear to be twins may be the remnants of all-female triplets.

The hypothalamic centers regulating GnRH release are sensitive to circulating steroids, which generally produce negative feedback on GnRH release. A notable exception to this pattern is the positive feedback effect on release of LH by estrogens (see Section V,B).

Several other hormones are involved in mammalian reproduction in addition to those of the hypothalamo–hypophysial–gonadal axis. PRL and, to a lesser extent, corticotropin (ACTH) from the adenohypophysis, as well as corticotropin-releasing hormone (CRH) from the paraventricular nucleus (PVN), oxytocin from the pars nervosa, and thyroid hormones all influence reproductive events. The placenta of eutherian mammals has assumed an endocrine role in pregnant females, producing steroids (primarily estrogens and progesterone) and polypeptide hormones [**chorionic gonadotropins (CGs); chorionic somatomammotropin (CS); chorionic corticotropin (CC); chorionic thyrotropin (CT),** GnRH, PRL, etc.]. In addition, the endocrine glands of the fetus may influence reproductive events; for example, contribution of the adrenal cortex to steroidogenesis by the placenta (see Chapter 9). The **pineal gland** may be a modulator of photoperiod and a source of antigonadotropic factors that influence gonadal function and prevent early onset of puberty (see Chapter 6). Finally, there are numerous reports of chemical agents termed **pheromones** that are produced by one sex to influence reproductive physiology, behavior, or both in the opposite sex.

II. Mammalian Life History Patterns

In the following sections, the generalized events of reproduction in mammals are presented along with some detailed information on gonadal and gonaduct differentiation.

A. Monotremes

The monotremes have retained the reptilian feature of laying eggs but have mammary glands for feeding the young after they hatch. The duckbill platypus lays its eggs in a nest, but the echidna places its eggs in a transitory pouch where they develop to hatching. On hatching, the new monotreme appears as a tiny, fetus-like creature with only a few well-developed features that enable it to attach itself to its mother's mammary gland to obtain nourishment. Development continues and the offspring, because of its primitive nature, can be considered an "exteriorized fetus" until it has completed development.

B. Marsupials

Technically, both marsupials and eutherian mammals have placentas. However, the marsupial placenta is rather primitive and apparently has no major endocrine function when compared to the eutherian placenta. The period of pregnancy or gestation is very short in marsupials (Table 11-2), and the young marsupial is born in an extremely immature condition. For example, among the macropodid marsupials (kangaroos), the "**joey**" must find its way essentially unaided to the mother's pouch, where it permanently attaches to the nipple of a mammary gland. The attached joey, like the newly hatched monotreme, continues its development. After a long period of pouch development (about 200 days in the red kangaroo), the young marsupial disengages itself from the teat and ventures outside of the pouch, returning first at regular and later at irregular intervals for milk.

C. Eutherians

Eutherian mammals employ the placenta not only as an endocrine organ to maintain gestation but also as a replacement for the mammary glands to supply nutrition to the fetus. Consequently, birth (**parturition**) is delayed considerably, and the newborn or neonate of most placental mammals is at a comparable stage of development to the young marsupial when it first leaves the pouch. Like the juvenile marsupial, the placental neonate relies at

Table 11-2. Comparison of Length of Estrous Cycle, Gestation Period, and Ratio of Body Weight of Neonate to Body Weight of Mother in Metatherian and Eutherian Mammals

Species	Length of estrous cycle[a] (days)	Length of gestation period (days)	Neonate: mother body weight ratio
Eutherial			
Rat	4–5	21	—
Sheep	16	148	1:14
Metatheria			
Virginia opossum	29	12	1:8,300
Long-nosed bandicoot	26	12	1:4,250
Brush possum	26	17	1:7,250
Dama wallaby	30	29	1:10,000
Swamp wallaby	31	37	—
Red kangaroo	35	33	1:33,400
Western gray kangaroo	35	30	—

[a]The estrous cycles are similar to the gestation period in metatherians and may not require a pregnancy-recognition mechanism. Based on Sharman, G. B. (1976). Evolution of viviparity in mammals. *In* "Reproduction in Mammals" (C. R. Austin and R. V. Short, eds.), Vol. 6, pp. 32–70. Cambridge University Press.

first on the mammary gland as the exclusive source of nourishment but gradually abandons it for other foods.

Three distinct reproductive patterns occur in sexually mature eutherians: one typically for males and two among females. Males of some domesticated species and humans are characterized by continuous secretion of gonadotropins and sometimes continuous spermatogenesis, and these males are capable of siring offspring at any time of year. Most eutherian species, however, exhibit seasonal episodes of spermatogenesis, sexual activity, or both, such as is displayed by nonmammalian vertebrates (see Chapter 12).

Although considered a continuous breeder, humans show some dramatic seasonal patterns of breeding activity correlated with photoperiod and latitude (based on the past 30–50 years of birth records). In Europe and Japan, there are more births in the early spring (March April), with the amplitude of the birth peak increasing in a south to north gradient. In South Africa, Australia, India, and New Zealand, the birth peak is associated with September–November. The amplitude of the birth peak decreases from north to south, which is the opposite of the latitudinal change observed in the northern hemisphere. However, in North America, the greatest number of births for both white and nonwhite people (many having their origins as transplants from Europe and Africa, respectively, in the past 300 years) occurs in August–October, with a decrease in amplitude from warmer to colder latitudes as occurs in countries of the southern hemisphere of the Old World. It has been proposed that North American patterns are a response to environmental temperature superimposed over the Old World patterns, which correlate more strongly with photoperiod. The variations with latitude are less clear for the northern hemisphere in the Old World, which are the opposite of that observed in the New World and the southern hemisphere of the Old World. Nevertheless, humans show seasonal patterns of reproductive activity that resemble those of other eutherians.

Females exhibit cyclic patterns of gonadotropin secretion that can be traced to a basic rhythmicity probably residing within the hypothalamus. There are two types of female cycles: the estrous cycle (usually seasonal) and the menstrual cycle (monthly and seasonal).

The **estrous cycle** is typical for mammals (except possibly humans) and consists of a series of precisely regulated endocrine events repeated in each cycle. Distinct stages within the estrous cycle can be readily distinguished. **Proestrus** is characterized by the hormonal changes that bring about ovulation. **Estrus** immediately follows or coincides with ovulation, and is a short period during which the female is receptive to the male and during which mating can occur. It is also the time when fertilization is most likely to lead to pregnancy and successful birth of offspring. The interim between estrus and the onset of hormonal changes characteristic of proestrus in cases in which pregnancy did not result is termed **diestrus.**

Carnivores and some other mammals may be classified as **monestrous.** If mating does not occur in a monestrous species or if mating occurs but fertilization and implantation are unsuccessful, the female will not return to estrus until the next breeding season (Gr. *mono,* one). Many mammalian species are **polyestrous,** however, and will return immediately to proestrus if mating does not occur or if mating is unsuccessful (Gr. *poly,* many). In some mammals (e.g., certain rodents), mating without successful fertilization and implantation may result in a short period of simulated or false pregnancy, termed **pseudopregnancy,** after which the female reenters proestrus.

Some mammals exhibit a sequence of events, known as the **menstrual cycle,** which is characterized by sloughing of the uterine lining if fertilization does not lead to pregnancy. Menstrual cycles are found in humans, monkeys, gibbons, the slender loris, marmosets, zebras, and several shrews and bats. The sloughing of the uterine lining results in a vaginal discharge of uterine epithelial cells and trapped blood. The blood is trapped due to constriction of special spiral-shaped arteries that supply most of the blood flow to the uterine lining. This stage of sloughing is known as the **menses,** or period of menstrual flow. In addition, a few mammals are known to exhibit **covert menstruation,** during which the sloughed tissues are resorbed and there is no uterine discharge. The onset of menses marks the end of one cycle and the beginning of the next. Although reproductive endocrinologists consistently focus on the cyclic nature of estrous and menstrual cycles and their restarting with the failure of pregnancy to occur, it is important to remember that the normal sequel to ovulation is pregnancy. In nature, it is probably unusual for a female to enter estrus and not become pregnant.

Most primates exhibit estrous behavior to some degree at about the time of ovulation, including species characterized as having menstrual cycles, although a well-defined estrus is not observed in the human female. Exhibition of menstrual or estrous cycles should not be considered as alternative strategies.

Although numerous hypotheses have been developed to explain why menstrual cycles evolved, most are discounted because they cannot be generalized to all mammals exhibiting menses. Some, for example, relate menstruation to a placental type common to menstruating mammals, yet other species with this same type of placenta do not have menstrual cycles. If it is simply a byproduct of endometrial function, then it is a costly one in terms of nutrient losses and should have been eliminated by natural selection, unless, of course, there were a special advantage that would offset the costs. One intriguing hypothesis suggests that menstruation evolved as an adaptation in mammals to neutralize and eliminate pathogens introduced during copulation and/or by sperm deposited in the vagina.

Uterine bleeding and discharge occur at other times in the estrous cycles of some mammals. For example, the cow and bitch (domestic dog, coyote) discharge blood prior to ovulation and the onset of estrous behavior. This discharge is estrogen induced and does not involve degeneration of the uterine lining. Periovolutary bleeding also occurs either overtly or covertly in several primates including humans, Sykes and vervet monkeys, and possibly in the cottontop tamarin.

D. Puberty

The achievement by the gonads of their full hormonal and gametogenic capacity is termed **puberty.** This process may be gradual or rather sudden, depending on the species. In rats, monkeys, and humans, puberty is associated with a marked increase in GnRH release from the hypothalamus, resulting in elevated gonadotropin secretion. In children, the frequency of these gonadotropin pulses increases twofold at night and later during the day as puberty progresses. The amplitude of the gonadotropin pulses also increases as the sensitivity of pituitary gonadotropes to GnRH is enhanced. Experimental studies suggest that the most critical factor in the induction of puberty is the increase in the frequency of GnRH pulses.

These changes in the gonadal axis are paralleled in other endocrine systems. Nocturnal release of prolactin also is elevated along with FSH and LH although its role in puberty is uncertain. Adrenal androgens increase in boys and girls prior to onset of puberty and continue to rise during puberty. This process is called *adrenarche* (see Chapter 9). These events are independent of the changes in the gonadal axis but are important contributions to puberty. For example, adrenal androgens stimulate the prepubertal growth spurt and the appearance of axillary and pubic hair.

GnRH mRNA in the hypothalamus rises markedly prior to the onset of puberty. The

GnRH neurons are interconnected so that release of their products is synchronous in a self-regulated rhythmic pattern (pulses). Thus, they are often called **pacemaker neurons,** analagous to the pacemaker cells of the heart. Evidence also suggests the existence of an external hypothalamic pacemaker that may drive the GnRH neurons.

There are several hypotheses concerning the mechanism(s) for the onset of puberty. The **gonadostat hypothesis** suggests that there is a development of decreased feedback sensitivity in the hypothalamus to gonadal steroids that brings about increased release of GnRH. Accelerated gonadotropin release from the pituitary activates gonadal steroidogenesis and gametogenesis or **gonadarche.** Gonadal responses also may involve receptor synthesis. For example, receptors for FSH are present in early ovarian follicles, but LH receptors are not. FSH stimulates production of LH receptors and increases the levels of aromatase enzyme. Now, the ovary can respond to both gonadotropes. The **missing link hypothesis** implies that some factor is missing and that it is the brain and not the gonads that is functionally incompetent prior to puberty. Data obtained from experimental and clinical observations supporting these hypotheses also support the **active inhibition hypothesis,** which currently is in favor. This hypothesis states that puberty occurs because of a progressive decrease in physiological inhibition. The well-known actions of the pineal gland and melatonin on inhibiting reproductive function (see Chapter 6) and the association of their reduction with precocial puberty (see Section VII,A) offer strong support to the inhibitory hypothesis. Furthermore, reduced activity of inhibitory GABA-secreting neurons on GnRH neurons has been documented to occur during puberty.

A separate hypothesis has been formulated to explain puberty and **menarche,** the onset of menstruation in girls. This **lipostat hypothesis** or **critical weight hypothesis** implies that attainment of puberty is in part a function of fat storage in the body. Consequently, girls who have more body fat reach menarche sooner than leaner girls. Furthermore, chronic strenuous exercise reduces body fat and may prevent young girls from reaching menarche just as it can block menstruation in women (see Section VII,B,2). Once the excessively lean girls are placed on a less severe regimen so that body fat increases, menarche usually is achieved. This mechanism probably is not related to puberty per se but rather may be an evolutionary mechanism to ensure that there is evidence of sufficient environmental resources (as evidenced in body fat) to support the energy demands of pregnancy and lactation for successful rearing of young.

III. Embryogenesis of Gonads and Their Accessory Ducts

The gonads, ducts, and associated glands are often termed the **primary sexual characters** because of their essential roles and direct participation in reproduction. Primary sexual characters include the oviducts and uterus of the female and the vasa deferentia and prostate gland of the male. **Secondary sexual characters** are often dependent on gonadal hormones and usually enhance reproduction but are not necessarily required. For example, the male physique and beard growth are secondary sexual characters.

A. The Gonads

The paired primordia of the mammalian gonads arise from the intermediate mesoderm as a genital ridge on either side of the midline in close association with the mesonephric

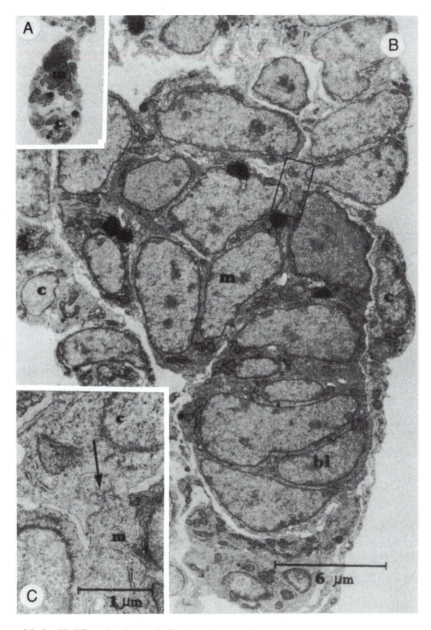

Figure 11-1. Undifferentiated gonad. Section of gonad from 25-mm tadpole of *Rana pipiens* showing cortical (c) and medullary (m) cells separated by a basal lamina (bl, basement membrane). (A) Total gonad; (B) enlargement of (A); (C) further enlargement, showing contact between cortical and medullary cells (arrow). [From Merchant-Larios, M. (1978).
Ovarian differentiation. *In* "The Vertebrate Ovary" (R. E. Jones, ed.), pp. 47–81. Plenum, New York.]

kidney. Numerous derivatives of the mesonephric kidney and duct system are retained as
functional portions of the vertebrate reproductive system. The gonadal primordium consists
of an outer **cortex** derived from peritoneum and an inner **medulla** (Figs. 11-1 and 11-2).

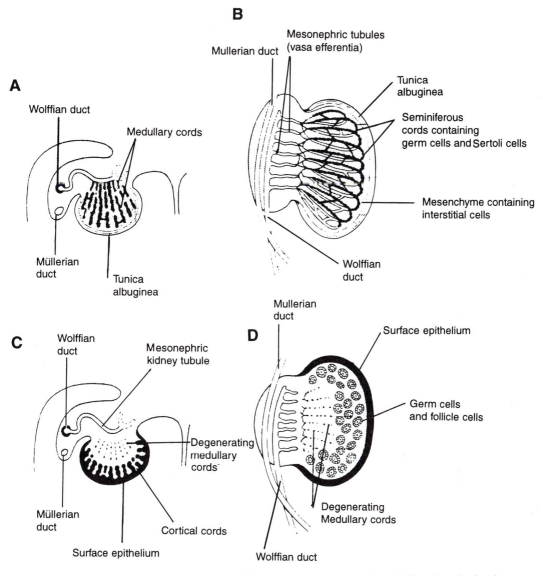

Figure 11-2. Development of the primitive testis and ovary. In the male, the medullary tissue develops into the testis cords (A), which give rise to the seminiferous tubules including the Sertoli cells. Mesonephric tubules give rise to the vas deferens and the vasa efferentia (B). In the female (C), the medullary cords degenerate and the cortical cords give rise to an ovary (D). Some mesonephric elements remain in the female as well. The wolffian ducts are retained in female amphibians to drain the kidneys, as they also do in males. Eventually the wolffian ducts degenerate in female reptiles, birds, and mammals, in which the ureters develop to drain the metanephric kidneys (not found in anamniotes). [Modified from Johnson, M., and Everitt, B. (1988). "Essential Reproduction," 3rd Ed. Blackwell, Oxford; originally from Metcalfe *et al.* (1967). *Physiol. Rev.* **47**, 782.]

Germ cells do not arise within the gonadal primordium itself but migrate from their site of origin to the yolk sac endoderm to either cortex or medulla, depending on which genetic sex is destined to develop.

Initially, the medullary component in males and females differentiates into **primary sex cords.** Differentiation of the primary sex cords and regression of the cortex result in a testis. Each testis consists of seminiferous tubules containing the **germ cells,** which eventually will produce sperm, and the **Sertoli cells,** which support gamete development. Steroidogenic **interstitial cells** are located between the seminiferous tubules. These interstitial cells arise from medullary tissue surrounding the primary sex cords and become sources of androgens. Germ cells that have migrated into the primary sex cords give rise to spermatogonia, which eventually will give rise to gametes.

In females, the primary sex cords degenerate, and **secondary sex cords** differentiate from the cortical region. These secondary sex cords become the definitive ovary. In the ovary, the germ cells give rise to **oogonia.** The ovaries contain **follicles,** which consist of one or more layers of **follicular cells** surrounding a single germ cell.

B. Accessory Ducts

In males, the central portion of each differentiating testis forms a network of tubules, known as the **rete testis,** that do not contain seminiferous elements. The rete testis connects the seminiferous tubules with a portion of the primitive mesonephric kidney duct called the **wolffian duct,** which, under the influence of testosterone, differentiates into the vas deferens and conducts sperm to the urethra. Most of the mesonephric kidney in mammals degenerates, with the exception of some of the anterior mesonephric kidney tubules. In the presence of testosterone, this tissue together with a portion of the wolffian duct forms two glandular structures, the **epididymis** and the **seminal vesicle** (Figs. 11-2 and 11-3).

A second pair of longitudinal ducts develops from the mesial wall of the mesonephric duct and lies parallel to the wolffian ducts. These structures are known as the **müllerian ducts.** In genetic females, the müllerian ducts develop into the oviducts, uterus, and part of the vagina (Fig. 11-3), usually fusing together to form a common vagina and, in some species, a single uterus as well. The wolffian ducts degenerate in female mammals. In males, it is the müllerian ducts that are suppressed in favor of wolffian duct development.

Müllerian-inhibiting substance (MIS) was first proposed by Jost in the 1940s to explain the inhibitory effect of the testes on development of müllerian ducts in rabbit embryos. It also has been called the **anti-müllerian hormone** or AMH. Implantation of a testis into a female embryo results in MIS secretion, which prevents development of the müllerian ducts. Evidence suggests that MIS is a glycoprotein. It not only blocks müllerian duct development but is capable of inhibiting growth of tumors from ovaries and müllerian duct derivatives (Fig. 11-3). It appears that MIS acts cooperatively with testosterone in producing its effects on the müllerian ducts.

IV. Endocrine Regulation in Eutherian Males

Maleness in eutherian mammals is dependent on secretion of androgens from the testis. In the absence of androgens, the male genotype (XY) will develop as a female phenotype. Conversely, treatment of newborn females with androgens destroys the cyclical secretory pattern of the hypothalamo–hypophysial–gonadal axis and replaces it with a noncyclical pattern like that of males. Becoming a male mammal, then, involves overcoming the basic

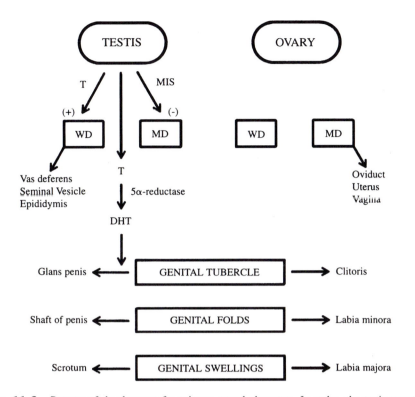

Figure 11-3. Patterns of development for primary sexual characters. In males, the testis secretes testosterone (T), which stimulates differentiation of wolffian ducts (WD), and müllerian-inhibiting substance (MIS), which causes regression of müllerian ducts (MD). Dihydrotestosterone (DHT) is either produced by the testes or converted from T in the genital tubercle, genital folds, or genital swellings, causing them to differentiate in the male direction. Estradiol from the ovary prevents MIS (also secreted by ovary) from causing müllerian duct regression and the absence of sufficient androgens determines the fate of the other structures.

tendency for mammalian embryos to develop as females. Sexual differentiation of the brain may occur prior to or soon after birth, depending on the species (Table 11-3).

A gene seemingly responsible for male sex determination, called **SRY,** has been localized on the short arm of the Y chromosome characteristic of genetic males. In mice, the SRY gene is activated in gonads of genetic males before they begin to differentiate into

Table 11-3. Critical Periods for Sexual Differentiation of the Brain in Mammals

Species	Gestation period (days)	Critical period (days)
Hamster	16	16–21
Laboratory rat	21–22	18–28
Laboratory mouse	19–20	20
Guinea pig	68	30–35
Human	270	84–126

testes. Insertion of the SRY gene into XX mice followed by its activation leads to formation of male-specific structures and regression of female ducts. The activated gonad secretes MIS, which causes regression of the müllerian ducts. Presumably, the SRY gene produces a factor that activates the MIS gene. Androgens secreted by the transformed gonad cause male-like differentiation of the external genitalia and the wolffian ducts as well as changes in the hypothalamus to suppress development of the surge center. This establishes the tonic secretory pattern for GnRH and gonadotropins that characterize males.

In some species, postpubertal males are capable of copulating with a female whenever she is receptive. Secretion of GnRH and hence of gonadotropins is more or less continuous but with daily fluctuations occurring in circulating levels of some gonadotropins. Daily secretory patterns for gonadotropins show considerable variation among different species. Hourly fluctuations of LH have been reported in bulls, and these variations in LH are correlated with following increases in circulating testosterone (Fig. 11-4). However, in human males, FSH shows no cyclic variation in blood levels although LH and testosterone exhibit obvious daily patterns with peak levels occurring during early morning hours and minimal values reported for the afternoon. Wild mammals exhibit distinct seasonal breeding, and spermatogenesis may be restricted to a few months or less.

A. Spermatogenesis

Each testis develops primarily from the medullary portion of an embryonic gonadal blastema (see Fig. 11-2). Differentiation of the medullary portion with concomitant regression

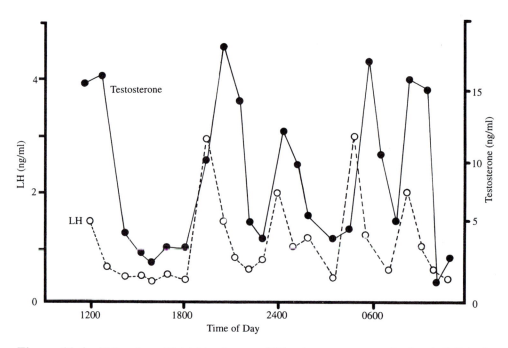

Figure 11-4. Male pattern of luteinizing hormone (LH) and testosterone secretion in a bull. Pulsatile release of GnRH (not shown) would precede each LH peak, which precedes each testosterone peak.

of the cortical components (progenitor of the ovary) appears to be controlled by embryonic androgen secretion. The medullary cords differentiate into seminiferous tubules and interspersed masses of interstitial cells. These interstitial cells are steroidogenic and are located between the seminiferous tubules. Androgens are synthesized and released into the circulation by interstitial cells.

The seminiferous tubules consist of large Sertoli cells, germ cells, spermatogonia, and cells derived from the spermatogonia (Figs. 11-5 and 11-6). **Peritubular myoid cells** develop early during puberty under the influence of androgens. They surround and provide support for the seminiferous tubules and are believed to be responsible for contractile activity of the tubules as well. The Sertoli cell has an extensive cytoplasm extending from the outer edge to the lumen of the tubule. The nucleus of the Sertoli cell is located at the outer edge. Sertoli cells form tight junctions with the peritubular myoid cells, and together they secrete the components that form the basement membrane. This basement membrane is the blood/testis barrier that isolates the seminiferous tubules from the blood and requires all chemical signals to pass through the Sertoli cells.

In addition to the Sertoli cells, the germ cells are present along the outer margins of the tubules and can differentiate into **spermatogonia.** Spermatogonial cells proliferate mitotically under the influence of FSH and eventually undergo differentiation characterized by nuclear enlargement to become **primary spermatocytes** that are capable of entering spermatogenesis (Fig. 11-7). Testicular androgens are somehow necessary for initiation of meiosis in primary spermatocytes. These cells undergo the first meiotic division to give rise to two smaller **secondary spermatocytes,** which are infrequently observed in histological preparations because, once formed, they quickly enter the second meiotic division to yield four haploid **spermatids,** which are transformed to sperm (spermatozoa) by concentrating the chromatin material into the sperm head and by elimination of the majority of the cytoplasm. The process of transformation of spermatids to sperm is termed **spermiogenesis.**

A given histological section of a seminiferous tubule may show varying numbers of spermatogonia, primary spermatocytes, secondary spermatocytes, spermatids, and sperm in sequence from the outer margin to the lumen. The tails of the sperm extend into the lumen, and the heads of the sperm typically are surrounded by highly folded margins of Sertoli cells (Fig. 11-6).

Millions of mature sperm may be sloughed off into the lumena of the seminiferous tubules each day. This process is termed **spermiation.** These sperm pass along through the tubules, which later will coalesce into larger ducts and eventually form the epididymis associated with each testis. Vast numbers of mature sperm are stored in the epididymis. Under the influence of androgens, the epididymis secretes materials into its lumen where the sperm are being held. Included in this secretion are protein-bound **sialic acids** (sialomucoproteins), **glyceryl phosphorylcholine,** and **carnitine.** These particular substances are involved directly in maintaining sperm in viable condition until ejaculation. Androgens and **androgen-binding protein (ABP)** produced by Sertoli cells in the seminiferous tubules are released along with sperm and travel to the epididymis (Fig. 11-6). Androgen molecules freed from ABP in the lumen, or androgen–ABP complexes, or both are absorbed by the epididymal cells. These androgens stimulate epididymal cells to secrete materials involved in maintenance of the sperm.

As a consequence of the forcible ejection (ejaculation) that occurs during copulation, the sperm leave the epididymis, enter the vas deferens, and travel to the urethra. The sperm traverse the length of the penis via the urethra and are deposited in the female's vagina

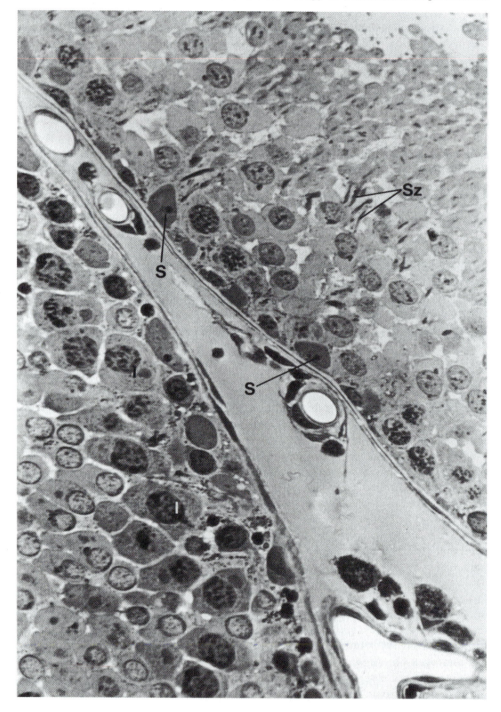

Figure 11-5. Section of rat testis showing edges of adjacent seminiferous tubules. S, Nuclei of Sertoli cells; I, primary spermatocytes; Sz, heads of sperm.

during coitus. Various glands such as the **prostate gland** add their fluid secretions to the sperm and epididymal secretions to form a watery mixture of sperm and various organic and inorganic substances known as **semen.** The entire ejaculatory event may be induced by

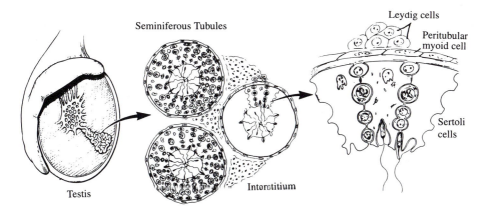

Figure 11-6. Cellular organization of the mammalian testis. Detail at right shows events of spermatogenesis and spermiogenesis in relation to three Sertoli cells. The endocrine roles of interstitial (Leydig) cells, peritubular myoid cells, and Sertoli cells are described in text. [Reprinted with permission from Skinner, M. K. Cell–cell interactions in the testis. *Endocr. Rev.* **12**, 45–77, 1991. © The Endocrine Society.]

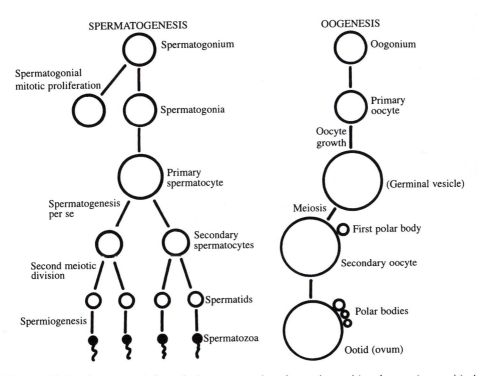

Figure 11-7. Gametogenesis in testis (spermatogenesis and spermiogenesis) and ovary (oogenesis). A major difference between males and females is the mitotic proliferation of gonial cells after birth in males, whereas the germ cells have all progressed to the primary oocyte stage in females and no new oocytes appear after birth. The primary spermatocyte undergoes meiosis to produce four spermatids, whereas meiotic division of the primary oocyte with unequal distribution of cytoplasm produces only one oocyte plus up to three polar bodies. The first polar body (a secondary oocyte with very little cytoplasm) often does not undergo the second division.

the release of oxytocin from the pars nervosa in response to a neural reflex initiated by mechanical stimulation of the penis.

B. Endocrine Regulation of Testicular Function

Spermatogenesis is apparently a temperature-sensitive process, and high temperatures such as are found within the body cavity of most eutherians can impair normal spermatogenesis and produce temporary sterility. Consequently, at some time prior to the attainment of sexual maturity or prior to the annual breeding season, the testes descend into the scrotum, where spermatogenesis can proceed at a slightly lower temperature. The failure of the testes to descend, a condition known as **cryptorchidism** (Gr. *crypto-,* hidden; *orchi,* testis), may cause irreparable damage to the seminiferous epithelium in most species. Some mammals lack a scrotum (for example, elephants, whales, seals), and the testes are permanently located within the abdominal cavity. In such species, spermatogenesis obviously does not exhibit the same temperature sensitivity characteristic of scrotal species, or these animals possess other mechanisms to reduce testicular temperature. Male elephants, for example, are capable of producing viable sperm and copulating with a female at any time of year.

Gonadotropins and a number of paracrine factors including testosterone (Table 11-4) control production of sperm. Many of the details of the endocrine regulation of these processes are not clear, but a generalized picture is emerging. Relatively separate roles have been defined for LH and FSH in males although FSH and testosterone work cooperatively in some cases. Spermatogenesis is initiated by FSH, which stimulates proliferation of spermatogonia and formation of primary spermatocytes. In addition, testosterone may initiate meiotic divisions of primary spermatocytes, resulting eventually in formation of spermatids. The production of ABP and cytoskeletal proteins (**actin, vinculin**) by the Sertoli cell also is stimulated by FSH. During development, FSH stimulates mitosis in Sertoli cells whose number in the adult testis is directly proportional to sperm abundance.

The synthesis and release of circulating androgens by the interstitial cells is controlled primarily by LH (Fig. 11-6). Steroidogenesis by the Leydig cell is enhanced by one of the actions of FSH on the Sertoli cell. FSH causes secretion of a protein complex called **steroidogenesis-stimulating protein (STP)** consisting of two separate proteins. STP is a paracrine regulator that enhances steroidogenesis in the Leydig cell in response to LH.

Testosterone is the major circulating androgen, although other androgens such as **androstenedione** or **5α-dihydrotestosterone (DHT)** may circulate in significant amounts (Table 11-5). Prior to attainment of puberty in bulls, androstenedione is the principal circulating androgen, but it is gradually replaced by testosterone at puberty. Testosterone within the testis, however, seems to be the most important androgen influencing sperm production.

Testosterone has several important paracrine effects on spermatogenesis and spermiogenesis. These actions probably are indirect since testosterone receptors (as well as FSH and LH receptors) appear to be absent or in very low numbers on germ cells. The attachment of Sertoli cells to spermatids involves cytoskeletal actin and vinculin interactions with the spermatids, followed by the indirect effects of testosterone. Peritubular myoid cells (Fig. 11-6) are stimulated by testosterone to release two proteins, **P-Mod-S$_A$** and **P-Mod-S$_B$**, which cause Sertoli cells to secrete several paracrine regulators (e.g., inhibin) that may alter spermiogenesis.

The interstitial cells of the testis also synthesize and release small quantities of estro-

Table 11-4. Possible Local Actions for Gonadal Regulators

Factor	Source	Proposed action
Gonadotropin-releasing hormone (GnRH)	Ovary	Working through IP_3 second messenger, GnRH alters steroidogenesis by granulosa cells in certain follicular stages; may signal atresia
	Sertoli cell	Alters androgen synthesis by interstitial cells; increases local permeability of capillaries
Testosterone	Interstitial cell	Regulates functions of Seroli cells; stimulates meiosis in primary spermatocytes
Transforming growth factor α (TFO-α)	Thecal cells	Facilitates proliferation of thecal and granulosa cells but slows their OTH-induced differentiation
	Sertoli and peritubular myoid cells	Causes EGF-like growth stimulation in interstitial cells, decreases steroidogenesis
Fibroblast growth factor (FGF)	Granulosa cell	Causes epithelial proliferation in early follicular development and conversion of thecal cells to ovarian interstitial cells after ovulation
	Testicular germ cells	Binds to receptors on Sertoli cell; function unknown
Nerve growth factor (NGF)	Ovarian cells	Stimulates follicular formation and organization as well as differentiation of ovarian interstitial cells
Epidermal growth factor (EGF)	Thecal/interstitial cells of ovary	Stimulates granulosa cell proliferation
Insulin-like growth factor I (IGF-I)	Granulosa cells	Increases number of LDL receptors on granulosa cells; stimulates cholesterol and inhibin synthesis
Activin	Granulosa cells	Unknown
	Sertoli and interstitial cells	Specific receptors shown on germ cells
Growth hormone-releasing hormone (GHRH)	Corpora lutea, oocyte	Promotes follicular development and ovulation
	Germ cells of testes	Stimulates Sertoli cells to make stem cell factor
Interleukin 6 (IL-6)	Ovarian T cells	Suppresses response of granulosa cells to FSH, i.e., decreased progesterone synthesis; induces apoptosis and atresia of granulosa cells
Interleukin 1 (IL-1)	Sertoli cells	Decreases steroidogenesis in interstitial cells
P-Mod-S protein	Peritubular myoid cells	Nonmitogenic factor that regulates differentiation and function of Sertoli cells

gens. Testicular estrogens reach dramatic levels in the stallion, and estrogens may have definite physiological roles in males, especially with respect to reproductive behaviors. Locally, estradiol can block androgen synthesis by interstitial cells and can influence the responsiveness of these cells to gonadotropins. The ratio of testosterone to estradiol in the general circulation may alter the ratios of FSH and LH being released from the pituitary through negative feedback. Finally, conversion of androgens to estrogen occurs in certain brain cells, and circulating estrogens themselves may influence male sexual behavior.

Table 11-5. Plasma Levels of Reproductive Steroid in Mammals

Species	Testosterone (ng/ml)	Dihydrotes-tosterone (ng/ml)	Estradiol (pg/ml)	Progesterone (ng/ml)
Mustela ermina (stoat)				
Male (annual range)	4.5–26			
Elaphus maximus (Asian elephant)				
Male (annual range)	0.7–45			
Female				
Not pregnant (peak)			26	0.153–0.195
Pregnant (peak)			26	0.263
Macaca fuscata (Japanese monkey)				
Male (annual range)	0.2–19.8			
Female				
Follicular phase (peak)			150	2
Luteal phase (peak)			250	5.3
Homo sapiens				
Male (mean)	7.9	0.4	50	
Female				
Follicular phase (range)	0.6	0.3	60–600	0.3–1.5
Luteal phase (range)	0.6	0.3	200	3–20
Pregnant (range)			5,500–30,000	45–210

C. Actions and Metabolism of Androgens in Males

Circulating androgens influence development and maintenance of several glands and related structures associated with the male genital tract, such as the prostate gland and seminal vesicles, and induce development of certain secondary sexual characters such as growth of the beard in men. Androgens also exert a negative feedback effect on the secretion of gonadotropins primarily through actions at the level of the hypothalamus (see Chapter 4).

The action of testosterone in some of its target cells is believed to involve first its conversion to DHT by the enzyme **5α-reductase.** It is DHT that has a greater affinity for the androgen receptor than does testosterone; it also binds better to the receptor. Timely development of prostate and bulbourethral glands, the penis, and scrotum are dependent on conversion of testosterone to DHT. Brain tissue also has been shown to employ 5α-reductase, and some effects of testosterone on behavior may involve prior conversion to DHT.

Many androgenic responses, however, are not mediated by DHT, and this conversion is not necessary for testosterone to produce these effects. For example, stimulation of wolffian duct derivatives (the epididymis, vas deferens, and seminal vesicles) is accomplished by testosterone without prior conversion to DHT. This action may occur because testosterone levels are sufficiently high to effectively stimulate the receptors, and DHT is typically effective as well. In some target tissues, androgens have been shown to undergo conversion to estrogens through aromatization of the A-ring and removal of the C-19 carbon atom (see Chapter 2). This process occurs, for example, in the central nervous system where aromatization as well as 5α-reduction may be essential to the mechanism of androgen actions. Induction of some male behaviors in castrates requires aromatization and cannot be induced by nonaromatizable androgens such as DHT, whereas others may be induced by either aromatizable androgens or by DHT.

Sertoli cells also convert androgens to estrogens, and this aromatization of androgens is stimulated by FSH. This action of FSH on aromatization of androgen is similar to that

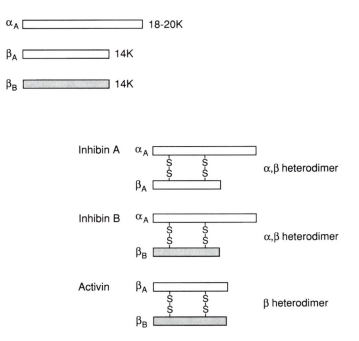

Figure 11-8. Inhibins and activin. The inhibin proteins are each heterodimers composed of one copy of α_A and one copy of either β_A (inhibin A) or β_B (inhibin B). Activin is a heterodimer composed of one β_A subunit and one β_B subunit. Other activin combinations are known but their biological importance, if any, is unclear.

proposed for the synthesis of estrogens in the ovarian follicle (see Section V,A). The role of these testicular estrogens is uncertain, however, but may be related to the action of FSH on the synthesis of ABP by the Sertoli cell.

The Sertoli cells, under the influence of FSH, secrete two forms of inhibin that selectively block FSH release from the adenohypophysis. Inhibin activity has been found in rete testis fluid, seminal plasma, testicular extracts, and ejaculate. Two forms of inhibin have been isolated: **inhibin A** and **inhibin B** (Fig. 11-8). These molecules are glycoprotein heterodimers of 31–35 kDa that possess a common α subunit combined with one of two β subunits (β_A or β_B). Inhibins are believed to be the major factor responsible for negative feedback in the regulation of FSH release in both males and females. In addition, a β-subunit heterodimer called **activin** has been isolated from gonads. Activin composed of one β_A subunit and one β_B subunit is a potent releaser of FSH from the pituitary gland in laboratory experiments, although its physiological role is still undetermined. It may be a local regulator since activin receptors occur on spermatogenetic cells in the testis. Alternative forms of activin also have been reported (β_A–β_A and β_B–β_B homodimers) but their importance is not established.

V. Endocrine Regulation in Eutherian Females

Although marked differences can be pointed out between endocrine events related to estrous and menstrual cycles, there are probably greater differences among species exhibiting one of these cyclic types than in comparison of the two cyclic types. Nevertheless there are several distinctive features that characterize females, regardless of whether they exhibit estrous or menstrual cycles.

In addition to the roles of gonadotropins and gonadal steroids described below, numerous other regulators (including GnRH, prostaglandins, and growth factors) are synthesized in the ovaries (as discussed earlier for the testes) and are suspected of playing important paracrine roles in ovarian events. A partial listing and some proposed autocrine and paracrine actions for these ovarian regulators are provided in Table 11-4.

A. The Follicular Phase

The basis for the cyclical nature of female reproductive events resides in the hypothalamus and is a genetically determined female characteristic. During the follicular phase of the ovarian cycle, the tonic hypothalamic center releases small quantities of GnRH into the portal circulation and relatively low but rather constant circulating levels of FSH, LH, or both are maintained. In general, prior to puberty, which is characterized by increased gonadotropin levels, the ovary contains **primary oocytes** invested with modified stromal cells forming the **primary follicle** (Figs. 11-9 and 11-10). In most mammals, there are no oogonia in the ovary because all of them differentiated into primary oocytes prior to or shortly after birth. The arrival of FSH at the ovary stimulates primary follicles to begin to enlarge and differentiate (Figs. 11-9 and 11-10). The growing oocyte becomes surrounded by two distinct layers of cells: the inner **granulosa cells** and the outer **thecal cells** (Figs. 11-9 and 11-10). Thecal cells further differentiate into inner and outer layers: the **theca interna** and **theca externa.** The granulosa cells are separated from the thecal layers by a connective tissue barrier or basement membrane similar to that described for the seminiferous tubules, and there is no penetration of capillaries into the granulosa. As the follicle grows, the granulosa cells secrete fluid contributing to the **liquor folliculi** or **antral fluid,** which is primarily an ultrafiltrate of blood plasma. Increasing production of liquor folliculi results in formation and progressive enlargement of a fluid-filled cavity within the follicle, the **antrum.** Most follicles exhibit apoptosis and degenerate to form **corpora atretica;** only a few will ever ovulate.

Under the influence of LH and FSH, ovarian follicles synthesize and release estrogens, predominantly estradiol (or estradiol-17β), into the general circulation (Table 11-5). The synthesis of estrogens in the ovary appears to be a cooperative effort between cells of the theca interna and the granulosa (Fig. 11-11). Studies support the interpretation that LH stimulates the thecal cells to produce androgens (principally androstenedione) that are aromatized by the granulosa cells to form estradiol. Conversion of androgens to estradiol by the granulosa cells is stimulated by FSH, which increases aromatase levels in these cells. In addition, FSH causes the granulosa cells to produce inhibin, which feeds back on the pituitary to selectively inhibit FSH release (as was described earlier for male). Inhibin also inhibits aromatase activity directly in granulosa cells whereas the related peptide activin increases aromatase activity.

Estradiol stimulates differentiation and proliferation of the uterine lining, or **endometrium,** in preparation for implantation of the **blastocyst,** a hollow embryonic stage formed from the first series of cellular divisions following fertilization. The blastocyst consists of an outer extraembryonic layer of cells, the **trophoblast,** which will form the fetal component of the placenta, and an **inner cell mass,** which will become the embryo proper. This interval of endometrial growth is termed the **proliferative phase** of the uterine cycle and

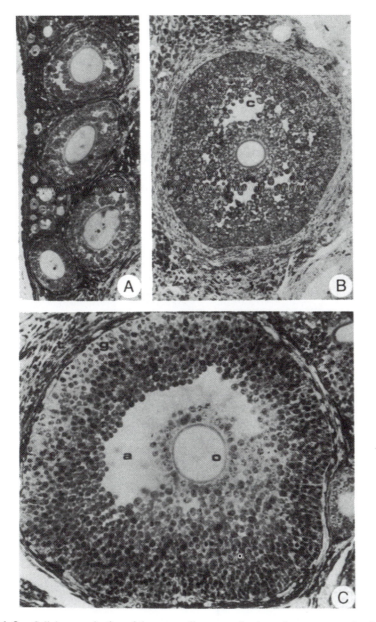

Figure 11-9. Cellular organization of the mammalian ovary. Sections of mouse ovary showing stages of follicular development. (A) Naked oocytes, primary follicles, and four early, growing follicles with the beginnings of formation of an antral cavity (c). (B) A later follicle with proliferated granulosa. Note the clear distinction of thecal–granulosa layers. (C) A young antral follicle. g, Granulosa; t, theca; o, oocyte; a, antrum. [From Natty, K. P. (1978). Follicular fluid. *In* "The Vertebrate Ovary" (R. E. Jones, ed.), pp. 215–259. Plenum, New York.]

is characterized by hyperplasia and development of secretory glands (see Fig. 11-12). In addition, there is a marked increase in the vasculature of the endometrium, which in higher primates is manifest in the development of special **spiral arteries.**

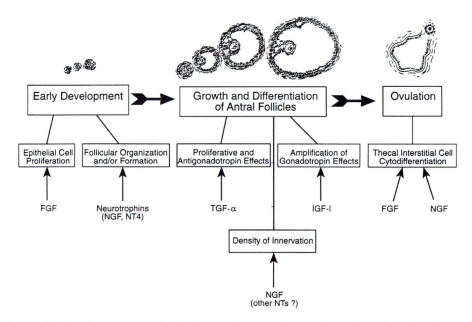

Figure 11-10. Summary of ovarian developmental stages and proposed paracrine actions of certain growth factors and neurotransmitters. FGF, Fibroblast growth factor; TGF-α, transforming growth factor α; NGF, nerve growth factor; NT, neurotropins; NT4, neurotrophin 4. [Reprinted by permission of the publisher from Ojeda, S. R., and Dissen, G. A. Developmental regulation of the ovary via growth factor tyrosine kinase receptors. *Trends Endocrinol. Metab.* **5**, 317–323. Copyright 1994 by Elsevier Science Inc.]

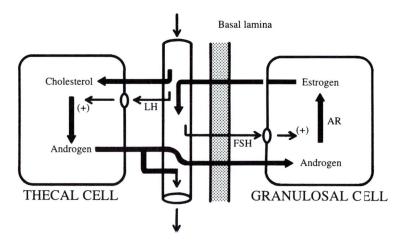

Figure 11-11. Two-cell model for steroidogenesis in the ovary. Binding of LH to receptors found only on thecal cells (or ovarian interstitial cells) stimulates androgen synthesis, most of which diffuses through the basal lamina (basement membrane) to the granulosa cell. FSH stimulates aromatase (AR) production, which transforms androgens into estrogens. A similar two-cell system is present in the testis with LH receptors associated with the interstitial cells (Leydig cells) and FSH receptors on the Sertoli cells. The low level of estrogen synthesis by testes may be accomplished in the Sertoli cell. Testosterone reaches the Sertoli cell by diffusion through the basal lamina surrounding the seminiferous tubule.

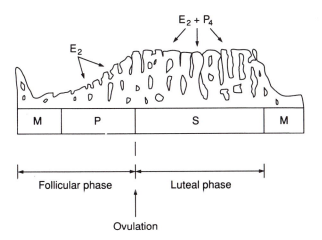

Figure 11-12. The uterine cycle. The menses (M) occupies the first 5 days of the cycle. The endometrium is stimulated by estradiol (E_2) during the proliferative stage (P). The follicular
phase in the ovary corresponds to M + P. Following ovulation the corpus luteum secretes
E_2 and progesterone (P_4) during the luteal phase, which maintains the vascularity of the
endometrium as well as secretion by exocrine glands during the secretory phase (S).
Following the death of the corpus luteum, the uterine lining degenerates and the uterus
reenters menses.

There may be two independent targets for estradiol in the uterus (see Chapter 2). One
target is the epithelial cell, which responds to estrogens with new RNA and protein synthesis. A second target may be the uterine eosinophil, which is a white blood cell that has
infiltrated the uterine lining. These eosinophils supposedly possess specific cytosol receptors for estradiol and appear to be responsible for the rapid uptake of water and release of
histamine that stimulates blood vessels and characterizes the uterine responses to estrogens.

Meiotic maturation of the oocyte apparently is influenced by the granulosa cells of
the **cumulus oophorus** that surrounds the oocyte in a growing or mature follicle (see
Figs. 11-9 and 11-10). However, the chemical mechanism for their influence is not known.

B. Ovulation

The levels of circulating estradiol increase progressively with the growth of the follicles
and the increase in the numbers of thecal and granulosa cells. A maximal or critical estrogen level in most cases activates the surge center in the hypothalamus, which releases a
large pulse of GnRH. The action of estrogens on GnRH release appears to be mediated via
norepinephrine-secreting neurons located in the brain stem with projections to the GnRH-
secreting neurons of the hypothalamus. Activation of these nonadrenergic neurons releases
norepinephrine and stimulates GnRH release. The pulse of GnRH released results in the
LH surge. (Fig. 11-13).

The LH surge causes ovulation of one or more follicles within a matter of hours (usually
12 to 24 hr regardless of the species). The number of follicles ovulating is species specific,
varying from a norm of one in women to a dozen or more in the sow. The physical mecha-

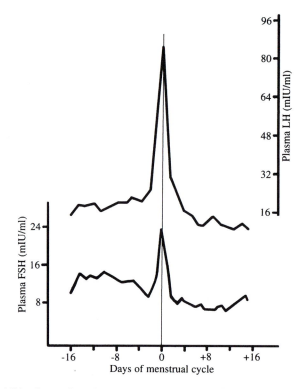

Figure 11-13. Midcycle gonadotropin surges in normal women. The greater release of LH is probably due to the presence of galanin released with GnRH at this time, as well as to negative feedback effects of inhibins on FSH release at the pituitary level.

nism by which LH causes the mature follicle to rupture and release the mature oocyte is not understood completely. It apparently involves changes in the follicular wall that can be attributed to the LH surge and the action of LH on the mature follicle.

Ovulation occurs as a consequence of a series of events (Fig. 11-14) that take place in the follicular wall at an avascular site called the **stigma.** At this location, locally produced estrogens induce the thecal cells to produce an enzyme, **collagenase,** that digests the intracellular matrix protein, **collagen.** This weakens the follicular wall. Enzymatic breakdown products bring about an inflammatory response and release of prostaglandins. Local blood vessels constrict in the presence of the prostaglandins, leading to local ischemia and cell death, weakening the follicular wall further. Pressure in the antral cavity causes the follicular wall to rupture at its weakest point, the stigma, and the ovum is expelled along with the antral fluid. It is speculated that the smooth muscle-like cells located in the follicular wall may be responsible for producing the increased antral pressure that triggers ovulation.

In addition to causing ovulation, the LH surge induces granulosa cells as well as some theca interna cells to differentiate into the **corpus luteum.** This process, known as **luteinization,** results in a corpus luteum that functions as an endocrine gland, secreting both estrogens and progesterone into the general circulation. One corpus luteum will form from each ovulated follicle. In addition, other developing follicles may undergo premature luteinization and function as **accessory corpora lutea** during pregnancy.

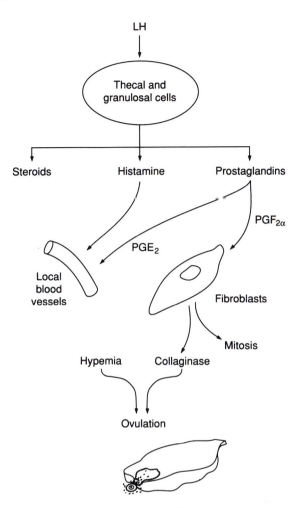

Figure 11-14. Endocrine regulation of ovulation. The LH surge causes secretion from ovarian cells that decreases local blood flow (hypemia) and destruction of collagen, which contribute to destruction of the follicular wall at the site of the stigma.

In some species, meiosis in the oocyte is not completed until after fertilization. Prior to fertilization, the ovulated cell is still technically an oocyte. If meiosis were completed prior to ovulation, this cell would be termed an **ovum.** The situation in mammals apparently varies from ovulation of oocytes to ova, but in the following discussions, the ovulated cell in every case is referred to as an ovum to simplify terminology.

Some mammals ovulate following coitus and are termed **induced ovulators.** Several carnivores (for example, ferret, mink, raccoon, and cat), rodents (for example, *Microtus californicus*), lagomorphs (for example, cottontail and domestic rabbits), at least one bat (lump-nosed bat), and several insectivores (for example, hedgehog and common shrew) are induced ovulators. Some other species are suspected to be induced ovulators including the elephant seal, the nutria, and the long-nosed rat kangaroo (a marsupial). Most mammals are believed to be **spontaneous ovulators,** in that ovulation is independent of coitus. However, even some spontaneous ovulators can be induced to ovulate following copulation under special conditions.

C. The Luteal Phase

Ovulation marks the onset of the **luteal phase** of the ovarian cycle. The corpus luteum begins secreting large quantities of progesterone, along with lesser amounts of estradiol-17β as well as other estrogens and progestogens. Progesterone maintains the proliferated uterine endometrium and stimulates the uterine glands to secrete a fluid called **uterine milk** or **embryotroph.** This interval is known as the **secretory phase** of the uterine cycle. Uterine milk is believed to be a source of nourishment for unimplanted blastocysts. The endometrium of both eutherian and marsupial mammals produce uterine milk. Progesterone also maintains the highly vascularized state of the uterus necessary for implantation and development of the embryo. Muscle layers of the uterus become desensitized by progesterone and are less responsive to stimuli that might produce rhythmic contractions and which thereby might dislodge an implanting blastocyst. Circulating progesterone and estrogens inhibit both the tonic and cyclic hypothalamic GnRH centers during the luteal phase so that additional follicular development is arrested and a second ovulatory episode is prevented. All developing follicles that do not ovulate undergo atresia. Some of the follicular cells will persist as **ovarian interstitial cells,** which are responsive to LH and synthesize androstenedione. These ovarian interstitial cells are considered to be important sources of androstenedione during the follicular phase that supplement the thecal supply of androstenedione necessary for the granulosa cells to secrete sufficient estradiol.

Depending on the species, regulation of corpus luteum function may require LH or be independent of LH once it has formed. In sheep, PRL together with LH apparently stimulate steroid secretion by the corpus luteum. However, only PRL is necessary to maintain the activity of the rat corpus luteum. Preovulatory estradiol-17β can produce a surge of PRL release in several species that might be related to corpus luteum function (Fig. 11-9). These actions of PRL on the corpus luteum explain the older name of *luteotropic hormone* for this molecule. However, PRL has no role in corpus luteum functions in primates and most other mammals, and the old name has been abandoned.

The corpus luteum secretes steroids for only a relatively short period in most species (5 to 8 days in women), after which it begins to degenerate. As the corpus luteum undergoes degeneration, steroidogenesis declines, and the uterus enters a regressive phase unless the animal is pregnant. In some species, the corpus luteum is relatively long-lived, especially in carnivorous species such as the dog, which will not reenter estrus until the next breeding season. Corpora lutea in the bitch are active for about 63 days after ovulation, which is equal to the normal gestation period.

The predetermined life span for the functional corpus luteum has provided one of the most intriguing mysteries of the ovarian cycle. Apparently, the corpus luteum sows the seeds of its own destruction. In female rats, mice, hamsters, rabbits, guinea pigs, and ewes, progesterone from the corpus luteum stimulates the synthesis and release of prostaglandins of the F series (PGFs; see Chapter 2) from the uterine endometrium. These PGFs, especially $PGF_{2\alpha}$, are **luteolytic factors,** that is, they cause destruction of the corpus luteum and cessation of steroidogenesis (Fig. 11-15). The mechanism of their luteolytic activity is not clear, although it may relate to an influence on the integrity of the blood vascular supply to the corpus luteum. In primates, the destruction of the corpus luteum toward the end of the luteal phase is not influenced by the uterus but appears to be caused locally by a luteo-

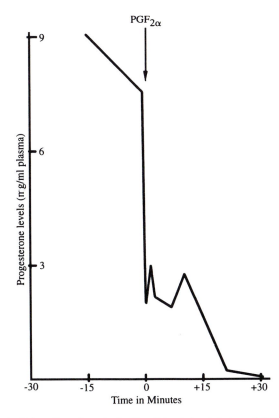

Figure 11-15. Luteolytic action of prostaglandins. Intrauterine infusion of $PGF_{2\alpha}$ (at time zero, arrow) causes an immediate drop in progesterone output from the corpus luteum of the cow as reflected in circulating plasma levels.

lytic factor (estrone, prostaglandins?), produced by the corpus luteum itself. Once fertilization has taken place, there are several mechanisms by which corpus luteum degeneration may be prevented and the life of the corpus luteum prolonged.

Should implantation not occur, the corpus luteum of many eutherian mammals will rapidly degenerate, resulting in a marked decrease in circulating levels of progesterone and estrogens. This decrease in circulating ovarian steroids brings about two major events.

First, the endometrium undergoes regressive changes following steroid withdrawal, becoming less secretory and less capable of supporting implantation of a blastocyst. In higher primates, spiral blood vessels constrict, preventing flow of blood to and from the endometrium and causing cell death. The degenerating tissue and trapped blood are sloughed into the uterine lumen and are discharged as the menstrual flow. Following the menses, considerable rebuilding of the endometrium must occur during the next follicular phase to prepare for implantation of blastocysts resulting from the next ovulation.

The second event is the freeing of the hypothalamic GnRH centers from the inhibitory influence of estrogens and progesterone, resulting in a moderate increase in circulating gonadotropins and consequently renewal of follicular development. In fact, increased FSH release occurs in many species during the later stages of the luteal phase so that follicular growth resumes even before regressive uterine events become obvious.

It should be emphasized that in wild, sexually mature mammals, fertilization and pregnancy are normal events, and coitus occurs frequently during estrus. The character of the ovarian cycle, with its rapid resumption in many species if fertilization and successful implantation do not occur, enhances the chances for successful reproduction. Either rapid reentry into estrus or rapid appearance of one or more new ova or both can occur and a second opportunity to produce offspring is made possible during that season.

The importance of the corpus luteum in maintaining pregnancy varies considerably, as does the role of pituitary hormones in stimulating corpus luteum function. For example, in rats, PRL is necessary for maintaining the first half of gestation through actions on the corpus luteum. However, in pigs, the corpora lutea secrete progesterone to maintain the uterine secretory phase during the early portion of the gestation period without the aid of any adenohypophysial hormones. In ewes, both LH and PRL are necessary for maintaining corpus luteum function during the first third of pregnancy, but maintenance of pregnancy actually resides in the ability of the conceptus to neutralize the uterine luteolytic factor $PGF_{2\alpha}$. Estrogens of placental origin apparently are responsible for prolonging the life span of corpora lutea in rabbits as well as for promoting progesterone synthesis. If the estrogen-secreting placental cells are damaged (for example, by X rays), pregnancy is abruptly terminated.

D. The Pregnancy Cycle

Estrus usually occurs just prior to ovulation and leads to mating. The recently ovulated ovum, still surrounded by some of the granulosa cells, the **corona radiata,** enters the upper end of the fluid-filled oviduct and is propelled toward the uterus by the action of cilia lining the oviduct. The possible role of muscular contractions of the oviduct wall in transport of the ovum has been suggested but not verified. Sperm deposited in the vagina by the copulating male during estrus are transported by peristalsis through the uterus and ascend into the oviduct in which recently ovulated ova are descending. Fertilization typically occurs in the upper third of the oviduct. Cleavage begins soon after fertilization, and the **zygote** or fertilized egg rapidly becomes a minute, multicellular **blastocyst.** The blastocyst consists of an **inner cell mass** that will become the embryo and a surrounding **trophoplast** that will give rise to the chorion. The trophoblast enables the blastocyst to erode the highly vascularized, secretory uterine endometrium and settle in for development. The gestation period may be as short as 12 days in the opossum (a marsupial) or as long as 22 months in elephants (eutherian).

In carnivores, such as the dog, the corpus luteum normally functions throughout gestation, which is equal to the length of the normal luteal phase. A central question that has puzzled reproductive physiologists for many years is related to the mechanism whereby other mammals "know" they are pregnant soon enough to prolong corpus luteal function and prevent premature regression or sloughing of the endometrium. The signal for prolongation of corpus luteal function in some species is the synthesis of LH-like chorionic gonadotropin (CG). Placental gonadotropins are structurally very similar to pituitary gonadotropins and generally produce LH-like effects (see Chapter 4). Their synthesis and release, however, are not influenced in a negative way by steroids in the manner of the steroidal feedback on pituitary gonadotropins. The trophoblast of the developing blastocyst secretes CG. Later, the trophoblast will become the fetal component of the placenta following implantation and will continue to secrete CG through most or all of pregnancy.

In the mare, only fertilized ova ever reach the uterus, implying some sort of early hormonal recognition that fertilization has occurred. This equine chorionic hormone is **pregnant mare serum gonadotropin (PMSG),** and it appears in large amounts circulating in the blood of pregnant horses.

Secretion of ovarian steroids brought about by extension of the life span of the corpus luteum or during the normal luteal phase of the ovarian cycle of carnivores inhibits hypothalamic centers controlling gonadotropin release so that follicular development and ovulation are blocked in pregnant animals. Having several embryos in the uterus at different stages of development might result in expulsion of the younger embryos and fetuses during parturition of the oldest one(s).

During the last third of pregnancy, another pituitary-like hormone, chorionic somatomammotropin (CS), is secreted by the placenta in a number of species (primates, mice, rats, voles, guinea pigs, sheep, chinchillas, and hamsters, but not bitches or rabbits). This placental hormone has both growth hormone (GH)-like and PRL-like activities. Antibodies to CS will cross-react with both GH and PRL in at least some of these species. The major roles for CS appear to involve effects on metabolism (GH-like) and stimulation of the mammary gland to begin milk synthesis during the later stages of pregnancy. This hormone is also known as placental lactogen.

The human placenta also secretes PRL that is identical to pituitary PRL. Placental PRL accumulates in the amniotic fluid during pregnancy, where it is thought to regulate volume and ionic composition of amniotic fluid. Levels of amniotic PRL are not affected by drugs that block maternal pituitary PRL release or even by hypophysectomy of the mother.

In humans, two additional adenohypophysial-like hormones have been identified in placental tissues: human chorionic thyrotropin (hCT) and chorionic corticotropin (hCC). Their functions are not clear. Perhaps they too serve to provide essential adenohypophysial hormones during pregnancy when maternal adenohypophysial function is inhibited. The placenta also synthesizes GnRH, which appears to increase placental CG production.

Relaxin, an insulin-like peptide unique to pregnancy, was discovered in 1932 to cause relaxation and softening of estrogen-primed pelvic ligaments, allowing the pelvis to stretch and expand (relax) during birth. This allows the relatively large head of the mammalian fetus to pass through the pelvis during parturition. Relaxin reaches peak levels prior to birth and rapidly disappears from the maternal circulation afterward. Spontaneous motility of the uterus may be inhibited by relaxin in some mammals. Relaxin, working with estrogens, progesterone, and prostaglandins, can alter the structural collagen of the uterine cervix, increasing its distensibility at parturition. There are also data supporting an action of relaxin in combination with steroids and PRL on the mammary gland and the onset of lactation.

The corpus luteum is the major source of relaxin in species where the corpus luteum is retained throughout gestation (pig, rat, and carnivores). Relaxin is produced by the human corpus luteum during early gestation and to some extent by the placenta. Only a little relaxin is found in the placentas of sheep, rats, cows, and rabbits, but in horses, the placenta is the major source of relaxin. In humans, the ovarian interstitial cells continue to be the major site for relaxin synthesis after death of the corpus luteum.

Relaxin is chemically similar to insulin and insulin-like growth factors (IGFs) and consists of two short A-chains (22–24 amino acids) and a longer B-chain (26–35 amino acids) joined together by disulfide bonds (see Chapter 2, Figure 2-11). The positioning of the disulfide bonds is the same as for insulin and the IGFs although there are many differences in amino acid sequences. Rat and pig relaxins exhibit numerous amino acid substitutions, although insulins from these species are similar. It has been suggested that the relaxin gene arose by duplication from the insulin gene although there has been considerable divergence in the relaxin genes among mammals.

1. Delayed Implantation

Several mammals, such as mink, bats, and skunks, have evolved a fascinating mechanism known as **delayed implantation,** whereby development of the blastocyst is arrested and the unimplanted blastocyst remains in the oviduct or uterus for an extended period prior to implantation. Among some eutherian mammals, delayed implantation appears to be an adaptation allowing copulation to occur at a particular time that is especially advantageous to the parent, while ensuring that the young are born at the most favorable time for their survival. Neither the basis for causing the blastocyst to remain in a healthy, arrested state nor the stimulus to bring about implantation are known. A similar phenomenon occurs in macropodid marsupials (Section VI,C), however, and its continuation is related to the presence of a young suckling on a teat. This is clearly not the mechanism involved in eutherians.

E. Lactation

The development of mammary glands, their synthesis of milk, and the ejection of milk to the suckling offspring are all regulated by hormones. Mammary glands in eutherian mammals usually occur as paired structures, from 2 to 18, and may be located on the thorax (human, elephant, and bat), along the entire ventral thorax and abdomen (sow and rabbit), in the inguinal region (horse and ruminants), along the abdomen (whale), or even dorsally (the nutria, a South American rodent). The internal structure is rather uniform and includes supporting stromal cells and a glandular epithelium that is organized into clusters of minute, saclike structures called **alveoli** (see Chapter 4, Figure 4-16). It is this glandular epithelium that is responsible for synthesis of milk. The alveoli are continuous with ducts and various duct-derived enlargements for storing milk. In addition, there are modified epithelial cells that contain muscle-like myofilaments parallel to the long axis of the cells. These cells are termed **myoepithelial cells** and are capable of contracting and causing ejection of milk from the alveoli into the duct system and out of the gland in the region of the nipple.

Information obtained from the mouse and rat indicates that differentiation of mammary glands from ectoderm involves specific induction by a particular underlying mesenchyme. These glands normally develop with the aid of estrogens during the last third of the gestational period. The fetal ovary is not the source of these estrogens since mammary glands develop in the absence of fetal ovaries. Androgens suppress mammary gland development and are presumably responsible for their altered development in the male fetus.

Postnatal mammary development involves hormones from the pituitary, ovaries, and adrenal cortex, at least in mice and rats. Growth of mammary ducts requires estrogens, GH, and corticosterone working in concert. However, expansion of the alveoli (lobuloalveolar growth) is dependent on the direct interactions of estrogens, progesterone, PRL, GH, relaxin, and corticosteroids.

Lactation can be separated into two basic processes or phases under separate endocrine control mechanisms. The first phase is milk secretion, or **lactogenesis.** This process is controlled primarily by pituitary PRL (or placental CS), growth factors, and corticosteroids. In primates, lactogenesis also is stimulated by GH. Lactogenesis involves synthesis

Table 11-6. Some Chemical Regulators Found in Milk[a]

Regulator type	Examples	Regulator type	Examples
Adenohypophysial hormones	PRL	Gastrointestinal peptides	VIP
	GH		CCK
	TSH		Gastrin
	FSH		GIP
	LH		
	ACTH		Substance P
			Neurotensin
Growth factors	IGF-I	Steroid hormones	Estradiol
	NGF		Progesterone
	EGF		Testosterone
	TGF-α		Corticosterone
	PDGF		Vitamin D
Neurohormones	TRH	Other regulators	Prostaglandins (PGE, PGF$_{2\alpha}$)
	GnRH		cAMP
	SS		Delta sleep-inducing peptide
	GHRH		Relaxin
	Oxytocin		Thyroid hormones (T$_3$, T$_4$, rT$_3$)
			Calcitonin
			Parathyroid hormone

[a]NGF, Nerve growth factor; PDGF, platelet-derived growth factor; TRH, thyrotropin-releasing hormone; SS, somatostatin; GHRH, growth hormone-releasing hormone; VIP, vasoactive intestinal peptide; CCK, cholecystokinin; GIP, glucose-dependent insulinotropic peptide; T$_3$, triiodothyronine; T$_4$, thyroxine; rT$_3$, reverse T$_3$. See text for other abbreviations.

of milk fat, milk protein, and milk sugar, typically lactose. The synthesis of lactose ultimately depends on protein synthesis; that is, the enzyme responsible for lactose synthesis, lactose synthetase, must be induced. Lactose synthetase is composed of two protein units, one of which is lactalbumin, which also is found in milk. Lactose, fat, and milk protein (largely casein) are secreted into the lumen of the alveolus. Water and numerous water-soluble substances enter the lumen by osmosis and result in a watery liquid known as milk. Many hormones are present in milk including hypothalamic peptides, pituitary hormones, growth factors, steroids, gastrointestinal peptides, and others (see Table 11-6).

The composition of milk produced by the mammary gland associated with suckling the young is very different at birth from what it will be shortly thereafter. This first milk is known as **colostrum** and is characterized by having a greater concentration of protein and less carbohydrate than does later milk. Colostrum contains antibodies and other substances that serve to protect the neonate against allergies and diseases while its own immune response system is developing.

The second phase of lactation is **milk ejection,** a simple reflex mechanism controlled by oxytocin from the pars nervosa. Mechanical stimulation of the nipple (suckling) evokes release of oxytocin from the pars nervosa via a spinohypothalamic neuronal pathway. Release of PRL also occurs when milk is ejected and stimulates further milk synthesis. Oxytocin stimulates contraction of myoepithelial cells that causes milk to be ejected from the alveoli into the ducts and storage channels of the mammary gland. The suckling young strips this milk from the gland by expressing it between the tongue and hard palate.

The milk ejection neurohormonal reflex exhibits classical conditioning responses as evidenced by stimulation of milk flow in the cow by sight and sounds of the milking parlor or in women by the cries of their hungry infant. This reflex can be influenced by other neural or chemical inputs to the hypothalamus. For example, stress or physical discomfort can inhibit ejection of milk in the presence of the stimulus that would normally elicit release of oxytocin.

F. Menopause

In nature, few animals live beyond their peak of reproductive activity owing to predation, disease, or other environmentally related phenomenon. In contrast, life after reproductive age is a common occurrence in human females. Whereas men may produce viable sperm most of their lives, the ovary becomes refractory to gonadotropins, usually during the mid- to late 40s. This transitional stage is called **menopause.** Cycles in these women become irregular and eventually they cease to ovulate and menstruate. This is accompanied by a marked depression in circulating levels of gonadal steroids as well as of adrenal androgens and by an elevation in gonadotropin levels. The transition from **premenopausal** (actively reproductive) to **postmenopausal** (nonreproductive) usually is gradual over several years and may be accompanied by additional symptoms including vaginal atrophy, hot flashes or flushes, reduced libido, and accelerated bone resorption leading to osteoporosis (see Chapter 15). Many studies have shown that heart disease and other cardiovascular disorders increase exponentially in postmenopausal women and deaths due to cardiac disease are severalfold greater than for uterine and breast cancer combined. In fact, heart disease is now the number one killer of women in the United States.

Estrogen replacement therapy, usually in combination with a progestogen, alleviates many of the symptoms of menopausal and postmenopausal women. When taken with calcium supplements and a regimen of weight-bearing exercise, steroid therapy also can prevent bone resorption. Large-scale studies currently are underway to verify or refute smaller studies reporting that estrogens reduce the risk of heart disease by as much as 50% in postmenopausal women. Apparently, estrogens or estrogens plus progestogens elevate high-density lipoproteins, which are associated with reduced cardiovascular risk. Other studies indicate that although estrogen-treated women have a higher incidence of breast cancer than women not being treated, the estrogen-treated women have a greater survival rate because the cancers are discovered earlier. The decision for a woman to elect estrogen therapy involves many complicating factors that must be weighed. For example, is osteoporosis, heart disease, or breast cancer a serious problem in her family? How severe are the symptoms and how do they affect her family life or her job?

VI. Reproductive Cycles in Selected Female Mammals

In this section, the reproductive cycles of the ewe (sheep), women, and the 4-day cycling rat are presented as being representative of eutherian mammals. This is followed by a generalized discussion of reproductive cycles for prototherians and macropodid marsupials, with an emphasis on a special version of delayed implantation in kangaroos.

A. Representative Eutherian Females

These cycles emphasize both the features described previously that are characteristic of eutherian mammals and some of the marked deviations that become obvious when different species are compared. The cycles of these three species are among the best known, and it may be argued whether they are truly representative of all mammals. Ewes and rats are polyestrous species, whereas women have no seasonal estrous behavior and exhibit a menstrual cycle. Both rats and women are continuous breeders, but ewes are distinctly seasonal breeders. Cows and pigs have cycles that are essentially like that of the ewe, although they differ somewhat in timing of the various events. None of these species exhibits delayed implantation, and all are believed to be spontaneous ovulators except under special conditions for the rat and probably the human.

1. The Ewe

Sheep estrous cycles occur seasonally, and the duration of one complete cycle is 16 days. The ewe may return to proestrus at least once if fertilization does not occur (Fig. 11-16). Reproductive cycles can be blocked in ewes by **genistein,** an antiestrogenic compound found in certain clovers. During the follicular phase (proestrus), there is a marked increase in both estrogen and androgen levels, a peak being reached about 24 hr after the onset of proestrus. About 12 hr later, a surge of plasma LH occurs, caused by the action of estradiol on the cyclic hypothalamic neurosecretory center. A high level of circulating androstenedione may be responsible for inducing estrous behavior. Usually, a single ovulation follows the LH surge by about 24 hr, and a corpus luteum forms from the ruptured follicle under the influence of LH. Low levels of LH following ovulation and the estrogen-induced surge of PRL stimulate the corpus luteum to secrete progesterone. Under the influence of progesterone, the uterine endometrium synthesizes a luteolytic PGF ($PGF_{2\alpha}$) that causes degeneration of the corpus luteum and resumption of proestrus.

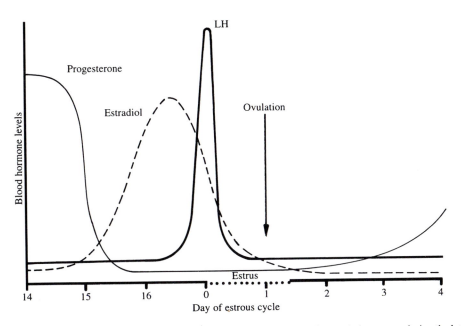

Figure 11-16. Hormonal events during the sheep estrous cycle. Note that ovulation occurs during the later part of estrus.

It appears that follicle growth, ovulation, and luteinization can be brought about by LH alone, although several studies have shown that FSH can stimulate follicular development, too. However, both postovulatory LH and PRL are necessary to induce progesterone synthesis by the sheep corpus luteum. Although FSH has been claimed to play no role in ovarian function for the ewe, several studies have reported a stimulatory effect of FSH on follicular development.

Fertilization followed by implantation delays degeneration of the corpus luteum. Although ovine CG (oCG) is produced, the conceptus apparently neutralizes the PGFs synthesized under the influence of progesterone so that the corpus luteum may secrete progesterone until the placenta is capable of producing sufficient steroids to maintain gestation. The placenta also secretes oCS.

As mentioned above, progesterone desensitizes uterine muscle to the stimulatory action of oxytocin. Just prior to birth, there is a marked reduction in circulating progesterone levels (Fig. 11-17), which presumably sensitizes the uterus to oxytocin. The contractions

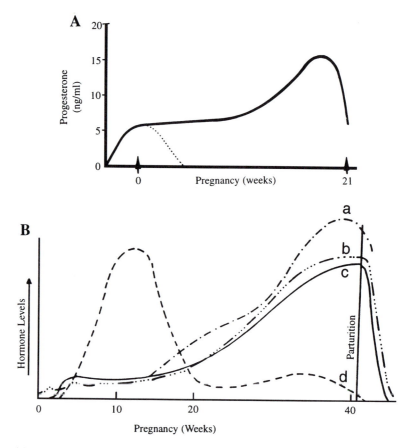

Figure 11-17. Patterns of hormone secretion during pregnancy. (A) Progesterone levels in ewes during the normal cycle (dotted line) and pregnancy. Note the drop in progesterone that occurs prior to birth. (B) Pregnancy levels of some hormones in women. Note that maternal progesterone levels do not decrease until detachment of the placenta. (a) hCS; (b) estrogens; (c) progesterone; (d) hCG. [Reprinted with permission from Bolander, F. F. (1989). "Molecular Endocrinology." Academic Press, San Diego.]

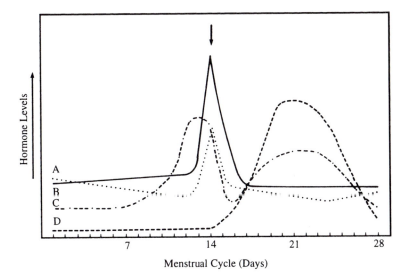

Figure 11-18. Hormone levels during the human menstrual cycle. (A) FSH; (B) LH; (C) estradiol; (D) progesterone. Arrow denotes time of ovulation. [Reprinted with permission from Bolander, F. F. (1989). "Molecular Endocrinology." Academic Press, San Diego.]

of the uterus initiated by oxytocin result in expulsion of the fetus as well as of the **afterbirth** (the detached placenta).

Experimental studies in sheep show that the fetal adrenal axis plays an essential role in the initiation of the birth process. Factors that interfere with adrenal function at any level retard the normal onset of labor, and premature birth can be induced by addition of ACTH or corticosteroids. Cortisol as well as adrenal androgens alter the ratio of estrogen to progesterone produced by the placenta, and this increase in the estrogen/progesterone ratio triggers parturition.

2. Women

The human female exhibits continuous cycling with a mean nonpregnancy cycle length of 28 days for most reproductively active women (Fig. 11-18). The rhesus monkey also has a menstrual cycle of 28 days, and there are many parallels in the menstrual cycles of these two primates. Studies of the rhesus monkey have provided insight into factors regulating the human menstrual cycle. Although this cycle is sometimes referred to as a lunar cycle because its periodicity is equivalent to approximately one lunar month, the human menstrual cycle is not correlated with any particular phase of the lunar month and should not be termed lunar. It could be that at one time the menstrual cycle was correlated more closely with moon phases but has become highly modified by numerous environmental and internal factors.

Cycles of women can vary from as short as 14 days to as long as 360 days, depending on both endocrine and psychological factors. Furthermore, considerable variation can occur in cycle length in a given woman at different times in her life history. In general, short cycles and irregular cycles are associated with the onset of puberty and with the end of the reproductive life prior to menopause, when the ovaries become refractory to pituitary gonadotropins and both estrogen synthesis and ovulation cease. The absence of circulating estrogens and progesterone releases the hypothalamus from negative feedback, and levels of circulating gonadotropins are exceptionally high in menopausal and postmenopausal women.

By convention, the menstrual cycle begins at the onset of the menses, which occupies the first 5 days of the cycle (Fig. 11-18). However, as soon as the corpus luteum of a previous cycle begins to regress prior to the onset of menses, there is a depression in plasma gonadal steroids and a resultant moderate elevation in plasma FSH that initiates growth of new follicles. Typically, only one follicle in one of the ovaries will reach maturity in a given cycle, with ovulation often occurring in the alternate ovary during the following cycle. The thecal cells of the growing follicles in both ovaries begin to secrete estrogens and progestogens, which peak on about day 14 of the normal menstrual cycle. Steroidogenesis is regulated by FSH and LH operating on the thecal and granulosal cells. The peak in estrogen level stimulates GnRH release from the cyclic center, causing a large surge of LH and a lesser surge of FSH. Ovulation follows the LH surge by about 24 hr, and the LH-stimulated granulosa cells and some thecal cells undergo luteinization. The resulting corpus luteum is independent of pituitary hormones and begins production of progesterone and estradiol. The hypothalamic centers governing LH and FSH release are inhibited by the high levels of these circulating steroids so that neither follicular growth nor ovulation can occur during the early portion of the luteal phase of the cycle. However, the corpus luteum functions for only a few days if the ovum is not fertilized. Unlike the case for sheep, the human uterus plays no active role in causing degeneration of the corpus luteum. Surgical removal of the uterus (hysterectomy) does not affect the duration of the luteal phase. Production of estrone by the corpus luteum of the rhesus monkey increases markedly prior to the onset of luteal degeneration. A similar mechanism employing estrone as a luteolytic factor may be operating in women. As the production of steroids by the corpus luteum decreases, the pituitary is released from steroid-induced inhibition, and a new cycle of follicular growth begins during the latter portion of the luteal phase. The outer portion of the uterine lining begins to slough off following the decline in progesterone levels, and the menses begins.

Should fertilization occur at the appropriate time during the menstrual cycle, the trophoblast of the blastocyst begins to secrete hCG prior to implantation, and the production of hCG continues at an accelerated rate during early pregnancy (Fig. 11-17). Under the influence of hCG, the corpus luteum continues to secrete both progesterone and relaxin for about 60 days, after which it degenerates even in the presence of exogenous hCG. However, the fetal adrenal–placental unit (see Chapter 9) produces sufficient estrogens and progesterone to continue the inhibition of pituitary gonadotropin release and to maintain the secretory condition of the uterus.

The stimulus for birth in humans is not related to a decrease in progesterone levels as shown for sheep (Fig. 11-17), but rather to a marked increase in estrogens relative to progesterone. Apparently, high levels of estrogens nullify the anesthetic properties of progesterone on uterine smooth muscle and allow oxytocin and/or prostaglandins to initiate uterine contractions and the onset of labor. This action of oxytocin seems to be a consequence of increased uterine receptors rather than a consequence of an immediate elevation in oxytocin levels. It is not clear if there is an adrenal component to initiation of birth as shown for sheep. However, premature birth often is correlated with enlarged fetal adrenals and delayed births are associated with fetuses that have underdeveloped adrenals.

Mechanical stimulation of the vagina, cervix, or uterus can release oxytocin in humans and induce a fetal ejection reflex. Administration of prostaglandins also can induce uterine contractions, and oxytocin may stimulate prostaglandin synthesis in the uterus. Synthetic oxytocin is normally used to induce labor in women and frequently is given to reduce postpartum bleeding following detachment of the placenta.

Pheromones may play important roles in human reproduction as inferred from studies of monkeys. Female rhesus monkeys produce a mixture of fatty acids of low molecular weight in vaginal secretions that stimulates sexual interest of males and may induce mounting behavior and ejaculation. The major volatile agents are fatty acids, including acetic acid, butanoic acid, propanoic acid, methylbutanoic acid, and methylpropanoic acid. Synthetic mixtures of these fatty acids in appropriate ratios stimulate male interest in females. Males are not interested in females receiving steroidal contraceptives, suggesting that the releasers of overt male sexual behavior are hormone dependent agents associated with the normal ovulatory cycle. Estrogens stimulate fatty acid secretions, and progesterone is inhibitory; observations that correlate well with levels of fatty acids observed in vaginal secretions throughout the menstrual cycle. Human vaginal discharges exhibit a similar cyclical variation in fatty acid composition (Fig. 11-19), although human females produce a much greater percentage of acetic acid than do rhesus monkeys. The use of oral contraceptives effectively obliterates the preovulatory increase in volatile fatty acids. However, the behavioral implications of these observations are not clear.

Although studies of pheromones in humans are complicated by a number of psychological and social considerations, some evidence exists for production of pheromones and their roles in reproduction. A "dormitory effect" of menstrual synchrony has been described by

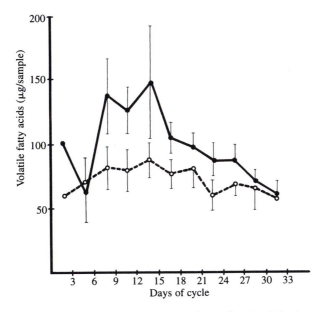

Figure 11-19. Lipid composition of human vaginal secretions collected at 3-day intervals during the menstrual cycle. Treatment with oral contraceptives (dashed line) reduced the normal midcycle rise observed in the volatile fatty acid content of vaginal secretions in 47 women (solid line). [Data originally reported by Michael, R. P., Bonsall, R. W., and Warner, P. (1974). Human vaginal secretion: Volatile fatty acid content. *Science* **186**, 1217–1219.]

McClintock for all-female living groups. Even though it is generally accepted that the ol-factory sense in humans is limited as compared to most mammals, several studies have shown definite sensitive olfactory discriminations, including sexually based differences, in the abilities to perceive certain odors. Trained perfumers can apparently distinguish olfac-torily between different skin and hair types, and some psychiatrists claim to be able to smell schizophrenics because of abnormal production and elimination of *trans*-3-methylhexanoic acid. The ability to detect some odors is sex dependent, such as the greater sensitivity of women to "boar taint" associated with spoiled pork. Some odors (e.g., licorice, lavender, doughnuts, and pumpkin pie) are claimed to induce sexual arousal. Clearly, controlled ex-perimental studies are needed to establish what, if any, roles pheromones have in human behavior and reproductive functions.

3. The 4-Day Cycling Rat

The presence of very short (4- or 5-day) cycles in laboratory rats (Fig. 11-20) is believed to be a consequence of the failure of rat corpora lutea to become functional during the normal luteal phase. The laboratory rat cycle is separable into **proestrus** (1 day), **estrus** (1 day), and **diestrus** (2 or 3 days) and is cued closely to environmental events.

On the morning of the day prior to estrus (that is, during proestrus), the levels of estro-gens in the plasma reach a peak, which stimulates an LH surge accompanied by a small surge in FSH. The gonadotropin surge occurs on the afternoon of proestrus and is followed rapidly by a marked surge of progesterone. Ovulation occurs a few hours after midnight on

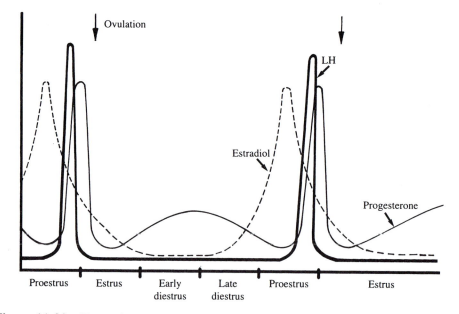

Figure 11-20. Hormonal events during the estrous cycle of the 4-day cycling female rat. The progesterone surge following the LH surge and prior to ovulation (arrow) is responsible for increased receptivity in the female for the male. There is a secondary, slow increase and decline in progesterone secretion from the corpus luteum that occurs during diestrus in the unmated female.

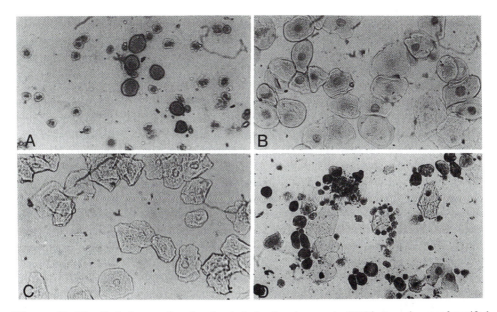

Figure 11-21. Vaginal smears from female rats during the estrous cycle. (A) Diestrus; absence of cornified (keratinized) cells and presence of small leukocytes. (B) Proestrus; many live epithelial cells with smooth margins. Leukocytes absent. (C) Estrus; large cornified cells with irregular margins. (D) Metestrus; leukocytes have infiltrated among the cornified cells. This stage is transitional between estrus and diestrus.

the day of estrus. Several follicles usually mature simultaneously, and multiple ovulations commonly occur. Estrus lasts about 9 to 15 hr, during which time the female is highly receptive to the male. Ovulation occurs during estrus. Cornified cells, which were produced by the actions of estrogens during proestrus, appear in the superficial layers of the vagina, and their presence in vaginal smears characterizes estrus (Fig. 11-21).

The third and fourth day of the cycle are termed **diestrus I** and **diestrus II.** Vaginal smears prepared during diestrus are characterized by the absence of cornified cells and a predominance of leukocytes in the smear. There is limited secretory function by the corpus luteum, as evidenced by a slight increase in plasma progesterone. Much of this progesterone is produced by ovarian interstitial cells, which constitute what has been called a permanent corpus luteum in the rat ovary. This interstitial tissue responds to LH by secreting both progesterone and 20α-dihydroprogesterone.

Many researchers recognize a short transitional period between estrus and diestrus in rats termed **metestrus.** The female in metestrus is no longer receptive to the male, but some cornified cells still appear in smears prepared from the vaginal mucosa.

Mating may stimulate gonadotropin release, and consequently the corpora lutea begin to secrete significant amounts of progestogens. This increased production of progestogens will inhibit hypothalamo–hypophysial function and delay the onset of the next cycle. If fertilization and implantation do not result from mating, the rat will not return immediately to proestrus but will delay resumption of proestrus for a few days. This period of copulation-induced **pseudopregnancy** often occurs in laboratory rodents and sometimes in other domestic mammals. The condition of pseudopregnancy is often accompanied by PRL-like effects on lactation and behavior presumably due to hypophysial release of PRL.

If implantation does occur the corpora lutea continue to secrete progestogens under the influence of placental CG and begin to secrete relaxin. The rat placenta also produces CS, which contributes to stimulation of mammary gland development and lactogenesis prior to birth.

Pheromones have been shown to play central roles in mating and successful pregnancy in rodents. Crowded female mice become anestrous when no males are present (called the Lee–Boot effect). However, simply the odor from a male mouse can cause them to enter estrus synchronously (Whitten effect). The endocrinological basis for these effects is suggested by observations that pheromones from female mice suppress pituitary release of FSH, whereas male pheromone stimulates FSH release that is followed in normal sequence by LH release and ovulation. A newly mated female mouse will abort if placed with a "strange" male (not the previous mate), and the incidence of spontaneous abortion increases with genetic dissimilarity of the strange male to the male with whom she was mated. If offspring result, they are always from the second mating (this is called the Bruce effect). This effect has also been observed in wild rodents and may not be peculiar to laboratory mice.

The sexual pheromones involved in the Lee–Boot and Bruce effects are probably modified steroids (steroid metabolites) and are transmitted via the urine of the male to the olfactory apparatus of the female. Male mouse urine induces and accelerates estrous cycles of females (Whitten effect), and the effect is most pronounced on Lee–Boot groups of females. The time of vaginal closing in females is also influenced by male urine. Anosmic females (animals whose nostrils have been blocked or whose olfactory bulbs have been removed surgically) do not respond to male urine.

Pregnant and lactating rats produce pheromones that influence other females. Odors from pregnant females shorten the estrous cycle of nonpregnant females so that more females will be pregnant at the same time. Presumably, it is advantageous to have many females give birth at the same time so that the females can cooperate in pup care. In contrast, odors from a lactating female with pups lengthens estrous cycles of nonpregnant females. Thus a socially dominant, lactating female may suppress the fertility of other females until she is again in estrus herself.

Males also may be influenced by female pheromones. Pairing of a previously paired male mouse with a strange female results in elevation of plasma testosterone in the male, indicating that endocrine responses of both males and females may be influenced through bisexual encounters.

B. Monotreme Reproductive Patterns

The egg-laying aplacental monotremes are reproductively more like reptiles than like other mammals, although they do possess some strictly mammalian features. For example, the developing monotreme follicle, like the reptilian follicle, does not form an antrum. Prior to ovulation, however, a thin layer of antral fluid is secreted by the follicular cells between the zona pellucida and the follicular cells. There is a well-developed theca interna that persists after ovulation. A corpus luteum is formed and the progesterone it secretes may stimulate production of nutrient substances by the uterine glands. These nutritive substances are incorporated by the growing oocyte while traveling through the oviduct and prior to application of leathery shells (like those found in reptilian eggs).

The duckbill platypus lays its eggs in a nest, but the female spiny echidna places her freshly laid eggs in a brood pouch that develops seasonally on her ventral surface. Formation of the pouch is probably dependent on estrogens, although no experimental data are available to confirm this. The eggs develop and hatch within the pouch, and after the breeding season the pouch regresses.

Corpora atretica may form from either large or small follicies in the monotreme ovary. The contents of a large follicle typically are extruded by rupture of the follicle prior to formation of the corpus atreticum. The extruded materials enter the peritoneal cavity or the prominent intraovarian lymph spaces that characterize the monotreme ovary.

The mammary glands of monotremes lack external teats. Consequently, milk is secreted onto a special common area, the **areola.** The newly hatched monotreme must attach itself to its mother with its forelimbs and suck or lick milk from the areola.

C. Marsupial Reproductive Patterns

Marsupials possess an estrous cycle with ovarian events similar to those described for eutherians, but the period of gestation is much shorter in relation to the estrous cycle than is seen in the eutherians (Table 11-2). As mentioned earlier, there is a short-lived placenta in marsupials.

Proestrus in females is characterized by follicular enlargement, estrogen-dependent uterine proliferation, and enlargement of elements of the vaginal complex. Peak uterine and vaginal development coincides with estrus and copulation. Ovulation occurs spontaneously one to several days after the onset of estrus, and postovulatory follicles transform into corpora lutea that maintain a short secretory uterine condition. Progesterone continues the secretory uterine phase in castrates and is undoubtedly the hormone responsible for maintaining gestation as it is in eutherian mammals.

The luteal phase is the same in mated and unmated females, and no "pregnancy-recognition signal" is necessary. Pregnancy does not affect ovarian function, and both pregnant and unmated females return to proestrus at about the same time in most species. Equivalent mammary gland development occurs during postestrus in both pregnant and nonpregnant females, and newborn foster young will develop normally if attached to virgin or nonlactating females at the equivalent postestrous state in relation to the time of parturition or cycle cessation. Circulating progesterone and urinary pregnanediol levels are similar in pregnant, unmated, and luteal-phase females. There are no endocrine differences between the pregnant and nonpregnant postestrous phases, which supports the conclusion that the marsupial placenta is not an endocrine organ.

Another unusual feature of marsupials is the ability to produce two kinds of milk simultaneously. As in eutherian mammals, the first milk differs markedly in composition from that produced later during lactation. Macropodids, which may have both a newborn joey and one that has already detached itself from a teat, will produce early and late milk simultaneously in the respective glands. The developmental state of a particular mammary gland would seem to be independent of endocrine conditions and strongly influenced by external conditions; that is, the joey.

A major developmental difference between eutherians and marsupials has had a profound influence on reproductive patterns in the latter group. Marsupials exhibit a primitive reptilian pattern of wolffian and müllerian duct origins. Instead of developing medially to the kidneys and ureters as in eutherians, these ducts develop laterally. Consequently, it is not possible for left and right müllerian ducts of females to fuse in the midline without placing considerable strain on the ureters. It is believed that the short gestation period of marsupials is a consequence of separate uteri and vaginas and limited space for uterine hypertrophy. A special birth canal must be formed so that parturition can occur, and in some species it forms anew each season. This development of separate vaginas has influenced evolution of the male reproductive system as well. Males of some species have a bifid penis, with left and right prongs apparently being inserted into the separate vaginas of the female during copulation.

1. Embryonic Diapause

Macropodid marsupials have developed a form of delayed implantation that has been termed **embryonic diapause** to distinguish it from delayed implantation as described earlier for eutherian mammals. Embryonic diapause has been reported for 14 macropodid species but does not occur in at least 1 species, the western gray kangaroo. One major difference from delayed implantation occurring in the eutherian mammals is the condition of the resting blastocyst. The macropodid blastocyst consists of about 70 to 100 cells of a uniform type termed the **protoderm.** It has not yet differentiated into embryonic and extraembryonic regions like that of the eutherians. The blastocyst is surrounded by a shell membrane and an albumen layer.

The presence of a joey suckling on a teat presumably evokes release of oxytocin from the pars nervosa. Oxytocin is believed to arrest corpus luteum functions while allowing lactation to occur. Removal of the suckling joey will allow the resting blastocyst to implant. Ovariectomy following ovulation induces diapause, but if ovariectomy is performed during diapause there is no effect on the duration of diapause. Progesterone administered to either intact or ovariectomized females stimulates cessation of diapause and reinstates blastocyst development. Estrogen is also effective, but continued embryonic development is not as successful as following progesterone treatment.

In the red kangaroo, embryonic diapause may be an adaptation to renew pregnancy immediately following the death of the joey living in the pouch. The gestation period for the red kangaroo is 33 days. After birth the newborn must find its way to the pouch virtually unaided. When it reaches the pouch, the joey attaches itself permanently to a teat and continues development as an exteriorized fetus. Soon after parturition, the mother kangaroo enters estrus again and mates. The presence of one joey in the pouch inhibits implantation of the new blastocyst resulting from the second mating. The new blastocyst remains in a suspended state of development for up to 200 days, at which time the first joey normally disengages itself from the teat and ventures into the outside world as a juvenile kangaroo. The newly liberated kangaroo will return at intervals to the teat to which it was formerly attached for nourishment. Meanwhile the detachment of the first joey from the teat either releases an inhibition to implantation or provides an endocrine stimulus for implantation of the waiting blastocyst.

In about 4 weeks, the gestation period terminates in birth of the second joey. The new joey enters the pouch and attaches to a teat. The mother kangaroo again enters estrus and mates, and another blastocyst enters embryonic diapause. Thus a female red kangaroo may have a young juvenile that requires occasional nourishment, a joey attached to a teat, and a blastocyst "waiting in the wings." During extensive periods of drought, the older joey could be denied milk and allowed to die. The implantation of the waiting blastocyst soon provides another joey whose demands on the mother's stored nutritional reserves and water supply would be very small in comparison to the demands of the larger joey. Next, the mother would reenter estrus, produce another diapausing blastocyst, and be ready to continue reproducing should conditions improve.

VII. Major Endocrine Disorders Related to Reproduction

Reproductive disorders have been studied extensively in order to prevent their occurrence as well as to correct them and increase reproductive capacities of both men and women. Some of the more common reproductive disorders are described here. Additional disorders related to reproduction include congenital adrenal hyperplasia (see Chapter 9) and osteoporosis (see Chapter 15).

The discussion of major endocrine disorders is separated into three major categories. The first two categories consider factors that influence the timing of puberty. The third deals with major genetically based disorders, many of which can result in ambiguous sexual determination.

A. Precocious Puberty

Puberty is a delayed period of development focusing on activation of the hypothalamo–hypophysial–gonadal axis and the functional integrity of sex accessory structures that may lead to successful sexual reproduction. **Precocity** is defined as the appearance of any one indicator of puberty at an age earlier than 2.5 to 3 standard deviations below the mean age at which the indicator normally appears in that population (see Table 11-7). The sequence and mean age for appearance of these indicators should be considered only as a guide. Implied precocity may not be evidence of an endocrine disorder, and variations in the sequence of these events are normal. Major deviations in a number of

Table 11-7. Mean Age for Normal Attainment of Certain Indicators of Puberty in Humans

	Mean age (in years ± SD)
Female	
Budding of breasts	11.2 ± 1.1
Sparse pubic hair	11.7 ± 1.2
Peak vertical growth rate	12.1 ± 1.0
Menarche[a]	
U.K.	13.5 ± 1.0
United States	12.9 ± 1.2
Male	
Enlargement of testes and scrotum	11.6 ± 1.1
Lengthening of penis	12.8 ± 1.0
Sparse pubic hair	13.4 ± 2.2
Peak vertical growth rate	14.1 ± 0.9
Adult genital size and shape	14.9 ± 1.1

[a]Ninety-five percent reach menarche between ages 11 and 15.

indicators may signal precocial endocrine activity of a pathological nature. **Isosexual precocity** involves early appearance of the genetically determined sex. It is termed **heterosexual precocity** if male features develop precocially in a female or if female features appear precocially in a male.

1. Precocity with Normal Endocrinology

Idiopathic (of unknown cause) precocity may be familial. Sexual development and body growth appear normal but are accelerated. Reproduction may be possible at an early age. For example, the youngest mother on record was 5 years 8 months of age at delivery!

2. Precocity and Pineal Tumors

Pineal tumors are not common but occur most frequently in young males. Precocious sexual maturation occurs in about one-third of these cases. Pineal tumors may impair release of antigonadotropic factors (melatonin?) from the pineal and allow sexual maturation to occur prematurely (see Chapter 6). In contrast, pineal tumors have been related to delayed puberty in a few cases. These tumors probably secrete more antigonadotropic factor than does a normal gland.

3. Precocity from Ectopic Gonadotropins or Gonadal Steroids

Rarely, pituitary tumors secrete excessive amounts of gonadotropins, and sometimes a nonpituitary tumor secretes chorionic gonadotropin causing premature gonadal maturation. Likewise, certain ovarian or testicular tumors can produce sufficient steroids to cause external evidence of puberty. Heterosexual precocity can occur from feminizing tumors in testes or androgen-secreting tumors of the female adrenals or ovaries. (See also the discussion of compensatory adrenal hyperplasia; Chapter 9).

B. Delayed Puberty

Numerous examples of delayed puberty are known. Since the causes and characteristics are unique for males and females, the following discussion is separated by gender considerations.

1. Causes of Delayed Puberty in Males

Prevalence of undescended testes or cryptorchidism is common at birth (10%) but is reduced to only 1% of males by 1 year of age. Only about 0.3% of adult males exhibit cryptorchidism, and the case of only one undescended testis is much more common than the bilateral condition. Testicular descent normally is initiated by local effects of DHT. Because of higher temperatures experienced by an undescended testis, the spermatogenetic tissue degenerates at about the time when spermatogenesis would normally begin (at about age 10). Androgen production usually is normal but may be reduced in some cases. The external testis of a unilateral cryptorchid develops normally, and these males are fertile.

There are several other causes for **hypogonadism** in males, including insufficient levels of LH and FSH due to hypothalamic or pituitary dysfunction. In some very rare cases, the interstitial cells may be unresponsive to gonadotropins.

2. Causes of Delayed Puberty in Females

Primary amenorrhea is the failure for menarche to occur at the normal time (Table 11-7). This condition can be related to many different causes including disorders of the hypothalamus, pituitary, and ovaries. Poor nutrition, stress, or rigorous athletic training programs can delay puberty through inhibitory actions on the hypothalamo–hypophysial–gonadal axis. For example, levels of LH and FSH are depressed in women suffering from **anorexia nervosa,** a disorder in which food intake is greatly reduced. Simple weight loss can depress FSH levels for a time but does not inhibit LH secretion.

Secondary amenorrhea occurs after menarche and can result from many endocrine disorders including thyrotoxicosis, drug therapy, and premature menopause as well as from a variety of hypothalamic, pituitary, and gonadal disorders. Two of the more common gonadal disorders associated with secondary amenorrhea are **polycystic ovarian syndrome (POS)** and **luteinization of atretic follicles (LAF).** The term *polycystic ovarian syndrome* is derived from the general thickening and simultaneous luteinization of several ovarian follicles, resulting in formation of numerous cysts in the ovaries. These cysts develop from thecal cells, and there is a corresponding decrease of granulosa cells in these follicles. Progesterone and estrogen production are diminished and gonadotropin secretion consequently is elevated. This resultant elevation of gonadotropins stimulates excessive production of androgenic steroids that cannot be aromatized to estrogens in the absence of granulosa cells. Uterine abnormalities and infertility result as well as obesity, hirsutism, and occasional balding from the androgens may accompany POS.

The LAF syndrome results from premature luteinization of ovarian follicles prior to formation of the cumulus oophorus. Gonadotropin levels are elevated, but masculinization usually does not happen. Numerous small ovarian cysts may be present, but these are easily distinguished from the large cysts that characterize POS.

C. Hereditary Disorders

Chromosomal rearrangements are common causes for deviations in sexual determination and/or expression. The most common disorders are the consequences of meiotic **nondisjunction** associated with the sex chromosomes; i.e., where the paired sex chromosomes fail to separate during the first meiotic division, resulting in gametes with either two sex chromosomes or none at all. Zygotes produced by the union of one of these gametes with a normal gamete would have only one sex chromosome or would have three sex chromosomes. In addition, numerous disorders have been linked to mutations in single genes.

Occasionally, people are born with a combination of ovarian and testicular tissue and are termed **hermaphrodites.** This term comes from Greek mythology, combining the names of Hermes, a god who sometimes played tricks on lovers, and Aphrodite, the goddess of love. Most hermaphrodites possess an ovotestis on one or both sides of the body. Rarely, **gynandromorphs** are discovered, individuals with an ovary and attendant müllerian derivatives on one side and a testis with its wolffian duct derivatives on the other. **Pseudohermaphrodites** have gonads of the genetic sex but externally resemble the other sex.

1. Klinefelter's Syndrome

Persons with **Klinefelter's syndrome** are born with an abnormal number of sex chromosomes: **47,XXY** (total number of chromosomes, sex chromosomes). This familial disorder occurs at fertilization and is present in about 0.2 to 0.3% of males. Klinefelter's syndrome can exist without obvious somatic abnormalities, although these persons are infertile and exhibit differing degrees of mental retardation. Similar syndromes have been described with additional sex chromosomes (48,XXYY, 48XXXY, etc.). Severity of the symptoms increases with the number of X chromosomes present.

2. Turner's Syndrome

Another disorder arising at fertilization is **Turner's syndrome,** in which there is a loss of one sex chromosome so that the resulting genotype is 45,XO (one X chromosome and no Y or no second X chromosome). Sometimes this condition occurs when a twin is found to exhibit Klinefelter's syndrome. Individuals with Turner's syndrome have a female phenotype but are infertile. They also exhibit a number of anatomical defects as well as cardiovascular and kidney disorders.

3. Galactorrhea

Secretion of a lactescent (milky) fluid from the breasts of either sex is called **galactorrhea.** It is usually caused by excessive secretion of PRL. Breast enlargement is not a prerequisite for its appearance. Galactorrhea frequently occurs in severe hypothyroidism characterized by elevated circulating levels of TSH and TRH. Prolactin release may be evoked by the high TRH levels.

4. Testicular Feminization Syndrome

A person with **testicular feminization syndrome** is a genetic male (46,XY). The testes are normal and secrete testosterone. Müllerian duct derivatives are absent because the embryonic testes also secreted MIS. However, the external appearance is that of a woman because of the congenital absence of androgen receptors in the tissues. This is an example of male pseudohermaphroditism.

5. 5α-Reductase Deficiency

A most unusual example of pseudohermaphroditism is the apparent shift of sex at puberty that occurs in a few people from several small villages of the Dominican Republic. This disorder is the result of a genetic deficiency in the ability to synthesize 5α-reductase. Males with this defect are born with undescended testes that synthesize testosterone like normal testes. However, these males cannot convert testosterone to DHT without 5α-reductase. Although testosterone also can bind to these receptors, DHT has a much greater affinity for them. Levels of testosterone in normal prepubertal males are too low to activate these receptors sufficiently and hence the need for 5α-reductase to convert testosterone to the more effective androgen, DHT. Apparently, in men with 5α-reductase deficiency, there is sufficient testosterone to stimulate other androgen-dependent structures, which develop normally (such as the vasa deferentia, epididymi) but the normally DHT-dependent structures such as the prostate gland, penis, and scrotum do not develop. These males understandably are raised as girls until these latter structures appear at puberty. The marked increase in circulating testosterone at puberty allows the penis and other structures to enlarge and causes facial hair to appear. Although these men usually change gender roles after puberty, they are infertile.

VIII. Summary

The reproductive system includes the hypothalamus, pituitary, gonads, and sex accessory structures. Primary control resides in pulsatile production of GnRH, which controls pituitary production of FSH and LH, which in turn causes gamete formation, gonadal steroid secretion, and ultimately the regulation of sexual characters and reproductive behaviors. FSH is involved primarily with gamete production whereas LH is responsible for initiating steroid secretion and release of mature gametes (ovulation and spermiation). FSH and testosterone work cooperatively to produce mature sperm. Androgens, progesterone, and estrogens stimulate other primary reproductive structures and secondary sex characters as well as mating behavior.

The male sex must be actively determined by the SRY gene, which is responsible for production of MIS by the testis. MIS causes regression of the potential female structures, and androgen production allows development of male sex accessory structures and reprograms hypothalamic reproductive centers to the male secretory pattern. The default sex in mammals is female.

Mammalian reproductive cycles show great variations among the major taxonomic groups. Monotremes are egg-laying but nourish their hatchlings with milk from their mammary glands. Marsupials allow their young to develop in a pouch after a short gestation period involving a nonendocrine placenta, relying on the mammary glands to support continued development of an exteriorized fetus. Eutherian mammals have a prolonged period of intrauterine fetal development supported by an endocrine placenta. Mammary glands are still used for extended nutritional support after birth.

Most female mammals exhibit an estrous cycle, called estrus, characterized by a period of enhanced receptivity of the female to the male. In certain species, including humans and the rhesus monkey, a special phase of the cycle occurs, the menses, that involves sloughing and discharge of the uterine lining and trapped blood. Because of this special feature, their cycles are typically called menstrual cycles and may (rhesus monkey) or may not (human) exhibit a distinct period of estrus. Species may exhibit one (monestrous), two (diestrous), or many reproductive cycles (polyestrous) during the year. In eutherians, the cycle can be separated into a follicular phase, during which ova develop in follicles, and a luteal phase, which prepares the uterus for implantation of the blastocyst. The luteal phase is named for the corpora lutea that develop from ruptured follicles and continue to secrete estrogens and progesterone. These phases are clearly separated by ovulation. During pregnancy, the eutherian placenta in cooperation with the fetal adrenal functions as an endocrine gland to maintain pregnancy, initiate birth, and prepare the mammary glands for postnatal functions. The placenta also functions in gas, nutrient, and metabolic waste exchanges. Numerous modifications of the placenta have evolved in different eutherian groups. Oxytocin, prostaglandins, and sometimes fetal adrenal steroids play important roles in the birth process. Postnatal functions of the mammary gland are controlled by PRL and oxytocin from the pituitary gland. Other modifications such as induced ovulation, delayed implantation (eutherians), and embryonic diapause (marsupials) are strategies that increase the effectiveness of reproduction.

Gonadal secretory activities involve two special cell types responsive to FSH and LH, respectively. Ovarian granulosa cells and testicular interstitial cells are responsive primarily to LH and synthesize androgens. Ovarian thecal cells and testicular Sertoli cells respond to FSH with conversion of androgens into estrogens (aromatase synthesis), synthesis of inhibin, activin, etc. Gonadal cells make many other local regulators (e.g., GnRH, IGFs, oxytocin, and prostaglandins) in response to gonadotropins or independently, and these regulators may have autocrine and/or paracrine actions that control local events in the gonads.

Suggested Reading

Books

Adashi, E. Y., and Leung, P. C. K. (1993). "The Ovary." Raven, New York.

Austin, C. R., and Short, R. V. (1982–1987). "Reproduction in Mammals." 5 Vols., 2nd Ed. Cambridge Univ. Press, Cambridge.

Burger, H., and DeKretser, D. (1989). "The Testis," 2nd Ed. Comprehensive Endocrinology Revised Series, Raven, New York.

Campbell, K. L., and Woods, J. W. (1994). "Human Reproductive Ecology: Interactions of Environment, Fertility, and Behavior." *Ann. N.Y. Acad. Sci.* **709**, 1–431.

de Kretser, D. (1993). "Molecular Biology of the Male Reproductive System." Academic Press, San Diego.

Ferin, M., Jewelewicz, R., and Warren, M. (1993). "The Menstrual Cycle: Physiology, Reproductive Disorders, and Infertility." Oxford Univ. Press, New York.

Findlay, J. K. (1994). "Molecular Biology of the Female Reproductive System." Academic Press, New York.

Gruhn, J. G., and Kazer, R. R. (1989). "Hormonal Regulation of the Menstrual Cycle." Plenum, New York.

Johnson, M., and Everitt, B. (1988). "Essential Reproduction," 3rd Ed. Blackwell, Oxford.

Jones, R. E. (1997). "Human Reproductive Biology," 2nd Ed. Academic Press, San Diego.

Knobil, E., and Neill, J. D. (1994). "The Physiology of Reproduction," 2nd Ed., Vols. 1 and 2. Raven, New York.

Krey, L. C., Gulyas, B. J., and McCracken, J. A. (1989). "Autocrine and Paracrine Mechanisms in Reproductive Endocrinology." Plenum, New York.

Tyndale-Biscoe, H., and Renfree, M. (1987). "Reproductive Physiology of Marsupials." Cambridge Univ. Press, Cambridge.

Wynn, R. M., and Jollie, W. P. (1989). "Biology of the Uterus," 2nd Ed. Plenum, New York.

Yokoyama, A. (1992). "Brain Control of the Reproductive System." Japan Scientific Societies Press, Tokyo, and CRC Press, Boca Raton, FL.

General Articles

Adashi, E. Y., Resnick, C. E., D'Ercole, A. J., Svoboda, M. E., and Van Wyk, J. J. (1985). Insulin-like growth factors as intraovarian regulators of granulosa cell growth and function. *Endocr. Rev.* **6**, 400–420.

Adashi, E. Y. (1990). The potential relevance of cytokines to ovarian physiology: The emerging role of resident ovarian cells of the white blood cell series. *Endocr. Rev.* **11**, 454–464.

Adashi, E. Y., and Rohan, R. M. (1992). Intraovarian regulation: Peptidergic signaling systems. *Trends Endocrinol. Metab.* **3**, 243–248.

Bilezikjian, L., and Vale, W. W. (1992). Local extragonadal roles of activin. *Trends Endocrinol. Metab.* **3**, 218–223.

Bourguignon, J.-P. (1995). The neuroendocrinology of puberty. *Growth Gen. Horm.* **11**, 1–6.

Bryant-Greenwood, G. D., and Schwabe, C. (1994). Human relaxins: Chemistry and biology. *Endocr. Rev.* **15**, 5–26.

Burrow, G. N. (1993). Thyroid function and hyperfunction during gestation. *Endocr. Rev.* **14**, 194–202.

Clarke, I. J. (1995). The preovulatory LH surge: A case of a neuroendocrine switch. *Trends Endocrinol. Metab.* **6**, 241–247.

Crowley, W. R., and Armstrong, W. E. (1992). Neurochemical regulation of oxytocin secretion in lactation. *Endocr. Rev.* **13**, 33–65.

Fraser, H. M., and Lunn, S. F. (1993). Does inhibin have an endocrine function during the menstrual cycle? *Trends Endocrinol. Metab.* **4,** 187–194.

Frisch, R. E. (1991). Body weight, fat, and ovulation. *Trends Endocrinol. Metab.* **2,** 191–197.

Giudice, L. C. (1992). Insulinlike growth factors and ovarian follicular development. *Endocr. Rev.* **13,** 641–649.

Giudice, L. C., and Saleh, W. (1995). Growth factors in reproduction. *Trends Endocrinol. Metab.* **6,** 60–69.

Griffiths, M. (1984). Mammal: Monotremes. *In* "Marshall's Physiology of Reproduction, Vol. 1, Reproductive Cycles of Vertebrates" (G. E. Lamming, ed.), pp. 351–385. Churchill Livingston, London.

Grosvenor, C. E., Picciano, M. F., and Baumrucker, C. R. (1993). Hormones and growth factors in milk. *Endocr. Rev.* **14,** 710–728.

Handwerger, S. (1991). The physiology of placental lactogen in human pregnancy. *Endocr. Rev.* **12,** 329–336.

Handwerger, S., Richards, R. G., and Markhoff, E. (1992). The physiology of decidual prolactin and other decidual protein hormones. *Trends Endocrinol. Metab.* **3,** 91–95.

Hawkins, J. R. (1993). The SRY gene. *Trends Endocrinol. Metab.* **4,** 328–332.

Hamum, E. (1991). Neuroendocrine peptides in milk. *Trends Endocrinol. Metab.* **2,** 2528.

Hiller, S. G., Whitelaw, P. F., and Smyth, C. D. (1994). Follicular oestrogen synthesis: The 'two cell; two gonadotropin' model revisited. *Mol. Cell. Endocrinol.* **100,** 51–54.

Hsueh, A. J. W., and LaPolt, P. S. (1992). Molecular basis of gonadotropin receptor regulation. *Trends Endocrinol. Metab.* **3,** 164–170.

Josso, N., Boussin, L., Knebelmann, B., Nihoul-Fekete, C., and Picard, J.-Y. (1991). Anti-mullerian hormone and intersex states. *Trends Endocrinol. Metab.* **2,** 227–233.

Kalra, S. (1993). Mandatory neuropeptide-steroid signalling for the preovulatory luteinizing hormone-releasing hormone discharge. *Endocr. Rev.* **14,** 507–538.

Kierszenbaum, A. L. (1994). Mammalian spermatogenesis *in vivo* and *in vitro:* A partnership of spermatogenic and somatic cell lineages. *Endocr. Rev.* **15,** 116–134.

Lee, M. M., and Donahoe, P. K. (1993). Müllerian inhibiting substance: A gonad hormone with multiple functions. *Endocr. Rev.* **14,** 152–164.

Leung, P. C. K., and Steele, G. L. (1992). Intracellular signalling in the gonads. *Endocr. Rev.* **13,** 476–498.

Licht, P., Frasnk, L. G., Pavgi, S., Yalcinkaya, T. M., Siiteri, P. K., and Glickman, S. E. (1992). Hormonal correlates of 'masculinization' in female spotted hyaenas (*Crocuta crocuta*). 2. Maternal and fetal steroids. *J. Reprod. Fert.* **95,** 1–12.

Matt, K. (1993). Neuroendocrine mechanisms of environmental integration. *Am. Zool.* **33,** 266–274.

McLachlan, R. I., Wreford, N. G., Robertson, D. M., and de Kretser, D. M. (1995). Hormonal control of spermatogenesis. *Trends Endocrinol. Metab.* **6,** 95–100.

Mellen, J. D. (1993). A comparative analysis of scent-marking, social and reproductive behavior in 20 species of small cats (*Felis*). *Am. Zool.* **33,** 151–166.

Moore, A., Krummen, L. A., and Mather, J. P. (1994). Inhibins, activins, their binding proteins and receptors: Interactions underlying paracrine activity in the testis. *Mol. Cell. Endocrinol.* **100,** 81–86.

Newman, J. (1995). How breast milk protects newborns. *Sci. Am.* **273**(6), 76–79.

O'Dell, W. D., Griffin, J., and Sawitzke, A. (1990). Chorionic gonadotropin secretion in normal nonpregnant humans. *Trends Endocrinol. Metab.* **1,** 418–421.

Ojeda, S. R., and Dissen, G. A. (1994). Developmental regulation of the ovary via growth factor tyrosine kinase receptors. *Trends Endocrinol. Metab.* **5,** 317–323.

Pescovitz, O. H., Srivastava, C. H., Breyer, P. R., and Monts, B. A. (1994). Paracrine control of spermatogenesis. *Trends Endocrinol. Metab.* **5,** 126–131.

Protet, M. (1993). Menstruation as a defense against pathogens transported by sperm. *Q. Rev. Biol.* **68,** 335–386.

Saez, J. M. (1994). Leydig cells: Endocrine, paracrine, and autocrine regulation. *Endocr. Rev.* **15,** 574–625.

Scheutz, A. W. (1985). Local control mechanisms during oogenesis and folliculogenesis. *In* "Developmental Biology, A Comprehensive Synthesis, Vol. I. Oogenesis" (L. W. Browder, ed.), pp. 3–83. Plenum, New York.

Schueler, L. A., and Kessler, M. A. (1992). Bovine placental prolactin-related hormones. *Trends Endocrinol. Metab.* **3,** 334–338.

Seron-Fere, M., Ducsay, C. A., and Valenzuela, G. J. (1993). Circadian rhythms during pregnancy. *Endocr. Rev.* **14,** 594–609.

Skinner, M. K. (1991). Cell–cell interactions in the testis. *Endocr. Rev.* **12,** 45–77.

Soares, M. J., Faria, T. N., Roby, K. F., and Deb, S. (1991). Pregnancy and the prolactin family of hormones: Coordination of anterior pituitary, uterine, and placental expression. *Endocr. Rev.* **12,** 402–422.

Tyndale-Briscoe, C. H. (1984). Mammals: Marsupials. *In* "Marshall's Physiology of Reproduction, Vol. 1, Reproductive Cycles of Vertebrates" (G. E. Lamming, ed.), pp. 386–454. Churchill Livingston, Edinburgh.

Walker, W. H., Fitzpatrick, S. L., Barrera-Saldana, H. A., Resendez-Perez, D., and Saunders, G. F. (1991). The human placental lactogen genes: Structure, function, evolution, and transcriptional regulation. *Endocr. Rev.* **12,** 316–328.

Clinical Articles

Bidart, J.-M., and Bellet, M. (1993). Human chorionic gonadotropin: Molecular forms, detection, and clinical implications. *Trends Endocrinol. Metab.* **4,** 285–291.

Imperato-McGinley, J., and Canovatchel, W. J. (1992). Complete androgen insensitivity: Pathophysiology, diagnosis, and management. *Trend Endocrinol. Metab.* **3,** 75–81.

Lufkin, E. G., and Ory, S. J. (1995). Postmenopausal estrogen therapy, 1995. *Trends Endocrinol. Metab.* **6,** 50–54.

Sunderland, M. C., and McGuire, W. L. (1991). Hormones and breast cancer. *Trends Endocrinol. Metab.* **2,** 72–76.

12

Comparative Aspects of Vertebrate Reproduction

AN UNDERSTANDING of reproductive patterns and their hormonal control in animals is central to our concerns about environmental quality and the future of aquatic and terrestrial ecosystems that are affected adversely by human activities. Levels of environmental contamination previously considered "safe" because they were not immediately toxic are now being seen to influence reproductive efforts through more subtle mechanisms than the dramatic thinning of bird eggshells by the pesticide DDT described some decades ago. Furthermore, documented declines in reproductive potentials in human males and dramatic increases in testicular cancer provide even more incentive to examine reproductive mechanisms. It is imperative that biologists learn more about the endocrine-regulated reproductive mechanisms that are most prone to disturbance, how these disturbances occur, and what remedies might be applied. In this respect, we need more information about the roles of natural environmental influences in reproduction as well as the influences of environmental contaminants.

Because of the great diversity among vertebrates and the important role of natural selection in reproductive phenomena, it is even more difficult to generalize about non-mammalian vertebrate reproductive patterns than it was for mammals in Chapter 11. The descriptions provided here are based on the discussions of mammalian reproduction. In the following accounts, many of the terms employed were introduced in Chapter 11.

I. Some General Features of Vertebrate Reproduction

Attainment of sexual maturity occurs at a time characteristic for each species and is followed by a series of reproductive cycles closely attuned to certain environmental factors. Bony fishes illustrate the full range of reproductive strategies known for vertebrates. Depending on the species, sexual maturity may be achieved during the first year of life (many teleosts), after more than 15 years of juvenile existence (Atlantic eel, sturgeon), or at some intermediate period. Some animals are **semelparous,** breeding only once after attaining sexual maturity and dying soon afterward (for example, Pacific salmon, *Oncorhynchus* spp); most species, however, are **iteroparous** and exhibit two or more reproductive cycles. Some of these may produce successive broods in a given year or season or may exhibit only one or two cycles per year. A few species may breed as one sex and then change to the opposite sex and breed again. Males may exhibit an **associated reproductive pattern,** in which gonadal steroids are highest during mating, or a **dissociated reproductive pattern,** where mating occurs when androgens are reduced (Fig. 12-1).

Natural environmental factors, such as temperature and photoperiod and the presence of suitable breeding or nesting sites, influence the central nervous system and the hypothalamo–hypophysial axis and regulate gonadal maturation and secretion of sex hormones. Steroid hormones, adenohypophysial hormones, or both determine development of various sex-dependent characters and influence courtship, breeding, and parental behaviors.

Like mammals, fishes, amphibians, and reptiles may be either **viviparous** or **oviparous,** with the exception of the cyclostomes and birds, which are exclusively oviparous. The term **ovoviviparity** has been applied in varying ways and will be avoided as suggested by Blackburn (1994; see Table 12-1). For simplicity, use of the term "viviparous" here will indicate live-bearing species regardless of whether there is a placental relationship or not. Oviparous species all lay eggs with protective coverings, from which a larval or juvenile form later will hatch regardless of the state of development at oviposition.

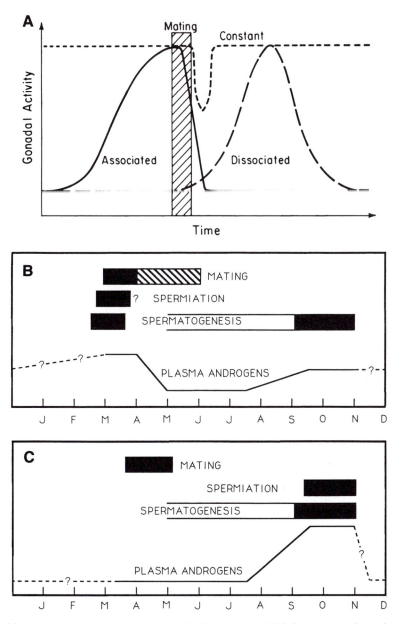

Figure 12-1. Associated and dissociated reproductive patterns. (A) Mating occurs at the peak of gonadal activity in species exhibiting the associated pattern, whereas mating occurs when gonadal activity is low in the dissociated pattern. (B) An associated reproductive pattern as seen in the newt, *Taricha granulosa*, in autumn. During spermatogenesis, the black bar represents the presence of sperm. Cross-hatched region is period of prolonged mating. (C) A dissociated pattern as seen in neotenic tiger salamanders, *Ambystoma tigrinum*. [(A) is modified from Whittier, J. M., and Crews, D. (1987). Seasonal reproduction: Patterns and control. *In* "Hormones and Reproduction in Fishes, Amphibians, and Reptiles" (D. O. Norris and R. E. Jones, eds.), pp. 385–410. Plenum, New York; (B) and (C) are from Houck, L. D., and Woodley, S. K. (1995). Field studies of steroid hormones and male reproductive behaviour in amphibians. From "Amphibian Biology," Vol. 2, Ed. H. Heatwole and B. K. Sullivan, Surrey Beaty & Sons, Chipping Norton, Australia.]

Table 12-1. Patterns of Reproduction That Have Been
Described as Ovoviviparity[a]

Pattern	Description
1	Internal fertilization; partial development of eggs within female reproductive tract; eggs at oviposition contain visible embryos
2	Restricted to anamniotes; site of metamorphosis is central to recognition; young at birth are premetamorphic larvae (amphibians)
3	Nutrients all supplied by yolk and not by placenta, oviductal secretions, or sibling embryos (including yolk)
4	A trace of egg shell appears whereas in viviparous species, no egg shell is shown
5	Includes anurans that brood eggs in vocal sacs, stomachs, dorsal skin pouches, etc.

[a] Based on Blackburn, D. G. (1994). Review: Discrepant usage of the term 'ovoviviparity' in the herpetological literature. *Herpetol. J.* **4**, 65–72.

Internal fertilization, a prerequisite for viviparity, requires evolution of a technique for transferring sperm from the male to the female. Some viviparous anurans (for example, *Nectophrynoides*) and birds transfer sperm through cloacal apposition or what has been termed the cloacal kiss. Aquatic fishes and urodele amphibians, which practice internal fertilization, rely on **spermatophores** for transfer of sperm. The spermatophore consists of a bundle of sperm that are aggregated and enclosed in a gelatinous substance that will not rapidly dissolve in water. This structure allows the male to transfer sperm directly or indirectly, without excessive dilution of the "semen." Selachians, viviparous teleosts, apodan amphibians, one anuran, and many reptiles possess **intromittent organs** that allow direct transfer of sperm from male to female. In contrast, an aquatic male urodele typically deposits spermatophores on the substrate and through a complicated behavioral ritual induces the female to pick one up with her cloaca. Frequently, the female receiving a spermatophore has a special storage site, the **spermatotheca,** that is capable in some species of storing viable sperm for months. The spermatotheca has special mechanisms to disperse and nourish sperm so that they can perform their destined functions at a later time.

A. Gonadal Features

Ovarian structures and events occurring in the gonads of nonmammalian vertebrates are similar to those described for mammals. Oocyte development is regulated by pituitary gonadotropins. The process of yolk protein formation is called **vitellogenesis.** Synthesis of these lipoprotein yolk precursors or **vitellogenins** by the liver is stimulated by estrogens. When released into the blood, these vitellogenins bind calcium ions and result in an elevation of total blood calcium in females undergoing vitellogenesis. Thus, marked increases in blood calcium can be used as an indicator of vitellogenesis and reflect circulating estrogen levels. Furthermore, vitellogenins are phosphoproteins, and an increase in plasma phosphoproteins can be monitored to provide information on reproductive status. Incorporation of vitellogenins by growing oocytes and their conversion to yolk proteins are controlled by gonadotropins.

There is a major difference in the structure of testes in most anamniotes and amniote vertebrates. Whereas testes of mammals, birds, reptiles, and anurans exhibit a tubular pattern of seminiferous elements with interspersed clumps of interstitial cells, the testes of urodele amphibians and most fishes consist of lobes or **lobules,** each of which is composed of large cellular **cysts.** Each cyst is derived from a **spermatogonial nest,** and all of the cells within a cyst and usually all the cysts within a lobule will be in the same stage of spermato-genesis. (Seminiferous tubules of anurans also exhibit a pattern of cystic spermatogenesis even though the lobular organization is absent.) **Sertoli (sustentacular** or **nurse) cells** are present in each cyst. The more posterior lobules may be in a more advanced stage of spermatogenesis in repeating breeders than are the more anterior lobules. Spermiation in these anamniotes is usually followed by complete evacuation of sperm from the mature lobules (more posterior ones) and regression of remaining cellular elements. Differentiation of lobules containing new cell nests occurs anteriorly from connective tissue elements and residual germ cells in the covering of the testis, the **tunica albuginea.** In some fish species, however, all lobules develop and discharge sperm more or less simultaneously, and if breeding recurs there must be extensive regeneration of spermatogonial nests and new cysts prior to the next breeding season. In urodeles, different lobules mature each breeding season and spent lobules do not regenerate. True interstitial tissue is lacking in most anamniotes, and the synthesis of androgenic hormones occurs in cells associated with the lobule walls. These steroidogenic cells are termed **lobule boundary cells.**

An additional steroidogenic tissue, the **interstitial gland,** may develop in the ovaries of gnathostomes. Interstitial glands develop from thecal cells derived from atretic previtello-genic follicles. It has been suggested that much of the estrogen synthesized during reproductive cycles is from the interstitial gland.

B. Endocrine Features

The endocrine factors in nonmammalian vertebrates are similar to and in many cases identical to those already described for mammals. The reader is reminded, however, that relatively few vertebrates have been examined with respect to endocrine factors and their involvement in reproduction. For example, about 50 of the 4600 mammals and about 15 of the more than 20,000 teleosts have been studied thoroughly. So, there is much to be learned before we can be certain that the patterns described here apply to all vertebrates or are even typical for a given group of vertebrates.

It is clear that the control of reproduction resides in the hypothalamus, which controls pituitary and ultimately gonadal functions. One or more gonadotropin-releasing hormones (GnRHs) have been identified in all vertebrate groups (see Chapter 5, Table 5-11) and GnRHs are responsible for release of gonadotropins.

There appear to be two distinct gonadotropins, which resemble follicle-stimulating hormone (FSH) and luteinizing hormone (LH) in their actions (except for possibly some teleosts and the squamate reptiles), and their release generally is under stimulatory hypothalamic control. Follicular development in females and spermatogonial mitoses in males are stimulated by FSH, with meiotic events in males being influenced locally by androgens produced by the Sertoli cells. Spermiation and ovulation are generally controlled by LH-like gonadotropins. However, in some teleostean fishes only an LH-like gonadotropin has been purified, and mammalian LH can regulate all aspects of reproduction in these fishes. Salmonid fishes exhibit two distinct gonadotropins (**GTH-I** and **GTH-II**), which are FSH-like and LH-like in their actions, respectively. GTH-I causes follicular steroid synthesis and maturation whereas GTH-II is responsible for release of mature gametes. Secretion of GTH-II occurs only late in the ovarian cycle of these fishes.

In contrast to teleosts and the other tetrapod vertebrates, reproduction in squamate reptiles requires only an FSH-like gonadotropin. Mammalian LH is ineffective in these reptiles (see Chapter 5).

The major circulating estrogen in nonmammals is estradiol, and testosterone or a closely related androgen is characteristic for males. Androgen and estrogen levels are higher than in mammals, possibly due to high circulating levels of steroid-binding globulin. Furthermore, relative levels of androgens and estrogens are not correlated with sex. For example, females may exhibit levels of androgens at certain times that exceed estrogen levels. The actions of gonadal steroids in nonmammals, including negative feedback effects on the hypothalamo–hypophysial axis, are similar to those described for mammals. Gonaduct differentiation and function, differentiation and maintenance of sex accessory structures, and induction of certain behaviors are regulated by gonadal steroids.

As in mammals, prolactin (PRL) exhibits in certain species some specialized functions that are closely linked to reproductive events. The specific involvements of PRL are discussed in some of the accounts that follow.

The relationships between thyroid hormones and reproductive events are discussed in Chapter 8, and actions of corticosteroids are described in Chapter 10. Briefly, thyroid hormones appear to enhance the onset of gametogenesis, especially in males. It is only in amphibians and certain avian species that a negative correlation has been reported between thyroid activity and the onset of sexual maturation. Stressful stimuli activate the hypothalamo–hypophysial–adrenocorticoid axis, causing an increased release of corticosteroids that is correlated with a reduction in or complete cessation of reproductive functions.

C. Sex Determination

There are several mechanisms of sex determination in vertebrates (Fig. 12-2). In mammals, birds, amphibians, and some fishes and reptiles, sex is associated with distinct differences

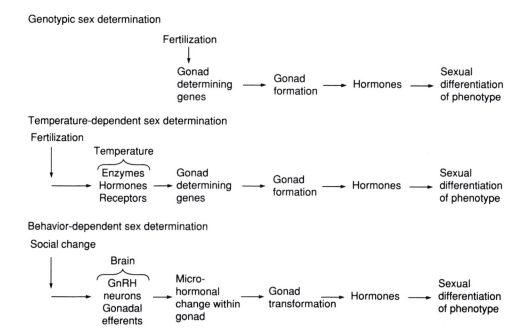

Figure 12-2. Mechanisms of sex determination. [From Crews, D. (1993). The organizational concept and vertebrates without sex chromosomes. *Brain Behav. Evol.* **42**, 202–214.

in one pair of chromosomes known as **sex chromosomes.** The remaining pairs are referred to as autosomes. In the **homogametic sex,** both sex chromosomes are identical (hence, all gametes produced by members of that sex are the same with respect to chromosome morphology; i.e., they are homogametic). The opposite sex has unlike sex chromosomes that separate during meiosis, producing two kinds of gametes with respect to the sex chromosomes; hence, the **heterogametic sex.** When the female is homogametic, as in mammals and most anuran amphibians, the common sex chromosome type is termed an X chromosome and her genotype is XX. The male is heterogametic with one X chromosome and one Y chromosome (his genotype is XY). When the male is the homogametic sex, as in birds and most urodele amphibians, he is designated as ZZ and the heterogametic female is ZW. Typically, the homogametic sex does not require gonadal steroids for early differentiation (i.e., it is the default sex), but gonadal steroids must be present for the heterogametic sex to overcome the default sex. The SRY gene on the Y chromosome of mammals and SRY-like genes in some other vertebrates are responsible for initiating heterogametic sexual differentiation. When sex is determined by this chromosomally based mechanism, it is called **genotypic sex determination (GSD).**

Many vertebrates do not have distinct sex chromosomes and may exhibit environmentally based mechanisms of sex determination. Although the lack of a morphologically distinct pair of sex chromosomes does not mean there is no genetic basis to sex determination, at least two mechanisms are well known that depend on environmental temperature or behavior of conspecifics. **Temperature-dependent sex determination (TSD)** occurs in all crocodilians, many turtles, and some lizards. It also occurs in a teleost, and there are correlative data that suggest it occurs in some amphibians. TSD is correlated with early incubation temperatures (Table 12-2). Incubation at a high temperature produces all one sex whereas at a lower temperature all the offspring are the other sex. These temperature effects are "all-or-none," and intermediate temperatures produce equal ratios of males and females (50:50) rather than intersexes. Hormones do not play an initiating role in TSD, although the levels of certain enzymes (aromatase and 3β-hydroxy-Δ^5-steroid dehydrogenase, 3β-HSD) are greatest in adrenocortical tissue and mesonephric kidney during sexual differentiation.

Estrogen treatment can reverse the effects of a male-producing temperature and aug-

Table 12-2. Temperature-Dependent Sex Determination in Vertebrates[a]

Group	Male-producing temperatures (°C)	Female-producing temperatures (°C)
Crocodilians	>34	<30
Turtles	23–27	30–33
Lizards	29–33	24–29
Teleosts	17–25	11–19

[a] Crocodilians, most turtles, and some lizards show TSD but snakes apparently do not. TSD also has been confirmed in one teleost (the Atlantic silverside, *Menidia menidia*) and may be a more common occurrence than is now known. Note that only in turtles is the low temperature male producing. Categories indicate an overwhelming predominance of one sex, if not 100%. Intermediate temperatures produce males and females in more or less equal numbers but no intersexes are reported.

ment the actions of a female-producing temperature, but androgens usually cannot override effects of a female-producing temperature. In turtle eggs incubated at a neutral temperature (one that should produce a 50:50 ratio), the sex ratio is sensitive to either androgens or estrogens. However, addition of both androgens and estrogens simultaneously at the neutral temperature will produce intersexes.

Social situations can initiate changes in gonadal sex. Simultaneous hermaphrodites (testes and ovaries in the same animal) and some parthenogenetic all-female species may alternate between male and female roles. (*Parthenogenesis* is the development of eggs without any genetic contributions from males.) However, in sequential hermaphrodites, sex change may be a result of **behavioral sex determination (BSD).**

Most cases of sex change (or sex reversal) occur in coral reef fishes. **Protogyny** (female-to-male sex change) is the more common and appears to be triggered by environmental cues. Some species are termed **diandric,** in that two types of males are found. One is a genetically determined or **primary-phase male,** and the **secondary** or **terminal-phase male** is derived through sex change of a female. In many cases, sex change is initiated by loss or removal of a dominant male in the social grouping and the largest ranking female takes his place. Often there are immediate behavioral changes in the dominant female that prevent sex reversal in subordinate females. Gonadal sex change usually follows the behavioral changes.

The bluehead wrasse, *Thalassoma bifasciatum,* is a diandric, sex-changing teleost living in the Caribbean. Treatment of females with human chorionic gonadotropin (hCG) induces gonadal sex change within 1 to 6 weeks. Spontaneous reversal is accompanied by a twofold increase in the number of GnRH-immunoreactive cells in the preoptic area of the brain. A similar increase in GnRH cells can be induced with implants of 11-ketotestosterone into females. How GnRH and gonadotropins produce a sex change in females previously employing these same hormones for female reproduction is unclear. Possibly the answer lies in changes in brain function that are initiated by environmental cues.

A similar pattern of sex change occurs in a closely related diandric species, the saddleback wrasse (*Thalassoma duperrey*), living among the coral reefs of Hawaii. A very different mechanism for triggering sex change operates in *T. duperrey,* a promiscuously mating fish that does not live in male-dominated groups. Long-term studies of these fish in underwater cages have determined that the cue for sex change is visual rather than chemical or tactile. The ratio of larger (usually males) to smaller fish (usually females) is perceived visually by the female, who undergoes sex reversal if sufficient males are not present. The physiological mechanisms underlying these changes are unknown at this time, but experimental studies suggest a role for biogenic amines that influence GnRH and gonadotropin release.

This early differentiation based on gonadal steroids or temperature resulting in a permanent designation of sex or of a sexual characteristic is known as an **organizational effect.** In the situation described for the coral reef fishes, sex and related structures and behaviors are not organized at an early age but show that sex determination and differentiation of the gonads is due to an **activational effect** that occurs later in life.

II. Reproduction in Agnathan Fishes: Cyclostomes

Lampreys (Petromyzontidae) are characterized by having no breeding cycle, and all individuals die after a single spawning. In contrast, the hagfishes (Myxinoidea) apparently breed more than once. However, little is known about the reproductive biology of the hagfishes, and the following account of agnathan reproduction by necessity must emphasize the lampreys.

It appears that the cyclostome gonad arises entirely from the embryonic cortex, whether it is destined to be a testis or an ovary. This singular embryonic origin may account for the common observations of what appear to be hermaphroditic gonads among the hagfishes. Males exhibit a single median testis with a cystic pattern of spermatogenesis. Because of fusion of the paired primordia early in development, the female lamprey has a single ovary. In contrast, the single ovary of myxinoids is due to the failure of one primordium to develop. Steroid-binding proteins are not present in cyclostome blood, and circulating levels of steroids are very low (Table 12-3). Gonaducts are absent in cyclostomes, and the gametes are shed into the coelom from which they exit via abdominal pores.

A. Male Lampreys

The mature lamprey testis exhibits the typical piscine pattern of lobules with germinal cysts. When the cysts have completed formation of sperm, they simultaneously rupture and release sperm into the body cavity. The testes of parasitic forms, such as *Lampetra fluviatilis,* contain only primary spermatocytes at the time of migration to the breeding grounds. These spermatocytes are transformed rapidly near the time of spawning into sperm masses. Typical interstitial cell masses can be identified cytologically between the lobules in testes of migrating lampreys. These cells accumulate cholesterol-positive lipids and have become densely lipoidal by spawning time. Cytologically, interstitial cells appear to be steroidogenic and exhibit maximal hydroxysteroid dehydrogenase (3β-HSD) activity

Table 12-3. Circulating Steroid Levels[a] in Selected Nonmammalian Vertebrates

Class and species	Testosterone	Estradiol (ng)	Progestogen (ng)
Agnatha			
Petromyzon marinus	UND[b]		
Eptatretus stouti	23.9 pg		
Chondrichthyes			
Torpedo marmorata (male)	15.6–35 ng		
Raja radiata			
Male	28–102 ng		
Female	0.2–6 ng		
Scyliorhinus canicula	2–6 ng		
Osteichthyes			
Oncorhynchus nerka			
Male	17 ng		
Female	78 ng		
Salmo trutta			
Male	2–33 ng		
Female	20–77 ng		
Oncorhynchus mykiss			
Prespawning female	52–235 ng	24–48	8–15
Spawning female	65–84 ng	2–3	354–416
Spent female	2–5 ng	1–2	8–19

[a] Per milliliter of plasma or serum.
[b] UND, Undetermined.
[c] 17β, 20α-dihydroxy-4-pregnen-3-one

in February and March, prior to the time of spawning. Development of lobule boundary cells during the later stages of spermatogenesis has been observed in *L. fluviatilis*. These cells are sensitive to hypophysectomy, and they may be homologous to Sertoli cells, although 3β-HSD activity has not been reported.

B. Female Lampreys

Oogenesis has been carefully examined in the parasitic sea lamprey *Petromyzon marinus* and in the river lamprey *L. fluviatilis*. In *P. marinus,* oogonia proliferate mitotically in the larvae to form the primary oocytes. By the time of metamorphosis of the larva to the juvenile, there are no oogonia remaining in the ovary. The primary ovarian follicles become more vascularized at this time. During the prolonged parasitic phase of body growth (about 10 to 20 months), the oocytes continue to enlarge slowly. The single follicular cell layer becomes thinner and less vascularized as spawning approaches, and the oocyte enters a period of rapid enlargement to reach the preovulatory condition. The mature follicles rupture immediately before spawning, and the eggs enter the coelom. Follicular atresia occurs throughout the history of ovarian development, and most oocytes undergo atresia, establishing this basic pattern early in the phylogeny of vertebrates. Phagocytes derived from the follicular cells ingest the yolk, and the follicle layers and surrounding stroma collapse into the area formerly occupied by the oocyte.

   Oocyte growth in parasitic *L. fluviatilis* accelerates markedly just prior to spawning. The granulosa contacts the oocyte only at the vegetal pole and reaches maximal development about 1 month prior to spawning. The thecal cells are greatly reduced and with the aid of an electron microscope can be seen covering the granulosa layer and the animal pole. The theca interna consists of a single layer of cells in which there is a marked increase in smooth endoplasmic reticulum and mitochondrial differentiation during vitellogenesis. These cells show maximal cytological activity prior to the time of most intensive vitellogenesis, following which they undergo progressive regression until the time of ovulation. The theca externa consists of fibroblasts, collagen fibers, and capillaries. 3β-HSD activity is apparently confined to the thecal cells, where peak activity is observed about 1 month prior to the appearance of secondary sex characters and the acceleration of follicular development.

   Vitellogenesis appears to be an estrogen-dependent event in lampreys involving cooperative action of the liver, which produces proteins that are secreted into the blood and are sequestered by the ovary to be incorporated in the developing oocyte. Estrogens stimulate liver hypertrophy and elevate plasma protein-bound calcium, suggesting the presence of a mechanism such as the one that has been documented so carefully in birds and other nonmammals (see Sections IV,B; V,D; VI,B; VII,B).

   In the free-living lampreys that do not feed after metamorphosis, the ovarian events occur over a much shorter time and are consequently more dramatic. In brook lampreys, the immediate postmetamorphic period is marked by the onset of both vitellogenesis and massive atresia. As many as 70% of the follicles present at metamorphosis may become atretic, and phagocytosis of the yolk may provide an essential nutritional source for growth of the remaining oocytes to maturity.

C. Endocrine Function in Lampreys

The importance of gonadal steroids to development of sex accessory structures has been demonstrated through classical experiments involving hypophysectomy, gonadectomy, and appropriate hormone therapy to either hypophysectomized or castrate animals. Lamprey GnRH or pituitary gonadotropins stimulate gonadal hormone secretions, which in turn stimulate formation of secondary sex characters. Spermiation also can be induced with lamprey GnRH. The gonads of hypophysectomized *L. fluviatilis* are less developed than those of sham-operated controls, although treatment of immature lampreys with mammalian gonadotropins has no effect on the gonads. These observations suggest a requirement for another factor or an endogenous maturation of the gonad to become responsive to gonadotropin.

D. Hagfishes

The reproductive biology of deep-water myxinoids is poorly known, and there is little literature available. The hagfishes apparently are not seasonal breeders and, unlike the lampreys, exhibit continuous reproduction. Only *Eptatretus burgeri,* which lives in the shallow coastal waters of Japan, shows a seasonal cycle of gonadal activity. There is a single gonad in adults similar to that described in lampreys. Although formation of both preovulatory "corpora lutea" (atretic follicles) and postovulatory "corpora lutea" has been described, the possible endocrine functions of such structures are unknown. Male hagfishes apparently lack true interstitial cells, and it is not clear whether steroidogenic lobule boundary cells are present.

Although almost no experimental work is available with respect to the function of the hypothalamo–hypophysial system in hagfishes, hypophysectomy of the Pacific hagfish *Eptatretus stouti* results in testicular degeneration in males, but hypophysectomy of mature females has no effect on either ovarian structure or circulating steroid hormone levels. Vitellogenesis is stimulated by treatment of *E. stouti* with estradiol.

III. Reproduction in Chondrichthyean Fishes

The selachians have been studied extensively, probably because of the incidence of viviparity in these species (present in 10 families of sharks and 4 families of rays). Unfortunately, no complete data are available on reproduction in ratfishes. Several different reproductive patterns have been described, extending from species that lay a precise number of eggs in a particular sequence, such as the oviparous clear-nosed skate *Raja eglanteria,* to the viviparous spotted dogfish, *Scyliorhinus canicula,* which is sexually active throughout the year and may have embryos in different stages of development *in utero* at the same time. Selachians are characterized by internal fertilization regardless of whether they are viviparous or oviparous. Some data on plasma levels of gonadal steroids are provided in Table 12-3.

A. Male Selachians

Spermatogenesis in paired testes (Fig. 12-3) is of the cystic type. Sertoli cells are present and possess 3β-HSD activity. These cells become densely lipoidal and cholesterol positive following spermiation and are eventually resorbed. Sertoli cells for the next cycle differentiate from connective tissue cells (fibroblasts) in the wall of the testis. Nests of spermatogonia proliferate from germ cells in the same regions, and they are responsible for producing sperm utilized during the next breeding period.

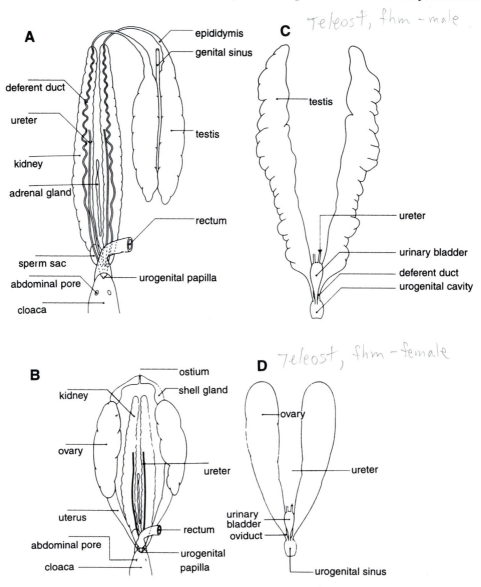

Figure 12-3. Reproductive systems in elasmobranch (A and B) and teleostean (C and D) fish. A dogfish
shark, *Mustelus manazo:* (A) Male with testes pushed to the side to reveal kidneys, reproductive ducts, and the interrenal; (B) female with ovaries obscuring the oviducts running between
the shell gland and uterus. A teleost, the carp, *Cyprinus carpio:* (C) Male; (D) female. [Modi-
fied from Matsumoto, A., and Ishii, S. (1987). "Atlas of Endocrine Organs: Vertebrates and
Invertebrates." Springer-Verlag, Berlin.]

Spermatophores produced by selachians are the result of secretory activities of male
accessory ducts. After spermiation, sperm pass through vasa efferentia and enter the coiled
tubules of the epididymis or **Leydig gland,** which is derived from the anterior portion of
the mesonephric kidney. Sperm and secretions of the Leydig gland pass on to an expanded
region of the vas deferens known as the sperm sac or **ampulla.** Here the sperm are con-

solidated and receive additional secretory material to form complex spermatophores typical for each species. Fertilization is internal, and spermatophores are transferred to the female by specialized structures termed **claspers.**

1. Endocrine Factors in Male Selachians

The importance of the ventral lobe of the selachian pituitary as the source of gonadotropin controlling spermatogonial proliferation (mitotis) has been demonstrated in the spotted dogfish. Degenerative changes in the testes appear 6 weeks after removal of only the ventral lobe, and 22 months later the testes contain only spermatogonia and mature sperm, indicating that removal of the ventral lobe blocks further differentiation of spermatogonia to spermatocytes, whereas all spermatocytes present at the time of surgery are able to complete meiosis and spermiogenesis. Removal of the rostral or neurointermediate lobes of the pituitary is without observable effect on the testes.

B. Female Selachians

The selachian ovary (Fig. 12-3) is covered by germinal epithelium and may contain a cavity derived from large lymph spaces within the stroma. Selachian follicles are similar to those of mammals in possessing several distinct layers of cells. The connective tissue near a nest of oogonia will differentiate into the theca. As each follicle begins to develop, some epithelial cells undergo hypertrophy and hyperplasia to become the granulosa. In some species, the granulosa may consist of only a single layer of cells. These cells are responsible for yolk deposition during oocyte growth as well as for yolk resorption should a given follicle become atretic. Granulosa cells also are thought to be the source of estrogens since they exhibit more 3β-HSD activity than do thecal cells. Most estrogen synthesis occurs in the mature follicle, which has a well-developed granulosa. During follicular development, a theca interna and theca externa can be discerned. However, both layers largely consist of connective tissue elements, and only a small amount of 3β-HSD activity has been observed in the theca interna cells.

The granulosa cells provide the source of both preovulatory (atretic) and postovulatory corpora lutea. The connective tissue layers surrounding these structures are derived from the theca. Corpora lutea of several species possess 3β-HSD activity, and corpora lutea persist during gestation in *Squalus acanthias,* the spiny dogfish. Furthermore, the corpora lutea from pregnant *S. acanthias* produce twice as much progesterone *in vitro* as do those from nonpregnant females, which possess only preovulatory corpora lutea formed from atretic follicles. These observations strongly support an endocrine role for postovulatory corpora lutea in the viviparous selachians.

Atresia is a common occurrence in selachian ovaries. Depending on the species being examined, either thecal or granulosa cells may contribute to formation of the preovulatory corpora lutea. Conflicting data have been published with respect to the question concerning whether these atretic follicles are steroidogenic or not, and final resolution awaits further study.

Selachian females have well-developed müllerian ducts that give rise to the oviducts as well as to the uterus of viviparous species. Oviducts have been examined in oviparous species that secrete horny shells to protect the eggs laid in the ocean as well as in viviparous species, and they possess a number of specialized features. **Oviductal** or **nidamental glands** secrete albumen and mucus in oviparious species. Villus-like structures may develop in the uterine portion of the oviducts of certain viviparous females, and they provide nourishment for their young. The oviductal glands of oviparous species are often differentiated into an anterior albumin-secreting area and a posterior shell-secreting region. An intermediate mucus-secreting zone may be found in some species. In one dogfish species, the shell-secreting portion of the oviduct serves also as a spermatotheca (a site for sperm storage in the female genital tract). The bonnethead shark is unusual in that it is viviparous, but it has a spermatheca, a feature typically found only in oviparous sharks.

1. Endocrine Factors in Female Selachians

Removal of the ventral lobe of the adenohypophysis blocks oviposition in female *S. canicula,* and all follicles containing oocytes larger than 4 mm in diameter undergo atresia. As in the male, removal of rostral or neurointermediate lobes has no effects on reproduction. Estradiol stimulates vitellogenin synthesis and oviduct growth. The possible role of progesterone from corpora lutea during gestation was suggested earlier. Additional studies with immature selachians, employing both gonadectomy and removal of the ventral lobe together with hormonal replacement therapy, are necessary to develop a complete understanding of endocrine factors responsible for reproduction in selachians.

The ovaries of viviparous sharks produce a relaxin molecule structurally more similar to mammalian insulins than to mammalian relaxin but with similar biological actions when tested for uterine-relaxing activity in the guinea pig. Mammalian relaxin causes dilation of the uterus of pregnant *S. acanthias* and premature loss of embryos. Relaxin appears to have evolved early in vertebrate evolution and performs a similar function in sharks as in mammals. This is the only group of nonmammalian vertebrates shown to produce a relaxin and to respond to mammalian relaxin.

IV. Reproduction in the Bony Fishes

Among the bony fishes, the teleosts exhibit almost every reproductive pattern and strategy known for vertebrates, including some that are unique to these fishes. Most of the information presented here is based on teleosts, but there are many similarities between teleosts and the other groups of bony fishes. Some features of these other bony fishes are illustrated, also. But first, some generalizations need to be made.

Like that of cyclostomes, the teleostean gonad develops only from a cortical primordium. Bony fishes may be dioecious or hermaphroditic. Among the hermaphrodites there are numerous examples of **protandry** (functioning first as males and later transforming to females), protogyny, and simultaneous hermaphroditism. Fertilization may be external or internal as in the viviparous teleosts and in the viviparous coelacanth, *Latimeria.* In some viviparous teleosts, the fertilized egg is known to develop within the ovary. Elaborate patterns of courtship, nest building, parental care, and other specific reproductive behaviors have been reported among diverse groups.

Breeding is cyclic, with each species exhibiting a well-defined spawning period regulated by environmental factors (seasonal changes in photoperiod, temperature, etc.). Although some species such as Pacific salmon are semelparous, many iteroparous species spawn several times during a single breeding season (for example, *Reprohanus melanochir,* an Australian garfish of the halfbeak family). In some viviparous species, such as the guppy *Poecilia reticulata,* ovulation is induced by treatment with prostaglandins but is unaffected

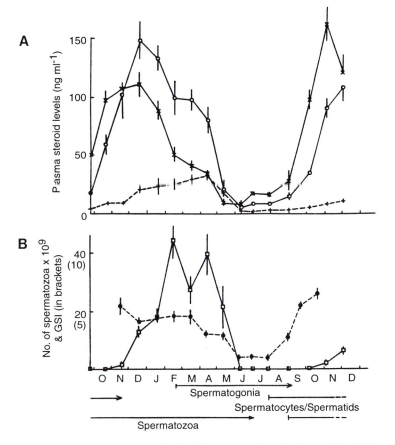

Figure 12-4. Annual reproductive cycle of male rainbow trout, *Oncorhynchus mykiss.* (A) Plasma levels of testosterone ($\times$), 11-ketotestosterone (open circles), and 17,20β-dihydroxy-4-pregnen-3-one (+). (B) Volume of sperm produced (open squares) and the gonadosomatic index, solid circles (GSI, gonad weight related to body weight). [From Scott, A. P. (1987). Reproductive endocrinology of fish. *In* "Fundamentals of Comparative Endocrinology" (I. Chester-Jones, P. M. Ingelton, and J. G. Phillips, eds.), pp. 223–256. Plenum, New York.]

by steroids or pituitary extracts. Soon after birth of the young, a new batch of oocytes is released and another brood is raised. Other viviparous species require a longer "interbrood period" for oocyte maturation and vitellogenesis [*Mollienesia* (molly) and *Gambusia* (mosquitofish)]. However, in *Quintana atrizoma,* oocyte development occurs during gestation so that a new batch of eggs can be fertilized as soon as the young are born.

The endogenous nature of seasonal or annual rhythms has not been demonstrated in fishes although seasonal reproductive cycles clearly are evident (see Figs. 12-4 and 12-5) even in tropical species. The effects of artificial lengthening and decreasing of the photophase may accelerate spawning in spring and fall spawners, respectively. A classical demonstration of environmental phasing of reproduction has been demonstrated by transporting a poecilid, *Jenynsia lineata,* from South America, where it normally spawns in January and February, to the northern hemisphere. In the new pond location, where photoperiod and seasons were reversed, the fish switched to spawning in July and August. However, the

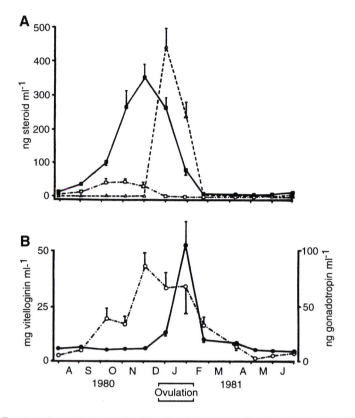

Figure 12-5. Annual reproductive cycle of female rainbow trout, *Oncorhynchus mykiss.* (A) Plasma levels
of testosterone (solid squares), estradiol (open squares), and 17,20β-dihydroxy-4-pregnen-3-
one (open triangles). (B) Plasma levels of vitellogenin (open circles) and gonadotropin (solid
circles). [From Scott, A. P., and Sumpter, J. P. (1983). Seasonal variations in sex steroids and
gonadotropin in females of autumn and winter spawning strains of rainbow trout *(Salmo
gairdneri). Gen. Comp. Endocrinol.* **52**, 79–85.]

possible importance of the temperature regimen, which was also switched, should not be
overlooked. In some species, temperature has been shown to be the critical factor in controlling recrudescence regardless of the light regimen imposed on the fish. In the tropics,
where photoperiod and temperature show little or no fluctuations, the reproductive cycles
of freshwater and brackish water fishes appear to be tuned to the wet and dry seasons as
occurs in many terrestrial vertebrates. Periodic flooding and drying cause marked changes
in water availability and also influence salinity and chemical composition of the aquatic
environment. Some tropical cichlid fishes that live in permanent bodies of water exhibit
successive breeding bouts over most of the year.

A. Male Bony Fishes

Spermatogenesis in bony fishes is of the cystic lobular type. In some teleosts, such as the
guppy, the testes have a tubular organization. Spermatogenesis in cyclic spawners resumes
soon after breeding. Nests of spermatogonia proliferate from germ cells near the margins
of the testicular lobules. Sertoli cells have been described in a number of species, appar-

ently arising from connective tissue elements of the spent lobule walls. 3β-HSD activity has been shown in Sertoli cells, implying a role for these cells in the synthesis of steroids.

The major circulating androgens in teleosts are testosterone and 11-ketotestosterone. In many teleosts, the androgen-secreting lobule boundary cells are located in the lobule walls and true interstitial (interlobular) cells are absent. These lobule boundary cells are believed to differentiate from fibroblasts in the lobule wall. Interstitial cells have been described in some species, and these cells possess 3β-HSD activity in teleosts, lungfishes, and in the coelacanth *Latimeria.*

A seasonal cycle of lipid accumulation and depletion has been described in detail for lobule boundary cells of pike, *Esox lucius,* and similar cells have been observed for other teleostean species. Viviparous teleosts, like their distant selachian relatives, produce spermatophores, employing secretions by the male gonaducts. The gonaducts and the endocrine control of their secretory activities have not been examined sufficiently.

Many species show marked seasonal development in gonaducts, accessory glands, and secondary sexual characters that are presumed to be under androgenic control. For example, testosterone induces formation of nuptial tubercles on the head of fathead minnows (Fig. 12-6), a distinctive male sexual characteristic in several cyprinid species. There are no wolffian ducts in bony fishes, and the sperm ducts are derived from the coelomic walls and are not homologous to the vasa deferentia of tetrapods or of selachians.

B. Female Bony Fishes

The teleostean ovary has been studied in considerable detail with respect to gonadal differentiation, oogenesis, vitellogenesis, and ovulation, both in oviparous and viviparous species. The ovary of most teleosts is hollow, whereas solid ovaries have been reported in most lungfishes and chondrostean fishes. A few teleosts also have solid ovaries. Unlike the hollow ovary of selachians and amphibians, in the ovary of teleosts, the cavity as well as the outer surface is lined with germinal epithelium. Each hollow ovary is continuous with an oviduct that is not homologous to the müllerian duct-derived oviducts of other verte-

Figure 12-6.   Nuptial tubercles on the snout of a minnow. These androgen-dependent secondary sexual characters are found on a number of cyprinid teleosts.

brates. Eggs are discharged from the ovary directly into the oviduct. In species with solid ovaries, the eggs are discharged into the body cavity, from which they pass to the exterior via oviducts or directly through temporary openings in the body wall. Mammalian or teleostean GnRH or LH induces ovulation in teleosts, but mammalian FSH has not been effective in stimulating reproduction.

Basically, the teleostean ovary consists of masses of follicles embedded in a rather sparse stroma. Each follicle begins as a single-layered epithelium surrounding the oocyte. As the follicle grows these epithelial cells undergo hyperplasia and hypertrophy to form the granulosa. Connective elements in the stroma near the follicular nest will differentiate into a theca, which may further differentiate into a theca externa and a theca interna. The granulosa cells are responsible for yolk deposition in the oocyte during follicular growth and resorption of yolk during atresia. In salmonids, receptors that will bind GTH-I are found on both thecal and granulosa cells, and GTH-I is responsible for synthesis of estrogens during follicular growth. The GTH-II receptors occur only on granulosa cells, where they are associated with synthesis of the steroid **17,20β-dihydroxy-4-pregnen-3-one (17,20β-P),** which is responsible for final oocyte maturation and ovulation in many species. The discovery and subsequent identification of 17,20β-P was accomplished by Yoshi Nagahama and colleagues in Japan. A different derivative of progesterone, **17,20β,21-trihydroxy-4-pregnen-3-one (17,20β,21-P),** has been found to be responsible for these events in some species. A corticosteroid, **deoxycorticosterone (DOC),** may be responsible for ovulation and final oocyte maturation in the catfish, *Heteropneustes fossilis*.

Three developmental patterns for ovaries can be identified in teleosts. In the **synchronous ovary,** all oocytes are in the same stage of development. Species with a synchronous ovary are semelparous (for example, *Anguilla* spp. and most members of the genus *Oncorhynchus*). Species such as rainbow trout and flounder have a **group-synchronous ovary,** with at least two populations of oocytes. These iteroparous species generally spawn once each year during a short breeding season. The last type is the **asynchronous ovary,** which has oocytes in all stages of development. These species spawn frequently each year during a prolonged breeding season.

Teleostean ovarian tissue synthesizes estrogens *in vitro* from radioactive precursors and also synthesizes androgens (testosterone and 11-ketotestosterone) and DOC. These three steroids have been identified in the peripheral plasma of several teleostean species. The levels of testosterone often are greater in prespawning females than they are in males, suggesting a behavioral role for androgens in females as reported for some female mammals.

All groups of bony fishes develop preovulatory, secretory corpora atretica as a result of atresia of developing follicles and develop corpora lutea following ovulation. However, a convincing endocrine function for corpora lutea is not yet established. Apparently, these atretic follicles do not have detectable 3β-HSD activity.

Vitellogenesis by the liver is stimulated by estradiol, and consequently total plasma calcium levels usually are elevated. Reproductively active oviparous females often exhibit significantly greater plasma calcium levels than do males or immature females (Table 12-4), presumably because of elevated estradiol levels and vitellogenesis. Calcium and vitellogenins are sequestered by the growing oocytes and are incorporated into the yolk.

C. Reproductive Behavior in Bony Fishes

Many aspects of reproductive behavior have been studied in teleosts, including migration, courtship, nest building, spawning, copulation, and parental care. Most of this work has

Table 12-4. Reproductive State and Serum Calcium Levels
during the Spring in Steelhead Trout (*Oncorhynchus mykiss*)
from a Natural Population

Sexual state	N	Mean body weight (g)	Mean serum calcium (mg/dl ± SE)
Immature males and females	12	38.4	11.6 ± 0.43
Sexually mature male prior to spawning	11	213.6	11.5 ± 0.74
Sexually mature females[a] after ovulation but prior to spawning	9	204.9	15.5 ± 1.44

[a] Sexually mature female trout differ significantly ($p < 0.01$) from sexually mature males and immature trout.

concentrated on roles of testis, testosterone, and synthetic androgens in males. Castration of males blocks breeding behavior and causes reversal to the nonbreeding condition of androgen-dependent characters. Some variations have been reported for **agonistic** (combative) **behavior,** which often accompanies breeding, depending on the time of castration. In 3-spined stickleback (*Gasterosteus aculeatus,* form *trachurus*) castration more than 1 week before building of the first nest abolishes all related behaviors. If castration is performed within the week prior to building of the first nest, however, agonistic behavior remains at a high level for 3 to 4 weeks. In some species, castration does not result in a decrease in agonistic behavior, and androgens may not be required to maintain the behavior once it has been induced in these fishes.

The roles of estrogens and androgens in relation to the breeding behaviors of females have been studied less than the roles of androgens in males. Castration of females may result in complete abolition of all reproductive behavior, loss of only some, or only a decrease in intensity. In one case, ovariectomized *G. aculeatus,* form *leiurus,* show more aggressive behavior than intact females, implying that ovarian steroids normally depress aggressive behavior in females. Estrogens have proven ineffective in inducing female behavior in females. Possible roles for androgens have not been studied, but the relatively high level of plasma androgens in prespawning females is suggestive of a behavioral role. Spawning behavior in females appears to be under the control of GnRH and gonadotropins, although in the killifish (*Fundulus heteroclitus*) neurohypophysial preparations or synthetic oxytocin induces reflexive spawning movements in hypophysectomized or castrated females (see Chapter 6). This spawning reflex is a behavior not dependent on shedding of ova. Similar observations have been reported for a few additional species. Spawning behavior in female goldfish is stimulated by ovarian prostaglandins.

Communication by pheromones is important in the reproduction of teleosts. Following ovulation, females of numerous oviparous species emit pheromones that attract and arouse sexual activity in males. For example, female *Bathygobia soporator* secrete a pheromone from the ovary that elicits courtship behavior by intact males but not by anosmic males (treated to prevent olfactory detection). Parental behavior is stimulated in jewel fish (*Heterochromis bimaculatus*) by chemical secreted by the young. Several species recognize their own young by using olfactory cues, and the offspring of some species use chemical recognition to identify their parents.

The goldfish has become a model system for investigations of pheromonal communication in teleostean reproduction, primarily due to the pioneering discoveries and subsequent work by Norman Stacey at the University of Alberta (Edmonton, Alberta, Canada) and collaborators. During the first stages of oocyte development, the ovaries synthesize a mixture of C_{21} steroids, including $17,20\beta$-P and $17,20\beta,21$-P. These steroids not only induce final oocyte maturation but act as a pheromone in males to induce GTH-II secretion, gamete release, and competence for spawning behavior. In addition, sulfated forms ($17,20\beta$-P-S and $17,20\beta,21$-P-S) are produced by goldfish ovaries and these too are potent stimulators of males. Evidence indicates that different regions of the male olfactory epithelium can detect $17,20\beta,21$-P and distinguish selectively between $17,20\beta$-P and $17,20\beta$-P-S. The physiological correlates of this ability have not been demonstrated, but should be forthcoming as this research continues.

In addition to releasing free and conjugated progestogens, goldfish ovaries release considerable quantities of the androgen androstenedione, which also plays a pheromonal role. Androstenedione release precedes release of $17,20\beta$-P and inhibits the responsiveness of males to $17,20\beta$-P. This mechanism may prevent premature gamete release in males.

Mammalian PRL has been shown to influence certain aspects of parental behavior that implies a role for endogenous PRL. Fanning behavior associated with aeration of the eggs can be stimulated in *Symphysodon aequifasciata* and *Pterophyllum scalars* by PRL treatment. However, similar treatment inhibits fanning behavior in sticklebacks. Stimulation of a mucous secretion that is fed to young *S. aequifasciata* is a PRL-dependent event, but it is not clear if the behavior of feeding the young is PRL dependent in these parent fish.

The implication of hormones in migratory behavior is largely circumstantial. The gonads and their secretions probably do not play a causative role since gonadal maturation usually occurs during migration. Thyroid hormones have been claimed to be causative factors of migratory behavior, and increased thyroid activity coincides with migratory behavior. It is possible that the increased activity of the thyroid gland is correlated to "permissive" effects related to metabolism and osmoregulation. Thyroid hormones may only enhance the physiological states favorable to migration, whereas the behavioral changes are neurally controlled through actions of environmental factors such as photoperiod and temperature or possibly by endogenous rhythmic neural cycles that are regulated by these environmental factors.

V. Reproduction in Amphibians

Many amphibians are characterized as terrestrial or semiterrestrial adults with an aquatic larval form. The origin of modern amphibians from their amphibian progenitors suggests that the primitive reproductive pattern for amphibians involved production of a small number of large, heavily yolked eggs. However, whether the primitive manner for development to proceed is directly to the adult form or indirectly via an aquatic larva is debatable. If direct development from the egg is primitive to amphibians, then it follows that the possession of an aquatic intermediate is a secondarily derived condition. The importance of the intermediate larval stage, regardless of whether this is a primitive or a derived feature, is that it allows for an aquatic feeding stage during which growth can be optimized without using energy stores or food resources of the terrestrial parents. If this is a derived condition, the presence of aquatic larvae would not be a primitive feature, and the notion that study of species exhibiting this life history pattern will provide evolutionary clues to the origin of terrestrial vertebrates from fishes would be false. Nevertheless, during metamorphosis of these fishlike amphibian larvae to terrestrial-type tetrapods, the same

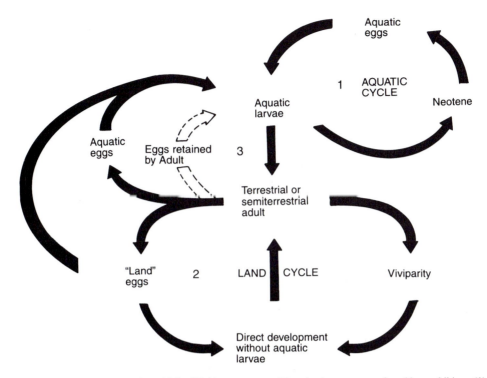

Figure 12-7. Summary of amphibian life history patterns. Three basic patterns are found in amphibians: (1) a totally aquatic cycle with sexually mature larvae (e.g., neotenes), (2) a totally terrestrial or land cycle, and (3) an aquatic–land pattern with terrestrial or semiterrestrial adults and aquatic larval stages. Within these patterns are some distinct variations such as viviparity vs. laying an egg on land within the land cycle.

problems are encountered and solved that confronted the evolution of terrestrial vertebrates from aquatic animals.

Within the modern Amphibia, several trends in reproductive evolution are evident (Fig. 12-7). All three extant orders (anurans, urodeles, and apodans) show a reduction in the use of the aquatic habitat with a tendency toward terrestrial development. This trend is accompanied by greater reliance on internal fertilization, a reduction in clutch size (number of eggs produced per breeding), and development of simple parental care of eggs and young. Mate selection and courtship patterns have become very elaborate in some species, often related to these complex patterns. Finally, oviparity apparently has given rise to viviparity independently on several occasions in each group: anurans, urodeles, and apodans.

Amphibians have gonaducts (Fig. 12-8) derived embryologically like those of amniote vertebrates and distinctly different from those of bony fishes. The male gonaducts are derived from the paired wolffian ducts of the pronephric and mesonephric kidneys. These ducts function as both urinary ducts and as sperm ducts or vasa deferentia. The oviducts are derived from müllerian ducts, which develop adjacent to the wolffian ducts. One hypothesis is that the müllerian ducts actually are produced from the wall of the wolffian ducts, which are retained in females for use as urinary ducts. Müllerian ducts appear in both male and female amphibians but degenerate in most anuran males and in some sala-

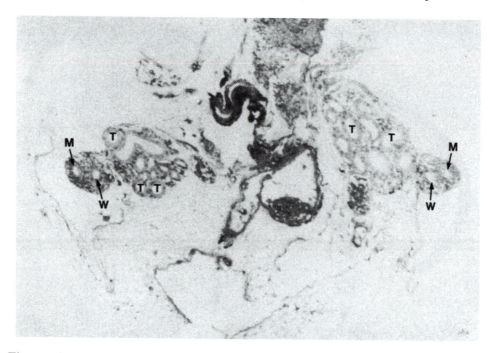

Figure 12-8. Müllerian and wolffian ducts of larval tiger salamanders. The wolffian ducts (W), which are adjacent to the kidney tubules (T), are medial to the müllerian ducts. (M). Magnification 55 ×.

manders. Müllerian ducts are retained in males of many species, and in bufonid toads can become functional oviducts if males are experimentally sex reversed (see Section V,D,1,a). Enlargement and differentiation of the wolffian ducts to the functional male condition is caused by either testosterone or dihydrotestosterone (DHT) but is antagonized by estrogens. In marked contrast, although normal ovarian differentiation of müllerian ducts can be stimulated by estrogens, both DHT and testosterone are effective at stimulating müllerian duct enlargement in tiger salamanders, although only testosterone is effective in frogs.

A. Oviparity in Amphibians

Most anuran amphibians studied are oviparous with external fertilization, although internal fertilization occurs in several species. Breeding in oviparous species is tied closely to a seasonal cycle involving photoperiod, temperature, availability of moisture, or a combination of these, although a few species are continuous breeders (the Indian frogs *Rana tigrina* and *R. erytrea* and the South American toads *Bufo arenarum* and *B. paracmenis*).

One predominant reproductive pattern is found in oviparous anurans. Spermatogenesis and ovarian follicular development are completed in the fall, and the animals simply "hibernate" until suitable breeding conditions occur in the spring. Many oviparous species lay their eggs in temporary or permanent ponds, with the egg developing into a free-swimming larval form. Tadpole larvae are the characteristic fishlike larval form of anurans and differ markedly from the larvae of urodeles, which possess external gills and four limbs. Anurans have internal gills like fishes and obtain their limbs later during metamorphosis. One anuran (*Ascaphus*) is known to lay its eggs in streams, and the tadpole larvae that result have special modifications to keep from being swept downstream. Some anuran and urodele species lay their eggs on land, usually in moist places such as under logs, in the axil of tree branches, or in temporary ponds held within the leaves of certain tropical plants. Terrestrial eggs that are heavily yolked develop directly into miniature adults, and no aquatic larval stage exists except within the egg.

Oviparous urodeles exhibit several reproductive patterns. In some species (*Triturus cristatus, Notophthalmus viridescens*), the pattern is similar to that of anurans. Spermatogenesis in the hellbender *Cryptobranchus alleganiensis* occurs in July, shortly before breeding in August and September. Other species such as the mudpuppy, *Necturus* spp., transfer sperm to the females in the fall, and oviposition and fertilization occur the next spring when males are not present.

A number of oviparous apodans have been described, all of which lay terrestrial eggs. In *Ichthyophis*, the eggs are laid in a burrow near a stream, and each newly hatched larva must emerge from the burrow and find its way to the stream. Apodans generally produce larger eggs than do the other amphibian groups, and clutch sizes are proportionally small.

B. Viviparity in Amphibians

Two European land salamanders, *Salamandra salamandra* and *S. atra*, give birth to live offspring that develop in the posterior portion of the oviducts. In *S. atra*, one young develops and undergoes metamorphosis in each oviduct during a 4-year gestation period. Gestation is shorter in *S. salamandra*, which gives birth to larval salamanders. The large size of these offspring indicates considerable nutrient contributions are made by the mother during these prolonged gestation periods. Details of these contributions and the mechanisms for their transfer have not been reported.

Viviparity in anurans typically involves a modification of a pouch that allows the eggs to develop into tadpoles on the body of the maternal animal. The South American tree frogs carry their eggs in a single mass on their back. A fold of skin may develop that completely covers the eggs in a pouch such as that found in the so-called marsupial frogs, *Gastrotheca* spp., of South America. In others, such as the African frog *Pipa pipa*, each egg develops in its own dermal chamber that forms on the back of the parent. Oviductal incubation of eggs occurs in *Nectophrynoides* and *Elutherodactylus*. One anuran incubates its young in its vocal sacs and at least one Australian species broods its young in its gut. Although this latter species was discovered very recently, its population was very small and may now be extinct.

It is estimated that viviparity occurs in the majority of apodan species. The contribution of maternal energy through oviductal secretion to support the developing young also is considerable. In *Typhlonectes*, one female may give birth to as many as nine larvae, each of which weighs about 40% of the mother's body weight at birth.

C. Reproduction in Male Amphibians

1. Male Urodeles

Spermatogenesis is of the cystic type in urodeles, and testicular structure and function are very similar to those of fishes. The urodele testis (Fig. 12-9) consists of one or more lobes, each containing several ampullae, which in turn are composed of several germinal cysts

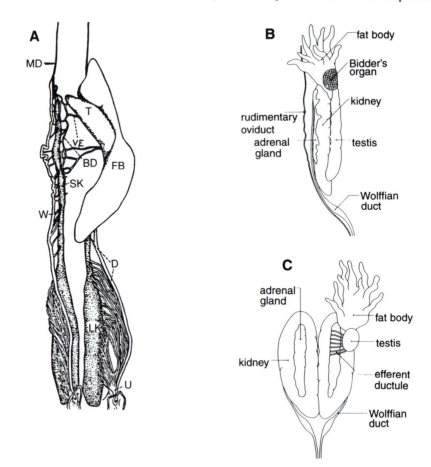

Figure 12-9. Amphibian gonads. (A) Reproductive system of male salamander, *Ambystoma tigrinum*. Note the vasa efferentia (VE) that carry sperm from the testis (T) to the wolffian duct (W), which doubles as both vas deferens and ureter (U). Many urinary collecting ducts (D) connect the lumbar portion of the kidney (LK) to the wolffian duct. One testis and its corresponding fat body (FB) have been removed. Even in the adult, a remnant of the müllerian duct persists (MD). BD, Bidder's canal; SK, sexual kidney. (B) The male toad *Bufo japonicus* with Bidder's organ anterior to the testis and a rudimentary oviduct, a remnant of the müllerian duct that persists in adults. This condition is common in all species of *Bufo* but is not found in other genera of anurans. Only one side of the reproductive system is shown. (C) In the male bullfrog, *Rana catesbeiana*, as in most anurans, there are no remnant female structures. One testis and its fat body have been removed. (D) In female tiger salamanders, the wolffian duct is reduced. The oviduct (O) typically is much longer and more folded in mature animals than is shown here. OV, Ovary. (E) Female bullfrog, *R. catesbeiana*. One ovary and its fat body have been removed to reveal the convoluted oviducts. [(A) and (D) from Rodgers, L. T., and Risley, P. L. (1938). Sexual differentiation of urogenital ducts of *Ambystoma tigrinum*. *J. Morphol.* **63**, 119–139. Copyright © 1938 John Wiley & Sons. Reprinted by permission of John Wiley & Sons, Inc. (B), (C), and (E) are from Matsumoto, A., and Ishii, S. (1992). "Atlas of Endocrine Organs: Vertebrates and Invertebrates." Japanese Society for Comparative Endocrinology and Springer-Verlag, Berlin.]

(Fig. 12-10). Germ cells associated with a germinal cyst divide mitotically to produce a cluster of secondary spermatogonia. These cells undergo synchronous differentiation to primary spermatocytes and enter meiosis. All of the cysts within an ampulla develop syn-

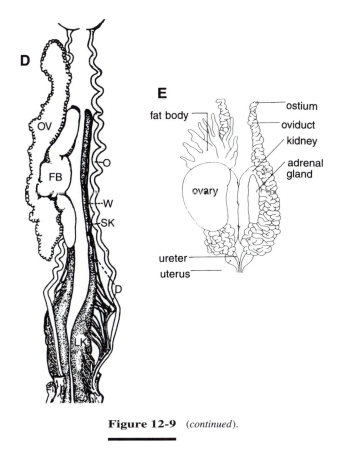

Figure 12-9 (*continued*).

chronously although it is typically only the more posterior ampullae that exhibit spermatogenesis prior to a given breeding season. The other ampullae represent the source of sperm for future breeding seasons. Sertoli cells develop from fibroblasts in the cyst walls while the spermatogonial divisions are taking place. As the ampullae mature, the posterior portion of the testis becomes swollen with sperm, whereas ampullae of the anterior-most portion consist primarily of spermatogonia. The posterior portion of the testis becomes dense and whitish because of the masses of sperm. After spermiation occurs the collapsed ampullae that have discharged their sperm into the male ducts are resorbed, and after breeding spermatogenesis is initiated in the anterior portion of the testis. If spermiation occurs in the fall, spermatogenesis will not be resumed until the next summer. New ampullae differentiate from connective tissue elements and germ cells in the tunica albuginea.

The urodele testis possesses lobule boundary cells in the ampulla walls. These cells exhibit a marked seasonal pattern of lipid accumulation, steroidogenesis, and lipid depletion. Lipid accumulation begins when the secondary sexual characters develop or hypertrophy. The lobule boundary cells become intensely lipoidal and cholesterol positive following breeding, when androgen-dependent accessory structures are regressing. Both lobule boundary cells and Sertoli cells possess 3β-HSD activity, and it is likely that both are influenced by pituitary gonadotropins.

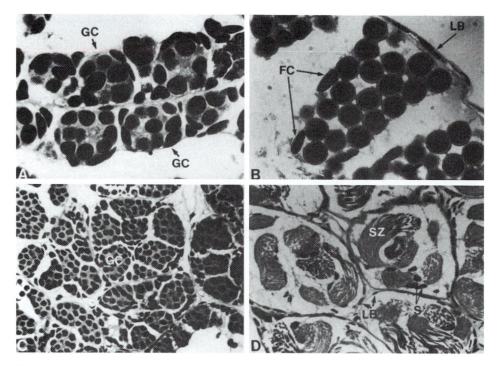

Figure 12-10 Testis of the newt, *Taricha granulosa.* (A) Early germinal cysts (GC). (B) Older cyst
containing secondary spermatogonia. Follicle cells (FC) have flattened nuclei. LB, Lobule
boundary cell. (C) Lower magnification showing several ampullae, each containing six
to eight cysts. (D) Enlargement of cysts from another region of the testis containing mature sperm (Sz) and prominent Sertoli cells (S) derived from follicle cells. (Courtesy of
Frank L. Moore, Oregon State University, Corvallis.)

Total androgen levels are comparable in most urodeles and anurans (Fig. 12-11 and
Table 12-5). Seasonal patterns of androgen secretion have been reported for *Taricha granulosa, Cynops pyrrhogaster,* and *Ambystoma tigrinum.* Androgen levels appear to be low
during breeding and high levels are more closely associated with development and maintenance of the vasa deferentia, which store sperm until breeding. Androgens also stimulate
development of cloacal glands in males that are associated with spermatophore production
and possibly with production of pheromones used in breeding. Contraction of the vasa
deferentia and discharge of sperm during mating and spermatophore production are caused
by arginine vasotocin (AVT).

Androgens also stimulate development of sex accessory structures called **nuptial pads**
in the newt *N. viridescens,* but maximal development is obtained by simultaneous treatment
with PRL and androgen. Development of skin glands also is influenced by androgens. In
urodeles such as *N. viridescens, T. cristatus,* and *A. tigrinum,* PRL is known to influence
the movement of land-phase animals to water for breeding and also induces heightening of
the tail fin, which is a male secondary sex character in some species.

Internal fertilization in both aquatic and terrestrial urodeles occurs through the production of an elaborate spermatophore produced through the actions of the cloacal glands of
the male. The spermatophore consists of a glycoprotein matrix to which a packet of sperm

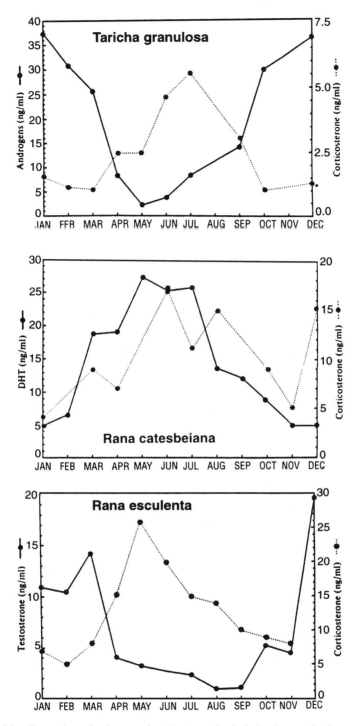

Figure 12-11. Comparison of androgen and corticosterone levels during the reproductive cycles of amphibians. Androgen levels appear to be negatively correlated with corticosterone levels in the newt *Taricha granulosa* and the frog *Rana esculenta*, but positively correlated in bullfrogs. [From Moore, F. L., and Deviche, P. (1988). Neuroendocrine processing of environmental information in amphibians. *In* "Processing of Environmental Information in Vertebrates" (M. Stetson, ed.), pp. 19–45. Springer-Verlag, Berlin.]

Table 12-5. Seasonal Extremes in Androgen Levels[a] in Selected Male Amphibians[b]

Species	Testosterone		5α-Dihydro-testosterone		Total androgens	
	High	Low	High	Low	High	Low
Urodeles						
Ambystoma tigrinum	32	0.4	7	0.3	39	0.7
Taricha granulosa					37	3
Triturus carnifex					28	2
Anurans						
Bufo japonicus	203	47	115	27	318	74
Bufo mauritanicus[c]	183		52			
Rana blythi[d]	4		1		5	
Rana catesbeiana					35	3

[a] In nanograms per milliliter of plasma.
[b] Note the much higher levels reported for *Bufo* as compared to other anurans and urodeles.
[c] For breeding animals only.
[d] Continuously breeding tropical frog.

is attached, the glycoprotein matrix acting as a base on which the sperm packet rests. Following a complex courtship procedure, the female is induced to pick off the sperm packet, using her cloacal lips. The sperm packet may then be stored in a specialized portion of the cloaca (spermatotheca) until ovulation and fertilization occurs.

2. Male Anurans

The anuran testis (Fig. 12-9) is structurally unlike that of urodeles and is more similar to that of amniotes, consisting of a homogeneous mass of seminiferous tubules with a permanent germinal epithelium and conspicuous interstitial tissue (Fig. 12-12). The interstitial cells are ultrastructurally like mammalian interstitial cells and possess a 3β-HSD activity. The lipid cycle within the interstitial cells and the degree of 3β-HSD activity closely parallel the development of androgen-dependent sex accessory structures such as the enlarged thumb pads of ranids. Interstitial cells of postspawning anurans exhibit considerable lipoidal accumulation but very low 3β-HSD activity, and thumb pads regress in ranids at this time.

During winter months, Sertoli cells of ranid testes lack lipid, but these cells elongate and exhibit small lipoidal granules as the breeding season approaches. Sertoli cells of breeding animals have a well-developed smooth endoplasmic reticulum, and 3β-HSD activity is detectable. After spermiation, the Sertoli cells detach from the tubule wall and degenerate. New cells for the next reproductive period differentiate from fibroblasts in the tubule walls.

Circulating testosterone, FSH, and LH vary seasonally, but estradiol levels are very low. Highest values for reproductive hormones occur during mating in some species (e.g., *Rana catesbeiana*) but not in others (*Rana esculenta, A. trigrinum*) where gametogenesis is dissociated from time of spawning and development of accessory structures is dependent on steroids.

3. Male Apodans

Male caecilians differ from urodeles and almost all anurans by possessing an elaborate intromittent organ, the **phallodeum,** associated with the posterior part of the cloaca. The

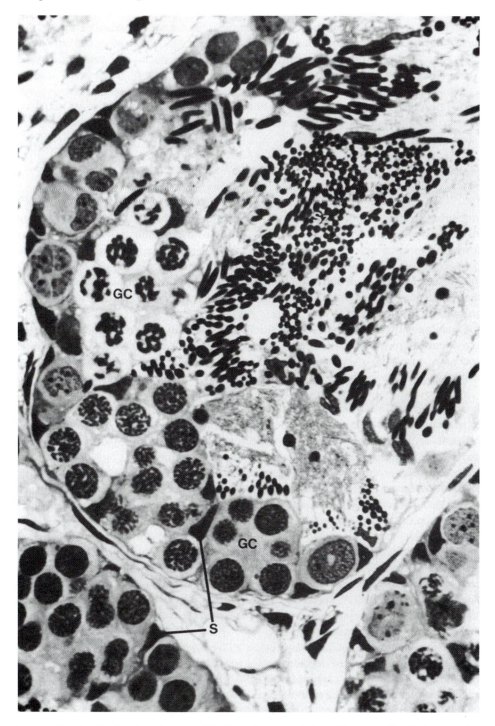

Figure 12-12. Section through testis of bullfrog, *Rana catesbeiana.* Note how all cells in a germinal cyst (GC) are in the same stage of development. Compare to urodele testis in Fig. 12-10. Sz, Spermatozoa; S, Sertoli cell. (Courtesy of Charles H. Muller, University of Washington, Seattle.)

phallodeum is employed for transferring sperm to the female reproductive tract. Consequently, fertilization is internal in all apodans. Another unique feature is the retention, in males, of only the most posterior portion of the müllerian ducts—the **müllerian glands.** These tubular apocrine glands are believed to produce seminal fluid and hence would be analogous to the prostate of mammals. The structure of the lobed apodan testes is similar to that of urodeles, but the cell nests within an ampulla are not synchronized and do not exhibit the same stage of spermatogenesis.

D. Reproduction in Female Amphibians

1. Female Anurans and Urodeles

Amphibian ovaries are hollow, saclike structures derived from the embryonic cortex and covered by germinal epithelium (Fig. 12-13). A derivative of embryonic medullary tissues forms the inner lining of the ovary. Oogonia are present in the germinal epithelium, and they give rise to nests of oocytes. The follicular epithelium consists of a single layer of granulosa cells throughout the maturation period. A very thin thecal layer does form around the follicle, but it is only obvious when viewed with the aid of an electron microscope (compare Figs. 12-13 and 12-14).

Granulosa cells and thecal cells contain 3β-HSD activity, but there is no steroidogenic activity in the ovarian interstitial cells of the ovary. Both thecal and granulosa cells may be sources of circulating ovarian steroids, but cytological evidence favors the granulosa as the major source. Gonadotropins stimulate these cells to synthesize estrogens.

At the end of a breeding season, the ovary typically contains young follicles that will become the next crop of mature oocytes, numerous cell nests that will become the young follicles of the next vitellogenic period, and primary germ cells that will give rise to new cell nests for future generations of oocytes. Progression from primary germ cells to mature oocytes may require three breeding seasons or more for completion.

Ovarian estrogens control development of sex accessory structures such as the hypertrophy of oviducts prior to sexual maturation and during each season prior to ovulation. In the marsupial frog *Gastrotheca riobambae,* the development of the brood pouch is dependent on estrogens secreted from preovulatory follicles.

Postnuptial ovaries frequently contain postovulatory corpora lutea, which are short-lived in oviparous species. Granulosa cells hypertrophy after ovulation and accumulate cholesterol-positive lipids. The follicle collapses and becomes a central mass of lipoidal cells surrounded by a fibrous capsule derived from the thecal layer. Postovulatory corpora lutea of *T. cristatus* and *R. esculenta* possess 3β-HSD activity and may be sources for steroids. However, no functional endocrine role for postovulatory corpora lutea has been demonstrated in these oviparous species.

In viviparous amphibians such as the anuran *Nectophrynoides occidentalis,* postovulatory corpora lutea are more persistent and appear to be functional throughout gestation. Furthermore, postovulatory corpora lutea from this species are capable of converting pregnenolone to progesterone. Corpora lutea are required for about the first 25 to 30 days of the 100- to 125-day gestational period in the pouch of the marsupial frog, *G. riobambae,* but ovariectomy after day 40 had no effect on gestation. The granulosa-lutein cells of the postovulatory corpora lutea in viviparous *S. salamandra* possess 3β-HSD activity and appear cytologically to be steroidogenic. Thirty or more corpora lutea persist in each ovary

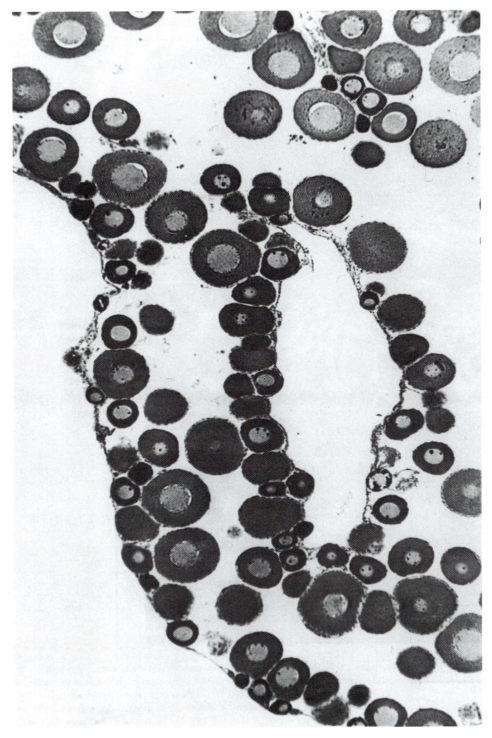

Figure 12-13. Amphibian ovary. This ovary from the cane toad, *Bufo marinus*, is typical of the hollow amphibian ovary with follicles attached to the germinal epithelium. (Courtesy of Charles H. Muller, University of Washington, Seattle.)

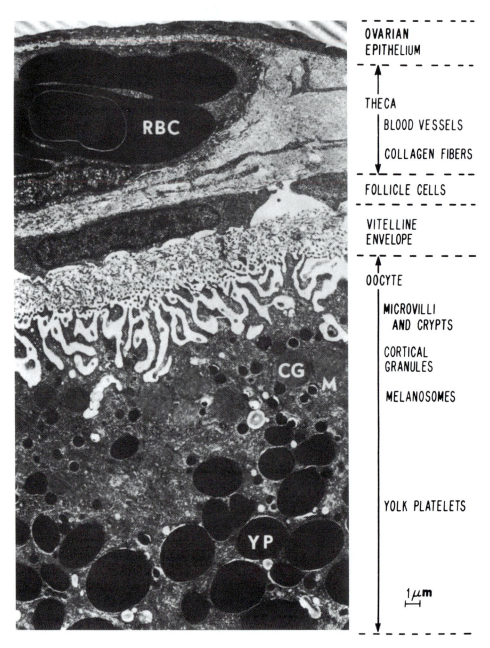

OVARIAN
EPITHELIUM

THECA

BLOOD VESSELS

COLLAGEN FIBERS

FOLLICLE CELLS

VITELLINE
ENVELOPE

OOCYTE

MICROVILLI
AND CRYPTS

CORTICAL
GRANULES

MELANOSOMES

YOLK PLATELETS

1 μm

Figure 12-14. Vitellogenic follicle of *Xenopus laevis*. This transmission electron micrograph shows the vascular theca and its relationship to microvillous processes projecting from the surface of the developing oocyte. YP, Yolk platelets. [From Dumont, J. N., and Brummett, A. R. (1978). Oogenesis in *Xenopus laevis* (Daudin). V. Relationships between developing oocytes and their investing follicular tissues. *J. Morphol.* **155**, 73–97. Reprinted with permission of the authors, publisher, and the Oak Ridge National Laboratory operated by Union Carbide Corporation for the Department of Energy.]

during the first 2 years of gestation and gradually decrease in both size and number over the next 2 years until the young are born.

Development of new oocytes may be arrested by the presence of corpora lutea. In *N. occidentalis* as well as in oviparous *Taricha torosa,* the succeeding crop of follicles begins development only after degeneration of the postovulatory corpora lutea. These latter data suggest an inhibitory action of progesterone produced in the postovulatory corpora lutea on release of gonadotropins from the adenohypophysis.

Atresia occurs frequently during follicular development in amphibians. Granulosa cells are responsible for phagocytosis of yolk and formation of the preovulatory corpora lutea (corpora atretica). Since no 3β-HSD activity has been identified in these structures, they should be termed corpora atretica.

The process of vitellogenesis in the liver and yolk deposition in oocytes of oviparous amphibians has been examined extensively. In *X. laevis,* gonadotropin stimulates micropinocytosis of vitellogenin by oocytes. Micropinocytotic vesicles of vitellogenin are hydrolyzed enzymatically in the yolk platelets to produce the yolk proteins **phosvitin** and **lipovitellin.** The yolk platelets containing these yolk proteins are utilized as an energy source during early embryogenesis. Estrogens will induce vitellogenin synthesis in both female and male livers when administered *in vivo.* Circulating vitellogenin binds free calcium ions, resulting in elevated total plasma calcium levels through release of calcium from storage sites.

Hypophysectomy results in atresia of all vitellogenic follicles whose diameters exceed about 0.4 mm, indicating the importance of endogenous pituitary gonadotropins in the process of vitellogenesis. Mammalian FSH will augment the growth of vitellogenic follicles and prevent atresia following hypophysectomy. The failure of mammalian gonadotropins to stimulate formation and growth of previtellogenic follicles (i.e., follicles less than 0.4 mm in diameter) coupled with their apparent insensitivity to hypophysectomy has led to the suggestion that these processes are completely independent of pituitary control. However, experimental studies have not ruled out completely a role for gonadotropin in the development of previtellogenic follicles, and it is possible that low endogenous levels of amphibian gonadotropin are necessary for even the earliest events in gametogenesis.

Amphibian ovaries *in vitro* produce progesterone, estradiol, estrone, testosterone, DOC (like teleosts), and DHT. The high levels of circulating testosterone reported for some anurans may be related to a precursor role for peripheral aromatization to estrogens or to a behavioral role of their own.

Ovulation is under the control of an LH-like gonadotropin and progesterone. *In vitro* studies of the ovary, initiated by Paul Wright about 50 years ago and subsequently elaborated by others, indicate that LH as well as a wide variety of steroids can induce oocyte maturation (completion of meiosis and breakdown of the germinal vesicle) and ovulation *in vitro.* Progesterone is the most potent maturational steroid, and it is reasonable to presume that progesterone or a closely related steroid plays a key role in normal ovulatory events. The synthesis of progesterone is under the control of LH, and the action of progesterone is believed to be indirect, operating through stimulation of a "maturation-promoting factor." Prolactin enhances the sensitivity of oocytes to gonadotropin or progesterone both *in vivo* and *in vitro.* This enhancement can be blocked by simultaneous *in vivo* treatment with thyroxine (T_4).

Ovarian maturation typically is completed in autumn, and ovulation is delayed over winter until favorable conditions occur in the spring. The endocrine basis for this diapause is not clear but may involve direct inhibition of ovulation by one or more pituitary hormones. Hypophysectomy of gravid anurans and at least one urodele results in ovulation and oviposition. Furthermore, hypophysectomy of gravid neotenic tiger salamanders increases their sensitivity to induced ovulation in response to a single injection of hCG. This "reflexive" ovulation could be due to inadvertent release of gonadotropins by the operation itself, or to removal of an active inhibitory substance of pituitary origin, or to both.

Growth of amphibian oviducts is stimulated by either estrogens or androgens. The feminizing action of androgens is often termed a "paradoxical" effect as would masculinization by estrogens. In mature females, contraction of oviducts is caused by AVT, which presumably is the hormonal stimulus in oviposition. Oviducts of breeding animals are more sensitive to AVT than are those of nonbreeding adults. Progesterone induces responsiveness to AVT in immature oviducts of salamanders, but estrogens are not effective. Possibly the pre- or postovulatory follicle releases sufficient progesterone in response to LH to alter receptor levels for AVT in the muscles of the oviducts. Androgens and estrogens do not affect the sensitivity of oviducts to AVT. The interrenal has been implicated as an additional progesterone source and may influence the responsivity of oviducts.

a. Bidder's Organ

Among both sexes of bufonid toads are found rudimentary ovaries, or **Bidder's organs,** that develop from cortical remnants of the embryonic genital ridge (Fig. 12-15). Histologically, Bidder's organ consists of a compact mass of small oocytes. In males, the bidderian oocytes undergo a limited seasonal growth and degeneration cycle correlated with the testicular cycle. These bidderian oocytes never reach the vitellogenic stage in males, however. The presence of 3β-HSD activity suggests they are steroidogenic during this time. After castration of male bufonids, Bidder's organ hypertrophies, presumably under the influence of increased gonadotropins, and forms a functional ovary. In castrated males, the rudimentary müllerian ducts may develop into functional oviducts and such sex-reversed animals may breed as females.

b. Fat Bodies

Conspicuous masses of adipose tissue called fat bodies are located adjacent to the gonads of amphibians (Fig. 12-9). In female anurans and urodeles, the size of the fat body is correlated inversely with gonadal weight, and it has been proposed that the lipoidal substances stored in the fat bodies are utilized for oocyte growth. Fat bodies of both male and female European newts (T. cristatus) can synthesize steroids and therefore may influence gonadal function, accessory sex structures, or both. In the European frog, R. esculenta, the steroidogenic function of fat bodies appears to be regulated by pituitary gonadotropins.

2. Female Apodans

Fertilization in apodans normally occurs in the upper portion of the oviduct following intromission by the male, and the fertilized eggs are either laid in burrows or, as may prove to be the case for most species, are retained in the oviducts until the developing larvae have completed metamorphosis.

Ovarian development is very similar to that described for anurans and urodeles, but in apodans the eggs tend to be larger and fewer. Postovulatory corpora lutea develop in the ovaries, and they appear to be important for maintaining oviductal secretion (even in

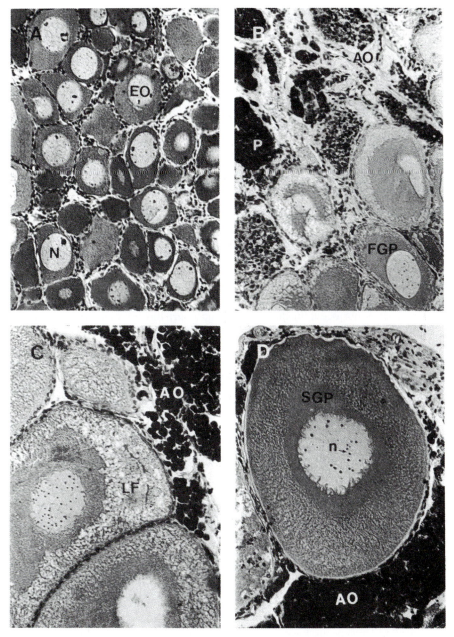

Figure 12-15. Sections through Bidder's organ from male *Bufo woodhousii*. These micrographs (all at the same magnification) show the effects of castration and gonadotropin treatment on oocytes of Bidder's organ. (A) Sham-operated toad, treated with saline injections for 26 days, shows unstimulated follicles. EO, Early previtellogenic follicle; N, nucleus. (B) Sham-operated toad, treated with mammalian gonadotropins for 26 days, shows moderately enlarged follicles. AO, Atretic follicle; P, pigment granules; FGP, first growth phase follicle (previtellogenic). (C) Castrated toad, treated with saline for 26 days, shows oocyte and follicle growth into the late previtellogenic stage (LF). (D) Castrated toad, treated with gonadotropins, shows vitellogenic or second growth phase oocyte (SGP). n, Nucleus. [From Pancak-Roessler, M. K., and Norris, D. O. (1991). The effects of orchidectomy and gonadotropins on steroidogenesis and oogenesis in Bidder's organ of the toad *Bufo woodhousii*. *J. Exp. Zool.* **260,** 323–336. Copyright © 1991 John Wiley & Sons, Inc. Reprinted by permission of John Wiley & Sons, Inc.]

oviparous species) and pregnancy. Oviductal secretions provide nutrition for the developing young, and these secretions may be controlled by hormones released from the corpora lutea.

Sufficient numbers of specimens have not been examined to determine the seasonality of breeding and ovarian cycles in caecilians. The endocrine factors involved in reproductive events can only be inferred at this time from studies on anurans and urodeles.

E. Reproductive Behavior in Amphibians

Numerous aspects of reproductive behavior have been described for many amphibians, but little is known about the mechanisms of endocrine control for most species. Reproductive behavior includes migration, calling, courtship, clasping, spawning, and various kinds of parental care. Studies involving castration, hypophysectomy, and/or injections of pituitary hormones support the conclusion that testicular hormones are involved in calling, courtship, and clasping. In neotenic *A. tigrinum,* a species with dissociated mating, androgens are not high at the time of mating. Although most species examined exhibit associated mating, attempts to stimulate reproductive behavior with androgens alone often have not been successful.

Plasma androgen levels may not tell the entire story. In the European crested newt, *Triturus carnifex,* inactive males have higher testosterone levels than courting males. Clasping of the female does not occur in these newts and courtship involves progressive stages: approaching, fanning, tail lashing, and spermatophore deposition. Aromatase activity increases in the brain and gonad during courtship, and courting males have significantly higher levels of estrogen in brain and plasma during the approach stage.

Neural peptides [AVT, GnRH, and corticotropin (ACTH)] trigger mating behavior in androgen-primed animals. For example, when a female frog that is not ready to spawn is clasped by a courting male, she croaks to signal her nonreceptivity. A receptive female will not emit this release call. This receptive female has accumulated water that will be used in ovulation and oviposition. Water retention is caused by AVT (see Chapter 6), and administration of AVT inhibits the release call, possibly through effects on the brain.

According to extensive studies by Frank Moore and associates at Oregon State University (Corvallis, OR), reproductive behavior by male rough-skinned newts (*T. granulosa*) also is stimulated by AVT, although androgens and other factors are involved (Fig. 12-16). During courtship, males in the breeding pond will attempt to clasp a female along the back. Attempts by several males to clasp the same female result in the formation of mating balls that persist for a time, but the unsuccessful males soon drop off, leaving only one clasping the female. A series of behavioral and chemical interactions between the clasping pair eventually will lead to spermatophore transfer. Androgens play a priming role that enhances the sensitivity of the newts to AVT. Stress or injection of corticosterone can rapidly repress clasping behavior within minutes. Clasping can be activated by cloacal stimulation and is controlled by neurons located in the rostral portion of the medulla. Furthermore, activity of the neurons is increased by AVT treatment but decreased by administration of corticosterone.

Pheromones are used in courtship by many urodeles that possess **hedonic glands** or **cloacal glands.** These glands are known to be sources of potent pheromones. In contrast, there is little evidence for chemical communication among anuran species and mating is accomplished presumably by using auditory, tactile, and visual cues. Male plethodontid

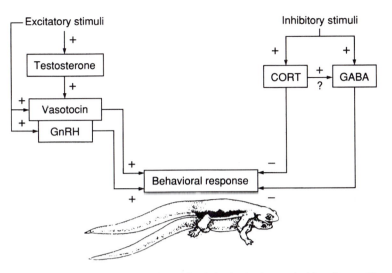

Figure 12-16. Neuroendocrine control of clasping behavior in the rough-skinned newt. Stress activates inhibitory pathways (CORT, corticosterone; GABA, γ-aminobutyric acid) that antagonize the actions of GnRH and arginine vasotocin, which are activated by external sexual parameters. [From Moore, F. L., and Orchinik, M. (1991). Multiple molecular actions for steroids in the regulation of reproductive behaviors. *Semin. Neurosci.* **3**, 489–496.]

and desmognathid salamanders employ a tubular **mental (chin) gland** for stimulating courtship behavior in females. Ambystomatids and salamandrid salamanders rely on pheromonal secretions from one of their cloacal glands, the **abdominal gland,** to stimulate females. Female attractants released into a stream readily attract reproductively active male *T. granulosa*. Furthermore, males of this species are attracted to females by airborne cues as well (see Table 12-6).

Table 12-6. Sex-Dependent Conspecific Odor Preferences in *Taricha granulosa*: Directed Locomotor Response of Newts to Newt and Nonnewt Odors[a, b]

No. of tests	Sex of test newt	Sex of stimulus newt	Ratio of S[c]/NS[d]	Probability that choices were random
10	Male	Male	7/3	0.0570
10	Male	Female	10/0	0.0003
20	Male	Male/female	17/3	0.0004
10	Female	Male	4/6	0.625
10	Female	Female	6/4	0.625
20	Female	Male/female	10/10	0.987

[a] Unpublished data of M. Schwartz. D. Duvall, and D. O. Norris (1978).
[b] Examined in a simple olfactometer.
[c] Subject chose air from stimulus animal.
[d] Subject chose air from nonstimulus source (no newt).

Nest building is a complex form of parental care that is exemplified by the foam nest frog, *Chiromantis xerampilina.* As many as 40 females may contribute to production of a huge, terrestrial foam nest in which fertilized eggs will be deposited. The females produce the foam by beating water with their hind limbs. Males are much smaller than females, who are attracted to males making soft clicking sounds. When a female arrives at a calling site, males drop from above onto her back. One male occupies the central position with up to seven peripheral males hanging on. This central position is considered to be the most advantageous position for fertilizing eggs emanating from the female's cloaca. In experiments where the cloacae of males were sheathed to prevent sperm release, no eggs were fertilized. When only the central male was sheathed, about half of the eggs were still fertilized but by the unsheathed peripheral males. In another study, 10 of 15 naturally breeding females were shown to produce clutches having 2 or more fathers. Thus, cooperation not only occurs in nest-building behaviors but also in mating, which ensures a good mixing of genetic material in the following generation.

VI. Reproduction in Reptiles

Living reptiles are members of diverse orders, and it is not surprising that considerable differences occur, making it difficult to generalize about reptilian reproduction. In many respects, the squamate reptiles possess features unique to their order, whereas the other orders may be more typical of reptiles as a group with respect to exhibition of primitive features. Most reptilian species are oviparous and exhibit well-defined annual reproductive cycles and breeding seasons. In addition, many examples of viviparity are known among snakes and lizards, and viviparity has been hypothesized to have evolved many times in these groups. Only a few, heavily yolked eggs are produced by most reptilian species, although clutch sizes may be relatively large among some turtles, crocodilians, and snakes. Fertilization is internal in all reptiles, and males have intromittent organs for placing sperm into the cloaca of a female. Mating frequently follows complicated behavioral patterns including male–male territorial and aggressive interactions. Following mating, females of many species can store sperm in the cloaca for months.

In males, the vas deferens develops from the wolffian ducts as described for amphibians (Fig. 12-17). The vas deferens conducts no urine, as the mesonephros kidney is completely replaced in reptiles by the evolution of the metanephric kidney, with its own ureter connecting to the urinary bladder. As elegantly demonstrated in the American alligator, müllerian ducts degenerate in males prior to hatching as a consequence of a müllerian-inhibiting substance (MIS) secreted by the testes (Figs. 12-18 and 12-19).

Crocodilians and turtles produce two distinct gonadotropins, but squamate reptiles rely on an FSH-like gonadotropin. The hypothalamus produces GnRH, which regulates gonadotropin release in response to environmental stimuli such as photoperiod and temperature (see Chapter 5).

Reptiles classically have been characterized by a lack of parental behavior and, in some cases, even lack of recognition of their offspring. More recent studies, however, have demonstrated that there is considerable investment in parental care even among oviparous species. Although members of the oldest extant group, the chelonians, typically abandon their nests once the eggs are laid, many squamates exhibit parental behavior. Crocodilian parents participate in the hatching process and in protecting the young. Evidence of nest

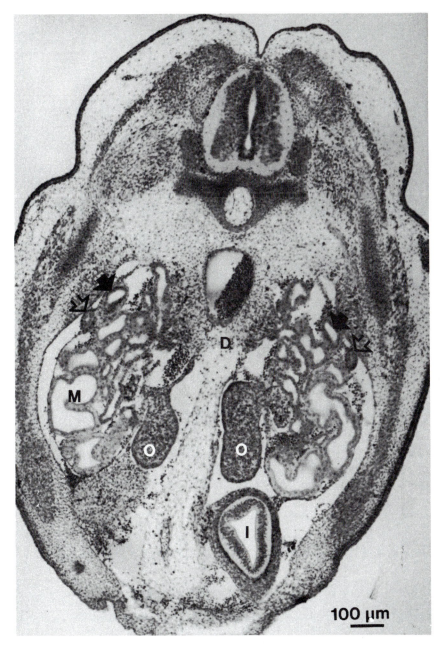

Figure 12-17. Cross-section through an embryo of the lizard, *Sceloporus undulatus.* Developing gonads (o), mesonephros (M) müllerian ducts (open arrows), wolffian ducts (solid arrows), intestine (I), and dorsal mesentary (D). [From Austin, H. (1988). Differentiation and development of the reproductive system in the iguanid lizard, *Sceloporus undulatus. Gen. Comp. Endocrinol.* **72**, 351–363.]

building and parental care has been unearthed for some extinct dinosaurs as well. Thus, complex parental care did not appear *de novo* in birds and mammals but already had become well established in reptiles.

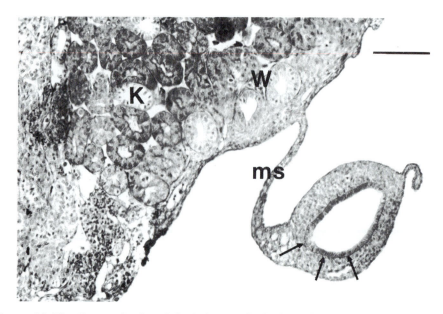

Figure 12-18. Cross-section through developing reproductive ducts of a male American alligator. The müllerian duct is attached to the mesonephros by a thin mesentary (ms). W, Wolffian duct; K, kidney tubules. (Photo courtesy of Harriet B. Austin, University of Colorado, Boulder.)

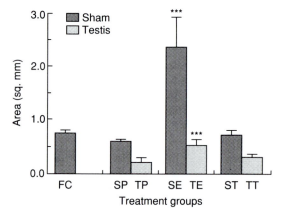

Figure 12-19. Effects of testis grafts and hormone treatment on müllerian ducts of female American alligator embryos. The source of müllerian inhibitory substance (MIS) appears to be the testis, and MIS is antagonized by the presence of estradiol (E) as evidenced by failure of SE group to exhibit regression. Shown here is the mean epithelial cell height of müllerian ducts in sham-operated (S) and testis-grafted (T) animals. The second letter indicates hormone treatment (P=placebo). Estradiol (E) treatment stimulated müllerian ducts (SE) and reduced the degree of regression (TE) caused by the presence of a testis (TP). Testosterone had no effect in either group (ST, TT). FC, Final control; ***, significantly different from treatments with same first letters. [From Austin, H. B. (1990). The effects of estradiol and testosterone on Müllerian-duct regression in the American alligator (*Alligator mississippiensis*). *Gen. Comp. Endocrinol.* **76**, 461–472.]

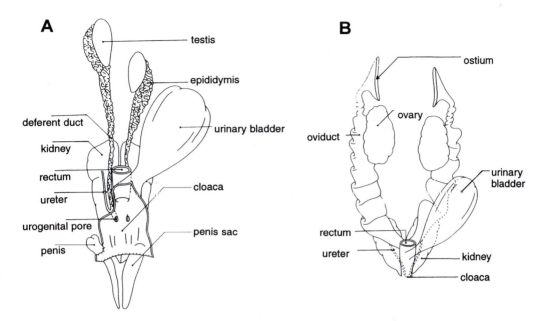

Figure 12-20. Reproductive system of a lizard. (A) Male: The top wall of the cloaca has been removed to show the duct openings and the penis. (B) Female. [Modified from Matsumoto, A., and Ishii, S. (1987). "Atlas of Endocrine Organs: Vertebrates and Invertebrates." Springer-Verlag, Berlin.]

A. Male Reptiles

The reptilian testis (Fig. 12-20) is typical of amniotes and consists of convoluted seminiferous tubules (Fig. 12-21), each surrounded by a connective tissue sheath, the tunica
propria. The entire testis is enclosed by a tunica albuginea. Spermatogenesis recurs soon
after the breeding season and is completed in most species prior to the onset of winter. Pituitary gonadotropins stimulate spermatogenesis in a variety of reptilian species.
Sperm may be stored in the vas deferens for up to several months prior to mating. Sertoli
cells are common, and they are steroidogenic. After spermiation and testicular collapse,
the Sertoli cells fill with cholesterol-positive lipid, which is depleted under the influence of FSH at the time mitosis resumes in the spermatogonia. Typical interstitial cells
have been described in the reptilian testis (Fig. 12-21), and they undergo cyclical changes
associated with androgen secretion and sexual changes in androgen-dependent sex accessory structures (Fig. 12-22). Representative plasma androgen levels are provided in
Table 12-7.

In sexually active squamates, a portion of the kidney tubules, known as the **sexual segment** of the kidney, undergoes hypertrophy under the influence of androgens. The sexual
segment appears to secrete materials that help maintain sperm that are stored in this region
prior to ejaculation, and it may be homologous to the seminal vesicles of male mammals.

B. Female Reptiles

Reptiles have paired hollow ovaries with little stromal tissue (Fig. 12-20). Oogonia are
present in the mature ovary as described for anamniotes and give rise to primary oocytes

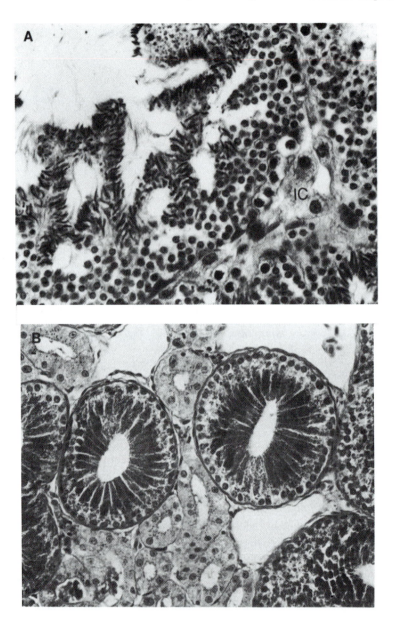

Figure 12-21. Spermatogenesis and sexual segment of kidney in sceloporine lizards. (A) Spermatogenesis
occurring in the testis of *Sceloporus jarrovi*. Portions of three tubule sections can be seen
separated by several large interstitial cells (IC). Note the dense, elongated sperm heads. (B)
Section through the kidney of *Sceloporus undulatus* showing small, lightly stained renal
tubules and large, darkly stained sexual segments of renal tubules modified for sperm storage. This enlargement is due to the action of testosterone. (Photomicrographs courtesy of Dr.
John Matter, Clemsen University.)

throughout reproductive life. The developing oocyte (Fig. 12-23) becomes invested with
granulosa cells, which are separated from the surrounding thecal cells by a connective
tissue layer, the **membrana propria.** The theca of fishes and amphibians has a simple

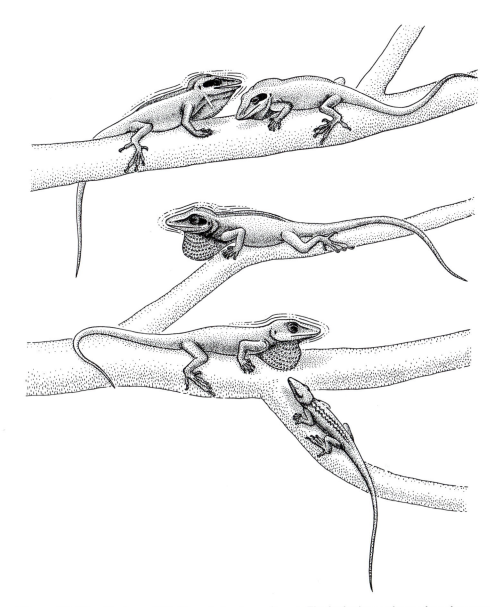

Figure 12-22. Displays exhibited by male *Anolis carolinensis*. The dewlap is an androgen-dependent structure that may be extended from the throat region. On the top limb, two males are performing
an aggressive bout. Head-bobbing and dewlap extension are common components of displays. Note the postorbital black pigment spot that appears in response to elevated epinephrine. The lone male on the middle limb is exhibiting an assertion-challenge display normally
performed by the dominant animal following an aggressive bout. Note the postorbital spot is
less intensely blackened. The male on the bottom main limb is making a courtship display to
the smaller female on the branch below. This display includes dewlap extension and head-
bobbing but there is no evidence of the postorbital pigment spot. [Reprinted with permission
from Crews, D. (1978). Integration of internal and external stimuli in the regulation of lizard
reproduction. *In* "The Behavior and Neurology of Lizards" (N. Greenberg and P. D. Mc-
Lean, eds.), pp. 149–171. NIMH, Rockville, Maryland.]

Table 12-7. Plasma Steroid Levels[a] in Reptiles

Species	Testosterone	Estradiol	Progesterone
Turtles			
Chrysemys picta			
Male	15–40 ng		
Preovulatory female	3.2–5.7 ng	0.79–1.37 ng	1.2–1.5 ng
Postovulatory female	0.2 ng	UND[b]	0.3–0.5 ng
Stenotherus ordonatus			
Female	250–1500 pg	500–5000 pg	700–4000 pg
Male	10–75 ng		
Lizards			
Lacerta vivipara	27–390 ng		
Uromastix hardwicki			
Preovulatory female	0.37 ng	0.18 ng	1.6 ng
Gravid female	1.57 ng	0.46 ng	13.41 ng
Iguana iguana			
Male	100 pg	79 pg	
Female	3 pg	270 pg	
Snakes			
Natrix fasciata (female)	50–1065 pg	10–540 pg	90–1445 pg
Naja naja			
Male	60–2300 pg		
Female	30–700 pg	10–310 pg	1.4–25 ng
Nerodia sipedon (male)	2–21 ng		

[a] Per milliliter.
[b] UND, Undetectable.

structure, but in reptiles it is differentiated into an inner, granular **theca interna** surrounded by a fibrous **theca externa.** The cells of the granulosa in most species are considered the primary source of follicular estrogen during ovarian recrudescence, although histochemical evidence in skinks (genus *Lamproholis*) implies that thecal cells rather than the granulosa cells are steroidogenic. Histochemical changes in ovarian cholesterol-positive lipid inclusions and 3β-HSD activity parallel estrogen-dependent oviductal growth and changes in other sex accessory structures as well as changes in the gonadotropes in the adenohypophysis. As the oocyte enlarges, it begins to project into the ovarian cavity and out from the surface of the ovary.

The squamate granulosa contains a unique flask-shaped cell type, the **pyriform cell** (Fig. 12-23), that is in direct contact with the developing oocyte. These cells apparently are involved with early steps in oocyte development as they either degenerate or transform into typical granulosa cells soon after the onset of vitellogenesis.

As ovulation approaches, the granulosa cells as well as some thecal cells accumulate cholesterol-positive lipids, and, following ovulation, proliferate and luteinize to form corpora lutea. These corpora lutea are well vascularized, exhibit 3β-HSD activity, and synthesize progesterone. They persist throughout egg laying in oviparous species or during gestation in most viviparous forms. Corpora lutea of viviparous species are said to synthesize greater amounts of progesterone than do those of oviparous species. Plasma progesterone levels are greatest following ovulation and are maintained at elevated levels throughout gestation in most viviparous lizards and snakes. Preovulatory peaks of progesterone are found in oviparous turtles, crocodilians, and lizards (Table 12-8).

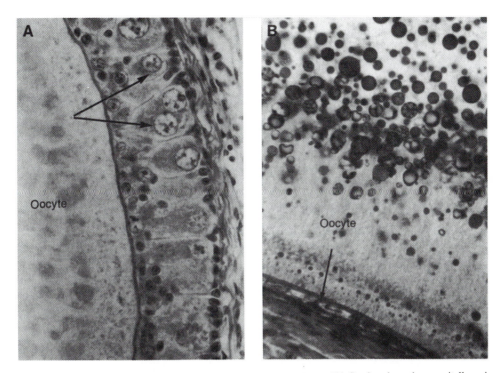

Figure 12-23. Ovary of the iguanid lizard, *Ctenosaura pectinata*. (A) Section through a pervitellogenic follicle with large pyriform cells (arrow) in the granulosa layer. (B) Lower power section through a vitellogenic follicle. Note how the granulosa (arrow) is flattened. (Courtesy of Dr. Mari Carmen Uribe.)

Only a few of the follicles that begin development reach maturity at a given time; the majority undergo atresia. Follicular atresia and formation of corpora atretica are a common occurrence in reptilian ovaries as in other vertebrates. The importance of corpora atretica is unknown, but the absence of 3β-HSD activity in corpora atretica makes it unlikely they would have an endocrine function. However, steroidogenic cells from atretic follicles may give rise to an "interstitial gland" that is believed to be a major source of ovarian estrogens.

Ovaries of reptiles show different patterns of follicular maturation and ovulation. Some produce several eggs simultaneously from each ovary (most reptiles). Others may alternate production of a single egg from each ovary (anoline lizards). Differences in follicular atresia rather than in the number of oocytes beginning development may be responsible for these patterns. Still others (e.g., certain turtles) may produce most of their eggs in one ovary during one season and most from the other ovary the next season.

Exogenous GnRH produces direct actions on ovarian follicles, which may reflect a paracrine role for ovarian GnRH. Treatment of lizards (*Podacris sicula sicula*) with salmon GnRH (sGnRH) increases secretion of prostaglandin ($PGF_{2\alpha}$) from mid- to late follicles and early corpora lutea and increases secretion of progesterone from follicles. Antagonists of GnRH produce opposite results. The physiological relevance of these observations is open to interpretation, but it does suggest paracrine effects of GnRH are not limited to mammals.

Table 12-8. Circulating Progesterone Levels[a,b] during Reproductive Cycles of Turtles, Lizards, and Snakes

Species	Period of early follicle growth	Pre-ovulatory stage	Early post-ovulatory	Mid pregnancy	Late pregnancy
Turtles					
Chrysemys picta (oviparous)	0.2 ± 0.06	5.0 ± 1.02	0.5 ± 0.01	—	—
Chelonia mydas (oviparous)	0.2 ± 0.08	1.8 ± 0.13	0.7 ± 0.88	—	—
Lizards					
Sceloporus cyanogenys	0.7 ± 0.15	0.9 ± 0.38	3.3 ± 0.48	—	3.5 ± 0.34
Chamaelo pumilis	0.9	1.0 ± 0.71	5.0 ± 3.90	2.3 ± 0.34	—
Snakes					
Natrix taxispilota	0.4 ± 0.04	0.9 ± 0.08	1.9 ± 0.24	—	—
Nerodia sipedon	1.3 ± 0.19	3.9 ± 0.83	5.0 ± 1.41	6.9 ± 0.78	2.8 ± 0.44
Thamnophis elegans	—	—	1.7 ± 0.30	6.2 ± 1.00	—

[a] Modified from Lance, V. and Callard, I. P. (1978). *In vivo* responses of female snakes (*Natrix fasciata*) and female turtles (*Chrysemys picta*) to ovine gonadotropins (FSH and LH) as measured by plasma progesterone, testosterone and estradiol levels. *Gen. Comp. Endocrinol.* **35,** 295–301.

[b] In nanograms per milliliter of plasma.

Alternation of ovulation in the ovaries of the anoline lizard *A. carolinensis* has been investigated extensively. There is a definite alteration in catecholamine activity in the hypothalamus that mirrors ovarian alternation. These observations may be explained by sensory neural connections between the ovaries and the hypothalamus that are transmitting information responsible for regulating the alternating pattern of ovulation.

Oviductal development apparently is under the influence of ovarian estrogens, and progesterone is without effect. In oviparous species, estrogens probably influence the secretion around the egg of albumin and shell from the anterior end of each oviduct. Estrogens also stimulate synthesis of vitellogenic proteins by the liver and cause increases in serum calcium of snakes, lizards, and turtles (Table 12-9). This increased availability of calcium is important for secretion of the shell by specialized cells in the lining of the oviduct. In most reptiles, measurement of calcium may be used as an indirect quantitative measurement of plasma vitellogenin. Undoubtedly the details of this process are very similar to those

Table 12-9. Effect of Estradiol-17β on Serum Calcium Levels of Ovariectomized Female Lizards (*Anolis carolinensis*)[a]

Treatment	N	Serum calcium (mg/dl ± SE)
Ovariectomized saline-injected females	5	14.8 ± 1.05
Reproductively active, sham-operated females	4	21.2 ± 3.47
Ovariectomized females injected with 1.0 μg of estradiol per day for 7 days	5	213.2 ± 9.88
Ovariectomized females injected with 10 μg of estradiol per day for 7 days	7	256.0 ± 22.13

[a] Unpublished data of K. Faber and D. Norris (1975).

described for amphibians and for birds, although vitellogenesis has not been investigated as extensively in reptiles. It is known that crocodiles produce yolk proteins that are biochemically similar to those of birds.

Oviposition or birth of live young is controlled by AVT, prostaglandins, and β-adrenergic innervation in turtles, lizards, and snakes. In the American chameleon, *A. carolinensis,* the sensitivity of the uterus to AVT is determined by the presence or absence of a corpus luteum in the adjacent ovary.

C. Environment, Behavior, and Reproduction in Reptiles

The role that physical and biological components of the environment play in sexual behavior and reproduction has been extensively studied in reptiles. Species living in temperate climates exhibit distinct seasonal patterns of hormonal secretion and reproductive events. Reproduction in tropical reptilian species varies from cyclic patterns to continuous breeding. There is a strong tendency for an observed increase in the incidence of viviparity among species inhabiting colder climates (altitude or latitude), but it is not clear which is cause or consequence.

Among temperate lizards, temperature is the dominant environmental factor influencing reproduction. Photoperiod, humidity, and nutritional status play decisive roles in some species. Other groups of reptiles have not been studied as extensively as lizards.

Although visual cues are the primary mechanism employed in reptilian courtship, evidence for pheromonal communication can be inferred from some experimental studies in all major reptilian groups. Male lizards of several families (Scincidae, Lacertidae, Teidae, and Gekkonidae) have androgen-dependent **femoral glands** located on the inner thighs as well as special cloacal glands that seem to play important roles in courtship and territorial behavior in association with breeding. Inguinal and axillary glands of chelonians have been implicated in reproductive behavior, too. Crocodilians appear to use chemical communication in courtship, but little detail is available.

When a female red-sided garter snake, *Thamnophis sirtalis,* emerges from her winter hibernaculum, she is immediately courted by a large number of males who had emerged previously. This behavior produces a mating ball of males, all attempting to copulate with a single female. The skin of the reproductive female produces methyl ketone, which apparently in the presence of estrogens attracts the males and signals them that she is ready to mate. As soon as one male successfully copulates with the female, she secretes another semiochemical that immediately turns off male mating behavior. Although injection of estrogens into adult males does not make them attractive to other males, some males with high testosterone and aromatase levels also are attractive to normal males, presumably because of the conversion of testosterone to estrogens. These "she-males" are more successful in achieving copulation with females than are normal males, presumably because the "she-males" confuse normal males who attempt to mate with them rather than with the true female, thus reducing their competition.

One of the more popular models for reptilian behavioral studies involves the work of David Crews and collaborators at the University of Texas, Austin. In general, estrogen and progesterone are responsible in reptiles as in most vertebrates for stimulating female receptive and mating behaviors whereas androgens, typically testosterone, control male behaviors. As previously mentioned, in a number of cases, androgens may be converted to estrogens by aromatase in order to produce behavioral effects. Among the species of whiptail lizards in the genus *Cnemidophorus,* about one-third are unisexual. These species consist only of females, which are further unusual in that these females are all triploid (3*n*). Studies by Crews have employed one 3*n* species, the desert grasslands whiptail *C. uniparens,* which apparently evolved from hybridization of two 2*n* species, the rusty rump whiptail *C. burti* and the little striped whiptail *C. inornatus.* Since *C. uniparens* has two sets of chromosomes derived from *C. inornatus* and only one from *C. burti,* Crews focused his behavioral studies on *C. uniparens* and *C. inornatus* (which exhibits normal sexual reproduction). A similar system occurs among the salamanders of the *Ambystoma jeffersonianum* complex, except that there the 3*n* females must mate with a male from a 2*n* species to activate cleavage in the egg, although no genetic material is contributed. In *C. uniparens,* mating does not occur with males of diploid species but rather the 3*n* females alternate between expressing female (receptive) and male (mounting) mating behaviors (Fig. 12-24). Although estradiol levels in the 3*n* females are five times less than in 2*n* females, this results in higher hypothalamic levels of estrogen re-

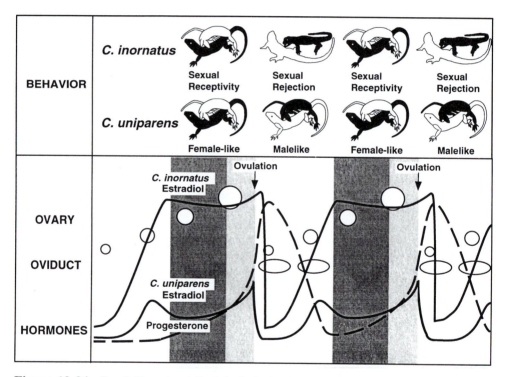

Figure 12-24. Female-like and malelike behavior in the desert grasslands whiptail lizard *(C. uniparens)* compared with female receptive behavior in a bisexual lizard, the little striped whiptail *(C. inornatus).* The differences in estradiol levels for the two species are illustrated. Female-like behavior in *C. uniparens* is elicited by a lower estrogen level and is followed by malelike copulatory behavior. The circles represent size of the ovarian follicles and the ovals indicate the presence of eggs in the oviduct. [Reprinted by permission of the publisher from Young, L. J., and Crews, D. Comparative neuroendocrinology of steroid receptor gene expression and regulation: Relationship to physiology and behavior. *Trends Endocrinol. Metab.* **6,** 317–323. Copyright 1995 Elsevier Science Inc.]

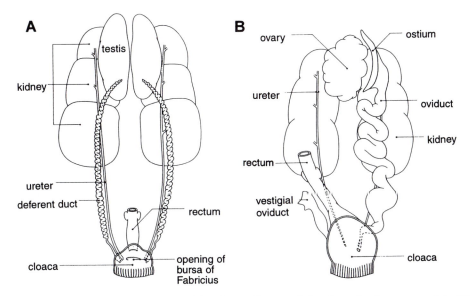

Figure 12-25. Reproductive organs of the pigeon. The top of the cloaca has been removed in both sexes. (A) The male. (B) The female: Note that the right ovary is absent in most birds (the left in others), as well as the corresponding oviduct. This regression is due to production of MIS in the embryo by the remaining ovary, which secretes estradiol locally and protects the müllerian duct on that side from MIS.

ceptor mRNA in the 3*n* brains and hence greater sensitivity to low plasma levels of estradiol. The postovulatory decrease in estrogen allows for the appearance of malelike mounting behavior or **pseudocopulation** although the exact chemical stimulus is not clear.

VII. Reproduction in Birds

Avian reproductive organs (Fig. 12-25) reflect the reductional anatomical adaptations to flight that characterize most bird systems. In the females of most species, only the left ovary and its attendant oviduct develop, whereas the right-hand components remain in a rudimentary state. This left–right asymmetry is reflected in the male, where the left testis usually is larger than the right although both are functional. Should the left ovary be removed surgically or destroyed by disease, the right rudiment may develop, but it will usually form an ovotestis or a testis rather than another ovary.

Avian gonads develop from a pair of undifferentiated primordia associated with the embryonic nephrotome. These primordia are invaded by primordial germ cells that migrate through the blood from the splanchnopleure and develop into the germinal epithelium. The embryonic gonad goes through a bipotential state in which both cortical and medullary components are present. Differentiation of cortical tissue is necessary for ovarian development and the medullary portion is suppressed. The reverse condition prevails in male birds. In contrast to mammals it is the male bird that is the homogametic sex (similar sex chromosomes), and it is the female that has unlike sex chromosomes. Developing a female phenotype requires estrogens, and castration of a young female may cause development of male plumage.

Development or suppression of the müllerian and wolffian ducts eventually depends on the direction of gonadal development, as it does in other vertebrates. In females, in which only the left half of the reproductive system usually develops, the left ovary receives the larger proportion of germ cells that migrate to the gonads. The mechanism behind this disproportionate distribution of germ cells is not known. Degeneration of the müllerian duct occurs only on the side of the smaller ovary and is induced by MIS produced in ovarian cells. However, local secretion of estrogens by the larger ovary is believed responsible for preventing oviduct degeneration on that side.

All birds are oviparous but display significantly more parental behavior than any other nonmammalian group of vertebrates. Birds are endothermic, like mammals, and use their body heat to support development of the embryo within the egg much as the mammal does *in utero.* Consequently, birds can breed successfully under conditions that are too cold for their reptilian relatives. The adaptation of long-distance flight coupled with high body temperatures allows utilization of polar and subpolar regions, where winter conditions are too severe for survival but where the summer months provide adequate warmth and food to breed and rear young birds to a size sufficient for successful fall migration to warmer latitudes for winter.

Reproduction is decisively cyclic in adult birds and is closely attuned to environmental factors (Figs. 12-26 and 12-27). Migratory and nonmigratory high-latitude and temperate species typically exhibit seasonal cycles, with breeding occurring in the spring and sometimes continuing through much of the summer. On the other hand, species occupying arid regions may show irregular cycles cued to the availability of water, which may not occur with any seasonal regularity. One tropical species, *Zonotrichia capensis,* has been reported to breed every 6 months regardless of rainfall as long as food is available.

Generally, ovaries and testes remain small in nonbreeding birds but may undergo tremendous hypertrophy in a very short time. This is especially advantageous for migratory species or species relying on particular stimuli for breeding where gonadal recrudescence can await arrival on the breeding grounds or appearance of suitable conditions such as abundant food.

A. Male Birds

Unlike the case for many mammalian species, bird testes are permanently located in the body cavity, and each testis consists of a mass of convoluted seminiferous tubules lined with a germinal epithelium and surrounded by connective tissue. Both developing germ cells and steroidogenic Sertoli cells as well as numerous fibroblasts can be seen in the germinal epithelium. As in mammals and other vertebrates the cytoplasm of the Sertoli cells completely envelops the germ cells. Typical steroidogenic interstitial cells occur between the seminiferous tubules as seen in reptiles and mammals.

In nonbreeding birds, testes are very small, and histologically these quiescent testes appear to be composed largely of interstitial cells. However, this is only an artifact produced by a marked postbreeding regression of spermatogenetic tissue. The onset of spermatogenesis (recrudescence) results in a rapid and marked increase in testicular size. Such rapid and extreme growth (to as much as 500 times the resting gonad weight) results in considerable strain and damage to the tunica albuginea surrounding the testis, and it must be replaced each year during the postnuptial phase of the testicular cycle. Replacement is accomplished through differentiation of fibroblasts and formation of a new tunica directly beneath the damaged one. It is often possible to distinguish histologically between juvenile birds and postnuptial birds by the presence of two connective tissue capsules around the testis in the latter.

Testicular recrudescence may involve a single synchronous spermatogenetic event or separate spermatogenetic waves, depending on whether a given species produces successive clutches during a particular breeding season. In either event, following spermiation, sperm migrate to expanded distal ends of the vasa deferentia known as **seminal sacs** from which sperm will be ejaculated forcefully during mating.

1. The Avian Testicular Cycle

The annual testicular cycle of temperate birds has three more or less distinct phases: (1) the regeneration or preparatory phase, (2) the acceleration or progressive phase, and (3) the culmination phase. Similar phases can be identified in all birds regardless of the seasonal nature of their reproductive cycles or of what environmental factors control testicular events.

The most common environmental factor influencing development of the avian testis is the **photoperiod** (see Fig. 12-26). Placing quiescent, temperate birds on long photoperiods will typically stimulate testicular recrudescence, whereas maintenance of these birds on short photoperiods even into the normal breeding season represses anticipated testicular events.

The **preparatory phase** of testicular development begins immediately after the reproductive period and is characterized by marked collapse of the testis. Animals in the preparatory phase are insensitive to effects of long photoperiod and are termed **photorefractory.** The end of the preparatory phase is heralded by restoration of **photosensitivity.** The endocrinological basis for the photorefractory period in birds is not clear and more than one mechanism may be involved in different species. Some studies suggest that feedback of testosterone on the hypothalamus is responsible for induction of the photorefractory period and for low levels of LH during the photorefractory period. Other investigations point to changes in hypothalamic sensitivity and/or steroid metabolism and not testosterone feedback.

Photosensitivity is separable into two phases, although there may be a continuum of the events associated usually with one phase. During the **progressive phase,** there is an increase in gonadotropin secretion brought about by actions of lengthening photoperiod on the hypothalamo–hypophysial axis. Increased circulating gonadotropins stimulate both spermatogenesis and androgen secretion by the interstitial cells. An increasingly intensive period of sexual activity and song occurs, and males of some species may begin exhibiting territorial behavior and mate selection. This effect of long photoperiod can be blocked by low temperatures.

The **culmination phase** coincides with the time of ovulation in females and includes the time of insemination. The male typically is ready for breeding before the female, and his testes will be bulging with sperm. Successful breeding involves a complex, hormonally dependent series of events involving precise male–female behavioral interactions.

a. Testicular Interstitial Cells

A characteristic lipid cycle occurs in avian interstitial cells similar to that described for other vertebrates. There is accumulation of lipid in young birds followed by rapid depletion coincident with onset of the first breeding season and spermatogenesis. The interstitial cells of adult birds are small and sparsely lipoidal in winter although they occupy a large proportion of the testis because of the regressed nature of the seminiferous tubules. There is gradual accumulation of lipids, including cholesterol, throughout the progressive phase as well as an increase in 3β-HSD activity. At the time of maximal sexual display, there is rapid depletion of interstitial cell lipid. Cholesterol disappears completely, but 3β-HSD activity remains strong, indicating lipid depletion is a consequence of rapid synthesis and secretion of androgens. The activity of 17α-hydroxylase is also high at this time (see Chapter 2 for its specific role in androgen synthesis). A massive disintegration of interstitial cells occurs during the preparatory phase, and new interstitial cells differentiate from fibroblasts.

b. Sertoli Cells

Cyclical changes in lipid content are characteristic of avian Sertoli cells, which ultrastructurally resemble steroidogenic cells. Both 3β-HSD and 17β-HSD activities have been reported for these cells. They become densely lipoidal following the breeding season, and no detectable 3β-HSD activity remains. The stored lipid is depleted with the onset of the next period of spermatogenesis.

2. Endocrine Control of Testicular Function

The hypothalamus contains two gonadotropic centers that separately control release of LH and FSH from the adenohypophysis. Hyperplasia of interstitial cells is caused by LH, and they become lipoidal and exhibit increased 3β-HSD activity. Avian testes are much more sensitive to avian LH than to mammalian LH. Androgens secreted by these cells stimulate sex accessory structures and secondary sexual characters. Similarly, purified mammalian FSH is less effective than avian FSH in stimulating spermatogenesis. Local effects of androgens from Sertoli cells are responsible for stimulating meiosis. Androgens are known to maintain spermatogenesis even in hypophysectomized birds.

Prolactin is present in the male pituitary and has been reported to inhibit FSH release and block spermatogenesis in some species. The formation of incubation patches on males of certain species is induced in part by PRL working cooperatively with testicular steroids.

3. Sex Accessory Structures in Male Birds

Wolffian ducts give rise to paired vasa deferentia, vasa efferentia, and the epididymides, all of which exhibit hypertrophy with the onset of sexual activity. These events are all

---

Figure 12-26. Plasma LH, plasma estradiol, and egg production (light bars) in female South African ostriches are affected by photoperiod. Plasma LH and testosterone levels in males are indicated by the black bars. [From Degen, A. A., Weil, S., Rosenstrauch, A., Kam, M., and Dawson, A. (1994). Seasonal plasma levels of luteinizing and steroid hormones in male and female domestic ostriches *(Struthio camelus)*. Gen. Comp. Endocrinol. **93**, 21–27.]

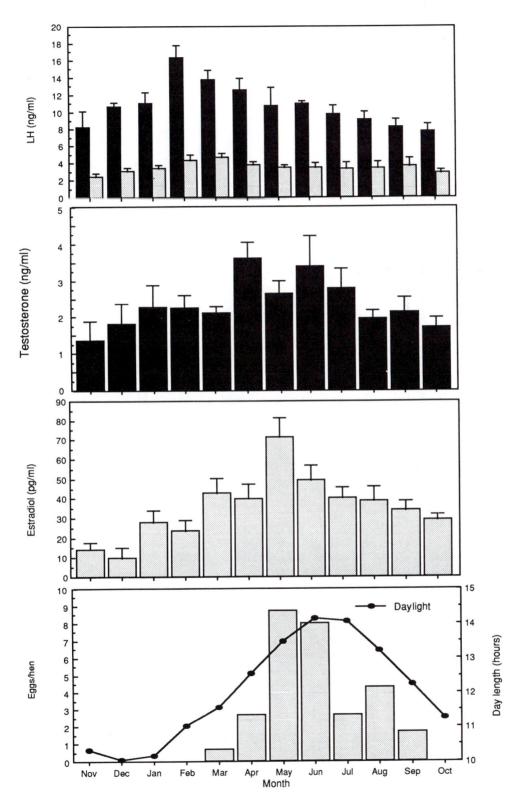

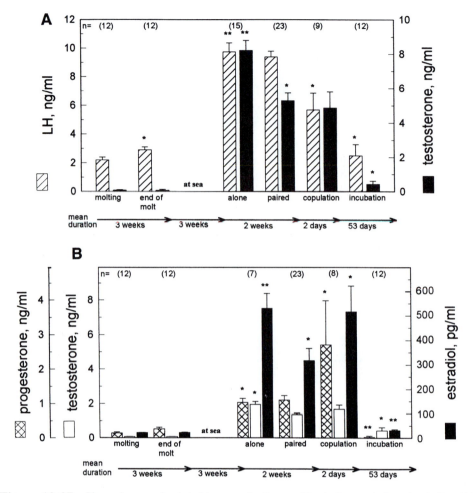

Figure 12-27.   Plasma hormone levels in king penguins from molting to the onset of egg incubation. (A)
Plasma LH and testosterone levels in males. Asterisks indicate a significant difference from
the previous data point. Numbers in parentheses indicate sample sizes. (B) Plasma progesterone, testosterone, and estradiol in females. LH levels paralleled that shown for the males
in (A). [From Mauget, R., Jouventin, P., Lacroix, A., and Ishii, S. (1994). Plasma LH and
steroid hormones in king penguins (*Aptenodytes patagonicus*) during the onset of the breeding cycle. *Gen. Comp. Endocrinol.* **93**, 36–43.]

prevented by castration. A testis is connected to the vasa efferentia by small rete tubules in
the tunica albuginea that become enlarged during the breeding season. The vasa efferentia
show increased secretory activity during the breeding season and coalesce to form a long,
coiled tube, the epididymis. Hypertrophy of the epididymis is accompanied by secretion of
seminal fluid. Mature sperm leave the epididymis and enter the enlarged vas deferens for
storage. The distal end of each vas deferens (seminal sac) fills with sperm. The posterior
walls of the seminal sacs protrude into the cloaca as erectile papillae that facilitate transfer
of sperm to the female. During copulation, the male's cloaca is everted and these erectile
papillae are brought into contact with the vagina of the female. In some species, the cloaca
actually is modified into a penis-like intromittent organ.

B. Female Birds

Considering the extensive literature available on domestic fowl, few studies of avian reproduction have employed wild species. However, these studies have provided relatively complete assessment of ovarian function in domestic species and provide a basis for comparison to wild species. In many respects, ovarian function in birds is like that of their oviparous ancestors, the chelonians, squamates, and especially the crocodilians.

The domestic hen differs most importantly from wild birds by being selected for continuous breeding. Prior to hatching, there is a proliferation of oogonia to produce thousands of primary oocytes that will serve the hen throughout her long and busy reproductive life. As in mammals, no new oocytes will be formed after hatching, unlike the situation for many anamniotes and reptiles. However, most of these oocytes will undergo atresia during early maturational stages. The primary follicle consists of an oocyte surrounded by a layer of granulosa cells. As the follicles grow, thecal layers are added, and the follicles become highly vascularized. Both granulosa and thecal cells are steroidogenic and possess 3β-HSD activity, and produce steroids as described for mammals. However, in birds, the steroidogenic process may involve three cells rather than two. In the domestic turkey, progesterone is synthesized by the granulosa cell while androgens are made in cells of the theca interna and converted to estrogens in the theca externa. This pattern has yet to be confirmed in wild species.

Estrogens secreted from the follicular cells cause the liver to produce large quantities of calcium-binding vitellogenin for the growing oocytes. This mechanism brings calcium to the oviducts, where it is used to construct the shell. Vitellogenin is enzymatically hydrolyzed to produce phosvitin and lipovitellin, which are stored in the yolk of the egg. The liver also synthesizes large quantities of triglycerides that are transported in the blood as β-lipoproteins and also are incorporated into growing oocytes. Micropinocytosis of these phosphoproteins and triglycerides by the oocytes is stimulated by FSH. Antrum formation does not occur in birds.

The large developing follicles of birds bulge conspicuously from the surface of the ovary, giving it the appearance of a bunch of grapes. The largest follicles are highly vascularized except for a rough, avascular spot, the **stigma,** where the follicle will rupture at ovulation.

Atresia of developing follicles may occur at any time during follicular development. These atretic follicles can be easily recognized by an influx of fibroblasts that phagocytize the yolk materials. Granulosa and thecal cells are lipoidal and contain cholesterol. There are many corpora atretica at all times in the ovary, but their importance, if any, is not recognized. As they disintegrate, some of the cells of the corpora atretica may become stromal interstitial cells and secrete estrogens.

Birds are characterized by the absence of persistent corpora lutea following ovulation. This feature is correlated with the absence of viviparity among the more than 8000 known species of birds. Following ovulation, collapsed follicles consist largely of granulosa cells containing progesterone, abundant smooth endoplasmic reticulum, and considerable 3β-HSD activity. The only evidence for a functional role, however, is the observation that surgical removal of these ruptured follicles increases the time that the ovulated egg is retained in the oviduct.

1. Endocrine Control of Ovarian Function in Birds

There is a close correlation between pituitary gonadotropin content and ovarian function in both domestic and wild birds (Fig. 12-26). Hypophysectomy causes ovarian regression and extensive follicular atresia, which can be prevented by gonadotropin replacement therapy. Follicular development is stimulated by FSH, and FSH will maintain oviducts in hypophysectomized but not in ovariectomized birds. Estrogen secretion is controlled by both LH and FSH. Mammalian gonadotropins, however, are not always as effective in birds as are avian pituitary or gonadotropin preparations. Furthermore, avian FSH is very effective at stimulating follicle development in lizards, emphasizing the close similarity between reptilian and avian pituitary hormones and the cautions necessary when interpreting the effects of mammalian hormones.

As is the case for certain reptiles, growth of follicles and ovulation is a continual process throughout the breeding season. Ovarian function is regulated so that typically only one egg is discharged at a time. This condition is reminiscent of the human and the lizard *A. carolinensis,* in which only one ovum is discharged, and the ovaries alternate in providing the ovum. However, birds have only one functional ovary, with a hierarchy of graded follicle size. The endocrinological basis for establishment and maintenance of this hierarchy is not known for reptiles, birds, or mammals, and it represents one of the major unanswered questions in reproductive biology.

The synthesis of vitellogenin by the avian liver is induced by estrogens as was described for other oviparous vertebrates. Total serum calcium concomitantly increases, which is related to the binding of calcium by vitellogenin (Table 12-10). In addition to incorporation of vitellogenic proteins into the oocyte, circulating calcium is sequestered by the shell glands of the "uterus" (an expanded region of the oviduct) for construction of the egg shell.

Pituitary LH is responsible for triggering ovulation of the fully mature follicle. As reptiles, progesterone levels may peak at about this time (Fig. 12-27). Plasma LH peaks about 6 to 8 hr before ovulation in domestic hens as well as in Japanese quail, but the magnitude of the avian LH surge is considerably smaller than that observed in mammals. This lower surge of LH might be an adaptation to ensure only sufficient LH for ovulating the largest follicle.

Calcium availability may be a potent factor regulating reproduction in female birds. Production of shelled eggs in domestic species directs as much as 10% of the body calcium stores per day into eggs. If large amounts of calcium are not available in the diet, the reproductive axis of the egg factory is shut down before damage to the skeleton occurs. When sufficient calcium becomes available, the birds resume laying. Although continuous egg laying does not occur in wild birds as it does in domestic fowl, it is possible that calcium depletion in wild birds contributes to cessation of breeding and induction of the refractory period.

Another pituitary hormone, PRL, plays essential roles in reproduction, including the development of a specialized, defeathered region in some species known as an **incubation**

Table 12-10. Effect of Mammalian Parathyroid Extract and the Influence of the Egg-Laying Cycle on Total Serum Calcium of Chicken[a]

Subject	Control (mg/dl ± SE)	Treated with parathyroid extract (mg/dl ± SE)
Rooster	10.1 ± 0.2	19.5 ± 3
Nonlaying hen	13.4 ± 2	19.5 ± 4
Laying hen	29.8 ± 11	47.7 ± 9

[a] Assenmacher, I. (1973). The peripheral endocrine glands. *In* "Avian Biology" (D. S. Farner and J. R. King, eds.), Vol. 3, pp. 183–286. Academic Press, New York.

patch, which aids in incubating eggs. Secretion of crop milk by the pigeon crop sac for use in feeding young birds also is induced by PRL and has resulted in development of a most useful biological assay for PRL activity in all tetrapod pituitaries (see Chapters 4 and 5). Prolactin does not affect steroidogenesis in cultured chick granulosa cells and probably has no effect on progesterone synthesis.

2. The Avian Oviduct

Estrogens secreted by the growing follicle stimulate hypertrophy of the oviduct and differentiation of secretory regions. Five differentiated regions can be identified in the mature avian oviduct that are like those of crocodilians: the **infundibulum, magnum, isthmus, shell gland,** and **vagina.** After ovulation, the ovum moves through the body cavity, enters the open end of the infundibulum, and is fertilized in the upper portion of the oviduct before the egg white protein **albumen** is added. The middle portion of the oviduct or magnum becomes highly glandular under the influence of estrogens, forming tubular glands and goblet cells. Estrogens stimulate synthesis of ovalbumen protein by these tubular glands, whereas progesterone stimulates the goblet cells to secrete the other major egg white protein, **avidin.** After accumulation of several coatings of albumen, the egg passes from the magnum to the muscular isthmus, where two shell membranes are applied. These membranes are composed of fibrous proteins cemented together with albumen. The shell consists largely of calcium salts supported by a fibrous protein matrix deposited on the outermost shell membrane by the shell gland or "uterus." After the shell has been applied, contraction of a powerful sphincter muscle causes the egg to rotate in the muscular vagina and enter the cloaca, pointed end first. Movement of the egg into the cloaca as well as its extrusion into the nest (oviposition) is controlled by AVT and prostaglandins. An increase in plasma AVT together with a concomitant decrease in neurohypophysial AVT coincides with oviposition. Treatment with AVT can cause premature oviposition.

3. Incubation Patches

In many avian species, a ventral region of skin called the **apterium** becomes defeathered, highly vascularized, and edematous just prior to or during egg laying. In addition, the epidermis of this region may exhibit hyperplasia. This specialized region is the incubation or brood patch, and when in contact with the eggs provides an efficient transfer of warmth from the parent bird to the eggs. Incubation patches may form in females, males, or both, depending on the species and which sex is responsible for incubating eggs. However, mere possession of an incubation patch is not proof of incubating behavior. Male house sparrows, *Passer domesticus,* have no incubation patch, yet exhibit incubating behavior. Conversely, male flycatchers (genus *Empidonax*) develop an incubation patch but do not show incubating behavior.

Formation of incubation patches involves cooperative actions of both estrogens and PRL. Estrogens seem to stimulate vascularization of the patch region, and PRL stimulates defeathering and epidermal hyperplasia. There typically is a concomitant transformation of pituitary lactotropes to a stimulated morphology known as "broody cells." Both PRL and estrogens are necessary for normal patch development. Furthermore, the response of epidermis in forming a patch is both site specific and tissue specific. After transplant to the dorsal surface ventral skin will still respond to PRL but not to estrogen. Vascularization of the ventral skin occurs only when it is in its normal location. On the other hand, dorsal skin transplanted to the normal patch site will not respond to either estrogens or PRL.

C. Androgen-Dependent Secondary Sex Characters in Male and Female Birds

Androgens play important roles in both male and female birds. In a number of species, a change in bill color is associated with breeding. Such changes are induced in both sexes by androgens but not by estrogens or progesterone. However, there is at least one case in which bill color change occurs only in the female, and, in that case, the color change is induced by estrogens.

Plumage color changes also may be controlled by androgens. This can occur even in the case of phalarope birds, among which the females possess the more colorful or **nuptial plumage.** One cannot presume androgens are responsible unless specific studies have been performed to verify this fact, because androgens are not always responsible for nuptial plumage. For example, development of nuptial plumage in castrated male weaver finches, *Euplectes orix,* has been used as the classic bioassay for LH (see Chapter 5). In some instances, estrogens actively inhibit formation of nuptial dress, and castration of females will cause development of male plumage. Assumption of nuptial dress in males also can be blocked with estrogens, verifying that it is the absence of estrogens that allows male plumage to develop and not the presence of androgens.

In some strains of chicken, both sexes have female-type plumage and castration causes development of male plumage. Treatment of castrated males with testosterone causes a return to female-type plumage, but growth of the comb and wattle are stimulated (a normal male trait). If castrated males are treated with DHT, the growth of the comb and wattle are stimulated but there is no reversion to female plumage. In these strains of chicken, the skin aromatizes testosterone to estrogens and stimulates female plumage. In the comb and wattle, 5α-reductase converts testosterone to DHT. Since DHT is not aromatizable, the plumage of DHT-treated castrated males does not revert to the female type.

D. Reproductive Behavior in Birds

Each avian species exhibits a precise sequence of endocrine-dependent behaviors such as migration, acquisition of territory, advertisement by song, attraction of mate, pairing, nest building, egg laying, incubation of eggs, and rearing of young birds. The actual sequence of events and their endocrinological bases are species specific and cannot easily be generalized. Successful breeding involves a complex interaction of male and female behaviors in precise sequences (that is, if male does A, then female does B, which stimulates male to do C, etc.) as well as the presence of suitable environmental cues such as proper nesting material, availability of water, etc. Little experimental work has been done with wild birds since it is difficult to get them to perform under laboratory conditions although several descriptive studies on hormone levels and behavior are available. However, much of our knowledge of wild birds comes from the work of John Wingfield at the University of Washington, Seattle, and his many associates.

Androgens appear to be responsible for territorial display and aggression in wild birds, as it is in domesticated species. Aggressive behavior also can be stimulated by FSH, but not LH, in males. Courtship appears to involve negative feedback of testosterone on FSH levels, which results in reduction in circulating androgens and allows for subsequent, less aggressive behaviors. In domestic ring doves, the initial aggressive behavior involves testosterone and copulatory behavior coincides with reduced androgens and increased aromatase activity. Androgens antagonize incubation patch development, and a reduction in circulating testosterone may be necessary for patch development in males of certain species.

Bowing behavior in feral pigeons coincides with maximal androgen synthesis but decreases prior to egg laying, coincident with an increase in progesterone levels. Progesterone is a well-known stimulus for incubation behavior in laying pigeons, and removal of the postovulatory follicle from chickens blocks nesting behavior.

VIII. Summary

Reproduction involves a precise integration of environmental factors (photoperiod, temperature, availability of nesting sites, etc.), physiological factors (nutritional state, general endocrine state with respect to thyroid hormones, adrenocortical functions, etc.), and specific endocrine secretions (FSH, LH, androgens, estrogens, progestogens, PRL, etc.). Reproductive patterns are finely tuned to environmental conditions in order to maximize evolutionary success, and this results in frequent observations of greater similarities in reproductive patterns between phylogenetically divergent species facing similar environmental problems than between closely related species living in diverse environments.

Environmental factors operate through the nervous system and specifically the hypothalamus to control release of gonadotropins and in certain cases PRL. Prolactin molecules or PRL activity as well as FSH and LH molecules have been identified in all tetrapods. Fishes have a PRL-like hormone, but some have only one LH-like gonadotropin that functions as both LH and FSH in other groups. In tetrapods, FSH initiates spermatogenesis in males and follicular development in females. Local androgens secreted from testicular cells under the influence of FSH appear to be necessary for initiating reductional division (meiosis) of primary spermatocytes. Luteinizing hormone induces androgen synthesis by interstitial (Leydig) or lobule boundary cells and spermiation in males and estrogen synthesis and ovulation in females. Androgen synthesis in female mammals also may be stimulated in thecal cells or ovarian interstitial cells by LH. Thecal androgens are thought to be converted to estrogens by granulosa cells.

Follicular atresia associated with formation of corpora atretica is a common occurrence in females. Atresia appears to be a mechanism for effectively reducing the biotic potential and placing reliance in production of a smaller number of offspring with better individual survival for evolutionary success. Corpora lutea form in many vertebrates primarily from granulosa cells of ruptured follicles, and corpora lutea synthesize progesterone that is related to gestation or behavior in many viviparous species. Many examples of autocine and paracrine regulation in the gonads are known.

Courtship and breeding behavior appear to be controlled primarily by gonadal steroids, although evidence is accumulating for participation of peptides. In addition, estrogens produce dramatic effects on vitellogenesis in nonmammalian liver and bring about a consequent disturbance in calcium metabolism. The basic oviparous mode of reproduction has become modified with respect to the development of viviparity in all nonmammalian groups except birds and agnathans.

Suggested Reading

Books

Balthazart, J. (1990). Hormones, brains and behavior in vertebrates. 1. Sexual differentiation, neuroanatomical aspects, neurotransmitters and neuropeptides. *In* "Comparative Physiology" (A. K. H. Kinne, E. Kinne-Saffran, and K. W. Beyenbach, series eds.), Vol. 8. Karger, Basel.

Balthazart, J. (1990). Hormones, brains and behavior in vertebrates. 2. Behavioral activation in males and fe-
    males—social interactions and reproductive endocrinology. *In* "Comparative Physiology" (A. K. H. Kinne,
    E. Kinne-Saffran, K. W. Beyenbach, series eds.), Vol. 9. Karger, Basel.
Becker, J. B., Breedlove, S. M., and Crews, D. (1992). "Behavioral Endocrinology." MIT Press, Cambridge, MA.
Duellman, W. E., and Trueb, L. (1986). "Biology of Amphibians." McGraw-Hill, New York.
Halliday, T. (1980). "Survival in the Wild—Sexual Strategy." Univ. of Chicago Press, Chicago.
Hoar, W. S., Randall, D. G., and Donaldson, E. M. (1983). "Fish Physiology, Volume IX. Reproduction, Part A.
    Endocrine Tissues and Hormones. Part B. Behavior and Fertility Control." Academic Press, New York.
Jones, R.E. (1978). "The Vertebrate Ovary." Plenum, New York.
Lamming, G. E. (1984). "Marshall's Physiology of Reproduction. Volume 1: Reproductive Cycles in Verte-
    brates." Churchill Livingston, Edinburgh.
Nelson, R. J. (1995). "An Introduction to Behavioral Endocrinology." Sinauer, Sunderland, MA.
Norris, D. O., and Jones, R. E. (1987). "Hormones and Reproduction in Fishes, Amphibians, and Reptiles."
    Plenum, New York.
Schreibman, M. P., and Jones, R. E. (1991). Reproduction. *In* "Vertebrate Endocrinology: Fundamentals and
    Biomedical Implications" (P. K. T. Pang and M. P. Schreibman, eds.), Vol. 4, Part A and B. Academic Press,
    San Diego.
Sharp, P. J. (1993). "Avian Endocrinology." The Society for Endocrinology, Bristol, UK.
Taylor, D. H., and Guttman, S. I. (1977). "The Reproductive Biology of Amphibians." Plenum, New York.
Van Tienhoven, A. (1983). "Reproductive Physiology of Vertebrates," 2nd Ed. Cornell Univ. Press, Ithaca, NY.
Wourms, J. P., and Callard, I. P. (1992). Evolution of viviparity in vertebrates. *Am. Zool.* **32,** 251–354.

Articles

General

Blackburn, D. G. (1991). Evolutionary origins of the mammary gland. *Mammal Rev.* **21,** 81–96.
Blackburn, D. G. (1994). Review: Discrepant usage of the term 'ovoviviparity' in the herpetological literature.
    Herpetol. J. **4,** 65–72.
Callard, I. P., Fileti, L. A., Perez, L. E., Sorbera, L. A., Giannoukos, G., Klosterman, L. L., Tsang, P., and Mc-
    Cracken, J. A. (1992). Role of the corpus luteum and progesterone in the evolution of vertebrate viviparity.
    Am. Zool. **32,** 264–275.
Crews, D. (1993). The organizational concept and vertebrates without sex chromosomes. *Brain Behav. Evol.* **42,**
    202–214.
Jones, R. E., and Baxter, D. C. (1991). Gestation, with emphasis on corpus luteum biology, placentation, and
    parturition. *In* "Vertebrate Endocrinology: Fundamentals and Biomedical Implications" (P. K. T. Pang and
    M. Schreibman, eds.), Vol. 4, Part A, pp. 205–302. Academic Press, San Diego.
Quay, W. B. (1984). Scent glands. *In* "Biology of the Integument" (J. Bereiter-Hahn, A. G. Matoltsy, K. S.
    Richards, eds.), Vol. 2, Vertebrates, pp. 357–373. Springer-Verlag, Heidelberg.
Schuetz, A. W. (1985). Local control mechanisms during oogenesis and folliculogenesis. *In* "Developmental
    Biology, A Comprehensive Synthesis, Volume I, Oogenesis" (L. W. Broder, ed.), pp. 3–83. Plenum,
    New York.
Young, L. J., and Crews, D. (1995). Comparative neuroendocrinology of steroid receptor gene expression and
    regulation: Relationship to physiology and behavior. *Trends Endocrinol. Metab.* **6,** 317–323.

Fishes

Callard, G. V. (1988). Reproductive physiology. Part B. The male. *In* "Physiology of Elasmobranch Fishes" (T. J.
    Shuttleworth, ed.), pp. 292–317. Springer-Verlag, Berlin.
Callard, I. P., and Klosterman, L. (1988). Reproductive physiology. Part A. The female. *In* "Physiology of Elas-
    mobranch Fishes" (T. J. Shuttleworth, ed.), pp. 277–291. Springer-Verlag, Berlin.
Davis, K. B., Goudie, C. A., Simco, B. A., Mac Gregor, R., and Parker, N. C. (1986). Environmental regulation
    and influence of the eyes and pineal gland on the gonadal cycle and spawning in channel catfish (*Ictalurus
    punctatus*). *Physiol Zool.* **59,** 717–724.
Dickoff, W. W., Yan, L., Plisetskaya, E. M., Sullivan, C. V., Swanson, P., Hara A., and Berrard, M. G. (1989).
    Relationship between metabolic and reproductive hormones in salmonid fish. *Fish Physiol. Biochem.*
    7, 147–155.

Dodd, J. M., and Sumpter, J. P. (1984). Fishes. *In* "Marshall's Physiology of Reproduction. Volume 1: Reproductive Cycles in Vertebrates" (G. E. Lamming, ed.), pp. 1–126. Churchill Livingston, Edinburgh.

Grober, M. S., Jackson, I. M. D., and Bass, A. H. (1991). Gonadal steroids affect LHRH preoptic cell number in a sex/role changing fish. *J. Neurobiol.* **22,** 734–741.

Koob, T. J., Laffan, J. J., and Callard, I. P. (1984). Effects of relaxin and insulin on reproductive tract size and early fetal loss in *Squalus acanthias. Biol. Reprod.* **31,** 231–238.

Kramer, C. R., Caddell, M. T., and Bubenheimer-Livolsi, L. (1993). sGnRH-A[(D-Arg^6,Pro^9,NEt)LHRH] in combination with domperidone induces gonad reversal in a protogynous fish, the bluehead wrasse, *Thalassoma bifasciatum. J. Fish Biol.* **42,** 185–195.

Mommsen, T. P., and Walsh, P. J. (1988). Vitellogenesis and oocyte assembly. *In* "Fish Physiology, Volume XI, The Physiology of Developing Fish. Part A. Eggs and Larvae" (W. S. Hoar and D. J. Randall, eds.), pp. 348–407. Academic Press, San Diego.

Nogahama, Y., Yoshikuni, M., Yamashita, M., and Tanaka, M. (1994). Regulation of oocyte maturation in fish. *In* "Fish Physiology. Molecular Endocrinology of Fish" (N. M. Sherwood and C. L. Hew, eds.), Vol. XIII, pp. 393–439. Academic Press, San Diego.

Parsons, G. R., and Grier, H. J. (1992). Seasonal changes in shark testicular structure and spermatogenesis. *J. Exp. Zool.* **261,** 173–184.

Ross, R. M., Hourigan, T. F., Lutnesky, M. M. F., and Singh, I. (1990). Multiple spontaneous sex changes in social groups of a coral-reef fish. *Copeia* **1990,** 427–433.

Scott, A. P. (1987). Reproductive endocrinology of fish. *In* "Fundamentals of Comparative Endocrinology" (I. Chester-Jones, P. M. Ingelton, and J. G. Phillips, eds.), pp. 223–256. Plenum, New York.

Sorenson, P. W., Scott, A. P., Stacey, N. E., and Bowdin, L. (1995). Sulfated 17,20β-dihydroxy-4-pregnen-3-one functions as a potent and specific olfactory stimulant with pheromonal actions in the goldfish. *Gen. Comp. Endocrinol.* **100,** 128–142.

Stacey, N. E., Sorenson, P. W., Van Der Kraak, G. J., and Dulka, J. G. (1989). Direct evidence that 17α,20β-dihydroxy-4-pregnen-3-one functions as a goldfish primer pheromone: Preovulatory release is closely associated with male endocrine responses. *Gen. Comp. Endocrinol.* **75,** 62–70.

Warner, R. R., and Swearer, S. E. (1991). Social control of sex change in the bluehead wrasse, *Thalassoma bifasciatum* (Pices: Labridae). *Biol. Bull.* **181,** 199–204.

Wourms, J. P., and Lombardi, J. (1992). Reflections on the evolution of piscine viviparity. *Am. Zool.* **32,** 276–293.

Amphibians

Del Pino, E. M., and Sanchez, G. (1977). Ovarian structure of the marsupial frog *Gastrotheca riobambae* (Fowler). *J. Morphol.* **153,** 153–162.

Houck, L. D., and Woodley, S. K. (1994). Field studies of steroid hormones and male reproductive behaviour in amphibians. *In* "Amphibian Biology" (H. Heatwole, ed.), Vol. 2. Social Behaviour. Surrey Beatty and Sons, Chipping Norton, Australia.

Jorgensen, C. B. (1992). Growth and reproduction. *In* "Environmental Physiology of the Amphibians" (M. Feder and W. W. Burggren, eds.), pp. 439–466. Univ. of Chicago Press, Chicago.

Lofts, B. (1984). Amphibians. *In* "Marshall's Physiology of Reproduction. Volume 1: Reproductive Cycles in Vertebrates" (G. E. Lamming, ed.), pp. 127–205. Churchill Livingston, Edinburgh.

Moore, F. L. (1987). Reproductive biology of amphibians. *In* "Fundamentals of Comparative Endocrinology" (I. Chester-Jones, P. M. Ingelton, and J. G. Phillips, eds.), pp. 207–221. Plenum, New York.

Moore, F. L., and Orchinik, M. (1991). Multiple molecular actions for steroids in the regulation of reproductive behaviors. *Semin. Neurosci.* **3,** 489–496.

Rose, J. D., Kinnaird, J. R., and Moore, F. L. (1995). Neurophysiological effects of vasotocin and corticosterone on medullary neurons: Implications for hormonal control of amphibian courtship behavior. *Neuroendocrinology* **62,** 406–417.

Wake, M. H. (1985). Oviduct structure and function in nonmammalian vertebrates. *In* "Functional Morphology in Vertebrates" (H.-R. Duncker and G. Fleischer, eds.), pp. 427–435. Gustav Fischer Verlag, Stuttgart.

Reptiles

Blackburn, D. G. (1992). Convergent evolution of viviparity, matrotrophy, and specializations for fetal nutrition in reptiles and other vertebrates. *Am. Zool.* **32,** 313–321.

Cree, A., Guillette, L. J., Jr., Cockrem, J. F., Brown, M. A., and Chambers, G. K. (1990). Absence of daily cycles in plasma sex steroids in male and female tuatara (*Sphenodon punctatus*), and the effects of acute capture stress on females. *Gen. Comp. Endocrinol.* **79**, 103–113.

Crews, D. (1983). Alternative reproductive tactics in reptiles. *Bioscience* **33**, 562–566.

Duvall, D., Guillette, L. J., Jr., and Jones, R. E. (1982). Environmental control of reptilian reproductive cycles. *In* "Biology of the Reptilia" (C. Gans and H. Pough, eds.), Vol. 13, pp. 201–231. Academic Press, New York.

Guillette, L. J., Jr. (1990). Prostaglandins and reproduction in reptiles. *In* "Progress in Comparative Endocrinology," pp. 603–607. Wiley-Liss, New York.

Guillette, L. J., Jr. (1993). The evolution of viviparity in lizards. *Bioscience* **43**, 742–751.

Jones, R. E., Propper, C. R., Rand, M. S., and Austin, H. B. (1991). Loss of nesting behavior and the evolution of viviparity in reptiles. *Ethology* **88**, 331–341.

Licht, P. (1984). Reptiles. *In* "Marshall's Physiology of Reproduction. Volume 1: Reproductive Cycles in Vertebrates" (G. E. Lamming, ed.), pp. 206–282. Churchill Livingston, Edinburgh.

Mason, R. T. (1993). Chemical ecology of the red-sided garter snake, *Thamnophis sirtalis parietalis. Brain Behav. Evol.* **41**, 261–268.

Owens, D. W., and Morris, Y. A. (1985). The comparative endocrinology of sea turtles. *Copeia* **1985**, 723–735.

Stewart, J. R. (1992). Placental structure and nutritional provisions to embryos in predominantly lecithotrophic viviparous reptiles. *Am. Zool.* **32**, 303–312.

Birds

Ball, G. F. (1993). The neural integration of environmental information by seasonally breeding birds. *Am. Zool.* **33**, 185–199.

Balthazart, J., and Ball, G. F. (1995). Sexual differentiation of brain and behavior in birds. *Trends Endocrinol. Metab.* **6**, 21–29.

Deviche, P. (1995). Androgen regulation of avian premigratory hyperphagia and fattening: From ecophysiology to neuroendocrinology. *Am. Zool.* **35**, 234–245.

Fivizzani, A. J., Colwell, M. A., and Oring, L. W. (1986). Plasma steroid hormone levels in free living Wilson's phalaropes, *Phalaropus tricolor. Gen. Comp. Endocrinol.* **62**, 137–144.

Follett, B. K. (1984). Birds. *In* "Marshall's Physiology of Reproduction. Volume 1: Reproductive Cycles in Vertebrates" (G. E. Lamming, ed.), pp. 283–350. Churchill Livingston, Edinburgh.

Lofts, B., and Murton, R. K. (1973). Reproduction in birds. *In* "Avian Biology" (D. S. Farner and J. R. King, eds.), Vol. 3, pp. 1–107. Academic Press, San Diego.

Wingfield, J. C., Ball, G. F., Dufty, A. M., Hegner, R. E., and Ramenofsky, M. (1987). Testosterone and aggression in birds. *Am. Sci.* **75**, 602–608.

Wingfield, J. C., O'Reilly, K. M., and Astheimer, L. B. (1995). Modulation of the adrenocortical response to acute stress in arctic birds: A possible ecological basis. *Am. Zool.* **35**, 285–294.

13

Regulators of the Gastrointestinal Tract

SURVIVAL OF any vertebrate requires ingestion of a nutrient source, enzymatic digestion of its macromolecules, and absorption of the end products of digestion. Once absorbed, these substrates may be used as energy sources, for synthesis of various molecules, or stored in some form for later utilization. The processes responsible for these events are closely regulated by the nervous system and by hormones arising from the gastrointestinal (GI) tract, the endocrine pancreas, and the hypothalamo–hypophysial system. In this chapter, we examine the major neural and endocrine players responsible for regulating digestion. In Chapter 14, we address the endocrine regulation of absorbed nutrients and the general regulation of metabolism.

Endocrinology began with identification in 1903 by Bayliss and Starling of the first hormone, **secretin,** produced by the duodenal mucosa. A few years later, Edkins proposed that **gastrin** from the antral stomach controlled acid secretion. Ironically, the GI hormones were the first endocrine messengers discovered but were among the last hormones to be chemically characterized. It was only after isolation of the first purified hormones in the 1960s that it became possible to identify the cellular types responsible for their secretion. Once purified peptides were available, we began to unravel the complex regulation of digestive events. However, advances associated with GI factors were slow compared to other areas of endocrine research. With developments of sensitive biochemical techniques, the sophisticated regulatory mechanisms operating to control nutrient procurement finally are being elucidated.

There has been rapid expansion in GI peptide research, and there are now many peptides that have been isolated and chemically characterized (see Table 13-1). Yet, there are still

Table 13-1. Mammalian Gastrointestinal Hormonal Regulators

Hormone	Number of amino acids	Cellular source	Primary function
Gastrin	17, 34	Antral G cell	Stimulates gastric acid secretion
Somatostatin (SS)	14, 28	Gastric D cell	Inhibits gastric acid secretion
Cholecystokinin (CCK)	8, 33, 39, 58	Intestinal I cell	Stimulates pancreatic enzyme secretion, gall bladder contraction, and bile release
Secretin	27	Intestinal S cell	Stimulates pancreatic HCO_3^- secretion
Vasoactive intestinal peptide (VIP)	28	Neurons	Relaxes arteriole smooth muscle and increases blood flow to intestines
Glucose-dependent insulinotropic peptide (GIP)	43	Intestinal K cells	Stimulates insulin release in presence of glucose
Motilin	22	Intestinal M cells	Stimulates migrating motor complex; causes intestinal contractions
Enteroglucagons: GLP-I and -II, glicentin	36 69	Intestinal L cells (EG cells)	Stimulates insulin release and inhibits glucagon release
Neurotensin (NT)	13	Intestinal N cells and neurons	May inhibit gastric acid secretion and motility
Gastrin-releasing peptide (GRP)	27	Gastric and intestinal neurons	Stimulates gastrin release
Calcitonin gene-related peptide (CGRP)	37	Gastric and intestinal neurons	May inhibit gastric secretion
Galanin	29/30	Intestinal neurons Brain neurons	May inhibit gastric acid secretion May stimulate appetite for fats
Peptide YY	36	Intestinal L cell	Inhibits pancreatic secretion
Peptide histidine isoleucine (PHI)	27	Intestinal neurons	Released with VIP and may have same actions

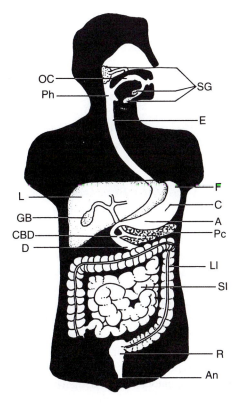

Figure 13-1. The human digestive system. OC, Oral cavity; SG, salivary glands; Ph, Pharynx; E, esophagus; F, C, and A, fundus, corpus, and antrum of stomach, respectively; L, liver; GB, gall bladder actually located beneath liver; CBD, common bile duct; D, beginning of the duodenum; Pc, pancreas; SI, rest of small intestine; LI, large intestine or colon; R, rectum; An, anus.

intestinal peptides whose physiological roles are not understood. Part of the confusion lies with the sources and modes of secretion for these factors. Some GI peptides appear to function as classic hormones, others are paracrine secretions, and still others are neurotransmitters associated with the autonomic nervous system. Many of these GI peptides later were identified within the central nervous system and in certain endocrine glands, where they function as neuromodulators, neurotransmitters, or even as neurohormones.

Endocrine regulation of digestion is an integral portion of the entire digestive process. In the following discussion, we emphasize processes occurring within the human digestive system, in particular the stomach and small intestine (Fig. 13-1). A brief comparative account of nonmammalian vertebrates follows. The metabolic fate of absorbed products of digestion and their endocrinological implications are discussed in Chapter 14. In addition to the secretions of the alimentary canal, digestion is aided by three essential exocrine glands: (1) the salivary glands, (2) the liver, and (3) the exocrine pancreas. Several pairs of salivary glands secrete saliva containing salts and a starch-digesting enzyme, **salivary amylase**. The liver and pancreas secrete enzymes and other substances directly into the small intestine via ducts (Fig. 13-1). In the following account, the reader should keep in mind the many anatomical and functional differences between the omnivorous human's digestive system and those of strictly carnivorous and herbivorous mammals. Many aspects of the human digestive system are designed for dealing with a carnivorous diet yet with some definite compensations for processing plant materials.

I. The Human Digestive System

The first detailed and systematic knowledge of human digestive processes came in the early 1800s from observations of Dr. William Beaumont on his patient Alexis St. Martin, a French Canadian who, while visiting Fort Mackinac, Michigan, was accidentally shot in the chest from close range with a shotgun. St. Martin had two fractured ribs, his lungs were lacerated, and his stomach was perforated. Beaumont assumed that St. Martin would not live the night, but he miraculously survived. The wound in St. Martin's stomach, however, never healed completely, resulting in a permanent opening to the outside (what we now call a **gastric fistula**) through which Beaumont was not only able to observe the progression of gastric digestion under varying conditions over a period of years, but he could remove samples of gastric contents for closer analysis. These pioneering observations by Beaumont stimulated much of the later interest in gastric physiology.

The accidental production of a gastric fistula in St. Martin provided the inspiration for a variety of surgical techniques in other animals, including production of gastric fistulas and gastric or intestinal pouches (isolated pouches no longer connected with the lumen of the gut). Transplantation of denervated pouches or pieces of digestive tract or pancreas to sites under the skin where revascularization can occur has enabled investigators to separate endocrine and extrinsic nervous regulatory mechanisms (Fig. 13-2). These isolated organs still contain functional elements of the enteric division of the autonomic nervous system, however, and may exhibit endogenous nervous regulation. Finally, the development of crossed circulatory systems between experimental animals confirmed the transfer of chemical factors (later to be called *hormones*) through the blood to target tissues. *In vitro* studies of pancreatic slices or mucosal tissues from various regions of the gut have been employed profitably to ascertain details of the actions of the various GI peptides and factors regulating their release. Table 13-1 summarizes the actions of some GI peptides in mammals.

A. Oral Events of Digestion

When food is ingested, it enters the mouth where it is first torn and ground by the teeth and mixed with saliva secreted from the exocrine salivary glands. Salivation is under direct neural control and can be stimulated by sight, smell, the thought of food, or by the presence of food in the mouth. Saliva is a basic fluid containing the hydrolytic enzyme salivary amylase, which hydrolyzes starches to disaccharides. The saliva and partially digested chewed food are mixed thoroughly to form a **bolus** that is pushed back into the pharynx by the tongue. After entry into the pharynx the bolus is swallowed by a reflexive series of events that propel it down the esophagus and into the stomach.

B. Gastric Digestive Events

After swallowing, the bolus of food enters an acidic environment of **hydrochloric acid (HCl)** secreted by **gastric glands** located in the **mucosa** (epithelial lining) of the **fundus** and **corpus** (body) of the stomach (Fig. 13-3). The specific source of HCl is the **parietal**

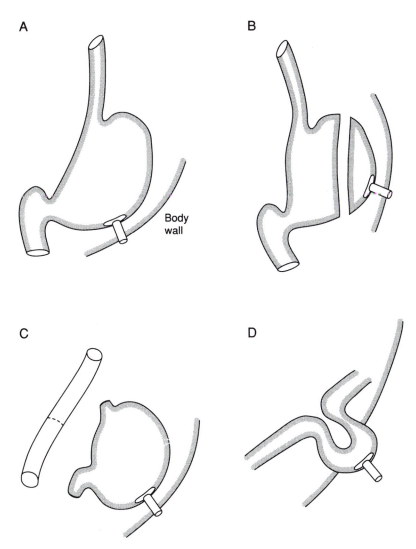

Figure 13-2. Some experimental preparations for studying digestive functions. (A) Equivalent to Alexis St. Martin, with a tube through the body wall to allow removal or addition to gastric contents. (B) A separate, enclosed Heidenhain's pouch for studying effects of blood-borne regulators without innervation or influence of stomach contents. (C) Entire isolated stomach pouch with intestine and esophagus connected. Pouch may be examined while still innervated or the nerves can be cut during the operation so that only hormonal and local factors can operate. (D) Similar approaches can be used for the small intestine as well.

cell of the mucosa. Hydrochloric acid reduces the pH of the stomach contents, thereby activating the proteolytic enzyme **pepsin,** which was secreted by the **chief cells** of the gastric glands. The acidity of the gastric contents also softens fibrous material, kills some bacteria ingested with the food, and extracts calcium salts from bone and cartilage. In addition, HCl brings about inactivation of salivary amylase as the acid penetrates the bolus and blocks further starch digestion.

The parietal cell secretes hydrogen ions (H^+), using an H^+-ATPase, with Cl^- leaving the

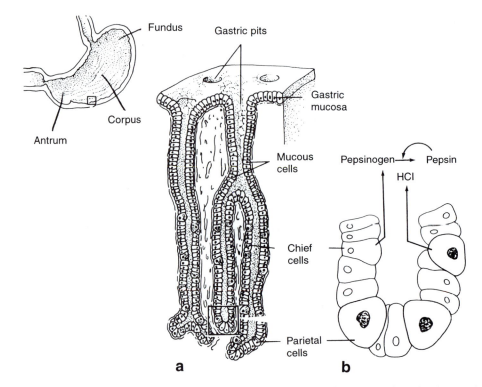

Figure 13-3. The gastric mucosa. (a) Enlargement of gastric mucosa from corpus (box) showing gastric glands. (b) Enlargement of box in a. [Modified from *Biology*, Third Edition, by Neil Campbell. Copyright (©) 1993 by The Benjamin/Cummings Publishing Company. Reprinted by permission.]

cell passively at the lumenal border. The presence of tight junctions between the mucosal cells and the high mitotic rate observed for gastric mucosal cells in part explain how the stomach contains such a caustic solution without causing irreparable destruction. These tight junctions prevent the penetration of acid and proteolytic enzyme (pepsin) into the mucosa. Meanwhile, as cells in this outer layer are damaged, they are replaced by mitotic activity to maintain the integrity of the mucosal barrier.

While in the stomach, the bolus becomes saturated with acidic gastric juices that are mixed thoroughly by peristaltic contractions of the stomach to form an acidic, viscous fluid called **chyme.** Motility of the stomach, which is stimulated by the parasympathetic system, is responsible for churning the food mass and mixing it with gastric juices to form chyme. If the pH of the chyme in the antrum is sufficiently low (that is, less than pH 4.5) and of the proper viscosity, the pyloric sphincter opens, and acidic chyme is squirted into the first segment of the small intestine, the **duodenum.** The exact mechanism controlling ejection of chyme from the stomach is not understood, and passage is normal even after surgical excision of the pyloric sphincter.

C. Intestinal Digestive Events

The intestinal phase of digestion involves secretions of the intestinal mucosal exocrine glands, which produce enzymes and alkaline mucus, as well as endocrine glands, which secrete a variety of peptide regulators. Secretions also enter the duodenum from the exocrine pancreas and liver, through the common bile duct, in response to events taking place in the duodenum. The liver produces bile containing bile salts that are important for emulsifying fats to aid their digestion by enzymes. The exocrine pancreas produces a basic bicarbonate secretion to help neutralize the acidic chyme as well as a variety of important digestive enzymes that attack proteins, complex carbohydrates, lipids, and nucleic acids.

D. Embryonic Origin of Gastrointestinal Endocrine Cells

Although the use of immunological and fluorescent techniques has enabled investigators to identify the actual cellular sources for many of the GI peptides, there is still considerable disagreement with respect to the embryonic origin or origins of these cells in mammals. Pearse proposed that all of these GI cellular types as well as calcitonin-secreting C cells of the thyroid gland, parathyroid chief cells, endocrine cells of the pancreas, melanin-containing cells, adenohypophysial cells, and the chromaffin cells of the adrenal medulla belong to the APUD cellular series (APUD, amine content and amine precursor uptake and decarboxylation) discussed in Chapter 4. These APUD cells are derivatives of neural crest or other neural ectoderm cells, suggesting that GI endocrine cells are of ectodermal rather than of endodermal origin, like the mucosa, and that they have migrated into the gastric or intestinal mucosa early during development. The neural origin for many of these endocrine cells has been confirmed although the majority do not come from neural crest cells as originally postulated by Pearse. The occurrence of many of the GI peptides in neural tissue argues strongly for a neural origin for these GI hormone-secreting cells. A single origin for GI endocrine cells is supported further by observations that immunoreactive gastrin, cholecystokinin (CCK), and glucagon appear to be colocalized in a single intestinal cellular type in an invertebrate chordate (amphioxus) and in the cyclostome fishes. An alternative view might be that at least some of these different cellular types have independently acquired APUD characteristics subsequent to or coincident with their differentiation from endodermal cells.

II. Regulation of Gastric Events

Gastric secretion is controlled at two levels. The cephalic phase involves stimulation of secretion via parasympathetic discharges elicited by the same stimuli that cause salivation; that is, sight, smell, taste, and thought or presence of food. In the gastric phase of secretory control, the presence of food in the stomach elicits secretion through vagovagal reflexes and/or through the gastrin mechanism. It has not been possible to determine which of these mechanisms is more important in controlling gastric secretion; probably all of these mechanisms operate in the normal digestive process.

A. The Gastrin Theory and Acid Secretion

Edkins in 1905 showed that extracts prepared from the most posterior portion of the stomach, the antrum, stimulated acid secretion by the gastric glands, and he suggested the name of **gastrin** for the active substance in these extracts. He found no gastrin activity in extracts prepared from the other portions of the stomach. Edkin's **gastrin hypothesis** temporarily lost credibility with the discovery of histamine, a potent stimulator of gastric acid secretion,

Peptide	Residues				
Human gastrin II	17	EEPWL	EEEEE	A Y*	G W M D F-NH$_2$
Caerulein	10		QQD	Y* T	G W M D F-NH$_2$
CCK$_8$	8		D	Y* M	G W M D F-NH$_2$

Figure 13-4. Peptides that stimulate gastric acid secretion. Caerulein is a peptide isolated from frog skin that has the common amino-terminal pentapeptide sequence (outlined) and hence activity similar to that of gastrin II. The same sequence occurs in all CCKs, although only CCK$_8$ is shown. Y*, A sulfated tyrosine residue. See Appendix C for an explanation of the single-letter code for individual amino acids.

and the demonstration of **histamine** in extracts of the gastric mucosa. It was almost 30 years before it was shown that histamine-free extracts from the mucosa of the antral portion of the stomach possessed the ability to stimulate acid secretion by the parietal cells. Neverthe-less, it was not until gastrin was finally isolated almost 60 years later and characterized chemically by Gregory and Tracey that the term "gastrin theory" finally was discarded, and the hormone gastrin was confirmed.

There are two peptide forms of gastrin, each composed of 17 amino acids: **gastrin I** and a sulfated form, **gastrin II,** which has a sulfate group attached at position 6 near the C-terminal end. A larger form of gastrin called **big gastrin** also has been found in the circulation. Big gastrin consists of gastrin I or II plus a different 17-amino acid peptide component. Only about 5% of the circulating gastrin occurs as big gastrin, however. The preprohormone for gastrin is composed of 104 amino acids and is cleaved enzymatically several times to release big and little gastrins. Most of the biological activity of these gas-trins resides in the four carboxy-terminal amino acids consisting of Trp-Met-Asp-Phe-NH$_2$. Several peptides that possess this terminal sequence (see Fig. 13-4) have been shown to stimulate acid secretion. A synthetic pentapeptide (**pentagastrin**) that incorporates the ter-minal tetrapeptide sequence is frequently used for experimental studies.

The control mechanism for acid secretion combines the observations that parasympa-thetic stimulation, acetylcholine (ACh), histamine, gastrin, and some other peptides all cause acid secretion (see Fig. 13-5). In contrast, atropine (an anticholinergic drug acting on muscarinic cholinergic receptors), procaine (an anesthetic), sympathetic stimulation, and certain antihistamines tend to reduce acid secretion under some experimental conditions. Cephalic stimulation through the parasympathetic system (ACh) can release gastrin from the **G cell** in the antral mucosa. Gastrin travels via the blood through the systemic circula-tion to the fundus and corpus of the stomach, where it stimulates the release of histamine from **tissue mast cells** situated in the mucosa. Gastrin also activates the synthesis of histi-dine decarboxylase, the enzyme responsible for histamine synthesis. Histamine in turn stimulates release of HCl from the parietal cell in the gastric glands. Gastrin may stimulate parietal cells directly without employing histamine as an intermediate, and some investi-gators conclude that histamine is not involved in endogenous acid secretion. Vagal stimu-lation or application of ACh may stimulate the parietal cells directly to secrete HCl.

The gastrin receptor is a G protein-linked molecule that works through inositol trisphos-phate (IP$_3$) and phosphokinase C (see Chapter 2) to activate H$^+$-ATPase and cause secretion of H$^+$, resulting in decreased pH of the stomach contents. Although the mechanism acti-vating H$^+$ secretion is not fully understood, an increase in cytosolic Ca^{2+} follows the bind-ing of gastrin by the parietal cell.

During active digestion the pH of the stomach may be between pH 1 and 2. Low pH in

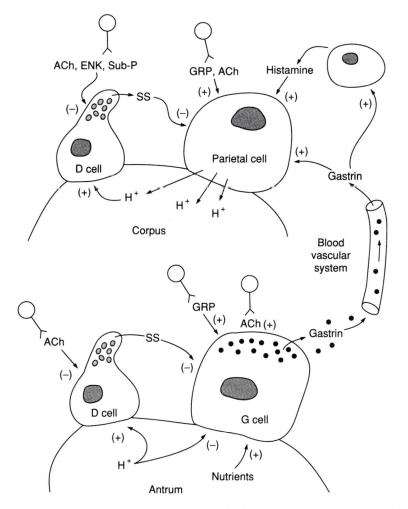

Figure 13-5. Neurocrine, endocrine, and paracrine mechanisms controlling acid secretion by the stomach. Release of gastrin from G cells in the antral stomach can be induced by nutrients in the gut lumen or by the neuropeptide gastrin-releasing peptide (GRP) or by acetylcholine (ACh). Gastrin travels through the blood and directly or indirectly (through release of histamine) stimulates the parietal cell in the corpus to secrete H^+ into the stomach lumen. The parietal cell is also stimulated by ACh and GRP. D cells in the antrum and corpus secrete somatostatin (SS), which blocks gastrin release and parietal cell secretion, respectively. D cells may be stimulated by H^+ and inhibited by neurotransmitters. H^+ in the stomach lumen also inhibits the G cell directly. ENK, Enkephalins; Sub-P, substance P.

the antral portion of the stomach (especially near the pyloric sphincter) reduces gastrin release, probably acting through release of somatostatin (SS) by **D cells** in the antral stomach.

In the 1930s, the term **"enterogastrone"** was coined to designate humoral inhibitors of intestinal origin that reduced gastric secretion and/or motility. Several peptides secreted by the small intestine may be candidates as enterogastrones either by evoking SS release or through more direct inhibitory actions. These peptides are identified in Section III.

B. Gastrin-Releasing Peptide

A novel peptide isolated from the mammalian stomach and intestine stimulates gastrin release and hence was named **gastrin-releasing peptide (GRP)**. This peptide of 27 amino acids bears a remarkable structural similarity to the 14-amino acid peptide **bombesin** that previously was isolated from the skin of frogs in the genus *Bombina*. In the stomach, immunoreactive GRP appears exclusively in neurons, but the axonal tips of these neurons do not contact the G cells directly, implying that this neurocrine behaves in a paracrine fashion.

In addition to its action on gastrin release, GRP stimulates secretion of pancreatic enzymes; contraction of gastric, intestinal, and gallbladder smooth muscle; and release of several gastrointestinal and pancreatic hormones. GRP also produces mitogenic effects resulting in hyperplasia of pancreatic, intestinal, and other tissues. At this time, it is uncertain whether all of these are physiological actions of GRP, and additional research is needed to ascertain the relevance of these varied observations to GI physiology.

Administration of GRP or bombesin into the ventricles of the brain causes a dramatic cessation of gastric acid secretion regardless of the stimulus used to evoke gastric secretion. This effect appears to be mediated via sympathetic nerves, and suggests another level for the involvement of this peptide in gastric function.

C. Secretion of Pepsinogen

The major gastric enzyme in adult humans and carnivores is the protease pepsin that is secreted by the chief cells in an inactive form called **pepsinogen.** Conversion of inactive pepsinogen to pepsin is accomplished by the presence of an excess of H^+ supplied by the parietal cells. The optimum pH for vertebrate pepsins lies between pH 1 and 2, the normal pH range observed in the stomach following stimulation of acid secretion. Pepsin is inactive above a pH of about 4.5. The presence of acid on the surface of the gastric mucosa may activate a cholinergic reflex that evokes pepsinogen release. Parasympathetic stimulation via the vagus nerve causes release of pepsinogen, and hormonal control of pepsinogen secretion may be absent. Gastrin causes release of pepsinogen only when applied in doses large enough to inhibit acid secretion by the parietal cells, implying that gastrin is not the normal factor causing pepsinogen release from the chief cells.

Several other GI peptides can invoke pepsinogen release but do so only when applied in pharmacological doses. However, one duodenal peptide (motilin; see Section III,D) has been implicated in regulating pepsinogen secretion at physiological levels. More research is needed to verify the physiological importance of motilin. Inhibition of pepsinogen release is caused by SS or by sympathetic stimulation.

III. Regulation of Intestinal Events

The duodenal mucosa contains many secretory cells, including cells responsible for digestive enzyme secretion as well as a variety of hormone-secreting cells and mucus-secreting cells. The intestinal mucosa is organized into thousands of tiny finger-like projections called **villi.** The presence of villi greatly increases the total surface area of the small intestine for releasing digestive secretions and for absorbtion of digestive products. The number of villi and secreting cells in the mucosa of the small intestine decreases progressively from the duodenum to the remaining regions, the **jejunum** and **ileum.**

Three stages or phases of intestinal regulation can be identified involving neural and endocrine mechanisms. There appears to be a distinct cephalic phase mediated via vagal stimulation that influences pancreatic secretion. The gastric phase involves vagal and vagovagal stimulation of gastrin release that appears to influence pancreatic secretion. Finally, the intestinal phase relies primarily on release of peptides stimulated by the composition of the intestinal contents.

A. Secretin

The presence of acidic chyme (pH less than 4.5) in the duodenum directly stimulates the **S cell** in the duodenal mucosa to release the peptide secretin into the blood. Although H^+ is the primary stimulus, secretin release also is stimulated by bile salts, fatty acids, sodium oleate, and several herbal extracts. Secretin stimulates the pancreas to secrete basic juice (rich in HCO_3^-) and helps to neutralize the acidity of the chyme that has entered the small intestine. Although secretin levels in the blood do not increase following ingestion of a meal, the action of secretin on the exocrine pancreas is potentiated by another intestinal peptide hormone that also increases in the blood following ingestion of a meal (see Section III,B).

Originally it was believed that secretin also was responsible for stimulating secretion of the digestive enzymes normally present in the pancreatic juice, including the proteases chymotrypsin and trypsin, pancreatic lipase, pancreatic amylase, and nucleases (DNase and RNase). After 40 years of controversy following the demonstration of secretin, it was confirmed finally by Harper and Raper (1943) that purified secretin stimulates secretion of pancreatic fluid that is rich in sodium bicarbonate and poor in digestive enzymes. A second duodenal peptide (they named it **pancreozymin**) was found to contaminate some secretin preparations. Since zymogen granules represent vesicles of stored enzyme within the acinar (exocrine) pancreatic cell, the peptide that causes extrusion of zymogen granules from pancreatic acinar cells logically could be called pancreozymin (pancreas–zymogen). It was postulated that the release of pancreozymin into the blood in response to the presence of peptides and amino acids in the chyme is due to direct actions of these molecules on pancreozymin-producing cells, similar to the action of H^+ on the S cell. Sometimes the term **secretagogue** is applied to substances present in food, products secreted from the mucosa into the gut lumen, or products of digestion that induce gastric or intestinal secretions.

Secretin consists of 27 amino acids and chemically is related to some other GI peptides (Fig. 13-6), several of which are described below. Secretin has been isolated from several mammalian species and evolutionarily appears to be very conservative (Fig. 13-7), although mammalian secretins differ markedly from avian secretin. Receptors for secretin are G protein linked, and secretin apparently operates through production of a cAMP second messenger to stimulate pancreatic HCO_3^- secretion.

B. Cholecystokinin

In 1928, Ivy and Goldberg proposed that fat in the chyme or some of the products from fat digestion stimulated release of yet another intestinal peptide that they named **cholecysto-kinin, CCK** [*chole* (bile), *kystis* (bladder), *kinein* (move)]. Cholecystokinin travels via the blood to the gallbladder, where it stimulates contraction of smooth muscles comprising the walls of the gallbladder. At the same time, it causes relaxation of the sphincter of Oddi, a

Position	Secretin	Glucagon	VIP	GIP
1	H	H	H	Y
2	S	S	S	E
3	D	Q	D	E
4	G	G	A	G
5	T	T	V	T
6	F	F	F	F
7	T	T	T	I
8	S	S	D	S
9	E	D	N	D
10	L	Y	Y	T
11	S	S	T	S
12	R	K	R	I
13	L	Y	L	A
14	R	L	R	M
15	D	D	K	D
16	S	S	Q	K
17	A	R	M	I
18	R	R	A	R
19	L	A	V	Q
20	Q	Q	K	Q
21	R	R	K	R
22	L	F	Y	F
23	L	V	L	V
24	Q	Q	N	N
25	G	W	S	T
26	L	L	I	L
27	V-NH$_2$	M	L	L
28		N	N-NH$_2$	A
29		T-NH$_2$		Q

Figure 13-6. The secretin family of peptides. There is considerable homology among secretin (bovine, porcine), glucagon (human, bovine, porcine), VIP (porcine), and the first 29 amino acids of porcine, rodent, and bovine GIP. [Data obtained from Walsh, J. H., and Dockray, G. J. (1994). "Gut Peptides." Raven, New York.]

muscle that controls exit of bile from the gallbladder. As a result, bile is expelled from the gallbladder, enters the bile duct, and is transported to the duodenum. Bile is a viscous, complex mixture consisting largely of bile salts and bile pigments. Bile salts are powerful emulsifiers (i.e., detergents). Bile pigments are breakdown products of hemoglobins, and

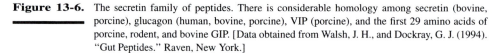

	1	11	21
Pig, cow	HSDGTFTSEL	SRLRDSARLQ	RLLQGLV
Dog	HSDGTFTSEL	SRLRESARLQ	RLLQGLV
Rat	HSDGTFTSEL	SRLQDSARLQ	RLLQGLV
Human	HSDGTFTSEL	SRLREGARLQ	RLLQGLV
Chicken	HSDGLFTSEY	SKMRGNAQVQ	KFIQNLM

Figure 13-7. Comparison of secretins from mammals and the chicken. Mammalian secretins are very conservative whereas more than half of the amino acids are different in the chicken. See Appendix C for the single-letter amino acid code. [Based on Leiter, A. B., Chey, W. Y., and Kopin, A. S. (1994). Secretin. *In* "Gut Peptides" (J. H. Walsh and G. J. Dockray, eds.), pp. 147–173. Raven, New York.]

they give bile and feces their characteristic colorations. Many other substances produced by the liver are present in bile, including metabolites of steroid hormones and inorganic iodide from deiodination of thyroid hormones. When bile enters the small intestine, the bile salts emulsify globules of fat, causing them to be dispersed as small fat droplets within the aqueous digestive fluids. The emulsification of fats (triacylglycerides, TGs) allows for marked reduction in the volume-to-surface ratio of the fat droplets, facilitating hydrolytic attack by pancreatic lipase to release monoacylglycerides and fatty acids for absorption by intestinal mucosal cells.

Isolation and purification of the first GI peptides was accomplished by Mutt and Jorpes in Sweden more than half a century after the discovery of secretin by Bayliss and Starling. Mutt and Jorpes discovered that the biological functions previously ascribed to pancreozymin and CCK resided in the same peptide consisting of 33 amino acids. Consequently, the rather cumbersome name **pancreozymin–cholecystokinin (PZCCK)** was proposed to designate the single peptide that for many years had been described in the literature under separate names. Because CCK was described first, Grossman (1970) proposed that we continue to use CCK for both actions and that we drop the designation PZ. In general, this proposal has been adopted. The **I cell** in the intestinal mucosa has been identified as the synthetic source for CCK.

Several CCK peptides have been identified, including molecules composed of 8 amino acids (CCK_8), 33 (CCK_{33}), 39 (CCK_{39}), and 58 (CCK_{58}). More recent studies indicate that the larger forms predominate and all forms are cleaved from the same gene product. In the central nervous system, CCK_8 is the common form. CCK activity is very low in the general circulation and it is difficult to obtain quantitative information on its various potential forms. However, CCK_8, CCK_{39}, and CCK_{58} are found in blood plasma of the dog, but it is not known if they are secreted in these forms or whether they are produced by peptidase activity in the blood following secretion of the larger form(s).

The separate functional roles of secretin and CCK have been demonstrated elegantly *in vitro* on slices of exocrine pancreas. Physiological levels of CCK do not evoke release of basic pancreatic juice except in the presence of secretin, implying a permissive role for CCK, enhancing the action of secretin. Purified CCK, but not secretin, causes extrusion of zymogen granules from the acinar cells. Acetylcholine also causes extrusion of zymogen granules, suggesting parasympathetic control of enzyme release from the exocrine pancreas. Parasympathetic influence via the vagus nerve and CCK probably operate via separate mechanisms since atropine blocks the effects of ACh but not of CCK on enzyme release. Furthermore, CCK does not require neural factors for its action.

Two G protein-linked receptors have been characterized for CCK. One type predominates in target cells of the digestive or alimentary tract (CCK-A receptor) and preferentially binds larger or sulfated forms of CCK. The second receptor type (CCK-B) occurs in the brain and has highest affinity for CCK_8. Once occupied, CCK receptors stimulate production of second messengers IP_3 and diacylglycerol (DAG). The former causes an increase in cytosolic Ca^{2+} and the latter activates phosphokinase C. There also is evidence suggesting CCK can work through a cAMP mechanism to produce some of its effects.

Fats, proteins, and amino acids in the gastric effluent entering the duodenum are the most potent, direct stimulators of CCK release. Conversely, active proteases such as trypsin in the duodenum inhibit CCK secretion (negative feedback). The presence of bile elements also may reduce CCK release. Basal levels of CCK are very low (about 1 pM per liter of plasma) but increase following eating to between 5 and 8 pM in about 10–45 min followed by a slow decline until the stomach is empty.

In addition to releasing enzymes in response to CCK, the rat exocrine pancreas also secretes a **monitor peptide** (61 amino acids) that stimulates CCK release, which in turn increases pancreatic secretion (positive feedback). The intestine also secretes a peptide that causes CCK release during early phases of digestion. During periods of fasting, this peptide is rapidly destroyed by the low levels of trypsin in the duodenum. However, when food enters from the stomach, the trypsin is diluted so that the intestinal peptide is not degraded so rapidly, which elevates CCK release.

In rats and dogs, bombesin and GRP are also potent releasers of CCK, which in turn stimulates pancreatic enzyme secretion. However, direct actions of bombesin and GRP have been demonstrated on pancreatic acinar cells as well. Furthermore, acinar cells are innervated by GRP-secreting neurons.

1. Additional Actions of Cholecystokinin

In several species, CCK slows the rate of gastric evacuation (emptying) through relaxation of gastric smooth muscle and contraction of the pyloric sphincter, thus slowing peristalsis and preventing passage of material from the stomach. These contradictory actions of CCK on gastric smooth muscle and the pyloric sphincter are similar to the effects observed on gallbladder smooth muscle and the sphincter of Oddi. CCK receptor antagonists accelerate gastric emptying in some human studies, suggesting a similar enterogastrone-like role for CCK. However, a physiological role for CCK on human gastric emptying is not established.

Intestinal blood flow may be altered by CCK. Certain neurons supplying blood vessels in the gut can release CCK, which causes relaxation of smooth muscle in arterioles and increases intestinal blood flow.

CCK may play an important role in regulating food intake. In the brain, CCK_8 secreted by neurons decreases food intake (appetite suppression) in rats. Furthermore, rats of an obese strain exhibit lower than normal brain levels of CCK_8. The use of CCK antagonists decreases satiety in humans, suggesting a similar mechanism is operational. Investigators also have reported increased food intake and accelerated weight gain in pigs treated with antibodies against CCK_8.

A peptide called **leptin,** which curbs appetite and increases the metabolic rate, has been discovered in rats. Leptin is secreted in greater amounts when an animal puts on fat, which then causes it to lose weight. In obese humans, however, leptin levels are about five times those of lean people, suggesting that either this protein is not regulating fat levels in humans or that obese people do not respond to this signal as do lean people.

C. Intestinal Regulation of Gastric Secretion

In 1930, Kosaka and Lim proposed the existence of an intestinal peptide that they called enterogastrone. This hypothetical hormone was released in response to the arrival of fat from the stomach and was believed to inhibit gastric motility (peristalsis) as well as reduce acid secretion by the gastric parietal cells. These actions would slow the entrance of fat into the small intestine and allow more time for proper processing of fat already present in the duodenum. There may be numerous enterogastrones, and several peptides with enterogastrone-like activity have been isolated from the small intestine (Fig. 13-8). The possible action of CCK on gastric secretion and peristalsis was discussed

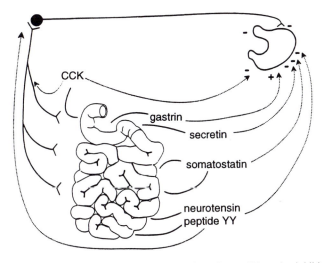

Figure 13-8. Enterogastrones. Several peptides released from the small intestine inhibit acid secretion and slow processing of food in the stomach. [From Lloyd, K. C. K., and Walsh, J. H., (1994). Gastric secretion. *In* "Gut Peptides" (J. H. Walsh and G. J. Dockray, eds.), pp. 633–654. Raven, New York.]

in Section III,B,1. Three additional intestinal peptides with enterogastrone-like actions are described below.

1. Gastric Inhibitory Peptide or Glucose-Dependent Insulinotropic Peptide

A peptide first called **gastric-inhibitory peptide (GIP)** was isolated from the intestine and shown to inhibit gastric function. GIP is produced by the **K cell** in the duodenal mucosa. Chemically, GIP is similar to glucagon and secretin (Fig. 13-6) but is considerably larger (42 amino acids). Introduction of fat or glucose into the duodenum evokes GIP release. Ingestion of a meal results in a five- to sixfold elevation in plasma GIP, which remains elevated for about 6 hr.

GIP has been extracted from several mammals, and its structure is rather conservative (Fig. 13-9). GIP blocks both gastrin-stimulated secretion of acid and enzyme by the stom-

	1	11	21	31	41
Porcine GIP	Y A E G T F I S D Y	S I A M D K I R Q Q	D F V N W L L A Q K	G K K S D W K H N I	T Q
Human GIP	Y A E G T F I S D Y	S I A M D K I [H] Q Q	D F V N W L L A Q K	G K K [N] D W K H N I	T Q
Bovine GIP	Y A E G T F I S D Y	S I A M D K I R Q Q	D F V N W L L A Q K	G K K S D W [I] H N I	T Q
Rodent GIP	Y A E G T F I S D Y	S I A M D K I R Q Q	D F V N W L L A Q K	G K K [N] D W K H N [L]	T Q

Figure 13-9. Comparison of mammalian GIP structures. These peptides are highly conservative and there have been few substitutions. See Appendix C for single-letter amino acid code. [Based on Pederson, R. A. (1994). Gastric inhibitory peptide. *In* "Gut Peptides" (J. H. Walsh and G. J. Dockray, eds.), pp. 217–259. Raven, New York.]

ach but is effective only in the experimentally denervated stomach. It may produce its inhibitory effects by stimulating release of SS, which in turn directly inhibits acid secretion by the parietal cell. In humans, large (pharmacological) doses of GIP are required to obtain even a weak effect on gastric acid secretion, implying that this is not a major role for GIP.

Other studies suggest the physiological role for GIP is to stimulate insulin release from the endocrine pancreas to facilitate absorption and distribution of glucose and amino acids (see Chapter 14). A dose-dependent release of insulin is caused by physiologically relevant amounts of GIP both *in vivo* and *in vitro*. Since it is believed this action represents the true role for GIP, the peptide has been renamed **glucose-dependent insulinotropic peptide,** which provides a more descriptive name but retains the original acronym, GIP. Insulin release requires concomitant elevation of blood glucose as well as GIP as evidenced by failure of elevated plasma GIP to release insulin following fat ingestion.

2. Peptide YY

Another potential enterogastrone may be **peptide YY (PYY),** which was first isolated from the pig intestine. It is also called peptide tyrosine–tyrosine, for it has both a C-terminal and an N-terminal tyrosine (the single-letter code for tyrosine is "Y"; see Appendix C). The peptide consists of 36 amino acids and is chemically very similar to neuropeptide Y (NPY) and pancreatic polypeptide (see Chapter 14). PYY inhibits secretion by the exocrine pancreas and reduces gastric acid secretion. Release of PYY into the circulation occurs following the introduction of fat into the duodenum, and physiologically relevant levels of PYY produce a dose-dependent inhibition of gastric acid secretion.

3. Calcitonin Gene-Related Peptide

Calcitonin is a calcium-regulating hormone produced in mammals by the thyroid C cells (see Chapters 7 and 15). **Calcitonin gene-related peptide (CGRP)** is formed in neurons by an alternative posttranslational processing of the calcitonin gene product that results in this 37-amino acid peptide. CGRP, which inhibits gastric acid secretion, has been isolated from human, rat, rabbit, bovine, and porcine sources (Fig. 13-10). Two forms have been isolated from humans and rats, each produced by a separate gene. A related peptide is **amylin** (also 37 amino acids), which has about 40–50% homology to CGRP (Fig. 13-10). Amylin, however, is cosecreted with insulin when pancreatic B cells are exposed to glucose.

Unlike most of the GI regulators, the distribution of CGRP-immunoreactive axons differs greatly among the species examined to date. For example, CGRP is virtually absent from gastric neurons in humans and pigs but is abundant in the stomachs of rat, hamster, mouse, mole, and ferret. Intermediate amounts of CGRP occur in stomachs of cats, dogs, and guinea pigs.

Most of the gastric CGRP-containing neurons are primary (sensory) afferent neurons. The cell bodies of these sensory neurons often are present in the dorsal root ganglia. CGRP neurons occur throughout the mammalian digestive tract, but endocrine cells secreting CGRP appear only in the human small intestine and the rat pancreas. CGRP-containing motor neurons from the brain stem have been described in the esophagus of monkeys and cats, and some CGRP-containing motor neurons as well as some CGRP-enteric neurons are seen in the stomach and intestine.

Figure 13-10. Amino acid sequences of CGRPs and amylins. Mammalian CGRPs are compared to chicken CGRP. Although there is considerable conservatism among the CGRPs and among the amylins, there are numerous differences between these groups. Amylin sequences homologous to CGRPs are underlined. See Appendix C for single-letter amino acid code. [Data from Holzer, P. (1994). Calcitonin gene-related peptide. *In* "Gut Peptides" (J. H. Walsh and G. J. Dockray, eds.), pp. 493–523. Raven, New York.]

CGRPS	1	11	21	31	37
Human CGRP-I	A C D T A T C V T H	R L A G L L S R S G	G V V K N N F V P T	N V G S K A	F-NH₂
Human CGRP-II	A C N T A T C V T H	R L A G L L S R S G	G M V K S N F V P T	N V G S K A	F-NH₂
Rat CGRP-I	S C N T A T C V T H	R L A G L L S R S G	G V V K D N F V P T	N V G S E A	F-NH₂
Rat CGRP-II	S C N T A T C V T H	R L A G L L S R S G	G V V K D N F V P T	N V G S E A	F-NH₂
Rabbit CGRP	G C N T A T C V T H	R L A G L L S R S G	G M V K S N F V P T	N V G S K A	F-NH₂
Bovine CGRP	S C N T A T C V T H	R L A G L L S R S G	G V V K S N F V P T	N V G S E A	F-NH₂
Porcine CGRP	S C N T A T C V T H	R L A G L L S R S G	G M V K S N F V P T	D V G S E A	F-NH₂
Chicken CGRP	A C N T A T C V T H	R L A D F L S R S G	G V G K N N F V P T	N V G S K A	F-NH₂

Amylins	1	11	21	31	37
Human amylin	K C N T A T C A T Q	R L A N F L V H S S	N N F G A I L S S T	N V G S N T	Y-NH₂
Rat, mouse amylin	K C N T A T C A T Q	R L A N F L V D S S	N N F G P V L P P T	N V G S N T	Y-NH₂
Guinea-pig amylin	K C N T A T C A T Q	R L T N F L V D S S	H N F G A A L P P T	D V G S N T	Y-NH₂
Cat amylin	K C N T A T C A T Q	R L A N F L I D S S	N N F G A I L S P T	N V G S N T	Y-NH₂

	1		11		21
Canine	F V P I F	T H S E L	Q K I R E	K E R N K	G Q
Porcine	F V P I F	T Y G E L	Q R M Q E	K E R N K	G Q

Figure 13-11. Amino acid sequences of canine and porcine motilins. See Appendix C for single-letter amino acid code. [Data from Poitras, P. (1994). Motilin. *In* "Gut Peptides" (J. H. Walsh and G. J. Dockray, eds.), pp. 261–304. Raven, New York.]

CGRP is released from the gastric mucosa of rats by the neurotoxin **capsaicin** (the active component in hot peppers) or by an increase in acidity of the duodenal contents to pH 6. CGRP probably stimulates SS release, which in turn inhibits gastric acid secretion.

D. Motilin

A small peptide (22 amino acids) that stimulates motility in the gastrointestinal tract and possibly the secretion of pepsinogen by the stomach has been isolated from the duodenum. This peptide is called **motilin** for its stimulatory action on smooth muscle, which hastens the passage of material through the small intestine. Motilin is structurally unlike the other GI peptides (Fig. 13-11). In most mammals, alkaline conditions in the duodenum cause release from the **M cells** of the small intestine. However, in humans motilin is released by acidic solutions.

In the fasting human, an endogenous rhythmic secretion of motilin activates a wave of muscular contraction called the **migrating motor complex (MMC).** This MMC begins at the antral end of the stomach every 75 to 90 min and moves along the intestine to sweep materials into the colon. This mechanism also prevents movement of bacteria from the large intestine into the small intestine, especially when food is absent.

E. Vasoactive Intestinal Peptide

Relaxation of vascular smooth muscle, which increases blood flow to the viscera, is caused by another peptide first isolated from the intestinal mucosa, **vasoactive intestinal polypeptide (VIP).** Increased blood flow into the small intestine may enhance absorption of digestion products into the blood. In addition, VIP relaxes gastric smooth muscle and slows gastric digestion.

VIP consists of 28 amino acids and is structurally similar to both glucagon and secretin as well as to GIP (Fig. 13-6). The immunoactive VIP found in the brain is structurally identical to intestinal VIP. In the intestine, VIP is produced typically in neurons, but a pyramidal-shaped **H cell** found in the small intestine and in the colon of some species also may secrete VIP. Posttranslational processing of the preprohormone produces not only VIP but another peptide consisting of 27 amino acids with an N-terminal histidine and a C-terminal isoleucine reflected in its name: **peptide histidine isoleucine** or **PHI.** Some of the actions previously attributed to VIP may be due to an interaction of PHI and VIP on the target cells. Sufficiently high doses of VIP have been shown to produce weak secretin-like effects on the pancreas as well as to induce a hyperglycemic response. These actions of VIP on the exocrine pancreas and blood sugar levels may be pharmacologic.

A major role for VIP is that of an inhibitory neurotransmitter produced by intestinal nerves. It inhibits vascular smooth muscle but stimulates glandular epithelia. It is the inhibitory action on vascular smooth muscle that increases blood flow into intestinal tissues.

Another proposed action for VIP is on the secretion of **Brunner's glands** in the duodenal mucosa. In response to the entrance of acidic chyme into the duodenum, Brunner's glands secrete a viscous, alkaline mucus that is believed to help in neutralizing gastric acid and protecting the duodenal mucosa.

F. Enteroglucagons

The **L cells** (also called EG cells) in the small intestinal mucosa produce several peptides that collectively can be called the **enteroglucagons.** These peptides are structurally and functionally like pancreatic glucagon (see Chapter 14). Enteroglucagons are extractable in a large form named **glicentin** (69 amino acids). Two other **glucagon-like peptides (GLP-I and GLP-II)** are also cleaved from the same preprohormone as glicentin. A truncated (shortened) form of GLP-I (**tGLP-I**) has been reported as well. One role for the enteroglucagons may be to stimulate insulin secretion while inhibiting pancreatic secretion of glucagon following ingestion of a meal. Furthermore, enteroglucagons may inhibit gastric acid secretion and delay gastric emptying. The physiological role for enteroglucagons and their relationship to pancreatic glucagon and glucose metabolism are considered in Chapter 14.

G. Other Gastrointestinal Peptides

Somatostatin (SS), a paracrine inhibitor produced by D cells in the stomach, also is produced by neurons and D cells located throughout the small intestine. Here, it seems to function in paracrine fashion by inhibiting release of virtually all other GI peptides. Somatostatin has been demonstrated in the arterial circulation of dogs following a meal and may act as a hormone to inhibit release of gastrin, insulin, glucagon, and pancreatic polypeptide, PP (see Chapter 14). Two forms of somatostatin have been isolated. The first is identical to the hypothalamic tetradecapeptide neurohormone (SS_{14}). The second consists of SS_{14} plus an additional 14 amino acids (SS_{28}).

Neurotensin (NT), a tridecapeptide, is another neurotransmitter first identified in the central nervous system that is found in the intestine. In addition to localization in neurons, NT also is found in epithelial **N cells** that directly contact the lumen of the small intestine. Although it has been shown to produce many effects, including that of an enterogastrone, no clear role has been established for NT.

Substance P (onadecapeptide) was the first peptide to be identified in both the intestine and brain (in 1931). It is not clear whether substance P and neurotensin function only as neurotransmitters or if they have paracrine functions in the intestine. Some of the substance P-containing neurons may be primary afferent (sensory), nociceptive (pain) neurons.

Dynorphin, a heptadecapeptide opioid whose role in the central nervous system is discussed in Chapter 4, is also present in the intestine. It is produced by neurons and probably acts as a local inhibitor of the neurotransmitter, substance P.

Galanin consists of 29 or 30 amino acids (Fig. 13-12) and is concentrated in the duodenum. It is capable of inhibiting gastric acid secretion and can block secretion of several other GI regulators including NT, enteroglucagons, SS, PYY, and pancreatic hormones. Galanin may be a modulator of GI functions, including motility, secretion, and blood flow.

	1	6	11
Common sequence (1–15)	G W T L N	S A G Y L	L G P H A -
Variable sequence (16–30)	16	21	26
Pig (29)	I D N H D	S F H D K	Y G L A -NH₂
Sheep (29)	I D N H D	S F H D K	H G L A -NH₂
Cow (29)	L D S H D	S F Q D K	H G L A -NH₂
Rat (29)	I D N H D	S F S D K	H G L T -NH₂
Chicken (29)	V D N H D	S F N D K	H G L T -NH₂
Human (30)	V G N H D	S F S D K	N G L T S -NH₂

Figure 13-12. Comparison of mammalian galanins with chicken galanin. All have the same 15 C-terminal amino acids, with only moderate differences among the N-terminal sequence. See Appendix C for single-letter amino acid code. [Data from Rokaeus, A. (1994). Galanin. *In* "Gut Peptides" (J. H. Walsh and G. J. Dockray, eds.), pp. 425–552. Raven, New York.]

Reports claim that galanin in the brain triggers a craving for fatty foods. Levels of galanin rise prior to lunch and dinner. Galanin is also thought to cause a weight increase in adolescent girls at puberty.

H. Complex Interactions of Gastrointestinal Peptides

Many studies have been published that involve observation of the effects of administering combinations of GI peptides as well as of the influences of one peptide on the release of another. Studies of this type indicate considerable overlap in the functional roles of the various peptides (for example, glucagon-like activity in secretin) although pharmacological doses usually were employed. The observation that many of these peptides double as neurotransmitters makes it even more difficult to interpret studies employing pharmacological doses. At the present time, it is difficult to sort out interactions due to structural similarities, pharmacological doses, or both from those interactions that might represent true synergisms, functional overlaps, or inhibitions. Slowly, a general pattern of regulation is beginning to emerge, but considerable research must be done before we will understand the myriad of interactions that already have been described.

IV. Comparative Aspects of Gastrointestinal Peptides

There have been relatively few studies concerning endocrine regulation of GI physiology in submammalian vertebrates, and most of these studies involve the demonstration of immunoreactivity to known mammalian peptides. The structural similarities among the different GI hormones make interpretation of these studies difficult without information on their physiological actions. Furthermore, the investigation into comparative regulation is hampered by the lack of basic understanding of the general physiology of digestion in submammalian species. For example, although extensive research in teleostean fishes of commercial importance has been accomplished with respect to diets, growth, and feeding ecology, few experiments have been concerned with physiological control mechanisms. There are many differences in digestive processes in nonmammals. For example, many fishes lack stomachs and may have other specialized structures (e.g., pyloric caeca) that imply major differences in control mechanisms. Also, digestion in fishes, amphibians, and reptiles may require days to accomplish what birds and mammals do in a few hours.

A. Invertebrates

Immunoreactive gastrin has been extracted from the GI tract of two molluscan species. The levels of extractable gastrin are comparable to those of mammals. Gastrin also has been demonstrated in the neuroendocrine cells of an insect, *Manduca septa,* supporting possible neural origins for this peptide. Galanin occurs in the nervous systems of *Bulla gouldiana,* a marine gastropod mollusk, and in the blowfly, *Phormia terraenorae.* These data suggest a broad phylogenetic distribution for these peptides that in vertebrates are associated with digestive functions. Peptides similar to or identical with other invertebrate systems are turning up in vertebrate guts and nervous systems as well. For example, bombesin from frog skin and the head-inducing substance of *Hydra* have been found in human brain and intestine.

B. Agnathan Fishes: Cyclostomes

Cyclostomes are stomachless fishes, and therefore studies have emphasized the intestinal and pancreatic tissues. Cytological studies by Ostberg and coworkers on the intestine of the Atlantic hagfish *Myxine glutinosa* have revealed the presence of primitive open-type endocrine cells. These cells extend from the basal portion of the intestinal epithelium to border on the lumen of the gut. Hagfish intestinal endocrine cells do not possess APUD characteristics, although APUD-type cells have been reported in the pancreatic islets that differentiate into insulin-producing B cells. These intestinal endocrine cells in the Atlantic hagfish do not resemble zymogen cells either, suggesting a separate origin for the endocrine and enzyme-secreting cells. Immunoreactive CCK, gastrin, glucagon-like peptides, SS, substance P, GIP, and VIP have all been reported in the hagfish intestine. Gall bladder strips prepared from a Pacific hagfish, *Eptatretus stouti,* however, did not respond *in vitro* with contractions to porcine CCK, although ACh caused contractions. Secretion of intestinal lipase in this same species is stimulated by porcine CCK.

In contrast to the hagfishes, the intestinal epithelium of larval and adult lampreys (*Lampetra* spp.) contains APUD-type cells. Immunoreactive glucagon-like peptides, NPY, PYY, SS_{14}, SS_{34}, and pancreatic polypeptide have been reported in the intestine of several lamprey species. Secretin-like and CCK-like activities are present in extracts prepared from intestines of river lampreys, *Lampetra fluviatilis,* and sea lampreys, *Petromyzon marinus.* Both secretin and CCK activities in these extracts were assayed by monitoring pancreatic secretions in the anesthetized cat.

C. Chondrichthyean Fishes

In their classical studies, Bayliss and Starling reported the presence of secretin-like activity in extracts prepared from dogfish shark and skate intestines and bioassayed in mammals. Intestinal extracts prepared from the holocephalan *Chimaera monstrosa* possess CCK activity. Porcine CCK stimulates contractions in strips of gallbladder prepared from dogfish sharks, and the intensity of the response is proportional to the dose of CCK. Immunoreactive SS and glucagon-like peptides have been reported in the gastric glands of

Pig	H S D D A V	F T D N Y	TRLRL	Q M A V K	K L Y N S	ILN-NH$_2$
Guinea pig	H S D D A L	F T D T Y	TRLRL	Q M A M K	K L Y N S	V L N-NH$_2$
Chicken	H S D D A V	F T D N Y	S R F R L	Q M A V K	K L Y N S	V L T-NH$_2$
Alligator	H S D D A V	F T D N Y	S R F R L	Q M A V K	K L Y N S	V L T-NH$_2$
Frog	H S D D A V	F T D N Y	S R F R L	Q M A V K	K L Y N S	V L T-NH$_2$
Rainbow trout	H S D D A I	F T D N Y	S R F R L	Q M A V K	K L Y N S	V L T-NH$_2$
Cod fish	H S D D A V	F T D N Y	S R F R L	Q M A A K	K L Y N S	V L A-NH$_2$
Bowfin	H S D D A I	F T D N Y	S R F R L	Q M A V K	K L Y N S	V L T-NH$_2$
Dogfish shark	H S D D A V	F T D N Y	S R I R L	Q M A V K	K L Y N S	LLA-NH$_2$

Figure 13-13. Comparison of vertebrate VIPs. This neuropeptide shows few substitutions have been made in the molecule through its long evolutionary history from fishes to mammals. [Modified from Wang, Y., and Conlon, J. M. (1995). Purification and structural characteristics of vasoactive intestinal polypeptide from trout and bowfin. *Gen. Comp. Endocrinol.* **98**, 94–101.]

five elasmobranch species. VIP has been isolated from the dogfish, *Scyliorhinus canicula* (Fig. 13-13).

D. Bony Fishes: Teleosts

Only a few species of the more than 20,000 species of teleosts have been investigated. The first observations were those of Bayliss and Starling, who reported that intestinal extracts prepared from salmon possessed secretin-like and possibly CCK-like activities. Similar activities were reported for pike, *Esox lucius,* and cod, *Gadus morhua,* using either avian or mammalian bioassays. CCK activity is present in the intestine of the Atlantic eel, *Anguilla anguilla,* and the pike. Isolated strips of gallbladder from Pacific salmon (*Oncorhynchus*) contract in the presence of porcine CCK, indicating sensitivity of the salmon gallbladder to the mammalian peptide. Intestinal strips from codfish contract in response to either gastrin or CCK$_8$.

A gastrin-histamine type of mechanism is present in the teleost stomach (cyprinid and labrid fishes lack stomachs). Extracts from the gastric mucosa of sunfish (*Lepomis machrochirus*) stimulate acid secretion in bullfrogs, and large doses of histamine (1015 mg/kg weight) induce acid secretion in the European catfish *Silurus glanis.* Histamine-induced acid secretion in cod (15 mg/kg) is blocked by certain antihistamines, supporting the existence of a mammalian-like regulatory system. Fish stomachs contain one cell type that secretes both pepsinogen and H$^+$ and probably secrete both products simultaneously.

Ten immunoreactive cell types were identified in *Sparus auratus,* including stomach cells reacting to NT, secretin, serotonin, substance P, or SS. Intestinal cells reacted to antibodies for gastrin, CCK, glucagon-like peptides, pancreatic polypeptide, substance P, or the opioid Met-enkephalin. Similar examination of the intestine of a stomachless fish, *Barbus conchonius,* exhibited all but substance P-immunoreactive cells. SS$_{14}$ is present in the cod stomach, where it inhibits acid secretion. Both SS$_{14}$ and SS$_{28}$ have been found in the intestines of several teleosts but not in the mudsucker, *Gillichthyes mirabilis.* VIP has been isolated and characterized chemically from the bowfin (a nonteleostean bony fish)and two teleosts (Fig. 13-13), but its physiological role is problematical.

E. Amphibians

Regulation of gastric mechanisms in amphibians has been studied almost exclusively in frogs. It appears that amphibians possess mechanisms very much like those of mammals, involving both neural and endocrine mechanisms. Stomachs of intact frogs or isolated gastric mucosa prepared from frogs (including *Rana pipiens, R. catesbeiana, R. temporaria,* and *R. esculenta*) respond with acid secretion when subjected to ACh, histamine, pentagastrin, or crude gastrin preparations from nonmammals or mammals. Similarly, gastric mucosa isolated from a urodele, *Necturus,* secretes acid in response to pentagastrin. Treatment with atropine or surgical vagotomy reduces acid secretion in frogs as it does in mammals. Supposedly, the release of pepsinogen in *R. esculenta* can be induced by increasing parasympathetic activity. Two skin peptides, caerulein and bombesin, have been associated with frog stomachs. Caerulein is known to stimulate acid secretion from stomach mucosa in a variety of vertebrates, and immunoreactive bombesin has been reported in frog stomachs. Chemical identification of GRP in the stomach of *Rana ridibunda* suggests that GRP may be an endogenous peptide stimulator of gastrin release.

Bayliss and Starling reported that extracts from frog intestines would evoke pancreatic secretion in dogs, providing evidence for the presence of secretin-like or CCK-like factors or both. Frog gallbladders will contract in the presence of porcine CCK *in vitro,* which supports the possible existence of a CCK-like factor in amphibians as well as a role for CCK in regulation of gastric processes. VIP also has been isolated from a frog stomach (Fig. 13-13).

F. Reptiles

Bayliss and Starling reported that a factor or factors capable of causing pancreatic secretion in mammals is present in the intestine of a tortoise. Immunoreactive somatostatin is present in the intestine of the lizard *Anolis carolinensis,* but motilin is not. In two chelonians, *Testudo graeca* and *Mauremys caspica,* immunoreactive bombesin, NT, gastrin, glucagonlike peptides, SS, PYY, and insulin were demonstrated in the intestinal mucosa, but no functional roles were indicated. The same study failed to find immunoreactive motilin, secretin, VIP, CCK, GIP, and opioid enkephalins. However, alligator VIP has been characterized chemically and found to be identical to chicken VIP (Fig. 13-13). It is remarkable that so few studies have been reported for reptiles, suggesting this is a group of vertebrates needing thorough examination.

G. Birds

Knowledge of the GI physiology of domesticated birds is more extensive than for any other nonmammalian group because of the emphasis by research in poultry science. However, little information on wild species is available. Gastrin can stimulate acid secretion in birds, and large doses of CCK cause release of enzymes from the avian exocrine pancreas. Extracts prepared from chicken intestines are strong stimulants of pancreatic secretion when assayed in turkeys, but these extracts are only weak stimulants in mammals (cat, rat), suggesting considerable molecular differences may exist in these avian factors. Mammalian glucagon and GIP appear to have no effects on pancreatic secretions in birds.

Other observations indicate there are some features in avian GI physiology that may be unique. For example, purified avian (chicken) or porcine secretin only weakly stimulates exocrine pancreatic secretion in turkeys, but purified mammalian VIP or PHI are potent stimulators. Chicken VIP has been isolated and differs structurally from porcine VIP at only four positions (Fig. 13-13). Immunoreactive SS and motilin have been demonstrated in the intestines of Japanese quail and chickens, but nothing is known about their possible functions.

V. Summary

The existence of three major GI hormones (gastrin, secretin, and CCK) postulated to be present in mammals at the beginning of this century has been established, and their primary chemical structures have been elucidated. Gastrin is produced by the G cell of the antral gastric mucosa in response to the presence of food. The parietal cell of the fundic and corpus portions of the stomach secret HCl in response to gastrin or direct neural (vagal) stimulation. Histamine may play a role as an intermediate in the action of gastrin on the parietal cell. Parasympathetic stimulation (vagal, ACh) stimulates acid secretion and also evokes secretion of pepsinogen from the chief cell in the gastric mucosa. Somatostatin produced in gastric D cells blocks gastric secretion. Release of gastrin also can be elicited by GRP.

Neural stimulation may be involved to a limited degree in the intestinal phases, but the endocrine factors predominate. Secretin is produced by the S cell of the duodenal mucosa in response to the presence of acidic chyme entering the small intestine from the stomach. The major hormonal action of secretin is to cause release of the basic juices from the exocrine pancreas. The presence of peptides, amino acids, or fats in the chyme causes release of CCK from the I cell of the intestinal mucosa, which in turn stimulates secretion of pancreatic enzymes and release of bile from the gallbladder. This pancreatic action of CCK originally was postulated to be due to a hormone called pancreozymin, which turned out to be molecularly identical to CCK.

Four additional peptides have been isolated from the mucosa of the small intestine and established as GI hormones. GIP from the K cell stimulates release of insulin from the pancreas and may inhibit gastric activity under certain circumstances. Motilin from the M cell stimulates gastric motility and possibly secretion of pepsinogen but does not influence acid secretion. VIP is a neurotransmitter that stimulates blood flow to the viscera. It is secreted along with PHI, which both are derived from the same preprohormone. Enteroglucagons (glicentin, GLP-I, GLP-II, and tGLP-I) produced by the L cells have been identified in the intestinal mucosa and appear to stimulate insulin release but inhibit pancreatic glucagon secretion. Additional GI peptides that appear to have regulatory roles include NT, PYY, dynorphin, CGRP, and galanin. Several peptides have been shown to have enterogastrone action (i.e., they inhibit gastric function), including enteroglucagons, NT, CGRP, galanin, and GIP.

Comparative studies of GI hormones are limited primarily to demonstrations of the presence of immunoreactivity to antibodies prepared against various mammalian GI peptides. Some putative regulatory peptides have been isolated that are closely related to mammalian counterparts but their physiological roles are not clear. It appears that intestinal peptides with GI hormone activities occurred early in vertebrate evolution (class Agnatha) and some counterparts are in evidence among invertebrates as well.

Suggested Reading

Books

Chivers, D. J., and Langer, P. (1994). "The Digestive System in Mammals: Food, Form and Function." Cambridge Univ. Press, Cambridge.
Holmgren, S. (1989). "The Comparative Physiology of Regulatory Peptides." Chapman & Hall, New York.

Schultz, G. M. (1989). "Handbook of Physiology, Section 6, The Gastrointestinal System." American Physiological Society, Bethesda, MD.

Stevens, C. E. (1988). "Comparative Physiology of the Vertebrate Digestive System." Cambridge Univ. Press, Cambridge.

Thompson, J. C. (1990). "Gastrointestinal Endocrinology." Academic Press, San Diego.

Walsh, J. H., and Dockray, G. J. (1994). "Gut Peptides." Raven, New York.

Articles

Conlon, J. M. (1995). Peptide tyrosine-tyrosine (PYY)—An evolutionary perspective. *Am. Zool.* **35,** 466–473.

Cooper, G. J. S. (1994). Amylin compared with calcitonin gene-related peptide: Structure, biology, and relevance to metabolic disease. *Endocr. Rev.* **15,** 163–201.

Fehmann, H.-C., Goke, R., and Goke, B. (1995). Cell and molecular biology of the incretin hormones glucagon-like peptide-I and glucose-dependent insulin releasing polypeptide. *Endocr. Rev.* **16,** 390–408.

McIntosh, C. H. S. (1995). Control of gastric acid secretion and the endocrine pancreas by gastrointestinal regulatory peptides. *Am. Zool.* **35,** 455–465.

Morely, J. E. (1995). The role of peptides in appetite regulation across species. *Am. Zool.* **35,** 437–445.

14

Chemical Regulators
of Metabolism

THE TERM **metabolism** represents the total of all enzymatic processes occuring within the cells of the body. Metabolism frequently is subdivided into specific pathways. One such subdivision is **intermediary metabolism,** which involves the metabolism of carbohydrates for energy and serves as the linkage between **protein metabolism** and **lipid metabolism.** Intermediary metabolism is especially important in the liver, which has the major responsibility for converting amino acids and fatty acids into carbohydrates for energy, especially during periods of fasting. In this chapter, we look first at the biochemical pathways involved in metabolism. We then consider the vertebrate endocrine pancreas as an endocrine system controlling metabolism. Finally, we examine the roles of extrapancreatic hormones in metabolism.

I. Major Elements of Metabolism in Vertebrates

This section is an overview of the major features of carbohydrate, lipid, and protein metabolism and their interrelationships. In addition, differences among tissues in terms of their metabolic capabilities are illustrated. This discussion sets the foundation for understanding how hormones regulate certain aspects of metabolism. Although the same metabolic enzymes and pathways are present in most vertebrates, the specific forms of these enzymes and their characteristics differ from group to group and from tissue to tissue within a group. The following account addresses a generalized pattern of metabolism for all vertebrates.

A. Intermediary Metabolism

The major pathway in intermediary metabolism involves the oxidation of glucose to produce ATP to drive energy-requiring reactions in cells. This oxidative process is known as **cellular respiration.** About 70% of the glucose used in cellular respiration is oxidized through **glycolysis** (Fig. 14-1), with the remainder being oxidized via the pentose phosphate pathway or **pentose shunt** (Fig. 14-2). These two pathways operate under both anaerobic and aerobic conditions. This alternative pathway produces **reduced nicotinamide adenine dinucleotide phosphate (NADPH),** which is necessary for the synthesis of fatty acids, for inactivation of steroids, and detoxification of many drugs. Some intermediates produced in the pentose shunt can feed into the glycolytic pathway (Fig. 14-1). Furthermore, some of the five-carbon sugars for which the pathway is named may be utilized for synthesis of ribose and deoxyribose, used for making the nucleotides found in RNA and DNA, respectively.

1. Glycolysis

The initial steps in glucose oxidation all occur in the cytosol. There are three limiting steps in glycolysis, each of which is catalyzed by a unidirectional enzyme (i.e., the reaction is irreversible because of thermodynamic considerations). Free glucose enters a cell by a protein carrier-mediated mechanism and is immediately phosphorylated to glucose 6-phosphate in most tissues by the first unidirectional enzyme, called **hexokinase** (recall that a kinase is a phosphorylating enzyme). In liver, this enzyme is called **glucokinase.** Once glucose has been phosphorylated, it can no longer be transported across the cell membrane and, in most tissues, is committed either to storage or to use for energy. However, in liver tissue (and in kidney to a limited extent), the enzyme **glucose-6-phosphatase** can remove the phosphate from glucose 6-phosphate, creating free glucose that can leave the liver cell. Glucose-6-phosphatase hydrolytic activity is lacking in almost all other tissues,

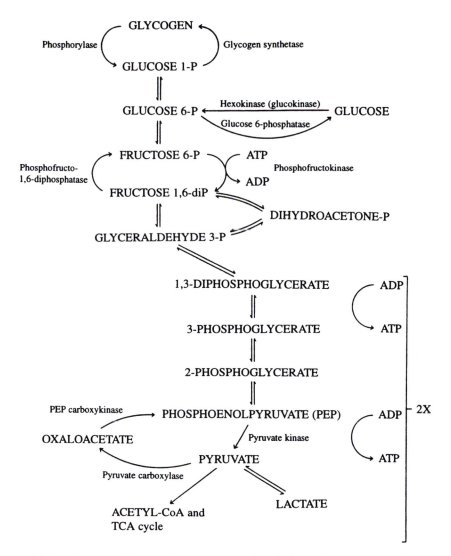

Figure 14-1. Glycolysis. This cytosolic process is called the Embden–Meyerhof pathway. It begins with glucose 1-phosphate (glucose 1-P) and ends with pyruvate (aerobic conditions) or lactate (anaerobic conditions). Lactate and pyruvate can be further oxidized under aerobic conditions via the TCA cycle or can serve as gluconeogenic substrates in the resynthesis of glucose (see text). Only the enzymes that perform regulatory roles in glycolysis are indicated.

so the liver is the only tissue that can readily supply glucose to the rest of the body under these conditions. In general, virtually all glucose in the blood is either absorbed following digestion of food or is contributed by the metabolic activities of the liver.

Glucose 6-phosphate can be metabolized through glycolysis or the pentose shunt, or it can be converted to glucose 1-phosphate and polymerized with the aid of the enzyme **glycogen synthetase** to a storage form of polysaccharide, **glycogen.** This process is called

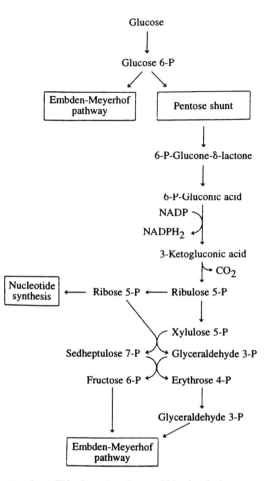

Figure 14-2. The pentose shunt. This alternate pathway within glycolysis generates substrates for fatty acid and nucleotide synthesis. See text for details.

glycogenesis. The enzyme **phosphorylase *a*** is responsible for hydrolyzing glycogen back to glucose as needed (see Fig. 14-3).

Following complete assimilation of a meal or during fasting, plasma glucose levels are maintained by the liver initially through hydrolysis of glycogen to glucose 6-phosphate (a process called **glycogenolysis**), which in turn is dephosphorylated to glucose and exits the liver cell. The liver also produces glucose from amino acids, lactate, fatty acids, and glycerol (see Section I,D).

The second one-way step occurs early in glycolysis by the action of phosphofructose kinase (PFK) with ATP and fructose 6-phosphate to form fructose 1,6-diphosphate. This commits the cell to further oxidation by splitting that C_6 sugar to two phosphorylated trioses through the glycolytic pathway, which yields a small amount of ATP.

In the third irreversible step, pyruvate kinase (PK) converts phosphoenolpyruvate (PEP) to pyruvate. In the presence of oxygen, PEP can be further oxidized to produce additional ATP or, in the absence of sufficient oxygen, converted to lactate (Fig. 14-1).

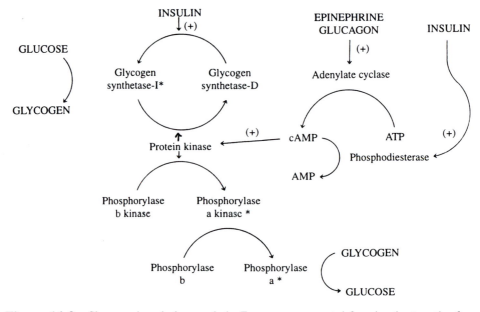

Figure 14-3. Glycogenesis and glycogenolysis. Enzymes are converted from inactive to active forms by various hormones that alter the availability of glucose for metabolism. Insulin favors glucose storage as glycogen whereas epinephrine and glucagon favor glucose oxidation or release into the blood. The asterisk (*) denotes the active form.

2. The Tricarboxylic Acid Cycle and the Electron Transport Chain

In the presence of oxygen and coenzyme A, pyruvate leaves the cytosol and enters a mitochondrion, where it interacts with coenzyme A and is converted to **acetyl-coenzyme A (acetyl-CoA)** plus CO_2. Acetyl-CoA combines with the four-carbon compound called oxaloacetate to produce citric acid or citrate (Fig. 14-4). Through a series of reactions that generates two additional CO_2 molecules and a number of important reduced compounds (NADPH, NADH, and FADH), the citrate is oxidized back to oxaloacetate that can combine with another acetyl-CoA generated from glucose. This series of cyclic events constitutes the **tricarboxylic acid** or **TCA cycle,** so named because citrate has three carboxyl groups in its structure. It also is known as the "citric acid cycle" since it begins with the combination of acetyl-CoA and oxaloacetate to form citric acid, as well as the "Krebs cycle" after the man who discovered it.

The reduced compounds generated by the TCA cycle are used to transfer electrons to a series or chain of electron-transfer molecules called cytochromes. This **electron transport chain** also is located in the mitochondrion. A final electron transfer is made to oxygen, which results in the generation of water. The energy released by the sequential transfers of electrons from higher to lower energy levels along the electron-transport chain is used to generate ATP. Thus, the complete oxidation of glucose to CO_2 and water requires glycolysis, the TCA cycle, the electron transport chain, and oxygen as the final electron acceptor in order to produce a large number of ATPs. This process is referred to as aerobic cellular respiration or simply as **aerobic respiration.** In the absence of oxygen, the oxidative pathway stops in the cytosol at pyruvate, with only a small amount of ATP produced. The excess pyruvate is converted by the enzyme **lactate dehydrogenase** to lactate. This process

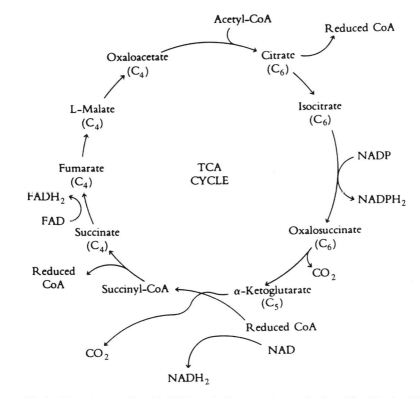

Figure 14-4. The tricarboxylic acid (TCA) cycle. Pyruvate is completely oxidized in the TCA cycle to CO_2, which is released from the cell. Electrons (plus hydrogen ions) are transferred to electron acceptors (FAD, NAD, and NADP) that will be used by the electron transport chain (cytochromes) to synthesize ATP and water. The TCA cycle and the cytochromes are all confined within the mitochondria.

is called **anaerobic respiration.** When oxygen is available, lactate can be converted back to pyruvate by the same enzyme. In some animals, a portion of the pyruvate is converted irreversibly to ethanol and CO_2.

B. Protein Metabolism

Proteins are composed of more than 20 different amino acids. In animals that have a protein-rich diet, such as most predators, ingested proteins are hydrolyzed to amino acids and absorbed through the intestine (see Chapter 13 for a more detailed description). These amino acids can be used to synthesize proteins and the excess can be converted to carbohydrate for energy use or to lipids for storage. Under conditions of starvation, hydrolysis of proteins located primarily in muscle cells provides amino acids for energy production in the muscle cells and amino acids for conversion to glucose by the liver (see Section I,D).

Some organisms can synthesize all the amino acids required for protein synthesis, but many vertebrates cannot. Humans and some other vertebrates must rely on a dietary source of what have been called **essential amino acids.** In humans, nine amino acids are considered to be essential and must occur in the diet.

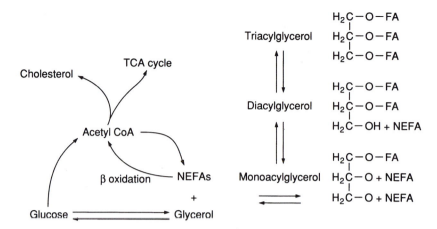

Figure 14-5.   Triacyglycerols: esterification and lipolysis. Fatty acids (FA) and glycerol produced from glucose may serve as the raw materials used to construct triacylglycerols or fats. Hydrolysis
of fats yields nonesterified fatty acids (NEFAs) and glycerol, which can be utilized for
energy. Triacylglycerides also may be used as a source for glucose (see Fig. 14-8).

C. Lipid Metabolism

Lipogenesis is the synthesis of lipids. The most important lipids are the **triacylglycerols**
(oils and fats), **fatty acids,** and **steroids.** Synthesis of a triacylglycerol begins with glycerol and one fatty acid and progresses stepwise to the triacylglycerol by the actions of
a series of acylglycerol enzymes. This synthetic process is usually called **esterification.**
The first step in esterification is the formation of a **monoacylglycerol** consisting of a
molecule of glycerol esterified to a fatty acid (Fig. 14-5). This step is catalyzed by a
monoacylglycerol synthetase and occurs more commonly in the liver. Addition of a
second esterified fatty acid produces a **diacylglycerol,** and a third addition produces a
triacylglycerol. If the attached fatty acids have many double bonds or are unsaturated,
the triacylglycerol is likely to appear as an oil at room temperatures. If there are few
double bonds in the esterified fatty acids, the compound is a solid at room temperature
and we call it a fat. We do not distinguish between these saturated and unsaturated triacylglycerols in the following discussions, although most of the natural compounds are
saturated.

   The liver provides triacylglycerols to other tissues (mainly muscle and adipose tissues)
for storage or utilization. Hydrolysis of triacylglycerols in storage tissues is called **lipolysis**
and occurs through the actions of **triacylglycerol lipase** and **diacylglycerol lipase.** Although lipolysis of stored triacylglycerides is initiated primarily in muscle and adipose
tissue, **monoacylglycerol lipase** is active only in the liver. Fatty acids freed through lipolysis are called **nonesterified fatty acids** (**NEFAs**). They can be released into the blood,
where they bind reversibly to serum proteins and are transported to the liver. NEFAs were
first called "free" fatty acids (FFAs), but NEFA is more appropriate since they are protein
bound as they circulate in the blood and are really not free.

   NEFAs can be synthesized in the liver and other tissues from acetyl-CoA. However,
acetyl-CoA is generated within the mitochondrion and cannot cross the mitochondrial
membrane and enter the cytosol where fatty acid synthesis occurs. Apparently citrate is
transported into the cytosol and used to regenerate acetyl-CoA. The rate-limiting enzyme,

acetyl-CoA carboxylase, initiates fatty acid synthesis from cytoplasmic acetyl-CoA. Glucose metabolism provides glycerol, which is enzymatically coupled to a fatty acyl–CoA complex to form a monoacylglycerol.

The steroid cholesterol may be absorbed following a meal or may be synthesized from cytoplasmic acetyl-CoA (see Chapter 2). This synthetic process involves activation of the rate-limiting enzyme **hydroxymethylglutaryl-CoA reductase (HMG-CoA reductase)** and ends with the synthesis of cholesterol in the cytosol. Cholesterol and cholesterol esters (cholesterol linked by an ester bond to a fatty acid) may be extruded in lipoprotein droplets and transported through the blood to sites of cholesterol utilization or may be converted to bile salts and stored in the gallbladder for use in digestion (see Chapter 13). Cholesterol synthesis may be in direct competition with fatty acid synthesis for the cytoplasmic substrate acetyl-CoA, and the regulation of the rate-limiting enzymes for these pathways is critical. Cholesterol is necessary for building cell membranes (especially in dividing cells) and is the precursor for the synthesis of all of the steroid hormones as described in Chapter 2.

Following absorption of a meal, excess carbohydrates and amino acids are converted into fats by the liver. These triacylglycerols plus cholesterol are surrounded by special proteins called **apoproteins** and are extruded into the blood as **very low-density lipoprotein (VLDL) droplets** (Fig. 14-6). The apoproteins apparently act as "docking proteins" for attachment to endothelial cell membranes containing **lipoprotein lipase.** This enzyme is most prevalent in the endothelial cells of capillaries in adipose tissue and to some extent in capillaries of skeletal muscle. Lipoprotein lipase converts triacylglycerols to monoacylglycerols and NEFAs, which diffuse into the adipose or muscle cell where they are reconstituted to triacylglycerols. The remaining lipoprotein droplet, which has lost some of its triacylglycerol, has a relatively higher concentration of cholesterol and protein, and therefore is now more dense. After the VLDLs have lost about 50% of their triacylglycerol content, they are classified as **intermediate-density lipoproteins (IDLs).** If the IDLs remain in the circulation, they continue to lose triacylglycerols. After they lose about 90% of their triacylglycerol content, as well as the special protein **apoprotein E,** they are renamed **low-density lipoproteins (LDLs).** Furthermore, without apoprotein E, they remain longer in the circulation and can transfer their remaining load of cholesterol to other cells, especially endothelial cells of arteries.

Apoprotein E, located on the VLDL and IDL surface, has a high affinity for special so-called **LDL receptors** found on dividing cells, steroidogenic cells, and liver cells. Although VLDLs and IDLs are readily bound by LDL receptors, an LDL has no apoprotein E and cannot be removed easily from the blood. However, **apoprotein B100,** still present on the surface of the LDL, also can bind to the LDL receptor but not as strongly as apoprotein E, which allows LDLs to enter liver cells at a slow rate. Thus, LDLs tend to accumulate in the blood and continue to deliver cholesterol to arterial sites for deposition. Once taken into a dividing or steroidogenic cell, the contents of the lipoprotein droplets are used up. In the case of liver cells, more triacylglycerols are added to the IDLs and LDLs, changing them back to VLDLs and shunting them back to the adipose tissues to unload additional triglycerides for storage.

Another group of lipoprotein droplets called **high-density lipoprotein (HDL) droplets** represents a special group of droplets that are high in cholesterol as well as apoproteins. A special role for HDLs is to remove cholesterol from tissues. In addition, HDLs can transfer excess apoprotein E to LDLs, allowing them to be removed more rapidly from the blood and recycled into VLDLs by the liver, thereby reducing their opportunity to interact with

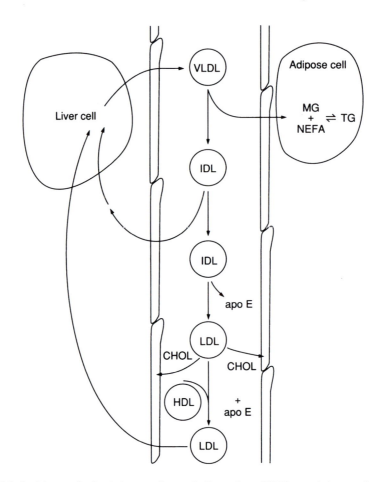

Figure 14-6. Lipoprotein shuttle between liver and adipose tissue. VLDLs consisting mostly of triacylglycerols (TG) leave the liver and travel through the blood to adipose cells. Lipoprotein lipase associated with the capillary cells hydrolyzes the triacylglycerols to monoacylglycerols (MG) and
NEFAs, which enter the adipose cells and are reformed into triacylglycerols. As they lose triglycerides, VLDLs become IDLs and either are removed by liver cells and recharged with triacylglycerols or lose more triacylglycerols and apoprotein E (apo E) and become LDLs. Interaction of LDLs with HDLs results in transfer of apo E to the LDLs so they can be removed more
rapidly by the liver cells. While in circulation, LDLs can transfer cholesterol (CHOL) to other cells.

other cells. It is presumed that a higher ratio of HDLs to LDLs represents a shorter turnover
time for LDLs and thus less opportunity for them to transfer cholesterol to arterial cells.

D. Gluconeogenesis

The synthesis of glucose from noncarbohydrate sources is termed **gluconeogenesis,** and
this term encompasses the conversion of amino acids, lactate, glycerol, or lipids into
glucose. Amino acids may simply be transaminated or deaminated to produce intermediates of glycolysis or the TCA cycle (Fig. 14-7). For example, transamination of the amino
acids alanine, serine, and glycine results in formation of pyruvate and transamination of
aspartate yields the TCA cycle intermediate oxaloacetate. These transaminations require

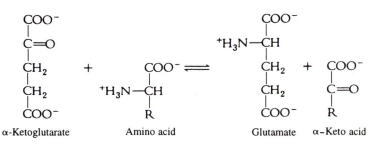

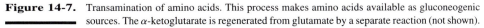

α-Ketoglutarate   Amino acid   Glutamate   α–Keto acid

Figure 14-7. Transamination of amino acids. This process makes amino acids available as gluconeogenic sources. The α-ketoglutarate is regenerated from glutamate by a separate reaction (not shown).

α-ketoglutarate as a recipient for the amino group of the amino acid. Regeneration of α-ketoglutarate from glutamate can be accomplished by urea cycle enzymes or by direct deamination of the amino acids arginine, lysine, or histidine. Of the amino acids mentioned above, only lysine cannot be resynthesized in humans.

Pyruvate, reconverted from lactate by liver cells in the presence of oxygen, can be a source of oxaloacetate. The rate-limiting enzyme for gluconeogenesis is **phosphoenolpyruvate carboxykinase (PEPCK),** which converts oxaloacetate to PEP, which in turn can be used to synthesize glucose (see Fig. 14-8). The ratio of PK to PEPCK can be used as an index to assess the gluconeogenic activity of a tissue.

Fats contribute to energy metabolism and glucose synthesis in two ways. Hydrolysis of

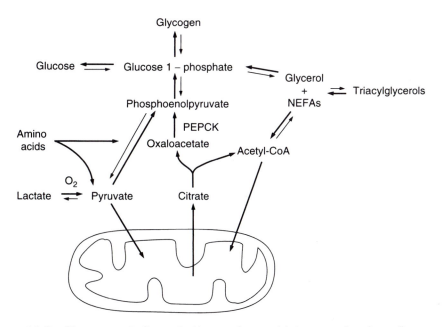

Figure 14-8. Gluconeogenesis. Summarized here are the potential gluconeogenic pathways (large arrows). Not all of these pathways are present in all cells. Some substrates must cycle through the mitochondria. Triacylglycerols, amino acids, and lactate can be used for gluconeogenesis. Note the central position of the gluconeogenic enzyme PEPCK.

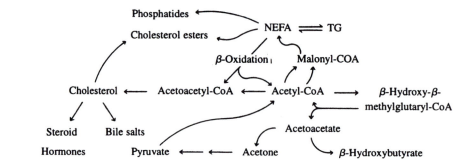

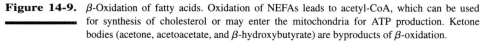

Figure 14-9. β-Oxidation of fatty acids. Oxidation of NEFAs leads to acetyl-CoA, which can be used for synthesis of cholesterol or may enter the mitochondria for ATP production. Ketone bodies (acetone, acetoacetate, and β-hydroxybutyrate) are byproducts of β-oxidation.

monoacylglycerol in the liver frees glycerol, which can be used to resynthesize glucose. NEFAs are metabolized through special pathways to cytoplasmic acetyl-CoA, which can be used for energy or to synthesize cholesterol and bile salts.

E. Fatty Acid Metabolism and Ketogenesis

NEFAs can be hydrolyzed through a pathway known as β-oxidation in liver cells (Fig. 14-9). This process results in formation of cytoplasmic acetyl-CoA, which can be transported into the mitochondria for ATP synthesis or converted to pyruvate and gluconeogenesis. However, certain byproducts of β-oxidation known as **ketone bodies** are not metabolized in the liver but are released into the blood. The most common ketone bodies are **acetone, acetoacetate,** and **β-hydroxybutyrate**. Under certain circumstances (e.g., during starvation), other tissues may utilize ketone bodies directly as gluconeogenic substrates. Excessive liver ketogenesis, however, can alter blood pH significantly and can lead to coma and eventually to death.

II. The Mammalian Pancreas

The mammalian pancreas is a mixed gland of exocrine and endocrine components that play essential roles in digestion and metabolism. The digestive role is accomplished by the exocrine portion of the pancreas, which produces digestive enzymes and an alkaline pancreatic fluid. The endocrine portion of the pancreas influences metabolism by secreting hormones that regulate carbohydrate, lipid, and protein metabolism.

The exocrine or acinar pancreas secretes critical digestive enzymes into the pancreatic duct, by which they reach the lumen of the small intestine. [An *acinus* is similar in structure to an endocrine follicle except that the lumen of the acinus makes contact with a duct through which the secretory products of the epithelium (acinar cells) are transported.] This pancreatic juice also contains bicarbonate ions that buffer the acidic material entering the small intestine from the stomach (see Chapter 13).

The endocrine pancreas consists of small masses (or islands, or islets) of endocrine cells scattered among the acinar tissue and known as the **islets of Langerhans** (Fig. 14-10). The pancreatic islets secrete at least four regulators: **insulin, glucagon, somatostatin (SS),** and

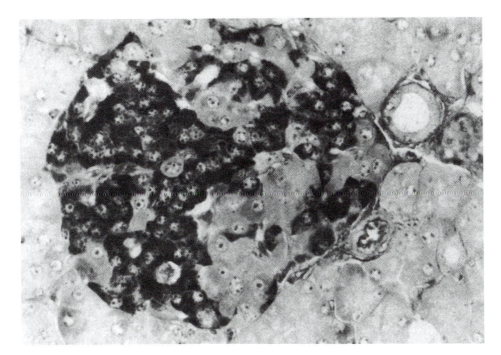

Figure 14-10. Pancreatic islet of Langerhans. This islet from a guinea pig pancreas contains mostly dark-staining B cells and light-staining A cells. [From Matsumoto, A., and Ishii, S. (1992). "An Atlas of Endocrine Organs." Japanese Society for Comparative Endocrinology and Springer-Verlag, Berlin.]

an additional peptide known as **pancreatic polypeptide (PP).** Insulin is primarily a hypo-glycemic agent that lowers blood glucose, whereas glucagon is hyperglycemic. Glucagon and insulin may produce opposing effects on lipid metabolism as well, with glucagon pro-moting lipolysis. The major role of pancreatic somatostatin may be as a paracrine agent that inhibits local release of the other pancreatic peptides. The role for PP is not well estab-lished, but its increase in the circulation following ingestion of a meal suggests a role in postabsorptive metabolism.

Historically, the endocrine pancreas is linked with blood glucose regulation. Recognition of this role originated from observations on the clinical syndrome **diabetes mellitus.** This disease was first described by the Greek physician Aretaeus in about 1500 B.C. as a condi-tion where "flesh and bones run together" and are siphoned into the urine. The term "dia-betes" comes from the Greek word *diabainein* (*dia,* through; *bainein,* to go). Indian phy-sicians reported the sweet taste of the diabetic's urine in the sixth century, but it was not until the eighteenth century that the word "mellitus," referring to the presence of glucose in the urine, was added. Diabetes mellitus was always fatal and even today shortens the lives of afflicted people. It is a leading cause of death in the United States. We now recog-nize two major forms of diabetes mellitus; one is related to insufficient production of insulin and the second involves an unresponsiveness of target cells to insulin (see Section II,D,1).

Diabetes mellitus is characterized by the production of large quantities of glucose-containing urine. Normally little or no glucose should appear in the urine, and its presence contributes osmotically to the larger urine volume produced by the untreated diabetic. It was the presence of glucose in the urine of pancreatectomized dogs, demonstrated by von Mering and Minkowski in 1889, that first linked diabetes mellitus to a possible disorder of the pancreas.

The occurrence of glucose in the urine results from excessively high circulating levels of glucose as a result of the failure by the endocrine pancreas to secrete adequate amounts of insulin. Blood levels of glucose become elevated, resulting in more glucose in the glomerular filtrate produced in the kidney. Under normal conditions, the proximal convoluted portion of each nephron transports all glucose that appears in the glomerular filtrate back into the blood. However, should blood sugar levels be sufficiently high, so much glucose enters the glomerular filtrate that the carrier-mediated transport mechanism in the nephron is saturated, and all of the glucose cannot be returned to the blood. The glucose remaining in the lumen of the nephron upsets the osmotic balance and less water can be reabsorbed from the filtrate, resulting in production of a greater volume of urine. Because of the increased water loss associated with the loss of glucose in the urine, this syndrome achieved the descriptive title of the "pissing evil" during the Middle Ages.

Other consequences of the inadequacy of insulin production contribute to metabolic disturbances and additional water losses. Low insulin levels allow increased utilization of fats and proteins through gluconeogenesis to replace the lost glucose. Proteins are hydrolyzed to amino acids that are in turn transaminated or deaminated to produce carbohydrates plus ammonia. Most of the highly toxic ammonia is converted to urea and excreted through the urine. Elevated urea in the urine also contributes to increased urine production and additional water losses. Lipid stores (triacylglycerides) are mobilized as the primary gluconeogenic source to compensate for carbohydrate shortages in the diabetic. As a consequence of β-oxidation of NEFAs and associated ketogenesis, ketone bodies increase in the blood (**ketonemia**) and consequently appear in the urine (**ketonuria**), contributing to increased urine production and additional water losses. Thus, the diabetic individual suffers from excessive water and nutrient (glucose, ketone bodies) losses. Several endocrine and metabolic alterations occur in an attempt to compensate for these nutrient losses.

The loss of glucose into the urine and increased production of nitrogenous wastes and ketone bodies all contribute to a process of desiccation or dehydration. Should the patient enter a coma, he or she would be unable to compensate for water losses through drinking. Severe dehydration may cause a marked decrease in blood volume and blood pressure that can lead to cardiac failure and death if proper treatment does not occur.

A. Development of the Mammalian Pancreas

The mammalian pancreas, like that of other vertebrates, develops from the endodermal lining of the primitive or embryonic gut. A dorsal bud from the embryonic intestine fuses with one or two ventral buds to form the definitive pancreas. Usually only the dorsal connection to the intestine is retained as the exocrine pancreatic duct. The exocrine tissue in the developing pancreas can be identified by the formation of small ductules. These ductules coalesce into larger ducts until they form one or more large pancreatic ducts connecting the exocrine pancreas with the lumen of the small intestine. Small buds develop from the ductules and become the islets of Langerhans. In addition, some of the endocrine cells may be scattered singly or in groups of only a few cells throughout various regions of the pancreas. Immunoreactivity of the various peptides occurs in humans at

Table 14-1. Appearance of Pancreatic Cell Types in Mammalian Fetuses[a]

Cell type	Human	Rat
A cells (glucagon)	8 weeks	12 days
B cells (insulin)	8 weeks	17 days
D cells (SS)	8 weeks	15 days
PP cells	10 weeks	—

[a] As determined by the presence of immunoreactive material (humans) and cell ultrastructure (rat).

about 8 to 10 weeks of development, marking the first synthesis of pancreatic hormones (Table 14-1).

Although the islets originate as outgrowths from the pancreatic ductules, the actual embryonic origin for these cells is not clear. It has been suggested that they arise from mesoderm or from neural crest cells of neuroectodermal origin or from endoderm of the gastrointestinal tract. However, experiments have shown that the islet cells are not derived from neural crest, at least in birds, and it is probable that mammalian islet cells are not derived from either neural crest or other neuroectoderm. Transplantation of quail neural crest cells containing a unique cytological marker into chick embryos confirms their incorporation into the avian pancreas, but they do not give rise to the pancreatic endocrine cells. Instead, they differentiate into parasympathetic ganglia that, as we shall see, are important for regulating islet cell activity. Until the origin of its secretory cells is confirmed, we shall refer to the endocrine pancreas as being of endodermal origin.

Several functional schemes have been proposed for the origins of the endocrine pancreatic cells that are divorced from their possible germ layer origins. These endocrine cells may represent modified gastrointestinal mucosal cells that initially synthesized "inducers." Exocrine cells in teleosts that are considered to be homologous to avian and mammalian islet cells often possess a concentration of protein granules (zymogen mantle), presumably induced by secretions of the adjacent pancreatic endocrine cells. Another suggestion is that these cells originally secreted hydrolytic enzymes related to digestion but that these "enzymes" lost their catalytic properties and acquired hormonal functions. In the fetal guinea pig pancreas, immunoreactive glucagon, PP, insulin, and somatostatin can be demonstrated in cells of the pancreatic tubules about 10 to 15 days before they appear in the islets, supporting the notion that they are derived from exocrine enzyme-secreting cells. These modified mucosal cells presumably lost their contact with the gut mucosa and specialized as centers for internal secretion. The similarity of peptides produced in the islet cells and those of the gastrointestinal tract (see Chapter 13) supports a common origin for these pancreatic and intestinal cells.

B. Cellular Types in Pancreatic Islets

At least five different cellular types have been identified in the mammalian endocrine pancreas: B, A, D, PP, and amphophils. Although first identified by their staining characteristics, they are now identified by their immunoreactivity to antibodies prepared against the specific pancreatic peptides. There are several general vertebrate patterns with respect to the proportions of these cellular types present in the endocrine pancreas (Table 14-2). However, careful analyses indicate that it is misleading to characterize classes of vertebrates as

Table 14-2. General Patterns of Islet Cellular Type Distributions in Vertebrates with Respect to A and B Cells[a]

Category	Dominant cytology	Representatives
I	Mostly B cells	Cyclostomes
II	More than 50% B cells	Teleosts, amphibians, mammals
III	Approximately 50% X cells[b]	Chimaeras
IV	More than 50% A cells	Most lizards, birds

[a] Based on Epple, A., and Brinn, J. E. (1987). "The Comparative Physiology of the Pancreatic Islets." Springer-Verlag, Berlin.
[b] X cells contain glucagon-like peptides but their function is unknown.

exhibiting a particular pattern (Tables 14-3 and 14-4). Immunocytochemistry reveals that several peptides may be localized in one cell type, and the pattern of colocalization differs markedly in different vertebrates (Table 14-5). The major pancreatic cellular types are illustrated in Fig. 14-11.

1. B Cell

The first cellular type identified in the pancreatic islets was the **B cell** (the β cell in an alternative scheme for naming pancreatic cellular types), which stains with aldehyde fuchsin [AF($+$)] and pseudoisocyanin [PIC($+$)]. The latter staining procedure applied following an oxidation step has been claimed to be specific for insulin granules, but it stains other intracellular structures in nonpancreatic tissues that do not contain insulin (for example, neurosecretory granules). Immunocytochemical techniques have verified that insulin is produced and stored in the B cell as granules about 300 nm in diameter.

Several drugs, such as **alloxan** and **streptozotocin,** selectively destroy the B cells and impair the ability of the pancreas to secrete insulin. The affected B cells undergo a process

Table 14-3. Comparative Cytology of the Endocrine Pancreas with Respect to Relative Predominance of A, B, D, and PP Cells[a]

Class	A cells	B cells	D cells	PP cells
Agnathans				
Hagfish		$++++$	$+$	?
Chondrichthyeans				
Elasmobranchs	$+++$	$++++$	$+$	$+$
Osteichthyeans				
Teleosts	$+++$	$++++$	$+$	$+$
Amphibians				
Anurans and urodeles	$+++$	$++++$	$+$	$+$
Reptiles				
Saurians	$++++$	$+$	$+$	$+$
Crocodilians	$+++$	$+++$	$+$	?
Birds	$+++++$	$+++$	$+$	$+$
Mammals	$++$	$+++++$	$+$	$+$

[a] The number of $+$ marks in each column represents only an attempt to show relative abundance and should not be construed as precise ratios.

Table 14-4. Percentage of B Cells
in the Pancreatic Endocrine Islets
of Selected Mammals

Species	Percentage of B cells
Metatherians	
Red kangaroo	8.1 ± 0.58
Gray kangaroo	15.9 ± 0.74
Euro	9.9 ± 0.82
Brush-tailed possum	52.7 ± 0.84
Eutherians	
Merino sheep	12.5 ± 1.16
Cattle	14.6 ± 0.58
White laboratory rabbit	76.9 ± 0.83
Laboratory rat	66.9 ± 0.43

of **hydropic degeneration** that is characterized by clumping of nuclear chromosomal material (formation of pyknotic nuclei) and eventual cell death. This technique of chemically induced degeneration is often used in experimental animals to selectively remove the insulin-secreting cells without altering the ability of the islets to secrete other pancreatic peptides.

2. A Cell

Cells of the pancreatic islets that are acidophilic and argyrophilic (affinity for silver-staining techniques) are called **A cells** (also α or α_2 cells). They do not stain with AF or PIC procedures. The A cell is the source of the second major pancreatic hormone, glucagon, which is shown by immunocytochemistry to be stored in secretory granules about 235 nm in diameter. In some species, the secretory granules of A cells may contain other peptides in addition to glucagon (see Table 14-5). The round secretory granules in the cytoplasm of the A cells are morphologically distinct from the angular insulin granules of the B cells, making these cells easy to distinguish with the aid of a transmission electron microscope. This difference in granule shape is not found in all mammals, however. Treatment with cobalt selectively impairs the ability of A cells to secrete glucagon, but such treatment does not lead to destructive degeneration such as alloxan produces in the B cells.

3. PP Cells

Pancreatic polypeptide (PP) has been localized by immunocytochemical techniques in 125-nm granules of cells found at the periphery of the endocrine islets as well as in cells scattered throughout the exocrine pancreas. These **PP cells** are distinct cytologically and immunologically from other cell types. The distribution of these PP-secreting cells (also called F cells) on the periphery of the islets varies greatly among different species, and it is difficult to generalize for all mammals.

4. D Cell

Immunoreactive somatostatin is localized in the cytoplasmic granules of **D cells** (also called α_1 cells or δ cells). Although D cells contain cytoplasmic granules similar in size to those of A cells, they are cytochemically distinguished from A cells by applying the toluidine

Table 14-5. Some Peptides Colocalized in Selected Vertebrates with Pancreatic Hormones[a]

Cell type and hormone	IGF-I[b]	IGF-II[b]	Gastrin	CRH[b]	GIP[b]	PP[b]	Endorphin	CCK[b]
A cell (glucagon)	Frog			Catfish	Human	Frog	Rat	Rat
				Toad	Rat	Rat		Human
				Lizard	Pig	Human		
				Chicken	Dog			
				Cat	Cat			
				Mouse	Guinea pig			
				Monkey				
				Rat				
				Human				
B cell (insulin)		Fish	Human					
		Frog						
		Lizard						
		Snake						
		Bird						
		Mammal						
D cell (somatostatin)	Lizard			Human				
	Snake			Rat				
	Bird			Guinea pig				
PP cell (pancreatic polypeptide)	Frog							
	Lizard							
	Bird							

[a]Colocalization was determined by immunocytochemistry.
[b]IGF, Insulin-like growth factor; CRH, corticotropin-releasing hormone; GIP, glucose-dependent insulinotropic peptide; PP, pancreatic polypeptide; CCK, cholecystokinin.

blue-staining procedure, and from both A cells and B cells by their staining with PIC following methylation but not following oxidation. Granule size (230 nm) is very similar to that of A cells.

5. Amphophils

Amphophilic cells have been demonstrated in the islets of many mammalian species as well as in sharks, teleosts, amphibians and reptiles, but no conclusions have been generated regarding their functional roles. They may represent either ungranulated and/or differentiating or granule-depleted degenerating forms of any of the four cellular types described above.

C. Hormones of the Mammalian Endocrine Pancreas

The two major pancreatic hormones are the hypoglycemic, antilipolytic, lipogenic hormone called insulin, and the hyperglycemic, lipolytic hormone glucagon. In addition, the pancreatic endocrine cells produce somatostatin and PP, but their roles in the regulation of metabolism are less well known than are those for insulin and glucagon.

1. Insulin: Hypoglycemic, Antilipolytic, Lipogenic

Von Mering and Minkowski in 1889 first observed the correlation between the pancreas and blood sugar regulation when they induced diabetes mellitus in dogs following pancreatectomy. The suspected hypoglycemic factor of the pancreas was later named "insuline" to emphasize its origin from the islets of Langerhans (L. *insula,* island). Purified insulin, however, was not isolated until a physician, Frederick Banting, and a graduate student in physiology and biochemistry, Charles Best, teamed up at the University of Toronto in the summer of 1921. Much of their success was due to the collaborative efforts of the biochemist J. B. Collip, who purified insulin, and the project's director, J. J. R. Macleod.

The presence of proteolytic enzymes in the exocrine pancreas had thwarted earlier attempts to extract insulin from a whole pancreas. Since it had been shown that tying off the pancreatic duct caused degeneration of the acinar pancreatic tissue but did not affect the islet tissue, Banting suggesting they employ this technique to avoid contamination of their extracts with digestive enzymes. The hypoglycemic factor they isolated was first named isletin, but they later changed its name to conform to the name proposed earlier by von Mering and Minkowski. The importance of the successful isolation of insulin and its use to alleviate the fatal symptoms of diabetes mellitus led to the awarding of the Nobel Prize in Physiology and Medicine for 1923 to Banting and Macleod. Best and Collip were not officially recognized in the award, although the recipients acknowledged them for their contributions. This pioneering research with experimental animals also is recognized for having saved millions of human lives.

a. Measurement of Insulin and Insulin-Secreting Capacity

Insulin may be bioassayed *in vitro* for its ability to stimulate uptake of glucose from the medium into cells. Muscle cells from the rat diaphragm are often employed in this bioassay. Radioimmunoassay (RIA), however, is routinely employed to measure circulating insulin levels, and many commercial RIA kits are available for performing these analyses. Insulin RIAs are highly specific, and some can detect the difference between porcine and human insulins, which vary by only one amino acid substitution.

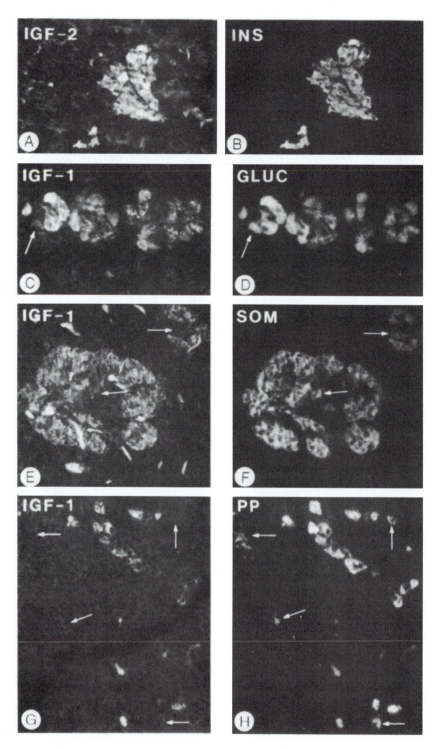

Table 14-6. Comparison of Human Responses to a Standard Glucose Tolerance Test[a]

Subject status	Venous glucose level (mg/dl)			
	Fasting	30 min	60 min	120 min
Normal person	<100	<160	<160	<100
Probable diabetic	<100 or 100–120	130–159	160–180	110–120
Diabetic	<100 or >100	>150	>160	>120

[a] An abnormal glucose tolerance test can also occur as a result of factors other than diabetes mellitus, including improper feeding prior to test, malnutrition, obesity, infection, fever, hyperthyroidism or hypothyroidism, acromegaly, kidney disease, and islet cell tumors.

The standard measure of the ability of the pancreas to secrete insulin is the **glucose tolerance test.** Following a period of fasting, a glucose load (excessive amount of glucose) is administered orally or intravenously, and its rate of disappearance (clearance) from the blood is measured. The clearance rate for a given glucose load is directly proportional to the secretion of insulin and hence reflects the ability of the pancreas to respond to hyperglycemia with elevated insulin secretion (Table 14-6).

b. Chemistry and Synthesis of Mammalian Insulins

Insulin is a very conservative and phylogenetically old molecule, occurring throughout the animal kingdom (Table 14-7) and even among unicellular organisms. The insulin molecule is a small protein composed of two different polypeptide chains linked together by disulfide bridges. It is structurally very similar to the insulin-like growth factors (IGFs) and relaxin described earlier, although each has its own receptor type (see Chapter 2). Secretion granules of B cells contain crystals of insulin and C-peptide fragments. The insulin fragment forms a complex with available zinc ions to produce a crystalline hexamer within the secretory granule. It is this crystalline structure that imparts the unique, angular structure to insulin secretion granules in the B cell.

c. Regulation of Insulin Release

Hyperglycemia can stimulate release of insulin through a direct action of glucose on the B cell. It is apparently not the presence of glucose but events associated with its uptake and metabolism by the B cell that stimulate insulin secretion. Decreases in circulating glucose cause a reduction in the secretion of insulin, and it appears that the basic regulatory control mechanism is directed through changes in blood levels of glucose. Any agent capable of elevating circulating glucose levels also will evoke insulin release. Amino acids, especially arginine, also can stimulate insulin release, although the mechanism is not well understood.

The discovery that **gastric inhibitory peptide (GIP)** released from cells of the small intestine after ingestion of a meal was a potent releaser of insulin was the basis for the

Figure 14-11. Pancreatic cell types. The four pancreatic cell types showing colocalization of IGF-I and IGF-II. (A, B) IGF-II and insulin in the B cells of *Scincus officinalis* (bird); (C, D) IGF-I and glucagon in A cells of *Psamorphis leniolatum* (snake); (E, F) IGF-I and somatostatin in the D cells of *Coluber ravergieri* (snake); (G, H) IGF-I and pancreatic polypeptide in PP cells of *Lacerta viridis* (lizard). Arrows indicate adjacent sections through same cell. (A,B) Magnification 410 ×; (C–F) 390 ×; (G,H) 440 ×. [From Reinecke, M., Broger, I., Brun, R., Zapf, J., and Maake, C. (1995). Immunohistochemical localization of insulin-like growth factor I and II in the endocrine pancreas of birds, reptiles, and amphibia. *Gen. Comp. Endocrinol.* **100**, 385–396.]

Table 14-7. Occurrence of Some Pancreatic Peptides in
Invertebrate Animals

Group	Insulin	Glucagon	PP	SS
Mollusk hepatopancreas	+	+		
Starfish	+			
Tunicate (urochordate)	+		+	+
Amphioxus (cephalochordate)	+	+		

alternative name for this peptide: **glucose-dependent insulinotropic polypeptide** (see Chapter 13). Elevated glucose appears in human hepatic portal blood in 2 min after ingesting 75 g of glucose, and elevated GIP and insulin follow within 5 min. Insulin, GIP, and glucose all reach maximal levels within 30 min and then decline to normal within 3 hr. Following ingestion of a "standard" mixed meal, maximal levels of circulating GIP are achieved within 45 min and remain elevated for 6 hr. A similar elevation of circulating glucose induced through direct intravenous injection of glucose evokes a smaller insulin response than does a meal. Although GIP release has been demonstrated following ingestion of long-chain but not short-chain fatty acids, the aumentation of insulin release by GIP apparently requires a simultaneous increase in glucose levels. Release of insulin by GIP occurs only when accompanied by hyperglycemia. This seems to relegate the insulinotropic role of GIP to that of a permissive effect, but a physiologically important action nevertheless.

The importance of inhibitory nervous regulation of insulin secretion has been demonstrated. Nonmyelinated autonomic axonal fibers are present in the islets. Similarly, acetylcholine or activation of parasympathetic nerves stimulates insulin release, whereas the anticholinergic drug atropine blocks insulin release. Norepinephrine from sympathetic neurons blocks insulin release. Epinephrine, which is a hyperglycemic agent affecting liver, muscle, and adipose tissue, also directly blocks the release of insulin. Thus, circulating epinephrine can potentiate its own hyperglycemic action by blocking the normal response of the B cell to increased blood glucose.

Somatostatin inhibits release of insulin by a direct action on the B cell, and it may play a paracrine role in regulating insulin release locally in the pancreas. It is not known what regulates somatostatin release from the pancreatic D cells, but studies show that both acetylcholine (parasympathetic) and α-adrenergic (sympathetic) agents can block somatostatin release whereas vasoactive intestinal peptide (VIP) and β-adrenergic agents stimulate its release.

The neuropeptide **galanin** has been identified in pancreatic islet neurons of dogs and humans, sometimes colocalized with other peptides such as neuropeptide Y. Release of insulin from the B cell by glucose or amino acids is blocked directly by galanin. Various stimulators of insulin release operate via different cellular mechanisms [e.g., blocking K^+ channels, activating inositol trisphosphate/diacylglycerol (IP_3/DAG)-mediated Ca^{2+} influx, inhibition of adenylate cyclase], all of which can be blocked by galanin. The physiological importance of this inhibitory mechanism and its relation to inhibition of insulin release by somatostatin are not clear.

A peptide of 49 amino acids was isolated from the porcine pancreas and named **pancreostatin (Pst)** for its ability to block glucose-induced release of insulin. Pst may be a product of a large acidic protein, **chromogranin A,** a member of a family of similar proteins associated with the secretory granules of catecholamine-secreting cells of the adrenal medulla (see Chapter 9), some gastrointestinal endocrine cells, and both catecholaminergic and peptidergic neurons. Although its physiological role is uncertain, a phylogenetic survey reveals Pst immunoreactivity in the gastric mucosa and pancreatic islets of all vertebrates examined including pig, human, chicken, Japanese quail, lizard, frog, shark, ratfish, and hagfish. No Pst immunoreactivity was observed in either the rat or in cephalochordates or urochordates.

d. Actions of Insulin on Target Tissues

After insulin binds to its membrane receptor, there is an increase in the transport of glucose, amino acids, fatty acids, nucleotides, and various ions into the target cell. Within a few minutes, enhancement of anabolic pathways and a decrease in catabolic pathways occur. Through delayed effects on nuclear transcription and protein synthesis, some cellular growth is stimulated. This growth-promoting action is due to an interaction between insulin and growth hormone (GH)-dependent, circulating IGFs, which operate through separate receptors but through similar intracellular pathways.

Insulin appears to bring about hypoglycemia through stimulating increased uptake of nutrients from the blood (glucose and amino acids) and by altering the activity of intracellular enzymes. Insulin facilitates transport of glucose into muscle and fat cells as well as transport of amino acids into muscle cells. Activity of hexokinase also is enhanced by insulin, which stimulates glucose oxidation and favors lipogenesis (adipose tissue) and protein synthesis (muscle). Insulin also prevents glycogen breakdown (glycogenolysis) and enhances glycogen synthesis following uptake of glucose in muscle and liver. This is accomplished through simultaneous inhibition of the cAMP-dependent enzyme phosphorylase *a* and stimulation of the enzyme glycogen synthetase (Fig. 14-3).

Glucose oxidation increases the intracellular pools of precursors for fat synthesis, that is, glycerol, acetyl-CoA, and fatty acids. In addition to indirectly enhancing esterification by stimulating acylglycerol synthetases, insulin inhibits triacylglycerol lipase and prevents lipolysis in adipose tissue. This reduces release of NEFAs and monoacylglycerol into the blood and hence reduces hepatic fatty acid oxidation and ketogenesis. These actions of insulin on adipose tissue and lipid metabolism are marked in carnivores, but there is little effect of insulin on either glucose metabolism or lipogenesis in herbivorous mammals. Consequently, carnivorous mammals are more sensitive to pancreatectomy than are herbivores.

The increase in intracellular glucose 6-phosphate and amino acids in both liver and muscle cells promotes protein, lipid, and glycogen synthesis. These are important actions of insulin immediately following ingestion of a meal and the entry of large quantities of digestive products such as glucose and amino acids into the blood.

e. Mechanism of Action for Insulin

The many actions of insulin on its target cells have prompted many different schemes to explain these diverse actions. At last, a unified picture is beginning to emerge that eventually may explain all of insulin's actions.

We know now that insulin binds to a tyrosine kinase transmembrane receptor that occurs as a dimer in the cell membrane (see Chapter 2). The occupied receptor undergoes autophosphorylation and then acts as a tyrosine kinase to phosphorylate various substrates. One of these substrates is involved in a chain of phosphokinase events called the kinase cascade (see Chapter 2). This mechanism activates glycogen synthetase (causing storage of glucose as glycogen) and alters levels of certain transcription factors responsible for the delayed actions of insulin on protein synthesis. The mechanism whereby insulin enhances glucose

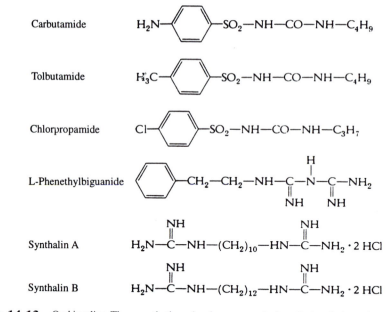

Figure 14-12. Oral insulins. These synthetic molecules cannot substitute for insulin but enhance the activity of insulin already present. These drugs are useful only in the treatment of NIDDM (type I diabetes).

transport into the cell still is not clear, but probably involves protein kinase-activated events. Presumably, the actions of insulin on adipose cells operate through similar mechanisms, but details have not been worked out.

f. Oral Insulins

The term **"oral insulin"** refers to the synthetic hypoglycemic drugs known as sulfonylureas and related compounds (Fig. 14-12). Because they are not proteins and thereby not subject to enzymatic degradation in the gut, the sulfonylureas may be taken orally and can relieve some of the symptoms of diabetes mellitus. Oral insulins thus have considerable therapeutic value in certain types of insulin-deficiency disorders. The effectiveness of these drugs may be related to their abilities in displacing insulin from nonspecific binding sites and noneffective linkages in plasma, connective tissues, and islet cells to provide marginally effective quantities of insulin for binding to effective (target) sites. If too high a dose is employed, the oral insulins also may block the effective sites and no hypoglycemic effect is observed. The oral insulins are useful only in disorders related to insulin insufficiency and would not be useful in totally pancreatectomized animals or in humans who cannot produce insulin. Furthermore, oral insulins are effective only if diet is carefully controlled.

2. Glucagon: Hyperglycemic Hormone

The discovery of glucagon was not marked by great publicity or special prizes, as was that of insulin, and this second pancreatic hormone has always been overshadowed by the clinical importance of insulin. Glucagon is only one of several hyperglycemic factors, and more attention has been given to growth hormone (GH) and epinephrine as hyperglycemic hormones. Glucagon is more important in fasting carnivores and in herbivores than it appears to be in fed humans or carnivores. It also is more important than insulin in birds and lizards for regulating metabolism.

a. Bioassay of Glucagon
Related to its role in fasting mammals, glucagon is often bioassayed in the fasting cat, in which it induces hyperglycemia. As in the case of insulin, however, rapid RIAs have been developed for measuring blood glucagon levels, and these are employed routinely.

b. Chemistry of Glucagon
Glucagon is a straight-chain polypeptide hormone consisting of 29 amino acid residues and with a molecular mass of 3485 Da. Structurally, glucagon is similar to the family of gastrointestinal hormones that includes secretin and cholecystokinin (see Chapter 13). Glucagon is a conservative peptide among mammals and has been identified immunocytochemically in all vertebrates as well as in a number of invertebrates. Nonmammalian glucagons are conservative, too.

In addition to the peptide glucagon, proglucagon also contains two **glucagon-like peptides (GLPs)** of unknown function. GLP-I is composed of 37 amino acids and has 48% homology to glucagon whereas GLP-II consists of 35 amino acids and has only 38% homology with glucagon. However, these peptides are not close enough structurally to affect binding of glucagon to its receptor.

Endocrine cells similar to the pancreatic A cells occur in the small intestine and produce **enteroglucagon** and related peptides, which may play important roles in digestion (see Chapter 13). Another extrapancreatic glucagon-like hyperglycemic factor has been characterized in extracts of submaxillary salivary glands obtained from rats, mice, rabbits, and humans. Release of this salivary glucagon is influenced *in vitro* in the same manner as pancreatic glucagon. Somatostatin, however, does not block release of salivary glucagon. To avoid confusion, the term *glucagon* in this chapter indicates glucagon of pancreatic origin.

c. Actions of Glucagon on Target Tissues
Glucagon promotes glycogenolysis in liver cells, its primary target with respect to raising circulating glucose levels. This effect appears to be mediated through stimulation of adenylate cyclase and production of intracellular cAMP. Increased glycogenolysis accompanied by decreased intracellular oxidation of glucose directs the movement of glucose from liver cells into the blood.

Lipolysis is stimulated by glucagon in fasting animals. The release of NEFAs into the blood further increases β-oxidation, ketogenesis, and gluconeogenesis in liver.

3. Pancreatic Polypeptide: Another Hormone?

Pancreatic polypeptide (PP) was discovered in the chicken pancreas and since has been purified from the usual mammalian sources (ovine, bovine, canine, porcine, rat, and human pancreatic tissue) as well as from wild equines (Przewalski's horse, zebra), rhino, tapir, several birds, and the alligator. The average molecular weight for PP is 4200, and all peptides isolated consist of 36 amino acid residues (Fig. 14-13). Structurally, PP is similar to neuropeptide Y (NPY) and peptide YY (PYY). The physiological role of PP is unknown, although there are several reasons for giving PP hormonal status. In humans, ingestion of a protein meal such as ground beef causes an increase in plasma levels of PP from preingestion levels of 57 pg/ml plasma to 229, 400, and 580 pg/ml, respectively, in 5, 10, and 240 min. Infusion or ingestion of glucose does not evoke this sort of increase in human PP (hPP), and it appears that the rapid increase in plasma hPP following ingestion of protein

	1	11	21	31
Human	APLEP V YPGD	NATPE QMAQY	AADLR RYINM	LTRPR Y-NH$_2$
Porcine/Canine	APLEP V YPGD	DATPE QMAQY	AAELR RYINM	LTRPR Y-NH$_2$
Tapir	APLEP V YPGD	NATPE QMAQY	AAELR RYINM	LTRPR Y-NH$_2$
Rhino	ASLEP V YPGD	NATPE EMAQY	AADLR RYINM	LTRPR Y-NH$_2$
Horse/Zebra	APMEP V YPGD	NATPE QMAQY	AAELR RYINM	LTRPR Y-NH$_2$
Bovine	APLEP E YPGD	NATPE QMAQY	AADLR RYINM	LTRPR Y-NH$_2$
Rat	APLEP MYPGD	YATHE QRAQY	ETQLR RYIN T	LTRPR Y-NH$_2$
Turkey	GPSQP T YPGD	DAPVE DL IRF	YNDLQ QYLNV	VTRHR Y-NH$_2$
Goose	GPSQP T YPGN	DAPVE DL?RF	YDNLQ QYRLV	VFRHR Y-NH$_2$
Alligator	TPLQP K YPGD	GAPVE DL IQF	YNDLQ QYLNV	VTRHR P-NH$_2$

Figure 14-13. Comparison of amino acid sequences for pancreatic polypeptide (PP). There is considerable homology within a taxonomic group. Note the similarity between avian and alligator PPs. See Appendix C for single-letter amino acid code.

is at least in part a neural response mediated via the vagus nerve. This release of hPP can be blocked by somatostatin, although it is not known whether somatostatin is involved in the normal regulation of PP release. A second phase of elevated PP occurs later and may last for several hours. The initial phase of PP secretion can be induced with experimental gastric distention, but the later response occurs only following protein- or fat-rich meals. The biphasic release of avian PP (aPP) in chickens apparently is independent of protein, fat, or carbohydrate content of the diet.

Bovine PP is reported to be a potent inhibitor of pancreatic exocrine secretion, but it does not seem to influence carbohydrate (glucose) metabolism. Avian PP has been reported to be a potent stimulator of gastric acid and pepsin secretion, but this effect has not been reported in mammals. It is too early to decide whether these reports are related to the physiological roles of PP or not.

Resting levels of plasma PP in humans increase with age, from about 50 pg/ml in 25 year olds to more than 200 pg/ml at age 70. This increase is paralleled by a progressive increase in the number of PP cells in the islets. In fasting birds, PP plasma levels are about 40 times greater than those observed in mammals.

4. Somatostatin: Paracrine Regulator of Pancreatic Secretion

The D cell is believed to be the source of pancreatic somatostatin that is chemically identical with hypothalamic somatostatin. In mammals, two somatostatin molecules have been identified in the pancreas, one of which consists of 14 amino acids (SS$_{14}$) and is identical to the hypothalamic peptide described in Chapter 4. The second somatostatin is the larger form composed of 28 amino acids (SS$_{17}$) of which the last 14 are identical to SS$_{14}$. Levels of pancreatic somatostatin in rats are equivalent to hypothalamic levels. Somatostatin is believed to inhibit the release of insulin, glucagon, and pancreatic polypeptide in a paracrine fashion.

Regulation of somatostatin release is not understood. The observed inhibitory actions of acetylcholine and norepinephrine mentioned earlier possibly mimic parasympathetic and sympathetic innervation, respectively.

D. Clinical Aspects of Pancreatic Function

1. Diabetes Mellitus

Diabetes mellitus is a complex, heterogeneous assemblage of disorders having the common feature of the appearance of glucose in the urine. Diabetes mellitus is always fatal if not treated, and even if treated, serious complications may develop including blindness, kidney disease, and circulatory or circulation-based problems. Approximately 5% of the U.S. population is afflicted with diabetes mellitus, and the incidence of the disease is increasing at the rate of 6% a year. Diabetes mellitus is directly responsible for about 38,000 deaths annually. However, cardio–renal–vascular complications resulting from diabetes mellitus are responsible for an additional 300,000 deaths annually, making diabetes a leading cause of death in the United States.

Most cases of diabetes mellitus have been characterized as either **maturity onset diabetes** (ketosis resistant; insulin independent) or as **juvenile onset diabetes** (ketosis prone; insulin dependent). "Maturity" and "juvenile" are not entirely appropriate terms since the former can occur occasionally in children and adolescents, whereas the latter can occur with low frequency in middle-aged or older adults. Consequently, these disorders were renamed first as **type-1 diabetes** and **type-2 diabetes** and more recently as **non-insulin-dependent diabetes mellitus (NIDDM)** and **insulin-dependent diabetes mellitus (IDDM),** respectively. Several features of these two types of diabetes are compared in Table 14-8. There are some additional forms of diabetes now recognized, but they are rare.

NIDDM, accounting for approximately 90 to 95% of the reported cases, typically occurs in overweight people (80% of cases) who are over 40 years of age. Much of the increase in diabetes observed in the United States in recent years is due to an increasing proportion of elderly people in the population coupled with an increasing tendency toward obesity.

Table 14-8. Comparison of Symptoms for Two Forms of Diabetes Mellitus

Features	IDDM	NIDDM
Age at onset	Usually during childhood or puberty	Frequently over 45 years
Type of onset	Abrupt onset of overt symptoms	Usually gradual
Genetic basis	Frequently positive relationship	Commonly positive relationship
Nutritional status	Usually undernourished	Usually obesity present to some degree (about 80% of cases)
Symptoms	1. Polydipsia, polyphagia, polyuria	1. May be none
	2. Ketosis frequently present	2. Ketosis uncommon except under stress
	3. Hepatomegaly rather common	3. Hepatomegaly uncommon
	4. Blood glucose fluctuates widely in response to small changes in insulin dose, exercise, and infection	4. Blood sugar fluctuations less marked than for juvenile onset
Fasting blood sugar levels	Elevated	Often normal
Requirement for insulin	Necessary for all patients	Necessary for 20–30% of patients
Effectiveness of oral agents	Rarely effective	Effective when diet is controlled

NIDDM has a slow onset with no obvious symptoms. Insulin levels in NIDDM are usually normal or in excess of normal. Apparently, the target cells become insensitive to insulin, possibly due to a decrease in receptor populations or to a deficiency in the mechanism of action of insulin after it is bound to the cell. Another possible cause might be an increase in nonspecific binding of insulin to nontarget cells, which reduces insulin availability for specific receptors. Whatever the defect, dietary regulation and the removal of excess weight will bring these patients back into normal carbohydrate balance, and the regulation of carbohydrate metabolism will return to normal. In more severe cases, oral hypoglycemic agents (e.g., sulfonylureas) are used to reduce blood glucose levels in these patients along with rigid control of dietary intake.

There is evidence to support a familial or genetic component in NIDDM. A series of studies on twins indicates that if one twin develops NIDDM after age 50, the other twin develops NIDDM in almost every case.

IDDM usually develops in persons under 20 years of age and is characterized by an abrupt onset of symptoms. A small percentage of people over 40 who contract diabetes mellitus also exhibit this form of the disease. Typically, the pancreatic B cells are reduced markedly (usually to less than 10% of normal), and insulin levels are very low or absent. Glycogen stores are depleted readily and fatty acid metabolism is accelerated, resulting in accumulation of ketone bodies in the blood, which in about 48 hr will produce ketoacidosis and eventually lead to coma and death unless insulin is administered. Patients with IDDM have an absolute requirement for exogenous insulin for the rest of their lives. Because of this reliance on exogenous insulin and difficulties in administering the appropriate dose when needed, the diabetic must always be prepared for the onset of hypoglycemia following excessive insulin. Hypoglycemia can also induce coma and usually can be counteracted by consuming a high glucose source.

There is some evidence to suggest a role of infection as a cause of IDDM. The symptoms of IDDM appear abruptly, and there is a higher incidence of onset during the fall and winter months when the incidences of infectious diseases also increase. In certain genetic strains of mice, IDDM has been linked to a specific virus. There also appears to be a correlation between the occurrence of certain viral diseases (e.g., mumps, rubella *in utero*) and the later appearance of IDDM.

The major cause of IDDM, however, seems to be an autoimmune reaction that destroys the B cells. Although characterized by an abrupt onset of severe symptoms soon leading to death if untreated, IDDM actually has a long incubation period (sometimes years). Only after about 80% of the insulin-secreting capacity of the pancreas is destroyed do the classic symptoms appear. The presence of lymphocytes and monocytes (white blood cells) in the islets at first seems to support the infection hypothesis. However, antibodies to components of B cells have been found in the blood of IDDM patients and in people believed to be at risk. One of these antibodies reacts against a ganglioside found within all B cells, and a second antibody reacts to a 64-kDa protein known only from the cell membranes of B cells. A third antibody reacts with insulin itself. The antibody against the 64-kDa protein is probably the one that initially damages the B cells. Apparently, these antibodies are present in the blood long before diabetes presents clinically, and administration of antiimmune system drugs to date has prevented the onset of IDDM in experimental subjects who would have developed the disease eventually. Although the circulating antibodies have been demonstrated, many scientists still believe that cell-mediated immunity is the cause of IDDM. Nevertheless, the ability to identify these circulating antibodies may prove to be an important diagnostic tool allowing early detection and possible prevention of irreversible damage to the B cells.

Dietary restrictions are similar for treating both NIDDM and IDDM patients, although a reduction in total caloric intake also is necessary for overweight NIDDM patients. Specific diets must be determined individually for every patient. Carbohydrates are still essential to the diabetic diet but need to be in the form of polysaccharides (e.g., starch). Mono- and disaccharides are usually avoided because of rapid uptake and resultant hyperglycemia soon after ingestion. The ingestion of simple sugars is less a problem for the NIDDM patient, as the pancreas can respond with some insulin release.

2. Extrapancreatic Tumor Hypoglycemia

Extrapancreatic tumor hypoglycemia (EPTH) sometimes appears in patients with cancer. Patients with extrapancreatic tumor hypoglycemia exhibit normal to low levels of insulin and normal IGF levels, but have high levels of circulating pro-IGF-II. This pro-IGF-II is of tumor origin and can bind to the insulin receptor and produce insulin-like hypoglycemia.

III. Comparative Aspects of the Endocrine Pancreas

A. Anatomical Features

Five anatomical arrangements for pancreatic islet systems can be distinguished readily by examination of the major vertebrate groupings. The **cyclostome type** is composed of aggregations of islet cellular types with no obvious relationship to the homologous intraintestinal equivalent of mammalian acinar cells. In hagfishes, islets surround the base of the common bile duct (Fig. 14-14), and in lampreys they are located on the surfaces of the intestine (epiintestinal), within the intestinal mucosa (intraintestinal; Fig. 14-14) and even within the liver (intrahepatic). Cyclostomes exhibit B and D cells but lack A cells and PP cells. GLP peptides have been identified in the gastric mucosa, however. The cyclostome pattern of islet distribution is considered to be the most primitive pattern in vertebrates. Exocrine pancreatic elements are embedded in the intestinal lining of these primitive fishes.

The **primitive gnathostome type** is found in the selachians (Fig. 14-15), holocephalans, and the coelacanth (sarcopterygian, *Latimeria*). In these fishes, the islet tissue occurs as layers surrounding the smaller ducts of a compact pancreas. In addition, some scattered cells are located throughout the exocrine pancreas. B, A, D, and PP cells occur in selachians and holocephalans.

In contrast to the coelacanth and the lungfishes (see below), we find the diffuse **actinopterygian type** among the other bony fishes, where the exocrine pancreatic cells are scattered along the bile ducts, abdominal blood vessels, and on the outer surfaces of the gastrointestinal tract, gallbladder, and liver. An intrahepatic pancreas has been observed in several species. The islet tissue often accumulates as clumps near the common bile duct but is not usually separated from the acinar tissues. Most teleosts possess isolated masses of pancreatic islet tissue called **Brockmann bodies.** A given species may have a number of Brockmann bodies, or the islet tissue may be concentrated largely in one "principal islet" as in the Atlantic eel, *Anguilla anguilla,* or in bullheads (*Ictalurus* spp.). Brockmann bodies of actinopterygian fishes contain A, B, and D cells, but PP cells occur only in those Brockmann bodies located in the pyloric region.

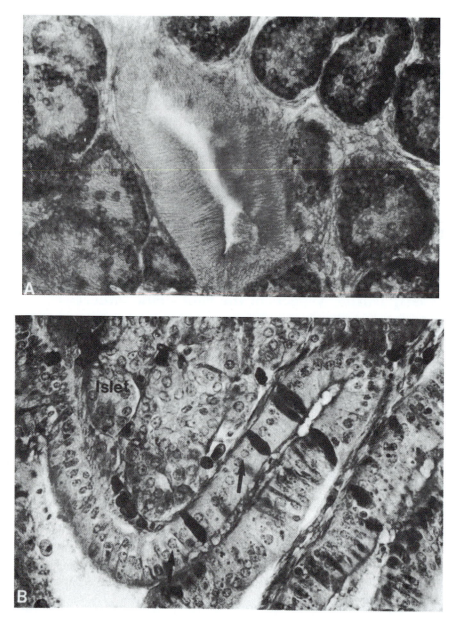

Figure 14-14. Islet tissues of hagfish and lamprey. (A) Follicles of islet tissue surrounding the lower bile
duct from the hagfish *Eptatretus burgeri*. (B) Islet located in intestinal submucosa of the sea
lamprey, *Petromyzon marinus*. (Courtesy of August Epple, Thomas Jefferson University,
Philadelphia.)

The pancreas of lungfishes forms a unique arrangement referred to as the **lungfish type.**
It is a compact, intraintestinal structure with a number of encapsulated islets. Only B, A,
and D cells have been identified in lungfish islets.

The **tetrapod type** of endocrine pancreas typically is a compact, extraintestinal structure containing scattered islets and occasionally some scattered endocrine cells. In the toad

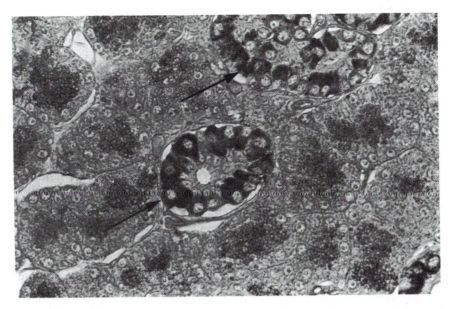

Figure 14-15. Pancreas of dogfish, *Scyliorhinus canicula.* These primitive islets consist of endocrine cells surrounding a small unstained ductule. (Courtesy of August Epple.)

Bufo (Fig. 14-16) and in birds (Fig. 14-17), all four cell types are present although their proportions often are very different from those in mammals. Islets may be concentrated in particular lobes of the pancreas and often are not evenly distributed.

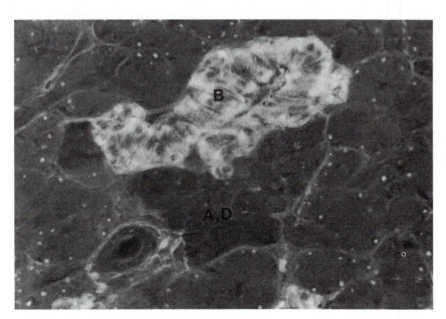

Figure 14-16. B cells in toad. The presence of insulin is indicated by the light fluorescence. The dark area immediately surrounding the B cells consists mostly of A cells with a few D cells. (Courtesy of August Epple.)

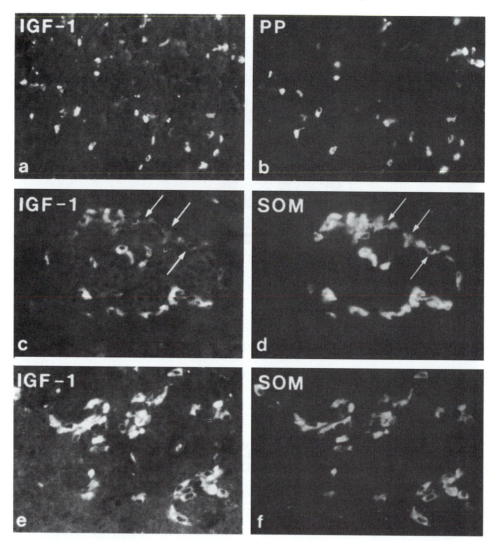

Figure 14-17. Avian cell types in the pancreas. (a–d) Colocalization of IGF-I in PP and D cells of the Japanese quail. (e and f) Colocalization of IGF-I and somatostatin in D cells of the domestic chicken. Arrows indicate adjacent sections through same cell. (a,b) Magnification 90×; (c,d) 300×; (e,f) 310×. [From Reinecke, M., Broger, I., Brun, R., Zapf, J., and Maake, C. (1995). Immunohistochemical localization of insulin-like growth factor I and II in the endocrine pancreas of birds, reptiles, and amphibia. *Gen. Comp. Endocrinol.* **100**, 385–396.]

Immunoreactive IGFs appear in the pancreatic islets or their homologs of protochordates, fishes, amphibians, reptiles, birds, and mammals. In cephalochordates and urochordates, there appears to be a single gene that produces a common insulin/IGF molecule associated with the enteroendocrine cells. These observations support a common gene origin for both insulin and the IGFs. IGF-II is colocalized with insulin in all vertebrates but

IGF-I may be colocalized in different pancreatic cells in amphibians, reptiles, and birds (see Table 14-5).

There are some generalizations we can make concerning the origin of the various components of the tetrapod pancreas and the relationship of their origin to the distribution of islets. The dorsal pancreatic bud gives rise to the tail and body of the pancreas (also known as the splenic portion) and the ventral buds give rise to the head or duodenal portion. Islets associated with the splenic lobe typically are larger and consist mainly of B, A, and D cells. The islets of the duodenal pancreas are smaller and may contain many PP cells.

B. Pancreatic Hormones in Nonmammals

Insulin is a conservative protein in that few amino acid substitutions have occurred at the critical sites related to its structure and biological activity. Insulin A-chains have 21 or 22 amino acids and B-chains have from 31 to 38. The C-peptide of proinsulin "tolerates" a greater number of substitutions, yet even portions of this peptide remain rather conservative; for example, the glycine-rich central core of the C-peptide. Single amino acid substitutions may have profound effects on the resultant molecule. One substitution in hagfish insulin is believed responsible for the observation that hagfish insulin does not bind Zn^{2+} and therefore does not form the typical hexamer crystals characteristic of most vertebrate insulins. The number of amino acids in the B-chain varies when insulins from different vertebrate classes are compared.

Mammalian glucagons that have been characterized chemically exhibit no amino acid substitutions. Nonmammalian glucagons also are structurally conservative, with the exception of teleosts. Turkey glucagon differs from mammalian glucagons only at position 28, where serine is substituted for asparagine. Chicken glucagon is identical to turkey glucagon, whereas duck glucagon has threonine at position 16 instead of serine. Glucagon extracted from eels (*A. anguilla*) also has 29 amino acids, with only 4 differences from mammalian glucagons. A second form with 36 amino acids resembles mammalian GLP-II. However, glucagon has 31 amino acids in salmon, 29 and 31 amino acids in angler fish, and a GLP with 34 amino acids has been isolated from catfish. Glucagon from the flounder, *Platichthyes flesus,* is identical to the 29-amino acid form of the angler fish. Furthermore, the variation in amino acid sequence among the teleosts is large. In contrast, glucagon isolated from the spiny dogfish shark *Squalus acanthias,* and from *Torpedo marmorata* are structurally more like mammalian glucagon than like that of teleosts.

B cells and insulin activity have been identified in the guts or digestive organs of many nonchordate invertebrate animals, and a number of actions of insulin have been described. For example, mammalian insulin reduces blood sugar levels and promotes glycogen synthesis in a gastropod mollusk, *Strophocheilus oblongus.* Furthermore, insulin-like molecules have been found in distantly related taxonomic groups including protozoans, bacteria, and fungi. Insulin is therefore a widely distributed factor that is capable of regulating carbohydrate metabolism in a great variety of organisms. The evolution of insulin must have involved a long and interesting process, most of which remains to be discovered.

Immunoreactive glucagon has been localized in the digestive glands of a crab, *Cancer pagurus,* and two gastropod mollusks, *Patella caerulae* and *Helix pomatia.* A glucagon-like molecule is present in extracts of invertebrate chordates, including tunicates and cephalochordates.

1. Agnathan Fishes: Cyclostomes

Insulin has been purified from hagfishes and three lampreys, *Lampetra fluviatilis, Petromyzon marinus,* and *Geotria australis.* The amino acid composition of hagfish and the northern hemisphere lampreys are more like mammalian insulins than like *Geotria* from the southern hemisphere, suggesting that *Geotria* diverged from the other lampreys very early in cyclostome evolution.

Hagfish insulin behaves similarly to other insulins in mammalian assays. The biological activity of purified hagfish insulin in mammals, however, is only about 4 to 7% of that reported for mammalian insulins, indicating the evolution of greater specificity in target cells as well as some changes in the insulin molecule itself. When injected into hagfish, both mammalian and hagfish insulins cause hypoglycemia. Hagfish insulin stimulates both glycogen and protein synthesis in hagfish skeletal muscle but stimulates only protein synthesis in liver. Amino acids are known to stimulate insulin release in lampreys, providing additional support for insulin's role as a regulator of protein metabolism. Known insulin antagonists, such as epinephrine, glucagon, thyroxine (T_4), corticotropin (ACTH), and prolactin (PRL), are all ineffective in altering blood sugar levels. Curiously, isletectomy of the hagfish *Myxine* does not cause hyperglycemia, and the response to a glucose load in isletectomized hagfish is normal. Hypophysectomy has no effect on islet cytology or blood sugar levels. There must be an explanation for the presence of insulin and the demonstration of its hypoglycemic action in the hagfish with the number of negative observations just mentioned, but it will require more research to reconcile these contradictory data. In contrast, total isletectomy of the lamprey *G. australis* resulted in a marked elevation in blood sugar within 24 hr.

Some large somatostatins have been isolated from lampreys and their amino acid sequences determined. As described for insulin, SS_{33} from *Geotria* is very different from the SS_{34} molecules that characterize the other lampreys, which more closely resemble hagfish SS. Most of the variations occur among the first 18 amino acids whereas little variation occurs in the rest of the molecules.

2. Chondrichthyean Fishes

Considerable variation in blood sugar levels has been found in chondrichthyean fishes. For example, values of 6 to 90 mg of glucose per 100 ml of blood have been reported for the dogfish shark. It appears that either blood sugar regulation is not very precise in these fishes or that blood sugar levels are very sensitive to capture and sampling procedures. Administration of mammalian insulin evokes hypoglycemia, and glucagon induces hyperglycemia in some species. Epinephrine also elevates blood glucose levels, and the effect of injected epinephrine occurs very rapidly. On a superficial basis, it appears that regulation of carbohydrate metabolism in these fishes is similar to the mammalian pattern.

There are many unusual features of metabolism in these fishes that require further study. Although liver glycogen levels in dogfish sharks conform with predicted responses to fasting and force feeding, the liver of holocephalans is either incapable of storing glycogen or is rapidly depleted of glycogen by the stress of capture and/or confinement. Injection of a glucose load does not alter tissue glycogen even though circulating glucose is increased. These observations may explain the inability of glucagon to increase blood glucose in the ratfish. Starvation forces sharks to rely on stored fats. Circulating levels of ketone bodies are elevated in fasting sharks as seen in mammals.

Protein and lipid metabolism with respect to the actions of pancreatic hormones has not been studied. Lipid metabolism in these fishes should be of special interest since hepatic fat content may be as high as 70% in the dogfish shark and 80% in the ratfish.

3. Osteichthyean Fishes

Insulins prepared from several teleosts also are structurally similar to mammalian insulins. Like the chondrichthyean fishes, blood glucose levels of teleosts are very sensitive to handling and sampling stress, presumably due to the hyperglycemic action of epinephrine from chromaffin cells associated with the interrenal system (see Chapter 10). Generally, both mammalian and teleostean insulins are hypoglycemic, and glucagon is hyperglycemic when tested in bony fishes. Injections of glucose or the amino acid leucine stimulate insulin secretion in the toadfish, *Opsanus tau,* and in the European silver eel, *A. anguilla.* Purified codfish insulin is a potent hypoglycemic agent in Northern pike, *Esox lucius.* Furthermore, mammalian insulin increases uptake and incorporation of labeled glucose and glycine into muscle lipids and protein in teleosts. Conversely, mammalian glucagon increases blood glucose, for example, in channel catfish. Glycogenolysis in isolated hepatocytes (liver cells) of goldfish is stimulated by either glucagon or epinephrine.

4. Amphibians

Mammalian insulin and glucagon produce effects in amphibians similar to those observed in fishes and mammals. Insulin induces hypoglycemia in both anurans and urodeles, stimulates incorporation of amino acids into proteins, and lowers fatty acid levels. Following administration of a glucose load, blood sugar levels return to normal in about 24 hr. This response to a glucose load and the effects of pancreatectomy indicate an endogenous role for insulin. Mammalian glucagon causes hyperglycemia although it is ineffective in some amphibians. Epinephrine is hyperglycemic in tadpoles and adults of *Xenopus laevis,* and administration of epinephrine or norepinephrine causes reduction in liver and muscle glycogen.

Early research by Bernardo A. Houssay about 50 years ago on the amphibian pancreas led to a widely employed procedure for research into mammalian diabetes mellitus. It was in the toad *Bufo arenarum* that the stimulation by pituitary hormones of blood glucose in pancreatectomized or insulin-deficient animals was discovered. This discovery led to the widespread use of hypophysectomy to alleviate extreme diabetic conditions in experimental animals, and such doubly operated dogs, cats, and others became known commonly as **Houssay animals.** This action of hypophysectomy in the pancreatic-deficient animal is termed the **Houssay effect.**

5. Reptiles

Reptiles exhibit mammalian-like responses to injected insulin and glucagon and respond typically to a glucose load, returning to preloading levels in about 48 hr. Lizards, however, are particularly insensitive to insulin, and large quantities of mammalian insulin are required to invoke hypoglycemic responses. It is virtually impossible to induce insulin shock even with massive doses to lizards irrespective of their diets (Table 14-9). However, pancreatectomy induces a hyperglycemic state in lizards as it does in snakes and alligators. These observations suggest that in spite of the predominance of A cells in lizard islets and

Table 14-9. Fatal Doses of Insulin for Various Vertebrates[a,b]

Animal	IU of insulin/kg body weight	Animal	IU of insulin/kg body weight
Nonmammals		**Mammals**	
Salamander	50	Rabbit	4–8
Lizard	>10,000[b]	Mouse	7–50
Canary	1000–4000	Rat	29–36
Pigeon	400–1200	Dog	16–300
Duck	50–500		

[a] Modified from Gorbman, A., and Bern, H. A. (1962). "A Textbook of Comparative Endocrinology." Wiley, New York.

[b] High doses of insulin cause intense hypoglycemia, followed by coma and death. Animals with fewer B cells are less sensitive.

[b] A dose of 10,000 IU of insulin had no effect on survival of lizards.

their demonstrated insensitivity to mammalian insulin, endogenous insulin may play an important metabolic role in lizards as well as in other reptiles.

Insulin has been purified from a rattlesnake, *Crotalis atrox,* and its amino acid composition is considerably different from bovine insulin, with some uncommon substitutions occurring in the B-chain. Alligator insulin is structurally very much like chicken insulin, and the alligator responds equally with hypoglycemia to bovine, alligator, or turkey insulin. Comparative analyses of reptilian insulins may shed some light on pancreatic regulation of carbohydrate metabolism in reptiles, especially the apparent insensitivity of many species to mammalian insulins. Curiously, cyclostome insulin is structurally more like mammalian insulins than is rattlesnake insulin, which points out the difficulties of reconstructing evolutionary events using molecular data obtained from one or a few species as being representative of a major taxonomic group.

Although glucagon has been isolated from the alligator, its role in carbohydrate metabolism in reptiles has been inferred largely from observations following mammalian glucagon treatment. Glucagon is hyperglycemic in squamate reptiles, as in birds, and circulating levels of glucagon often are high. For example, fasting lizards, *Varaus exanthematicus,* have high glucagon levels whereas glucagon is lowest during times of food intake.

Pancreatic polypeptide has been isolated from alligators. The amino acid composition of alligator PP is more like that of birds than like mammals (Fig. 14-13).

Of particular interest is the absence of studies in which lipid and protein metabolism of reptiles was examined. Insulin and other "metabolic" hormones have not been examined for their roles in lipid or protein metabolism. Until all aspects of the hormonal regulation of metabolism in reptiles are integrated with the observations of effects of insulin and glucagon, a complete picture for the roles of reptilian pancreatic hormones will not emerge.

6. Birds

Hyperglycemia in birds, initially observed following pancreatectomy, is now believed to have been a result of overlooking the splenic lobe (rich in A cells) during pancreatectomy. Total pancreatectomy results in hypoglycemia rather than hyperglycemia and suggests that glucagon plays a major role in avian blood sugar regulation. This suggestion is supported further by observations that a high glucose load is required to induce insulin release in the duck and by the relative insensitivity of avian tissues to injected mammalian insulin

(Table 14-9). Elevated glucose levels depress glucagon release, whereas high circulating levels of NEFAs induce glucagon release.

Avian tissues respond with hyperglycemia, glycogenolysis, and lipolysis following administration of mammalian glucagon. The avian pancreas contains about 5 to 10 times the extractable glucagon per gram of pancreas as compared to mammals, emphasizing its probable importance to metabolic regulation in birds. In one of the few studies performed in wild birds, mammalian glucagon stimulated both plasma levels of free fatty acids and glucose in penguins, whereas insulin caused a reduction in blood glucose. Additional studies with wild avian species representing diverse taxonomic groupings would be welcome in order to confirm or alter generalizations based largely on observations of domestic species.

Insulin has been extracted from the avian pancreas, and its structure has been examined. Chicken insulin is more effective in inducing hypoglycemia in birds than it is in mammals, and it is more potent in birds than is mammalian insulin. Glucagon has been extracted and its primary structure determined for duck, turkey, and chicken hormones. These hormones are almost identical structurally to mammalian glucagon (see Chapter 13).

Like mammalian PP, avian PP (aPP) consists of 36 amino acid residues and has a molecular mass of approximately 4200 Da (Fig. 14-13). Avian PP, unlike its mammalian counterpart, appears to have a strong stimulatory influence on gastric acid secretion in chickens. The physiological role for aPP, however, is uncertain at this time.

IV. Hormones Regulating Mammalian Metabolism

Metabolism is affected by many hormones that directly or indirectly affect key enzymes in glycolysis, gluconeogenesis, glycogen synthesis or degradation, lipogenesis, lipolysis, and in the synthesis and utilization of proteins (Table 14-10). These hormones include insulin and glucagon from the endocrine pancreas as well as epinephrine from the adrenal medulla (Chapter 9), glucocorticoids from the adrenal cortex (Chapter 9), GH from the adenohypophysis and the IGFs from liver (Chapter 4), thyroid hormones (Chapter 7), and gonadal steroids (Chapter 11). A hormone may be classified according to its general actions (Table 14-11) as being **glycolytic** (favoring glycolysis), **glycogenolytic** (favoring glycogen hydrolysis to glucose phosphate), **gluconeogenic** (favoring conversion of lipids and/or proteins or their intermediates into glucose production), **lipogenic** (favoring synthesis of triacylglycerols or steroids), **lipolytic** (favoring hydrolysis of triacylglycerols to NEFAs and monoacylglycerols), **protein anabolic** (favoring protein synthesis), or **protein catabolic** (favoring breakdown of proteins to amino acids). A hormone may affect directly or indirectly many metabolic processes and several of the preceding descriptors may be ap-

Table 14-10. Regulation of Some Key Liver Enzymes in Metabolism

Name of enzyme	Process	Stimulatory factor	Inhibitory factor
Phosphorylase *a*	Glycogenolysis	Glucagon, E, NE	Insulin
Glycogen synthetase	Glycogenesis	Insulin	
Triacylglycerol lipase	Lipolysis	Glucagon	Insulin
Triacylglycerol synthetase	Lipogenesis	Insulin	
Phosphoenolpyruvate carboxykinase (PEPCK)	Gluconeogenesis	Glucagon, E, NE	

Table 14-11. Generalized Actions of Hormones with Respect to Protein, Carbohydrate, and Lipid Metabolism

Hormone	Protein	Carbohydrate	Lipid
Insulin	Anabolic	Hypoglycemic, glycogenic, antiglycogenolytic	Lipogenic, antilipolytic
Epinephrine		Glycogenolytic, hyperglycemic, gluconeogenic	Lipolytic
Glucagon		Hyperglycemic, glycogenolytic	
Growth hormone	Anabolic	Hyperglycemic, gluconeogenic	Lipolytic
Glucocorticoids	Antianabolic	Hyperglycemic, gluconeogenic, inhibits peripheral utilization	
Thyroxine	Anabolic or catabolic	Hyperglycemic, glycogenolytic	Lipolytic
Androgens	Anabolic		

plied to the same hormone (see Table 14-11). These metabolic actions may be due to effects on enzyme activity or effects on transport at the cell membrane and hence affect the availability of substrates for various pathways. Although a detailed analysis of this topic is beyond our scope, we briefly describe some hormonal interactions in the following sections that occur under some common conditions in mammals. A brief review of nonmammals can be found in Section V.

A. Endocrine Regulation Following Feeding

In carnivores and omnivores, the most important hormones following ingestion of a meal and following its absorption into blood are insulin and GH as well as GIP. This time period is termed **postprandial** and is characterized by elevated blood levels of glucose, amino acids, and lipids in proportion to the composition of the meal. Release of insulin is evoked directly by an increase in circulating glucose in the blood supplying the pancreatic islets. In addition, there is evidence of early enhancement of the glucose-stimulated release of insulin caused by release of GIP from the small intestine in response to the presence of glucose in the small intestine (see Section II,C,1,c). It is not clear what the role of circulating PP is in humans and dogs after a meal. No convincing evidence for a direct metabolic action of PP has been demonstrated, and this peptide may be limited to regulating digestive functions (see Chapter 13). Direct neural stimulation of insulin release also may occur following or during ingestion, but this does not appear to be a major regulatory pathway.

Insulin stimulates the uptake of glucose by muscle cells, liver cells, and adipocytes. In liver, insulin promotes glycogenesis and formation of VLDLs from glucose whereas in muscle cells it favors glycogenesis and glycolysis. These effects are mediated largely through reduction in the amount of enzyme phosphorylase *a* (which normally would cause hydrolysis of glycogen) and an increase in the enzyme glycogen synthetase. In liver, there also is a reduction in glucose 6-phosphatase and glucokinase that also favors glycogenesis. In adipose cells, synthesis and storage of triacylglycerols as well as inhibition of lipolytic pathways are favored by the intracellular actions of insulin. Increased lipoprotein lipase activity also is favored in adipose tissue by insulin, which enhances incorporation of NEFAs and monoacyglycerol from VLDLs and LDLs. In addition, insulin enhances transport of amino acids into muscle cells, thereby enhancing protein synthesis. In liver

Table 14-12. Metabolic Actions of Glucocorticoids in Different Mammalian Tissues[a]

Process	Liver	Adipose	Heart, diaphragm	Skeletal muscle
Glucose uptake		−	−	
Glycogenesis	+		+	−
Amino acid uptake	+		+	+
Gluconeogensis	+		+	
Protein synthesis			−	
Protein catabolism				+
Lipogenesis	+			
Lipolysis		+P		

[a]Note that the responses of each tissue to the same hormone are specific to the tissue.
+, Stimulatory; −, inhibitory; P, permissive effect.

cells, insulin favors protein synthesis and reduces both deamination of amino acids and gluconeogenesis.

Release of GH may occur much later after a meal and together with insulin causes stimulation of protein synthesis in muscle cells. During sleep, insulin levels typically are low and GH release is maximal. GH stimulates additional protein synthesis as well as glycogenolysis and lipolysis.

B. Effects of Acute and Chronic Stress on Metabolism

The details of the stress response are described in Chapter 9. Acute stress, such as the "flight-or-fight" response and exercise, causes release of epinephrine from the adrenal medulla. Epinephrine brings about important changes in energy metabolism resulting in increased glycogenolysis and hyperglycemia. Glucagon levels also may increase, contributing to hyperglycemia. Chronic stress involves the gluconeogenic and hence hyperglycemic actions of the glucocorticoids (see Table 14-12). In cases of fasting or starvation, GH and glucagon as well as glucocorticoids play important metabolic roles.

1. Acute Stress

An early response to stress is the elevation in epinephrine release from the adrenal medulla. Epinephrine activates adenylate cyclase in muscle cells, resulting in increased glycogenolysis even in the face of elevated insulin. Because glucose 6-phosphate resulting from glycogenolysis cannot be converted into free glucose and leave the muscle cell, it is used directly in glycolysis. This reduces the need for muscle cells to remove glucose from the blood and spares blood glucose for use by the nervous system during stressful times. Another cAMP pathway is stimulated by epinephrine in adipose cells and enhances the activity of triacylglycerol lipase. This in turn increases the availability of NEFAs and glycerol for gluconeogenesis and a trend toward ketogenesis during stress.

2. Chronic Stress

If stressful conditions persist beyond the acute phase, the hypothalamo–adrenal axis is stimulated such that additional glucocorticoid is released. The most important role for glucocorticoids is to prevent utilization of amino acids for protein synthesis and favor conversion of amino acids to carbohydrate via gluconeogenesis. In liver, gluconeogenesis from amino acids provides glucose that can be added to the circulation, producing a subsequent hyperglycemia. This action also protects circulating glucose as an energy source for the nervous system by forcing muscle cells to employ amino acids for energy production and preventing use of circulating glucose by these cells.

During starvation, GH and glucagon have important roles in metabolism. Although glucagon normally stimulates glycogenolysis in liver, the absence of significant glycogen stores means that glucagon can have little direct effect on circulating glucose levels. However, GH and glucagon both stimulate lipolysis in adipose tissue and increase the levels of NEFAs for energy production as well as ketogenesis. Meanwhile, glucocorticoids continue to encourage muscle cells to catabolize protein stores and to use the resulting amino acids as a major gluconeogenic source. Protein catabolism in muscle cells also provides amino acids for liver gluconeogenesis and thereby can contribute indirectly to blood glucose. Gluconeogenesis from amino acids produces toxic nitrogenous wastes (ammonia, urea, uric acid) that must be eliminated by the kidney with attendant loss of water. In turn, muscle cells will utilize some of the circulating ketone bodies produced during the metabolism of NEFAs as an energy source. The rest of the ketone bodies are excreted in the urine along with the nitrogenous wastes from protein catabolism. As described earlier, addition of ketone bodies to the urine contributes to water loss and dehydration.

Chronic starvation can result in severe protein deficiencies as a consequence of the emphasis on gluconeogenesis from proteins to meet the glucose demands of the nervous system. Accompanying ketogenesis from use of NEFAs can produce disturbances in acid–base balance as well. Calculations based on analyses of energy metabolism associated with acute starvation indicate that most humans could survive only about 21 days of total starvation before they would die of protein deficiencies. However, as early as 3 days after the onset of total starvation, the brain undergoes a remarkable decrease in its metabolic requirement for glucose and begins to utilize ketone bodies produced from lipid metabolism to provide the bulk of its energy needs. This change not only enhances lipolysis as a source of ketone bodies for brain metabolism and for gluconeogenesis in the liver but also spares the need for amino acids as an energy source and reduces the rate at which proteins are catabolized. Consequently, the anticipated protein deficiencies do not occur until much later. However, once the body's lipid stores are exhausted, protein remains as the only source to sustain life.

What causes this shift in brain metabolism during total starvation is unknown at this time, but we do know it occurs only under conditions of total starvation. If there is any caloric intake, the switch to ketone utilization is prevented. Consequently, people experiencing semistarvation (where caloric intake is less than required to balance energy expenditures) commonly suffer from protein deficiencies because they employ protein stores to make up the energy deficit in preference to lipid stores. Humans who undertake self-regulated semistarvation (commonly called "dieting") run the risk of developing protein deficiencies and reducing body weight more in terms of lean muscle mass than of fat. The losses of water involved with additional excretion of nitrogenous wastes account for most of the initial weight losses associated with "dieting" unless balanced carefully by increased water intake. Furthermore, it appears that protein stores can be protected only by simultaneously stimulating protein synthesis. Periodic bouts of strenuous exercise result in enhanced protein synthesis during nonexercise periods, which is directed in part by episodes of GH release.

C. Protein Anabolic Hormones

Alterations in nitrogen retention and excretion reflect episodes of protein synthesis and protein catabolism for gluconeogenesis, respectively. Many hormones have been shown to enhance nitrogen retention through actions on protein synthesis. These **protein anabolic** agents include GH, androgens, estrogens, thyroid hormones, and insulin. We have already considered the actions of GH and insulin above. Since thyroid hormones are protein anabolic in thyroidectomized animals but not in hypophysectomized animals (lacking GH as well as the other tropic hormones), it is safe to conclude that the thyroid effect on protein synthesis is brought about by its interactions with GH (see Chapter 7).

The earliest suggestion of a protein anabolic role for androgens was the claim for their rejuvenating power by Brown-Sequard in 1889, which he reportedly discovered after treating himself with extracts prepared from animal testes. In the 1930s, Charles D. Kochakian and coworkers began 40 years of research that clearly established the unique role of androgens as **protein anabolic steroids.** Although estrogens were shown to stimulate protein synthesis in a few tissues (e.g., uterus, mammary gland, skin, skeleton), only the **androgenic–anabolic steroids** had such a profound effect on skeletal muscles that they determined an overall nitrogen retention by the body. Progesterone is protein anabolic on uterus but overall causes increased nitrogen excretion.

The greater muscle masses of male mammals as compared to female mammals long has been recognized and is attributed to the actions of androgens. In nonmammalian vertebrates, androgens have been shown to have sex-specific effects on certain muscles in males. Many analogs of the naturally occurring androgens have been prepared, but it has not been possible to separate their androgenic action from their protein anabolic effect. Androgenic–anabolic steroids do not influence transport of amino acids into muscle cells but apparently affect the levels of amino acid-activating enzymes employed in protein synthesis.

The potential effect of anabolic–androgenic steroids on muscle strength and endurance resulted inevitably in their use by humans for athletic training programs. Although their effect on muscle mass is easily demonstrated, the improvement of athletic performance is mostly anecdotal and is not well supported by controlled studies. Regardless, the established effects of anabolic–androgenic steroids on reduced reproductive performance (through negative feedback on the hypothalamo–hypophysial–gonad axis), increased liver dysfunction (80% of subjects in one study), and correlations with hepatitis, liver failure, and fatal liver cancer should serve to warn people to avoid their use for unsubstantiated, short-term benefits on athletic performance.

V. Nonpancreatic Hormones and Metabolism in Nonmammalian Vertebrates

In contrast to mammals, the endocrine regulation of metabolism has been neglected in vertebrate groups that have relatively little economic importance. The differences in body temperature, reliance on anaerobic metabolism, and great variety of activity levels make it difficult to generalize even within a reasonably defined taxonomic unit. A brief review of metabolism in nonmammalian vertebrates is provided below.

A. Fishes

Lampreys exhibit hyperglycemia following treatment with either cortisol or epinephrine. Cortisol elevates liver glycogen, suggesting a gluconeogenic action. Insulin is hypoglycemic and a single injection can decrease plasma glucose levels significantly for several days. Hagfishes do not exhibit hyperglycemia in response to either cortisol or catecholamine treatment.

Chondrichthyean fishes have been studied less frequently than have the jawless fishes, and few generalizations are possible. Blood sugar levels are relatively low in selachians but can be elevated with treatment by either ACTH or glucocorticoids, implying metabolic actions similar to those described in mammals.

Among teleosts, gluconeogenesis occurs mainly in the liver and kidney and is accelerated by high-protein diets or by starvation. Cortisol is the principal glucocorticoid secreted by the interrenals. In addition to its role as a mineralocorticoid, cortisol depletes lipid and protein stores and promotes gluconeogenesis. High cortisol levels, for example, are associated with up to a 96% reduction in body fat and loss of more than half the body protein in spawning sockeye salmon. Catecholamines stimulate glycogenolysis, lipolysis, and gluconeogenesis. The possible role(s) of thyroid hormones (see Chapter 8) is unclear at this time, but they do not appear to be important for regulating metabolism.

B. Amphibians

Corticosterone and, to a lesser extent, aldosterone are gluconeogenic and hyperglycemic in amphibians, but the primary role of these steroids as seen in fishes appears to be the regulation of ion and water balance. Specific gluconeogenic enzymes are activated by corticosteroids as well as are urea cycle enzymes associated with elimination of nitrogenous wastes generated from amino acid degradation. Epinephrine is a potent stimulator of glycogenolysis in both liver and muscle and therefore is hyperglycemic. Thyroid hormones can increase oxidative metabolism, but apparently only at higher temperatures (for example, 25°C; see Chapter 8); any metabolic role is highly speculative.

C. Reptiles

The metabolism of reptiles and the roles of corticosteroids and catecholamines in regulating reptilian metabolism are poorly studied. Although there is little evidence to establish a gluconeogenic action in reptiles, there is some indirect evidence. Corticosteroids are hyperglycemic and favor deposition of glycogen in lizards, turtles, alligators, and snakes. Epinephrine is hyperglycemic in representatives of all major reptilian groups and promotes glycogenolysis in liver and glycogen deposition in alligator muscle. As reported for amphibians, thyroid hormones stimulate oxygen consumption in lizards at warmer temperatures (30°C) but not at lower temperatures (see Chapter 8).

D. Birds

Domestic birds exhibit all of the glycolytic, glycogenic, and glycogenolytic pathways for carbohydrate metabolism described for mammals. Because of the dependence of developing birds on lipid-rich yolk, the normal glycolytic and oxidative pathways are not important until after hatching.

Corticosterone is hyperglycemic and gluconeogenic primarily owing to stimulation of protein catabolism. Corticosteroids also stimulate synthesis in the liver of triacylglycerols, which are transported to adipose tissue for storage, an important action during fattening prior to spring or fall migrations. This lipogenic action is in marked contrast to the effects of glucocortioids in mammals and other vertebrates and may represent a special adaptation in long-distance migrators. It would be interesting to examine long-distance migrating whales in this regard.

Epinephrine and norepinephrine are potent glycogenolytic and lipolytic hormones in birds. Catecholamines cause hepatic and muscle glycogenolysis, which can lead to hyperglycemia. Thyroid hormones affect oxygen consumption and therefore have direct actions on metabolism and heat production.

VI. Summary

Overall metabolism is the sum of all anabolic and catabolic reactions occurring in the body. Intermediary metabolism consists of pathways, involved in energy production from carbohydrates, that are connected to pathways for protein and lipid metabolism such that excess carbohydrates can be converted to protein or lipid for storage; the latter can also be used to supply carbohydrates for energy production via gluconeogenesis. The carbohydrate pathways include glycolysis and the pentose shunt (anaerobic), the tricarboxylic acid cycle, and the cytochrome system (which together with glycolysis and the pentose shunt constitute aerobic metabolism). Lipids (fatty acids, triacylglycerols, cholesterol, and phospholipids) are synthesized from carbohydrates and can be converted to carbohydrate via gluconeogenesis from glycerol and NEFAs (β-oxidation). Ketone bodies (ketogenesis) are formed during the β-oxidation of NEFAs. Amino acids derived from protein catabolism can be converted (by deamination or transamination) to carbohydrates through gluconeogenic events. However, in many animals it is not possible to resynthesize all of the necessary amino acids from carbohydrates and some (the essential amino acids) must be obtained through the diet.

The endocrine pancreas secretes hormones that are directly involved in the regulation of metabolism. The predominant cellular types in the vertebrate endocrine pancreas are the A cell, B cell, and D cell. The A cell and B cell have been identified as the sources for glucagon and insulin, respectively. The D cell produces somatostatin. A fourth cellular type located at the periphery of the mammalian pancreatic islet produces PP. These four cellular types plus sometimes a fifth type of unknown function are found in most vertebrate groups. However, only A and B cells have been confirmed in cyclostomes. It is not certain what the germ layer origins are for the pancreatic endocrine cells. They do not come from the neural crest in spite of their APUD characteristics, and it is not clear whether they are of endodermal or ectodermal origin.

Five patterns of islet morphology can be distinguished in vertebrates: (1) cyclostome type, (2) primitive gnathostome type, (3) actinopterygian type, (4) lungfish type, and (5) tetrapod type. It is assumed that the most primitive anatomical arrangement for the endocrine pancreatic tissue occurs in lampreys as the follicles of Langerhans embedded in the gut lining. The equally if not more primitive hagfishes, however, have a definite endocrine pancreas. Bony fishes tend to have discrete islets but lack an acinar (exocrine) pancreas. Elasmobranchs, holocephalans, and *Latimeria* all exhibit the primitive gnathostome type of endocrine pancreas. Tetrapods have distinct pancreatic organs containing both acinar and islet tissues.

Glucagon is one of several hyperglycemic regulatory factors in vertebrates, and it stimulates glycogenolysis and lipolysis in liver and adipose cells while inhibiting glucose oxidation. These actions of glucagon appear to be mediated through activation of adenylate cyclase and formation of cAMP. Low blood sugar levels stimulate glucagon release, and somatostatin blocks glucagon release. High levels of circulating fatty acids (NEFAs) stimulate glucagon release in birds and herbivorous mammals. Glucagon may be the primary pancreatic regulator of carbohydrate metabolism in birds and lizards, with insulin playing only a minor role. Glucagon seems to be hyperglycemic in amphibians and fishes. Immunoreactive glucagon also is present in several invertebrate phyla including arthropods, mollusks, and the invertebrate chordates.

Insulin is the only naturally occurring hypoglycemic factor in vertebrates. It is present in many invertebrate species as well and may even function there as a hypoglycemic factor. The many actions of insulin include permeability effects on the plasmalemma, resulting in the increased uptake of glucose and amino acids into muscle cells and the uptake of glucose by adipose cells. Lipogenesis increases after binding of insulin to the plasmalemma of adipose cells because of the increased availability of substrates as a consequence of glucose uptake and oxidation as well as effects on enzymes that increase lipogenesis and inhibit lipolysis. Growth effects occur in cooperation with IGFs produced under the influence of GH.

High circulating glucose levels stimulate release of insulin from the B cells of the pancreatic islets. It is both the uptake and the metabolism of glucose that actually are responsible for causing insulin release. Neural factors (parasympathetic) and other hormones (for example, GIP) also may stimulate insulin release. Somatostatin, epinephrine, and gastrin, as well as low blood glucose levels, all inhibit insulin release.

The most primitive function for insulin may be its effect on amino acid incorporation into protein. Glucose administration does not invoke insulin release in lampreys, lizards, or birds, and insulin stimulates only protein synthesis in hagfish liver. These observations suggest limited involvement of insulin in carbohydrate metabolism of nonmammals and that its hypoglycemic actions may be largely due to effects on protein and/or lipid metabolism. Furthermore, the effects of insulin on lipid metabolism may be largely a mammalian occurrence.

Metabolism is controlled by the actions and interactions of many hormones under different nutritional conditions, which maintain a healthy organism with energy reserves necessary for maintenance of the organism and for reproduction. The major metabolic hormone is insulin, which is protein anabolic, lipogenic, antilipolytic, glycogenic, antiglycogenolytic, and hypoglycemic. These actions are most important following ingestion of a meal and show lesser importance under postprandial conditions and during starvation. The actions of insulin are opposed primarily by six other hormones, including GH (protein anabolic, lipolytic, gluconeogenic, hyperglycemic), glucagon (lipolytic, glycogenolytic, hyperglycemic), epinephrine (lipolytic, glycogenolytic, gluconeogenic, hyperglycemic), glucocorticoids (protein catabolic, gluconeogenic, hyperglycemic), thyroid hormones (lipolytic, glycogenolytic, hyperglycemic), and anabolic–androgenic steroids (protein anabolic). Epinephrine directly inhibits insulin release. Glucocorticoids also prevent peripheral use of glucose, especially by muscle.

Suggested Reading

Books

Campaigne, B. N., and Lampman, R. M. (1994). "Exercise in the Clinical Management of Diabetes." Human Kinetics, Champaign, IL.

Epple, A., and Brinn, J. E. (1987). "The Comparative Physiology of the Pancreatic Islets." Springer-Verlag, Berlin.

Kochakian, C. D. (1984). "How It Was: Anabolic Action of Steroids and Remembrances." Univ. Alabama School of Medicine, Birmingham, AL.

LeRoith, D., Olefsky, J. M., and Taylor, S. I. (1996). "Diabetes Mellitus: A Fundamental and Clinical Text." Lippincott-Raven, Harestown, MD.

Samols, E. (1991). "The Endocrine Pancreas." Raven, New York.

General Articles

Baskin, D. G., Figlewicz, D. P., Woods, S. C., Porte, D., Jr., and Dorsa, D. M. (1987) Insulin in the brain. *Ann. Rev. Physiol.* **49,** 335–347.

Bliss, M. (1989). Special lecture: J. J. R. Macleod and the discovery of insulin. *Q. J. Exp. Physiol.* **74,** 79–96.

Fehmann, H.-C., and Habener, J. F. (1992). Insulinotropic glucagonlike peptide-I(7-37)/(7-36) amide: A new incretin hormone. *Trends Endocrinol. Metab.* **3,** 158–163.

Geary, N. (1990). Pancreatic glucagon signals postprandial satiety. *Neurosci. Biobehav. Rev.* **14,** 323–338.

Hazelwood, R. L. (1993). The pancreatic polypeptide (PP-fold) family: Gastrointestinal, vascular, and feeding behavioral implications. *Proc. Soc. Exp. Biol. Med.* **202,** 44–63.

Heidenreich, K. A. (1991). Insulin in the brain: What is its role? *Trends Endocrinol. Metab.* **2,** 9–12.

Larner, J. (1988). Insulin signaling mechanisms. Lessons from the Old Testament of glycogen metabolism and the New Testament of molecular biology. *Diabetes* **37,** 262–275.

McDermott, A. M., and Sharp, G. W. G. (1993). Mini review: Inhibition of insulin secretion: A fail-safe system. *Cell. Signal.* **5,** 229–234.

Quon, M. J. (1994). Insulin signal transduction pathways. *Trends Endocrinol. Metab.* **5,** 369–376.

Sargeant, R., Mitsumoto, Y., Saraba, V., Shillabeer, G., and Klip, A. (1993). Hormonal regulation of glucose transporters in muscle cells in culture. *J. Endocrinol. Invest.* **16,** 147–162.

Thissen, J.-P., Ketelslegers, J.-M., and Underwood, L. E. (1994). Nutritional regulation of the insulin-like growth factors. *Endocr. Rev.* **15,** 80–101.

Trenkle, A. (1981). Endocrine regulation of energy metabolism in ruminants. *Fed. Proc.* **40,** 2536–2541.

Comparative Articles

Conlon, J. M., Reinecke, M., Thorndyke, M. C., and Falkmer, S. (1988). Insulin and other islet hormones (somatostatin, glucagon and PP) in the neuroendocrine system of some lower vertebrates and that of invertebrates—a minireview. *Horm. Metab. Res.* **20,** 406–410.

Conlon, J. M., Nielsen, P. F., Youson, J. H., and Potter, I. C. (1995). Proinsulin and somatostatin from the islet organ of the southern-hemisphere lamprey *Geotria australis. Gen. Comp. Endocrinol.* **100,** 413–422.

Gapp, D. A. (1987). Endocrine and related factors in the control of metabolism in nonmammalian vertebrates. *In* "Fundamentals of Comparative Vertebrate Endocrinology" (I. Chester-Jones, P. M. Ingleton, and J. G. Phillips, eds.), pp. 509–660. Plenum, New York.

Reinecke, M., Hoog, A., Ostenson, C.-G., Efendic, S., Grimelius, L., and Falkmer, S. (1991). Phylogenetic aspects of pancreastatin- and chromogranin-like immunoreactive cells in the gastro-entero-pancreatic neuroendocrine system of vertebrates. *Gen. Comp. Endocrinol.* **83,** 167–182.

Reinecke, M., Broger, I., Brun, R., Zapf, J., and Maake, C. (1995). Immunohistochemical localization of insulin-like growth factor I and II in the endocrine pancreas of birds, reptiles, and Amphibia. *Gen. Comp. Endocrinol.* **100,** 385–396.

Sheridan, M. (1994). Mini review: Regulation of lipid metabolism in poikilothermic vertebrates. *Comp. Biochem. Physiol.* **107B,** 495–508.

Clinical Articles

Atkinson, M. A., and Maclaren, N. K. (1990). What causes diabetes. *Sci. Am.* **263(1),** 62–71.

Bach, J.-F. (1994). Insulin-dependent diabetes as an autoimmune disease. *Endocr. Rev.* **15,** 516–542.

Bretherton-Watt, D., and Bloom, S. R. (1991). Islet amyloid polypeptide: The cause of type-2 diabetes? *Trends Endocrinol. Metab.* **2,** 203–206.

Eisenbarth, G. S., and Ziegler, A. (1995). Type I diabetes mellitus. *In* "Molecular Endocrinology: Basic Concepts and Clinical Implications" (B. D. Weintraub, ed.), pp. 269–282. Raven, New York.

Fajans, S. S. (1987). MODY—A model for understanding the pathogenesis and natural history of type II diabetes. *Horm. Metab. Res.* **19,** 591–599.

Hoberman, J. M., and Yesalis, C. E. (1995). The history of synthetic testosterone. *Sci. Am.* **Feb. 1995,** 76–81.

Macintyre, J. G. (1987). Growth hormone and athletes. *Sports Med.* **4,** 129–142.

Weindruch, R. (1996). Caloric restriction and aging. *Sci. Am.* **Jan. 1996,** 46–52.

15

Regulation of Calcium and Phosphate Homeostasis

CALCIUM AND phosphate are important for the regulation of many basic body functions. They are involved in the structure of bones and teeth, which serve as the major reservoirs for maintaining plasma and interstitial fluid levels of these ions. Although inorganic calcium and phosphate are not related closely with respect to most of their essential roles in vertebrate physiology, they are regulated by some of the same hormones. Consequently, a discussion of the endocrine regulation of calcium homeostasis cannot be divorced entirely from the regulation of phosphate. The succeeding discussion of calcium and phosphate homeostasis in mammals sets the standard for making comparisons to other vertebrates.

I. Importance of Calcium and Phosphate

A. Calcium Homeostasis

Calcium levels in extracellular fluids and in the cytosol of cells must be regulated precisely if normal body functions are to be maintained. In addition to its role in bone and tooth construction, calcium ions have other important roles to play (Table 15-1). They are responsible for excitation and contraction of muscle cells and are important in the induction of spontaneous excitations of cardiac pacemaker cells. Calcium ions are essential for exocytosis of secretion granules in neurons and glandular cells and serve as second messengers in many target cells. Certain key metabolic enzymes are activated by calcium ions, which can also serve as cofactors for several blood-clotting proteins (factors VII, IX, and X).

The calcium level in adult mammalian blood plasma is maintained at about 10 mg/dl or approximately 2.5 mM. About half of this calcium is free in the plasma in ionic form (Ca^{2+}), and the remainder is bound to circulating proteins (40%) or occurs in other chemical complexes (10%). It is this free ionic calcium that is essential to so many important life processes because it can be exchanged readily with extracellular fluids and cells. About 90% of the protein-bound calcium is linked to albumin, and this binding is pH sensitive. Acute acidosis decreases binding and elevates ionic plasma calcium. Acute alkalosis increases binding and reduces free plasma calcium.

After birth, calcium is obtained in the diet, absorbed through the small intestine, and deposited in bones and teeth or excreted via urine or feces. Urinary excretion of calcium is

Table 15-1.   Some Physiological Roles for Calcium and Phosphate

Roles for calcium	Roles for phosphate
Structural component (with phosphate) of bones, teeth	Structural component (with calcium) of bones, teeth
Necessary for contraction of skeletal, smooth, and cardiac muscle	Activator of enzymes from inactive form
	Buffer in blood and other body fluids
Maintenance of membrane potentials in some neurons	Essential component for energy storage in chemical bonds of ATP, creatine phosphate, etc.
Unique role in membrane potentials of pacemaker cells	Essential component of nucleic acids
	Essential component of phospholipids in cell membranes
Second messenger in hormonal and neurocrine mechanisms of action	Involved in second-messenger formation
Role in exocytosis of secretory products from cells	Necessary for metabolism of glucose and other carbohydrates
Necessary cofactor for certain enzymes	
Component of intercellular matrices (e.g., basement membrane)	

Table 15-2. Calcium Content of
Human Body[a]

Age (years)	Body weight (kg)	Calcium content (g)
1	10.6	100
5	19.1	219
10	33.3	396
15	55.0	806
20	67.0	1078

[a] Irving, J. T. (1973). "Calcium and Phosphorus Metabolism." Academic Press, New York.

directly proportional to plasma levels, and little calcium is excreted unless plasma calcium levels exceed normal. Calcium deposited in bone serves as a reservoir to provide adequate plasma Ca^{2+} for minute-to-minute regulation of body needs and during acute or chronic periods of dietary deprivation. Human body calcium levels vary according to size and age (Table 15-2).

1. Calcium Regulation

Blood calcium homeostasis is maintained by the cooperative actions of the bones and teeth and the intestine, which together serve as internal and external sources of calcium (see Fig. 15-1). The kidneys prevent loss of calcium to the urine or can allow excretion of excess calcium. The intestine also excretes calcium. Calcium homeostasis is maintained by three

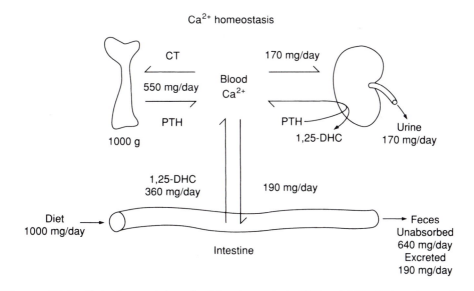

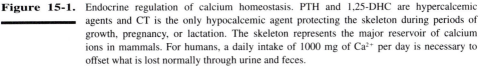

Figure 15-1. Endocrine regulation of calcium homeostasis. PTH and 1,25-DHC are hypercalcemic agents and CT is the only hypocalcemic agent protecting the skeleton during periods of growth, pregnancy, or lactation. The skeleton represents the major reservoir of calcium ions in mammals. For humans, a daily intake of 1000 mg of Ca^{2+} per day is necessary to offset what is lost normally through urine and feces.

hormones in mammals. **Parathyroid hormone (PTH)** secreted by the **parathyroid glands** causes calcium release from bones and favors calcium reabsorption by the kidneys. **Calcitonin (CT)** from C cells embedded in the thyroid gland (see Chapter 7) or the parathyroids (dogs) promotes deposition of calcium into bone. A steroid derivative of vitamin D known as **1,25-dihydroxycholecalciferol (1,25-DHC)** released from the kidney stimulates intestinal uptake of calcium.

B. Phosphate Homeostasis

Phosphate, like calcium, is an essential component of bones and teeth. Approximately 80% of the total body phosphate is sequestered in the skeleton as calcium phosphate. In addition, many essential molecules contain phosphate, including structural phospholipids in cellular membranes, nucleic acids, nucleotides, and hexose phosphates. Furthermore, phosphate is indispensable for energy storage within cells in the form of ATP or creatine phosphate. Hydrolysis of ATP or guanosine triphosphate (GTP) to form cyclic adenosine $3',5'$-monophosphate (cAMP) or cyclic guanosine $3',5'$-monophosphate (cGMP), respectively, is necessary for mediating the actions of many hormones, for neural transmission, and for many other cellular processes. The actions of protein kinases determine the presence or absence of phosphate, which activates or inactivates key enzymes in metabolic pathways (see Chapter 14). Although phosphate ions play a minor role as buffers of hydrogen ions in most body fluids, they are the major buffer system in urine.

It is rare that phosphate becomes a limiting factor for an organism, and the usual phosphate disturbances in mammals are a result of excessive levels of phosphate. Inorganic phosphate (P_i) in mammalian plasma is generally about 3.1 mg/dl (about 1.0 mmol/liter). Most of this phosphate (about 80%) is in the form of HPO_4^{2-}, with almost 20% occurring as $H_2PO_4^-$ and only a trace as PO_4^{3-}. Henceforth, the chemical formula HPO_4^{2-} is used to represent all of the phosphate ions.

In normal plasma about 90% of the P_i is free ionic (filterable) phosphate, and about 10% is bound to the plasma proteins. In addition to P_i, plasma contains considerable amounts of lipid-bound phosphate and esterified phosphate so that total plasma phosphate is actually about 12.5 mg/dl. Phosphate values vary considerably with diet, age, and metabolic state, however, and it is difficult to provide a "normal" value without specifying the conditions under which this "normal" occurs.

C. Interrelationship of Ca^{2+} and HPO_4^{2-}

Calcium and phosphate are regulated in such a way that the product of the free plasma concentrations of Ca^{2+} and HPO_4^{2-} equals some constant ($[Ca^{2+}][HPO_4^{2-}]) = k$). This constant however, may change according to differing physiological states or pathological conditions. For example, k is greater in growing mammals than it is in adults. This relationship between Ca^{2+} and HPO_4^{2-} implies that if there is an increase in Ca^{2+}, a corresponding decrease in HPO_4^{2-} should follow. Likewise, an increase in HPO_4^{2-} should cause a decrease in Ca^{2+}. This generalization is useful to illustrate some of the relationships that exist between the regulatory mechanisms governing these ions. Minute-to-minute adjustments of Ca^{2+} and HPO_4^{2-} levels in extracellular fluids are accomplished primarily through a combination of bone destruction (resorption) or formation, absorption of dietary calcium by the small intestine, and renal excretion of phosphate.

D. Bone Formation and Resorption in Mammals

In bones and teeth, calcium phosphate occurs in the form of submicroscopic crystals deposited on an organic matrix composed primarily of **collagen fibers.** These crystals assume a uniform structural and complex chemical form known as **hydroxyapatite crystals.** Construction of bone through formation of calcium phosphate is not understood completely, but some of the major features are well accepted. Bone formation may involve **apatite formation** (deposition of new hydroxyapatite crystal) or simply **mineral accumulation** (the additional growth of existing crystals). Exchange of Ca^{2+}, HPO_4^{2-}, and water can occur between the surface of these crystals and the extracellular fluids. This exchange is inversely proportional to the size of the crystal. Thus, larger crystals contain considerable amounts of calcium phosphate that cannot engage in free exchange with the extracellular fluids. About 99% of bone calcium phosphate is found in these larger, nonexchangeable, stable or **diffusion-locked crystals.**

The specific process of bone formation and growth also involves a number of local chemical factors that are responsible for collagen matrix formation, cartilage matrix deposition, and formation of hydroxyapatite crystals as well as cellular replication and differentiation. Cells known as **osteoblasts** (literally, bone-forming cells) are responsible for bone formation. (The homologous cell in teeth is termed an *odontoblast.*) The osteoblasts comprise the **endosteal membrane** that lines the cavities within bone, and they synthesize the collagen matrix on which apatite formation occurs. The factors controlling osteoblast activities are poorly understood. Some osteoblasts give rise to **osteocytes** that become completely surrounded by bone except for minute channels through which the osteocytes communicate with one another. There seems to be little agreement on the role for osteocytes in calcium–phosphate metabolism, but they may be important targets for hormonal regulation.

Resorption of bone may involve either removal of the collagen matrix and/or solubilization of hydroxyapatite crystals with consequent release of Ca^{2+} and HPO_4^{2-}, but both processes usually occur. Another bone cell, the **osteoclast** (literally, bone destroying) is primarily responsible for bone resorption. The osteoclast (Fig. 15-2) is a large, multinucleate cell and is easy to distinguish from uninucleate osteoblasts or osteocytes. Osteoclasts may arise from either osteoblasts or bone marrow cells, but the details of their formation are not clear.

II. Endocrine Regulation of Calcium and Phosphate Homeostasis in Mammals

Three hormones regulate calcium and phosphate homeostasis in mammals, as mentioned earlier. Parathyroid hormone is a **hypercalcemic factor;** that is, its actions cause an elevation in the level of plasma calcium. Release of PTH is increased by low plasma calcium levels and is decreased by elevated plasma calcium levels, but PTH release is not directly influenced by fluctuations in phosphate levels. Calcitonin is a **hypocalcemic factor,** and its release is related primarily to changes in plasma calcium. One major site of action for PTH is bone, where it may stimulate calcium release from bone (bone resorption) and release both calcium and phosphate ions into the circulation (see Fig. 15-1). Parathyroid hormone also increases calcium reabsorption by the nephron as well as the secretion of phosphate into the urine, resulting in a decrease in plasma phosphate levels and a concomitant increase in urinary phosphate.

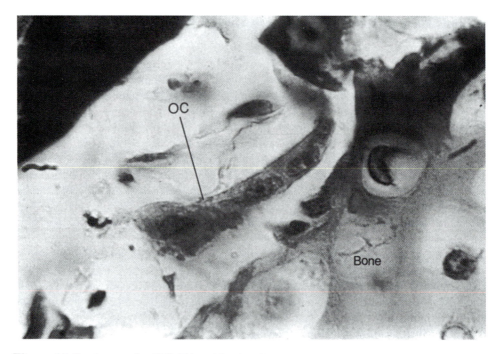

Figure 15-2. An osteoclast (OC). This multinucleate bone-destroying cell is a target for PTH, and the ratio of osteoclasts to osteoblasts is increased by PTH. This sample is from a turtle bone. (Courtesy of Nancy B. Clark, University of Connecticut, Storrs.)

The effects of PTH on calcium homeostasis are linked closely to the actions of the third regulatory hormone, 1,25-DHC. Absorption of calcium from the gut lumen is stimulated by 1,25-DHC, which also influences the actions of PTH on bone and kidney. Furthermore, as discussed below, PTH influences the production of 1,25-DHC by the kidney.

A. The Parathyroid Glands and Parathyroid Hormone

Although the parathyroid glands had been observed previously, Sandstrom rediscovered and named them in 1880. Frequently the parathyroids, which develop like the thyroid from pharyngeal tissues (Fig. 15-3), are embedded within the thyroid glands (Fig. 15-4), for example, in the mouse, cat, and human. In other mammals, such as goats and rabbits, they are separate glands located near the thyroid. Some mammals have more than four separate parathyroid glands, and accessory parathyroid tissue is not uncommon. Thyroidectomy may result in lowered plasma Ca^{2+} levels in some species because of simultaneous removal of the embedded parathyroids, whereas parathyroidectomy in another species may not alter plasma Ca^{2+} markedly unless all of the accessory parathyroid tissue was removed. Decreasing plasma calcium by artificial means can increase PTH secretion severalfold. This can be accomplished by infusion of calcium-chelating (calcium-binding) agents such as ethylenediaminetetraacetic acid (EDTA).

The parathyroid glands developmentally arise from pharyngeal endoderm. However, the parathyroid **chief cells** that secrete PTH arise from neuroectoderm, as shown first in the frog *Rana temporaria*. Chief cells are part of the amine precursor uptake and decarboxylation (APUD) series of peptide-secreting cells (Chapter 1). Indirect evidence also supports

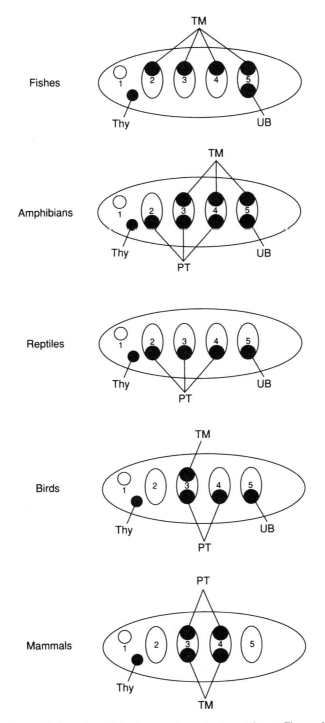

Figure 15-3. Embryonic derivatives of the pharyngeal pouches in vertebrates. The numbers refer to pouch number, with anterior to the left. Pouch 1 remains only as the spiracle (selachians) or the eustachian tube. The thyroid (Thy) actually develops from the pharynx floor between the first and second pouches. Parathyroids (PT) appear first in amphibians. In reptiles, the origin of the thymus (TM) may be from pouches 2 and 3 (lizards), 3 and 4 (turtles), or 4 and 5 (snakes). In mammals the origins of parathyroids and thymus are reversed. The ultimobranchial body (UB) is absent in mammals and the calcitonin-secreting cells migrate to the thyroid instead of the UB.

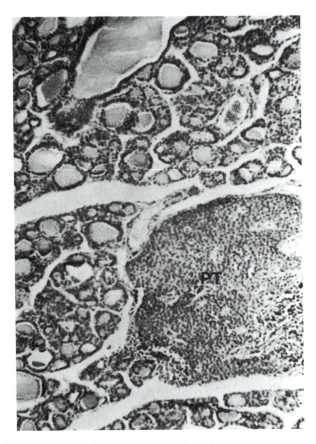

Figure 15-4. A mammalian parathyroid gland (PT) located in the thyroid, surrounded by thyroid follicles.

a similar origin for PTH-secreting cells in birds and mammals. Chief cells are cuboidal with no unique cytological features other than the presence of granules containing immunoreactive PTH. They are the dominant cellular type and comprise about 99% of the parathyroid's cellular population in most species. Chief cells exhibit β-adrenergic receptors, suggesting some neural influence on PTH release may exist. Glucocorticoid receptors also are present and when activated result in enhanced PTH release. Estrogens may diminish the response of the chief cells to lowered plasma calcium.

In a few species, such as deer, the parathyroids are composed exclusively of chief cells. A second cellular type called an **oxyphil** is usually present in the parathyroids. Oxyphils are eosinophilic cells rich in mitochondria (hence their name). The function of the oxyphil and the significance for its large number of mitochondria are unknown. Oxyphils represent only about 1% of the total parathyroid cell population.

1. Parathyroid Hormone

Mammalian PTH has been isolated and characterized from several species. It is a large polypeptide consisting of 84 amino acids (Fig. 15-5). Apparently all of the biological ac-

	1	6	10	16	21	26	31
Bovine PTH	AVSEI	QFMHN	LGKHL	SSMER	VEWLR	KKLQD	VHNF---
Rat PTH	AVSEI	QLMHN	LGKHL	ASMER	VEWLR	KKLQD	VHNF---
Human PTH	SVSEI	QLMHN	LGKHL	NSMER	MQWLR	KKLQD	VHNF---
Human PTHrP	AVSEH	QLLHD	KGKSI	QDLRR	RFFLH	HHIAE	IHTA---

Figure 15-5. Amino acid sequences of mammalian PTHs and PTH-related peptide (PTHrP). Only the first 34 amino acids are shown. Chicken PTHrP differs from human PTHrP at positions 23, 25–27, and 29–32. See Appendix C for single-letter amino acid code.

tivity resides in the first 34 amino acids. Synthesis of PTH occurs in two steps. **Prepropar-athyroid hormone (PreProPTH)**, consisting of 115 amino acid residues, is synthesized on the ribosomes of the rough endoplasmic reticulum (RER) but is rapidly cleaved at the NH_2 terminus to **proparathyroid hormone (ProPTH)** as it enters the cisterna of the endoplasmic reticulum. ProPTH consists of 90 amino acids and travels through the cisterna to the Golgi apparatus, where the remaining 6 NH_2-terminal residues of the prohormone are removed. The resulting PTH is then packaged into storage vesicles to await release.

The classic bioassay for PTH is an increase in plasma calcium following administration of parathyroid gland extracts or PTH to normal or parathyroidectomized dogs. Current bioassays employ parathyroidectomized rats maintained on a calcium-deficient diet. Owing to the activity of plasma endopeptidases, parathyroid hormone activity may reside in numerous peptide fragments in the circulation, varying from 4500 (34 amino acids) to 9000 Da (84 amino acids). The half-life for PTH biological activity due to these various peptide fragments is about 20 min in cattle or rats. However, the complete 84-amino acid PTH peptide is cleared from the blood much more rapidly (2 to 4 min), indicating that the remainder of the bioassayable activity is due to persistence of fragments produced by partial hydrolysis of the parent PTH peptide.

2. Regulation of Parathyroid Hormone Secretion

In Chapter 2 we emphasized the role of increased calcium in endocrine cell secretion of hormones through exocytosis. However, unlike most secretory cells, parathyroid cells release PTH when extracellular and intracellular calcium levels are minimal. In fact, higher levels of extracellular or intracellular calcium actually inhibit PTH secretion (for example, Fig. 15-6). Although the precise mechanisms for regulating PTH secretion have not been worked out completely, several cellular events are established.

Low intracellular calcium is correlated with activated protein kinase C (PKC), which is a known participant in cellular regulation (see Chapter 2). High extracellular calcium induces formation of inositol trisphosphate (IP_3) and diacylglycerol (DAG), but the latter is not involved with PKC activation in chief cells as it is in other secretory cells. DAG may activate some as yet undiscovered mechanism that inhibits PTH release. IP_3 releases intracellular stores of calcium ions that activate neutral proteases called **calpains** and/or activate lysosomal hydrolytic enzymes. Degradation of intracellular PTH is accomplished by the actions of these enzymes. Furthermore, calpains can cause down-regulation of PKC, which would prevent PTH secretion. Both calcium ions and 1,25-DHC have been shown to reduce PTH synthesis, possibly through direct effects on transcription.

Extracellular calcium may open calcium channels in the chief cell membrane by binding

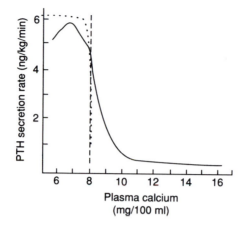

Figure 15-6. Relationship of calf parathyroid secretion of PTH to experimental manipulation of plasma calcium levels. Plasma levels above 8 mg/100 ml (about 1 nM) suppress PTH secretion. [Modified from Mayer, G. P., and Hurst, J. G. Sigmoidal relationship between parathyroid hormone secretion rate and plasma calcium concentration in calves. *Endocrinology (Baltimore)* **102**, 1036–1042, 1978. © The Endocrine Society.]

to a G protein-linked membrane receptor system. There is also evidence to support a direct action of calcium ions on membrane calcium channels.

3. Parathyroidectomy

In all cases, parathyroidectomy causes a reduction in plasma Ca^{2+} leading to detrimental muscular effects. As Ca^{2+} levels decrease, hyperexcitability of motor neurons and skeletal muscles occurs, resulting in twitches, spasms, and, in extreme cases, violent convulsions. This condition is known as **low calcium-induced tetany.** If prolonged contractions (tetany) of the respiratory muscles occur, death due to asphyxiation may result. The severity of these neuromuscular effects, however, differs markedly with respect to the species involved and under various physiological conditions. For example, exercise raises the body temperature and increases the breathing rate, causing a reduction of blood CO_2. Reduced blood CO_2 in turn alters blood pH (alkalosis) and retards ionization of Ca^{2+}. This further reduction in Ca^{2+} in a parathyroidectomized animal may precipitate a tetanic seizure. Animals on diets low in Ca^{2+} and high in HPO_4^{2-} exhibit tetany following parathyroidectomy more readily than do animals on normal diets. Hyperventilation reduces CO_2 levels and can induce tetany especially when calcium levels are already low.

4. Parathyroid Hormone-Related Protein

A protein with hypercalcemic activity first was isolated from patients exhibiting human hypercalcemia of malignancy (HHM) as a result of cancer. Because this protein has considerable N-terminal overlap with PTH (Fig. 15-5), it was named **parathyroid hormone-related protein (PTHrP)**. This protein has proven to be a natural product of many cell types; the gene encoding this protein is believed to have been derived from an ancestral gene that also gave rise to the PTH gene. This gene can be active in more than 20 tissues in humans, including the lactating mammary gland, the uterus, and the amnion and parathyroid glands of the fetus.

During fetal development, PTHrP is synthesized by the amnion and accumulates in amniotic fluid. It is also a major product of the fetal parathyroid. PTHrP may modulate transfer of Ca^{2+} across the fetal–placental unit and may be responsible for the high Ca^{2+} levels observed in fetal as compared to maternal plasma.

PTHrP gene activity appears in the rat lactating mammary gland within 24 hr after birth and remains elevated during suckling. Removal of the pups causes a drop within 1–2 hr and PTHrP production ceases by 4 hr. Resumption of nursing reinstates lactating levels of PTHrP.

In humans, most but not all (12 of 19) breast-feeding woman exhibit elevated plasma PTHrP (2–8 pmol/liter), but PTHrP is absent from the plasma of bottle-feeding mothers (0 of 16 examined). There are considerable amounts of PTHrP in human and bovine milk, suggesting it may be important for maintaining calcium homeostasis in newborns. Much lower levels of PTHrP are present in commercial, milk-based infant formulas, and PTHrP is absent from soy-based formulas (see Table 15-3).

PTHrP causes relaxation of both vascular and nonvascular smooth muscle. During pregnancy, PTHrP may be an important relaxant of smooth muscle, and its level decreases in the uterus prior to birth. It also has been shown to relax smooth muscle of vascular tissue, urinary bladder, and the stomach although a physiological role for PTHrP in these tissues is not established.

In nonlactating adults as well as in fetuses, PTHrP may be an autocrine growth regulator. It is expressed in more than 20 human tissues and stimulates growth in several cell lines including renal carcinoma, fibroblasts, osteoblasts, chondrocytes, and lymphocytes.

B. The C Cells and Calcitonin

The C cells (parafollicular cells) of the mammalian thyroid gland have been identified as the source of CT in most mammals. However, the actual origin of these cells is unclear. The most generally accepted origin is that proposed by Godwin, suggesting that these cells originated from the ultimobranchial body that develops from the sixth pharyngeal pouch

Table 15-3. Approximate Levels of PTHrP in Milk and Infant Formulas[a,b]

Type of milk/formula	PTHrP (ng eq/mL)
Fresh milk	
Human	50
Bovine	96
Commercial milk	
Whole milk	81
Nonfat milk	118
Buttermilk, chocolate milk	5–7
Milk-based formulae	
Six brand names	1–30
Soy-based formula	Undetectable

[a] Using these contrived units, plasma PTHrP in an HHM patient would be 0.08 ng eq/L. This means that PTHrP in breast milk is about 1000× more concentrated than in plasma and about 10,000× greater than normal plasma PTHrP.

[b] Data from de Papp, A. E., and Stewart, A. F. (1993). *Trends Endocrinol. Metab.* **4**, 181–187.

(endoderm). These ultimobranchial cells become incorporated into the thyroid gland of mammals just as parathyroids frequently do. The ultimobranchial body remains as a distinct separate structure in other vertebrate groups.

Like the chief cells of the parathyroids, the C cells of the mammalian thyroid exhibit APUD characteristics. Some elegant studies employing cellular chimeras of chicken and quail embryonic tissues have confirmed the origin of the CT-secreting cells of the ultimobranchial body from neural crest cells. C cells also contain somatostatin (SS) and ultrastructurally resemble the D cells found in the intestinal lining. It has been suggested that SS released along with CT may play an autocrine or paracrine role.

1. Calcitonin

Production of a potent mammalian hypocalcemic factor by cells of the parathyroid gland was first reported by Copp and associates, who named it calcitonin (CT). Other investigators provided evidence that the source of this hypocalcemic factor was the C cells of the thyroid gland and suggested the alternative name of **thyrocalcitonin** to reflect its origin. It was soon discovered that ultimobranchial glands of sharks and chickens contained CT and that C cells of the thyroid were of ultimobranchial origin.

Mammalian CT is a small, single-chain peptide of 32 amino acids with a disulfide bond linking residues 1 and 7. There are no active fragments of CT, and the entire molecule is necessary for biological activity. The amino acid sequence for some CT molecules is provided in Fig. 15-7. It is cleaved from a prohormone of 136 amino acids. Circulating levels of CT in humans are reported to be from 5 to 100 pg/ml, with a short biological half-life (5 min) related to the fact that all fragments of CT are inactive.

The CT gene product may be processed through several mRNA pathways. This results

	1	6	11	16	21	26	31
Mammals							
Porcine	CSNLS	TCVLS	AYWRN	LNNFH	RFSGM	GFGPE	T P-NH$_2$
Bovine	CSNLS	TCVLS	AYWLD	LNNYH	RFSGM	GFGPE	T P-NH$_2$
Human	CGNLS	TCMLG	TYTQD	FNKFH	TFPQT	AIGVG	A P-NH$_2$
Rat	CGNLS	TCMLG	TYTQD	LNKFH	TFPQT	SIGVG	A P-NH$_2$
Teleosts							
Salmon-I	CSNLS	TCVLS	KLSQE	LHKLQ	TYPRT	NTGSG	T P-NH$_2$
Salmon-II	CSNLS	TCVLS	KLSQN	LHKLQ	TFPRT	NTGAG	V P-NH$_2$
Salmon-III	CSNLS	TCMLS	KLSQN	LIIKLQ	TFPRT	NTGAG	V P-NH$_2$
Birds							
Chicken	CASLS	TCVLS	KLSQE	LHKLQ	TYPRT	DVGAG	T P-NH$_2$

Figure 15-7. Amino acid sequences of CTs. The sequence of the first 10 residues is highly conserved when comparing teleosts, birds, and mammals. The presence of valine (V) at position 8 increases the biological activity of the molecule four- to fivefold. Hence, salmon-I is more potent in humans than is human CT. Note also that human and rat CTs are more like each other than they are like bovine, porcine, or ovine CTs (not shown), which are all similar. See Appendix C for single-letter amino acid code. [Modified from Matsumoto, A., and Ishii, S. (1992). "An Atlas of Endocrine Organs." Japanese Society for Comparative Endocrinology and Springer-Verlag, Berlin.]

in formation of several peptides known as calcitonin gene-related peptides (CGRPs). CGRPs have no CT-like activity but may play other physiological roles (see Chapter 13).

C. Skin, Liver, Kidney, and 1,25-Dihydroxycholecalciferol

The actions of PTH on bone and possibly kidney require the steroid 1,25-DHC, as does intestinal absorption of calcium. The first steps in the formation of 1,25-DHC involve the conversion of cholesterol to 7-dehydrocholesterol and then to **cholecalciferol** (**vitamin D₃**) by the skin. Conversion of cholesterol to 7-dehydrocholesterol occurs in the presence of adequate sunlight. The chemical transformation of 7-dehydrocholesterol to cholecalciferol is a two-step process. The first step occurs quickly in the presence of ultraviolet (UV) light, and the second step is a temperature-dependent isomerization that requires several days. Penetration of the skin by certain wavelengths of UV light is strongly dependent on the angle of incidence of sunlight (influenced by time of day, season, and latitude) as well as by cloud cover and air pollution. Furthermore, window glass absorbs UV light effectively and reduces cholecalciferol synthesis. The brown skin pigment melanin also absorbs UV light and decreases cholecalciferol synthesis. The lack of melanin pigment that evolved in northern Europeans may have served as an adaptation to increase UV penetration and hence enhancement of cholecalciferol synthesis, thereby allowing them to inhabit northern lattitudes. Prolonged exposure of skin to sunlight not only increases melanin synthesis but causes cholesterol to be converted primarily to inert steroids known as lumisterol and tachysterol instead of to cholecalciferol.

Transport of cholecalciferol to the liver is facilitated by a binding protein in the plasma that has a high affinity only for the isomerized cholecalciferol. Cholecalciferol is converted by the liver to **25-hydroxycholecalciferol** (**25-HC**) and then by the kidney to 1,25-DHC (see Chapter 2, Fig. 2-29). Circulating levels of 25-HC reported for humans are between 7 and 42 ng/ml, with an extensive biological half-life of about 15 days.

The kidney converts 25-HC primarily to 1,25-DHC. Some 24,25-DHC is made as well but is of little importance because 1,25-DHC is 100 to 1000 times more potent. This final step is stimulated by PTH. Circulating levels of 20–50 pg of 1,25-DHC per milliliter are reported for humans. These low levels are largely a product of its biological half-life of only 3 hr. During pregnancy, the placenta also will synthesize 1,25-DHC, which augments uptake of dietary calcium for fetal growth.

III. Interactions of Parathyroid Hormone, Calcitonin, and 1,25-Dihydroxycholecalciferol

The major disturbance to calcium homeostasis is the influx of Ca^{2+} following ingestion of a meal. Periods of rapid growth cause an additional demand for dietary calcium, as do pregnancy and lactation. The small intestine, kidney, and bone are the primary sites where these regulatory hormones (PTH, CT, and 1,25-DHC) produce their actions during times of calcium excess or deficiency to maintain calcium homeostasis (see Fig 15-1).

A. Calcium and Phosphate Regulation in Bone

Minute-to-minute regulation of plasma calcium and indirectly of plasma phosphate may reside in the activity of **osteocytes** embedded permanently in the bone matrix. PTH may cause **osteocytic osteolysis** (one form of bone resorption) through direct action on the osteocytes. This effect may be dependent on an interaction between PTH and 1,25-DHC. In contrast, major disturbances such as occur during growth, pregnancy, or lactation involve the interactions of PTH and CT on osteoclasts and osteoblasts.

Parathyroid hormone stimulates osteoclast activity, and chronic elevation of PTH causes an increase in the number of osteoclasts. The action of PTH on the osteoclast first involves intracellular synthesis of prostaglandins, which in turn activate adenylate cyclase, resulting in an increase in cAMP within the osteoclast and release of lysosomal enzymes from the osteoclast. It is the action of these hydrolytic lysosomal enzymes that destroys both mineral and bone matrix components of bone and is responsible for bone resorption and release of Ca^{2+} and phosphate to the plasma. Calcitonin apparently prevents bone resorption by interfering with the activation of adenylate cyclase by the prostaglandins.

Another hypothesis concerning PTH action on bone resorption is focused on PTH effects on the bone-forming osteoblasts. According to this hypothesis, the osteoblasts possess a pumping mechanism for calcium transport. PTH, through activation of adenylate cyclase and cAMP formation, increases the flux of Ca^{2+} into the osteoblast from the bone surface and out of the osteoblast on the blood side of the endosteal membrane. There is evidence to suggest that, at normal physiological levels, PTH stimulates bone formation rather than bone resorption through an increase in osteoblast activity. The observed effects of PTH on osteoclast activity require considerable elevation of PTH levels and may occur only under conditions of calcium deprivation. Although calcitonin may influence osteoblast functions to a limited extent, its major action appears to be inhibition of the osteoclast.

B. Calcium and Phosphate Regulation in Kidney

The stimulation of calcium reabsorption by the kidney may be the most important physiological action of PTH, although some investigators believe the increased excretion of phosphate is more important. Parathyroid hormone also increases the enzymatic formation of 1,25-DHC in kidney and thus may enhance intestinal calcium absorption indirectly. Estrogens and prolactin (PRL) also enhance formation of 1,25-DHC, and these actions may be essential in pregnant and lactating mammals, respectively. Although CT antagonizes the action of PTH on bone resorption, it apparently does not influence renal processes in normal animals.

C. Regulation of Calcium Uptake in the Intestine

Calcium uptake by the intestinal mucosal cell is dependent on a calcium-binding protein within these cells that is linked to a calcium-activated ATPase. Calcium is actively absorbed by this binding protein–ATPase complex at the mucosal surface that is in contact with the lumen of the gut. Once in the cell, these calcium ions are transported to the opposite (serosal) border of the cell, where calcium ions diffuse into the interstitial fluid and then into the blood capillaries. Phosphate ions passively follow the movements of Ca^{2+}. Synthesis of both a calcium-binding protein and a calcium-activated ATPase are stimulated by 1,25-DHC.

No role has been hypothesized for CT in this intestinal mechanism for calcium uptake. Following the influx of Ca^{2+} ions, however, calcitonin is released partly in response to the increase in plasma Ca^{2+} but may also be released by gastrin, a gastrointestinal hormone released from the gastric mucosa during the early phase of digestion of a meal (see

Table 15-4. Hormones That Influence Calcium Metabolism in Mammals

Hormone	Action
Parathyroid hormone (PTH)	Stimulates bone resorption, renal calcium reabsorption, and synthesis of 1,25-DHC
Calcitonin (CT)	Antagonizes action of PTH on bone
1,25-dihydroxycholecalciferol (1,25-DHC)	Facilitates calcium absorption from intestine
Growth hormone (GH)	Stimulates cartilage and bone growth
Thyroid hormones	Permissive effect on GH secretion and action
Insulin-like growth factors (IGFs)	Mediators of GH action on bone
Estrogens/androgens	Effects closure of epiphysial plate, blocking further long bone growth; protects adult skeleton from resorption; may reduce PTH release and prevent hypercalcemia
Glucocorticoids	High levels stimulate PTH release, causing increased bone resorption and resultant hypercalcemia

Chapter 13). This increased level of CT inhibits the action of PTH on bone osteoclasts and favors addition of the absorbed dietary Ca^{2+} to the bone matrix.

Calcitonin becomes indispensable during pregnancy and lactation with respect to Ca^{2+} mobilization. Its role is apparently to protect the maternal skeleton from excessive destruction in meeting the calcium requirements of fetus or neonate and to allow the diversion of dietary Ca^{2+} to it. The relative importance of CT in different species varies markedly according to the precise demands for maternal calcium.

D. Other Hormones and Calcium–Phosphate Homeostasis

Estrogens, androgens, glucocorticoids, and thyroid hormones have direct and indirect effects on mineral homeostasis (see Table 15-4). In addition, GH has indirect effects through production of insulin-like growth factors (IGFs).

Although the effects of estrogens and androgens bring about cessation of long bone growth at puberty, these hormones also have stimulatory effects on osteoblast activity. Estrogens not only protect the skeleton from resorption but enhance the reabsorption of calcium by the kidney and increase the production of 1,25-DHC. Androgens are thought to play a similar role in males.

In addition to their ability to stimulate PTH secretion, chronic excesses of glucocorticoids, such as occur during prolonged stress, can reduce intestinal uptake and kidney reabsorption of calcium. This could result in a significant calcium loss and could lead to low calcium-induced tetany.

Estrogens are hypothesized to alter the calcium set point in chief cells so that greater reduction in plasma calcium is required to elicit PTH release. This action might explain, at least in part, the effects of estrogen withdrawal on loss of skeletal calcium during and after menopause.

Growth hormone causes an increase in secretion of IGFs by the liver and stimulates IGF synthesis in bone. IGF-I stimulates bone proliferation as well as collagen synthesis by osteoblasts. Increased osteoblast activity in turn may activate osteoclasts, which promote resorption and can stop or reverse bone growth.

Hypothyroidism delays bone growth, probably indirectly through its adverse effects on GH secretion and action. Delayed ossification of cartilage also is observed in hypothyroidism. In the hyperthyroid animal, thyroid hormones cause bone resorption and can lead to weakening of the skeleton (see Section IV,C).

IV. Major Clinical Disorders Associated with Calcium Metabolism

In general, plasma calcium does not vary greatly or death results. Consequently, it is not easy to diagnose parathyroid abnormalities by examining plasma levels of calcium. **Hypercalcemia** (elevation of plasma calcium) is not excessive but often is accompanied by a reduction in bone density and possibly increased calcium in the urine. Similarly, **hypocalcemia** may show few overt symptoms.

A. Hypercalcemia

Excessive plasma calcium levels (greater than 10 mg/dl) can result from a variety of causes. **Primary hypercalcemia** is characterized by chronically elevated levels of PTH. The most common cause is a single parathyroid adenoma (90% of cases). Carcinomas of the parathyroids are very rare and may account for less than 1% of primary hypercalcemic cases. Primary hypercalcemia is difficult to diagnose since approximately one-third of these cases are without overt symptoms, and the remainder exhibit rather generalized and nonspecific symptoms such as weakness, nausea, and anorexia. Serum calcium levels are moderately elevated (10.2 to 11.0 mg/dl) and are usually accompanied by lowered phosphate levels. Often, elevated serum levels of calcium are not seen (i.e., there is no hypercalcemia per se) because the kidney compensates with increased calcium excretion or **hypercalcuria.** Among the many causative factors that can induce **secondary hypercalcemia** are carcinomas that spontaneously secrete PTHrP, leading to HHM. Breast carcinomas also may produce vitamin D-like sterols that increase calcium absorption. Chronic immobilization of an experimental animal or of humans can bring about extensive bone resorption and cause hypercalcemia.

B. Hypocalcemia

There are many different conditions that can lead to hypocalcemia. In most cases, there is a reduction in PTH secretion. This may be caused by abnormal development of the parathyroid glands, accidental damage, or removal by surgery. Hypomagnesemia impairs PTH secretion and indirectly can cause hypocalcemia. In some cases (e.g., pseudo-hypoparathyroidism), the target organs do not respond to PTH and in others an abnormal PTH is secreted that will not activate tissue receptors. Reductions in 1,25-DHC due to failure to convert 25-HC or to synthesize vitamin D also lead to hypocalcemia. Most cases are without serious overt symptoms in adults. Tetany may be inducible under calcium stress but otherwise is absent.

C. Osteoporosis

Osteoporosis is characterized by decalcification and loss of bone matrix from the skeleton, resulting in shrinkage, distortion, and increased brittleness of the bones. Bones become subject to easy fracture as a result of falls, blows, or lifting (stress fractures). Approximately

1.2 million bone fractures each year in the United States are attributed to osteoporosis. Of these, 530,000 vertebral and 227,000 hip fractures occur. Osteoporosis accompanies aging and is most common among postmenopausal women. Osteoporosis is eight times more common in women than men largely as a consequence of women having smaller bone calcium reserves. Maximal adult bone mass is achieved in women at about age 35, after which calcium losses exceed calcium gain. Beginning at menopause (see Chapter 11), there is a gradual reduction in estrogen levels, causing a disproportionate decrease in bone mass by allowing increased bone resorption to take place. Weight-bearing exercise and careful attention to dietary calcium are important for postpubertal, premenopausal women to ensure maximal bone density prior to onset of menopause.

Type I osteoporosis may occur in women soon after the onset of menopause and is characterized by vertebral crush fractures or fracture of the arm just above the wrist. It is attributed to the marked reduction in estrogen levels accompanying menopause. **Type II osteoporosis** appears later in life and can affect both men and women. It is mainly a consequence of decreased ability to absorb sufficient dietary calcium with advancing age.

Estrogen replacement (women only), calcium supplements (usually fortified with extra cholecalciferol or a related compound to facilitate intestinal uptake), and exercise are commonly prescribed. Frequently, estrogen is given to women with a low dose of a progestogen (e.g., medroxyprogesterone acetate), which prevents uterine bleeding and reduces uterine hyperplasia and the risk of uterine cancer. When a history of breast cancer is known, estrogen therapy may cause increased risk of breast cancer although the mortality rate caused by breast cancer is lower in women receiving estrogen. This apparent anomaly is a result of earlier detection and hence higher cure rates in these women. Although early studies suggested these therapeutic efforts only prevented or slowed additional skeletal losses, more recent studies show improvement in mineral density of bones even if estrogen replacement therapy is not begun until after age 70.

Hormonal manipulations are ineffective without close attention to diet and exercise. Emphasis should be placed on weight-bearing activities (e.g., walking). Swimming, while an excellent aerobic exercise, does not stimulate bone deposition. Weightlessness, like immobilization, accelerates bone resorption. Bone density in athletes is related directly to the type of exercise, with weight lifters having the most dense bone and swimmers having the least dense bone; runners are intermediate. The racquet arms of tennis players are 35 and 28% more dense than the other arm, respectively, in men and women. Even mild activity for nursing home patients averaging 82 years of age not only prevented further bone loss but resulted in bone buildup over a 36-month period.

D. Paget's Disease

Paget's disease is caused by increased osteoclast activity resulting in accelerated bone resorption. It occurs in 3% of the population over age 40, and occurs with greatest frequency in people of western European or Mediterranean descent. About one-third of the people afflicted with Paget's disease do not exhibit any overt symptoms, but the bones become brittle as the disease progresses. Serum levels of calcium and phosphate are usually normal because the excess ions are excreted in the urine. Salmon CT has been used with some success to treat Paget's disease, because it is more effective (Table 15-5), possibly due to its much longer biological half-life in mammals compared to mammalian CT (Table 15-6). However, therapeutic treatment is thwarted in part by down-regulation of CT receptors on osteoclasts after a few days of treatment.

Table 15-5. Comparison of Relative
Activity of Purified Human, Salmon, and
Porcine Calcitonin as Determined by a
Standard Bioassay

Source	Activity in MRC[a] units/mg hormone
Porcine	200
Human	120
Salmon	5000

[a]MRC, Medical Research Council of England.

V. Calcium and Phosphate Homeostasis in Nonmammalian Vertebrates

An attempt to discuss the evolutionary aspects of this topic is complicated by major differences, indeed a distinct dichotomy, between the fishes and the tetrapods. Some fishes lack true bone (agnathans, chondrichthyeans), and many of the bony fishes possess acellular bone (no osteocytes) rather than cellular bone. Scales may provide a major physiological reserve of stored calcium to bony fishes, and the surrounding waters (especially sea water) may be an important source of calcium as well. Furthermore, parathyroid glands are lacking in fishes although immunoreactive mammalian PTH has been shown in trout and goldfish plasma as well as in brain, pituitary, and the **corpuscles of Stannius** embedded in the kidneys of several species (Fig. 15-8). Calcium metabolism in fishes appears to be regulated by one or more hypercalcemic factors from the pituitary and a hypocalcemic factor from the corpuscles of Stannius and/or the ultimobranchials. Definitive parathyroid glands and PTH first appear fully differentiated in the amphibians. The comparative approach also is hampered by the lack of detailed information concerning calcium and phosphate regulation

Table 15-6. Effects of Calcitonin in Selected Nonmammalian Vertebrates

Species	CT source	Effect
Agnathan fishes		
Myxine glutinosa	Mammalian	Decreased urine flow and electrolyte content of urine
Bony fishes		
Periophthalmus schosseri	Eel	Hypocalcemia
Carassius auratus	Goldfish	Hypocalcemia
Cyprinus carpio	Salmon	Hypocalcemia
Pacific salmon (*Oncorhynchus*)	Salmon	Decreased gill uptake of calcium
Amphibians		
Rana tigrina	Salmon	Transient hypocalcemia; increased calcium deposition in paravertebral calcium sacs
Ambystoma mexicanum	Eel	No effect on blood calcium; decreased calcium influx
Reptiles		
Dipsosaurus dorsalis	Salmon	No effect on kidney or basal salt gland handling of calcium

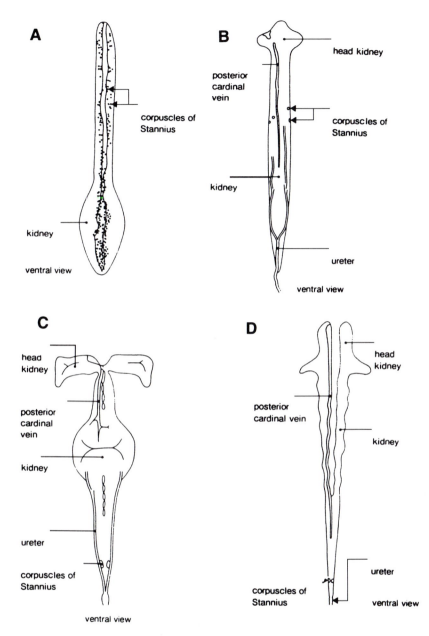

Figure 15-8.   Location of corpuscles of Stannius in selected bony fishes. (A) The bowfin, *Amia calva*; (B) a salmonid, *Salvelinus pluvius*; (C) the crucian carp; (D) *Mugil cephalus*. [Modified from Matsumoto, A., and Ishii, S. (1992). "An Atlas of Endocrine Organs." Japanese Society for Comparative Endocrinology and Springer-Verlag, Berlin.]

in fishes, amphibians, and reptiles. The distribution of parathyroid and ultimobranchial glands in vertebrates is provided in Figs. 15-9 and 15-10.

Estrogenic hormones elevate circulating calcium and phosphate levels indirectly in females of most nonmammalian species during the process of vitellogenesis associated with

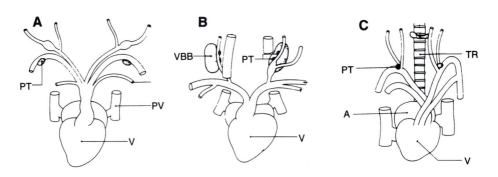

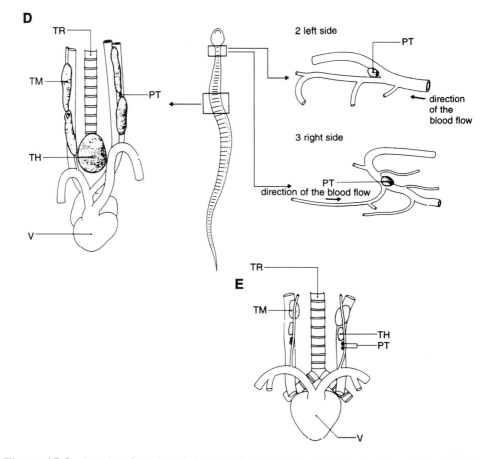

Figure 15-9. Location of parathyroid glands in selected tetrapod vertebrates. A, Atrium; PT, parathyroid; PV, precaval vein; TH, thyroid; TM, thymus; TR, trachea; V, ventricle; VBB, ventral branchial body. (A) Newt, *Cynops pyrrhogaster*. Urodele amphibians typically have one pair. (B) Frogs and other anurans typically have two pairs of parathyroids. (C) Lizards have one pair of parathyroids, except for *Anolis*, which has two. (D) Snakes, such as *Agkistrodon halys*, have two pairs of parathyroids, with the posterior pair associated with the thymus. (E) Many birds, such as the seagull *(Larus argentatus*; shown here), and domestic species have two pairs of parathyroids located near the thyroid, although some (e.g., stork, quail) have one pair. [Modified from Matsumoto, A., and Ishii, S. (1992). "An Atlas of Endocrine Organs." Japanese Society for Comparative Endocrinology and Springer-Verlag, Berlin.]

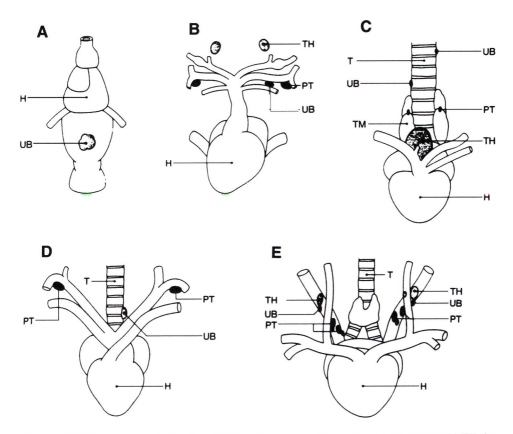

Figure 15-10. Location of ultimobranchial tissue in nonmammalian vertebrates. PT, parathyroid; TH, thyroid; TM, thymus; T, trachea; UB, ultimobranchial; H, ventricle. (A) Goldfish; (B) newt; (C) snake; (D) lizard; (E) bird. [Modified from Matsumoto, A., and Ishii, S. (1992). "An Atlas of Endocrine Organs." Japanese Society for Comparative Endocrinology and Springer-Verlag, Berlin.]

oocyte growth. Synthesis of phosphate-containing, calcium-binding, vitellogenic proteins by the liver is stimulated by estrogens. These proteins are released into the blood, through which they travel to the ovaries, where they are incorporated into growing oocytes as yolk protein. Vitellogenic proteins readily bind Ca^{2+}, indirectly decrease free plasma Ca^{2+} levels, and consequently stimulate release of more Ca^{2+} from reservoirs such as bone. The presence of vitellogenic proteins elevates total plasma levels of Ca^{2+}, although free Ca^{2+} remains about the same. The relationship between reproductive hormones and vitellogenesis is discussed in more detail in Chapter 12.

A. Agnathan Fishes

Cyclostomes do not appear to have regulatory mechanisms specific for calcium and phosphate metabolism although mammalian CT does decrease urinary excretion (Table 15-6). Whatever regulatory mechanisms are employed, they appear to be not as efficient as those found in other vertebrates as evidenced by circulating levels of calcium. Marine species

Table 15-7. Biological Half-Life for
Purified Vertebrate Calcitonins When
Incubated in Either Mammalian or Avian
Blood Plasma

Source	Half-life (min)
Porcine	2
Salmon	20
Chicken	90

exhibit an intermediate level of plasma Ca^{2+} between that of sea water and plasma of tetrapods, whereas the plasma Ca^{2+} level in freshwater lampreys is intermediate between that found in plasma of freshwater teleosts and fresh water. These observations raise some interesting questions about the roles of calcium and its regulation in agnathans.

B. Chondrichthyean Fishes

Most selachians and other chondrichthyeans live in sea water, where calcium availability is not as serious a problem as it is for freshwater species. Although these fishes lack true bone tissue, calcium salts are added to their cartilaginous skeletons for additional strength, especially in larger species. Selachian ultimobranchial glands contain a potent hypocalcemic factor when assayed in mammals, but extracts of shark ultimobranchial glands, salmon CT, and porcine CT are all ineffective in sharks. Curiously, estrogens do not produce any appreciable alteration in plasma calcium as they do in bony fishes and nonmammalian tetrapods. No experimental evidence has been reported for an adenohypophysial factor that would directly affect plasma calcium. Studies of sharks that are either freshwater residents or that penetrate long distances up major rivers, such as the Mississippi, might provide some interesting insight into calcium homeostasis in these fishes.

C. Bony Fishes: Teleosts

Calcium regulation appears to be under the control of a hypercalcemic adenohypophysial factor and a hypocalcemic factor associated with the corpuscles of Stannius. Hypophysectomy of either *Anguilla* (which has cellular bone) or *Fundulus* (acellular bone) results in a decrease in serum calcium and induction of tetany. One candidate for the pituitary hypercalcemic factor is PRL, which would be similar to its action on Na+ balance in freshwater fishes.

Teleost pituitaries also produce somatolactin (SL) that can influence Ca^{2+} levels (see Chapter 5). In addition, PTHrP has been demonstrated in the brain, pituitary, and blood of several species. In the sea bream, *Sparus auratus,* two populations of pituitary cells are found that contain immunoreactive PTHrP. In the anterior group located in the rostral pars distalis, PTHrP is colocalized with β-thyrotropin (β-TSH). The second group of immunoreactive PTHrP cells occurs at the borders of the proximal pars distalis and the pars intermedia. These cells do not react with antisera for β-TSH, corticotropin-like peptide (CLIP), or corticotropin (ACTH). Although plasma levels of PTHrP in sea bream are 10 times greater than in normal human plasma, no specific role for PTHrP is evident. The name **hypercalcin** has been applied to the pituitary hypercalcemic factor of teleosts, but this activity could be due to PTHrP, SL, and/or PRL.

The corpuscles of Stannius have been examined for their possible involvement in calcium homeostasis. Little experimental work has been performed with species other than teleosts, and only a few teleosts have been examined. Stanniectomy of the saltwater eel *Anguilla anguilla* results in an increase in serum calcium, and the administration of corpuscle extracts reduces calcium to normal levels. Stanniectomy of the closely related freshwater *Anguilla japonica* is followed by decreased urinary Ca^{2+} levels and increased urinary phosphate levels. It is possible that the corpuscles of Stannius are mainly active in reducing Ca^{2+} levels in fish adapted to high-calcium environments such as sea water. The active hypocalcemic agent in the corpuscles may be a peptide called either **hypocalcin** or **stanniocalcin.**

Treatment with eel, salmon, or goldfish CTs lowers calcium influx across the gill, and the gill may be a major site where CT regulates Ca^{2+} transport (Table 15-6). Salmon CT is a powerful hypocalcemic agent when tested in avian or mammalian systems (Tables 15-5 and 15-6), and it is now employed clinically to reduce pathological hypercalcemia (Paget's disease). This potency of salmon CT is related, at least in part, to its persistence in the circulation, that is, its long biological half-life as compared to mammalian CT (Table 15-7).

Some teleost kidneys are capable of hydroxylation of cholecalciferol to form 1,25-DHC. This hormone is apparently not necessary for Ca^{2+} uptake by the intestine, but large doses of 1,25-DHC enhance Ca^{2+} uptake. Treatment with 1,25-DHC increases plasma calcium and phosphate in the freshwater catfish, *Heteropneustes fossilis*. Calcium-binding proteins (calbindins) are present in the intestinal mucosa of teleosts (Table 15-8), and their synthesis

Table 15-8. Calbindin Immunoreactivity in Vertebrate Tissues[a]

Species	Intestine	Kidney
Bony fishes		
Carassius auratus (goldfish)	+	
Salmo trutta (brown trout)	+	
Oncorhynchus mykiss (rainbow trout)		0
Amphibians		
Bufo bufo (toad)	+	0
Rana esculenta (frog)	+	0
Rana temporaria (frog)	+	
Xenopus laevis (frog)	+	
Triturus alpestris (newt)	+	
Reptiles		
Chrysemys (turtle)	+	
Psuedemys (turtle)		0
Gekko gecko (lizard)		0
Birds		
Japanese quail	+	
Chicken	+	+
Pigeon	+	
Mammals		
Rat	0	
Monkey	0	
Human	0	

[a] All tetrapods examined have two forms of brain calbindin, but the fish examined have only one. +, Present; 0, absent; blank, no data.

Table 15-9. Serum Calcium Levels in Radiothyroidectomized
Steelhead Trout, *Oncorhynchus mykiss*

Treatment	Total serum calcium ± SEM (mg/100 ml)	Range (mg/100 ml)
Radiothyroidectomized as fingerlings	10	4–12
Intact controls	11	9–15
Radiothyroidectomized prior to yolk sac	11.8 ± 1.01	
Intact controls	14.9 ± 0.51	

may be controlled by 1,25-DHC as it is in mammals. In females, total plasma phosphate and calcium increase markedly during oogenesis, owing to the actions of estrogens on liver vitellogenic protein synthesis and secretion.

1. Thyroid State and Calcium Homeostasis in Teleosts

Serum calcium may be influenced by thyroid state, but the physiological importance of these observations has not been established. Serum calcium levels are decreased in juvenile steelhead trout (*Oncorhynchus mykiss*) that were radiothyroidectomized prior to complete resorption of the yolk sac (Table 15-9). It is not possible to distinguish between general effects of radiation that may have damaged some calcium-regulating mechanism and effects due to the absence of thyroid hormones, however. Growth of the skeleton is also abnormal following radiothyroidectomy, supporting an involvement of thyroid hormones at some level. Abnormal skeletal growth observed in these fish could be related to ineffective action of growth hormone in the absence of thyroid hormones (see Chapter 8).

D. Bony Fishes: Lungfishes

The lungfishes lack parathyroid glands and are relatively insensitive to tetrapod calcium-regulating hormones. It is both surprising and somewhat disappointing that they do not exhibit some tetrapod-like feature. Parathyroid extracts, PTH, and salmon CT are all ineffective in altering Ca^{2+} levels in the South American lungfish *Lepidosiren paradoxa*. Surprisingly, however, CT and PTH are diuretic and antidiuretic, respectively, in these fishes. The physiological importance of these observations is not clear.

E. Amphibians

Studies of amphibians have shed little light on the origin of the parathyroid glands and the evolution of calcium regulation in tetrapods. Ultimobranchial and parathyroid glands are present in the apodans, anurans, and urodeles, and calcium regulation is similar to that observed in other tetrapods.

Parathyroid glands are not present in some urodeles until after metamorphosis, and parathyroids never develop in some permanently neotenic aquatic species such as *Necturus*. These glands are more important in the calcium balance of early terrestrial urodeles and especially of anurans. Ultimobranchial bodies are well developed and produce a hypocalcemic CT-like factor. As in fishes, a hypercalcemic factor (probably PRL) is present in pituitaries of urodele amphibians, and it is of greater importance for calcium regulation in the more aquatic species. PTHrP has not been reported.

1. Amphibian Ultimobranchial Glands

The amphibian ultimobranchial glands develop as in all tetrapods from the fifth pharyngeal pouches. Most urodeles have only one ultimobranchial gland (usually on the left side). In apodans, anurans, and some urodeles (for example, *Necturus* and *Amphiuma*), the ultimobranchial glands are paired. Immunoreactive CT is present in the ultimobranchial glands of *Rana temporaria* and *Rana pipiens*. Cytologically, the ultimobranchial gland consists of one or more simple follicles composed of C cells, which may appear in either a "dark" form (relatively electron dense) or in a less electron-dense "light" form, at least in anurans. In addition to the C cells, the ultimobranchial gland of the apodan *Chthonerpeton indistinctum* contains cholinergic and purinergic neuronal endings. Although the presence of sympathetic neurons has been demonstrated in the frog *R. pipiens,* most anurans investigated do not exhibit innervation of the ultimobranchial glands. The condition of the urodele ultimobranchial gland with respect to innervation has not been described. Salmon CT produces a transient decrease in plasma calcium and phosphate in anurans. Eel CT decreases calcium influx in aquatic axolotls (*Ambystoma mexicanum*) as it does in fishes.

2. Amphibian Parathyroid Glands

As in mammals, parathyroidectomy results in lowered calcium and usually leads to tetany and death. The parathyroid glands appear to develop from the third and fourth pharyngeal pouches as they do in mammals. Chief cells responsible for PTH secretion actually arise from neuroectodermal pharyngeal components rather than from endoderm. In apodan parathyroids, only chief cells are present whereas two cellular types have been described for anurans. However, these appear to be two forms of chief cells that occur in light and dark phases as described above for the ultimobranchial bodies. Two distinct cellular types have been reported in parathyroids of the urodele *Cynops pyrrhogaster.* One of these cell types is considered to be only "supportive," however.

3. Endolymphatic Sacs

In addition to cellular bone, the endolymphatic sacs, located at the base of the skull and/or along the vertebral column, appear to be major targets for factors regulating calcium homeostasis. These sacs contain large amounts of calcium carbonate and may be important reservoirs of these ions, particularly during metamorphosis and subsequent ossification of bones. These structures also may provide bicarbonate ions for buffering the blood following the dissociation of calcium carbonate.

4. Amphibian Calcium and Phosphate Homeostasis

In general, the amphibians regulate plasma calcium and phosphate similarly to mammals. The parathyroids, pituitary gland, and ultimobranchial glands control calcium and phosphate metabolism, and the vitamin D complex seems to be related to at least some actions of PTH in amphibians. A calcium-binding protein occurs in the intestinal mucosa (Table 15-8), and this may be related to 1,25-DHC activity.

Removal of the ultimobranchial glands in frogs generally causes an increase in osteoclast activity and a consequent increase in blood calcium levels. The ultimobranchial gland secretions also prevent removal of calcium from the endolymphatic sacs and block uptake of calcium through the gut. Parathyroidectomy causes a decrease in plasma Ca^{2+} in anurans and in the newt *C. pyrrhogaster.* However, no changes in serum Ca^{2+} occur following parathyroidectomy of immature giant salamanders, *Megalobatrachus davidianus.* Administration of bovine PTH to frogs increases plasma Ca^{2+} and decreases plasma phosphate, implying that the kidney may also be a target for PTH.

F. Reptiles

The regulation of calcium and phosphate in lizards has been well studied, and these investigations have been extended to include snakes and turtles. Ultimobranchial glands and parathyroids have been described, and regulation of calcium is essentially like that in other tetrapods. The major sites for endocrine regulation of calcium metabolism are the kidney, cellular bone, and the endolymphatic sacs, which in lizards (as in amphibians) are important reserves of calcium and carbonate ions. The presence of PTHrP and its possible roles in reptilian calcium metabolism have not been addressed.

1. Reptilian Ultimobranchial Glands

The ultimobranchial glands of reptiles are located near the thyroid and parathyroid glands. They are small glands consisting primarily of follicles. All the reptilian ultimobranchials are innervated, although the nature of these neuronal endings has not been examined extensively. In crocodilians, chelonians, and some snakes the ultimobranchial glands are paired. In lizards only the left gland persists. Seasonal changes in the cytology of ultimobranchial glands have been reported for a few species, but the relationship between these changes and physiological and environmental parameters has not been ascertained.

2. Reptilian Parathyroid Glands

Four reptilian parathyroids develop from the third and fourth pharyngeal pouches as described for the amphibians, and they resemble mammalian parathyroids cytologically (Fig. 15-11). Adult lizards and crocodilians have only one pair of parathyroid glands, whereas snakes and turtles have four glands. In addition to cellular cords, the presence of follicles containing PAS(+) material is a common feature of reptilian parathyroids although their functional significance is not known.

3. Calcium and Phosphate Homeostasis in Reptiles

Parathyroidectomy of lizards and snakes causes a marked decrease in plasma Ca^{2+}, accompanied by tetany. However, there is little or no change in circulating Ca^{2+} in turtles following a similar operation, and tetany does not occur in turtles. This insensitivity of turtles to parathyroidectomy is apparently a consequence of the immense calcium reservoir represented by the shell. Treatment with mammalian PTH causes increased plasma Ca^{2+} and urinary phosphate as well as decreased urinary Ca^{2+} and plasma phosphate in both lizards and turtles. The renal action of mammalian parathyroid extracts on phosphate excretion is marked in parathyroidectomized snakes (four species of *Natrix*), although calcium excre-

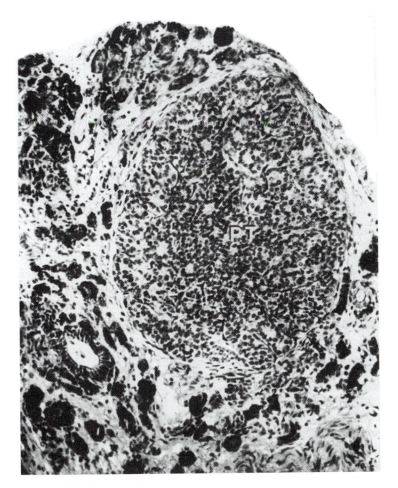

Figure 15-11. Parathyroid of the false map turtle, *Graptemys pseudogeographica*, surrounded by ultimo-branchial tissue. (Courtesy of Nancy B. Clark, University of Connecticut, Storrs.)

tion is probably not affected significantly. Thus, it appears that reptiles possess basically the same regulatory control of calcium and phosphate homeostasis as exhibited by mammalian parathyroids.

The role of CT in reptiles is uncertain. Extracts of reptilian ultimobranchial glands produce hypocalcemia when injected into rats, and salmon CT has been shown to lower Ca^{2+} in the green iguana. As with fishes and amphibians, when reptilian ultimobranchial factors are tested in mammals the presence of hypocalcemic factors is noted, but their endogenous physiological roles are not known.

No information is available on 1,25-DHC production or actions in reptiles. A calcium-binding protein has been demonstrated in the intestinal mucosa of three species (Table 15-8).

G. Birds

There are typically four parathyroid glands in birds, and separate ultimobranchial glands are present. Cytologically, the parathyroid glands resemble those of mammals, and the ultimobranchial cells are like the mammalian C cell. The regulation of calcium and phosphate homeostasis is typically mammalian with only minor differences. It should be noted, however, that few avian species have been investigated thoroughly, and most of the physiological studies have been conducted with domestic birds (e.g., chickens, pigeons, and Japanese quail).

1. Avian Parathyroid Glands

Avian parathyroid glands are separate and usually distinct, except for a few species, such as the domestic chicken, in which some fusion of the separate parathyroids may occur. Cytologically the parathyroid glands contain only chief cells; no oxyphils have been reported.

2. Avian Ultimobranchial Glands

The ultimobranchial glands are usually separate structures although some fusion with the thyroid gland occurs in the pigeon. There are both light and dark cells in chicken ultimobranchials. The light cell is more abundant and resembles the mammalian C cell cytologically. A rich vagal innervation (parasympathetic) has been described for the chicken, but its importance has not been elucidated.

3. Calcium and Phosphate Homeostasis in Birds

Parathyroidectomy in birds usually causes marked hypocalcemia accompanied by tetanic seizures and death within 24 hr. This degree of sensitivity to parathyroidectomy is not manifest on this time scale in amphibians or reptiles and may be related to the much higher body temperature and metabolic rate of birds. Treatment with mammalian parathyroid extracts produces marked increases in plasma Ca^{2+}. Dietary deprivation of Ca^{2+} or vitamin D causes marked hypertrophy and hyperplasia of the parathyroid glands, whereas high Ca^{2+} diets result in regression of the parathyroids. As in mammals, estrogens and parathyroid extracts both increase levels of 1,25-DHC.

Mammalian parathyroid extract stimulates renal excretion of phosphate in normal starlings but does not seem to alter renal treatment of calcium. The effect of mammalian parathyroid extract or PTH on avian bone is similar to that described for mammals.

PTHrP has been isolated from chickens and is present in both embryos and adults. Numerous embryonic tissues express PTHrP, and it may be an important growth stimulator in bird embryos. Expression of the PTHrP gene in the isthmus and shell gland of the adult oviduct is related to entrance of the egg into the oviduct and the calcification of the shell. PTHrP relaxes vascular smooth muscle and increases blood flow to the shell gland during calcification. It is not known whether it alters Ca^{2+} transport as well. Relaxation of oviductal smooth muscle by PTHrP allows the entrance and passage of the egg through the oviduct.

Calcitonins that are structurally similar to mammalian CT have been isolated from chickens and turkeys and are potent hypocalcemic agents in both birds and mammals. Unlike the condition for fishes, amphibians, and reptiles, the avian kidney specifically binds CT, suggesting a renal action for CT in birds. Dietary levels of calcium are directly related to ultimobranchial activities, and high levels of Ca^{2+} cause marked stimulatory changes. An immunoreactive calcium-binding similar to that of other vertebrates is present in the intestinal mucosa of birds (Table 15-8).

VI. Summary

Precise regulation of calcium and phosphate homeostasis is necessary for many processes in vertebrates. In birds and mammals, this regulation is accomplished through secretion of hypercalcemic PTH from the parathyroid glands and hypocalcemic CT produced by the C cells embedded between the follicles of the thyroid gland (mammals) or by the ultimobranchial glands (birds). PTHrP may be an important regulator of calcium homeostasis and growth. Its presence in milk may help newborn mammals regulate calcium as well. PTHrP is involved in both avian embryonic development and shell formation in adults.

Parathyroid hormone increases plasma calcium levels through direct and indirect action on bone, kidney, and intestine. Parathyroidectomy invariably brings about tetany and usually death as a consequence of the decrease in circulating Ca^{2+} following this operation. This condition can be alleviated through administration of PTH or calcium. The actions of PTH on bone may be anabolic (bone formation mediated through the osteoblasts) or catabolic (bone reabsorption via the osteoclasts). In addition to the effect of PTH on calcium, PTH stimulates renal excretion of phosphate ions. Both the resorption of bone and the kidney action of PTH may be dependent on a derivative of vitamin D, 1,25-DHC. The uptake of calcium through the intestinal wall requires 1,25-DHC. Synthesis of 1,25-DHC involves sequential intermediates produced in the skin, altered in the liver, and finally converted to 1,25-DHC in the cortical portion of the kidney. This enzymatic conversion in the kidney may be influenced by PTH as well as by estrogens and PRL.

Calcitonin has been isolated from mammalian parafollicular thyroid tissue and from bird ultimobranchial glands. The major site of action for CT is bone, where it antagonizes the action of PTH on osteoclasts.

The comparative aspects of calcium and phosphate regulation are complicated by the absence of parathyroid glands in fishes and the presence of the unique corpuscles of Stannius in teleosts. The pituitary of teleosts produces PRL, SL, and PTHrP, which may be involved in calcium homeostasis. Cyclostomes and elasmobranchs have not been studied sufficiently and mineral regulation is not well understood. Among bony fishes, calcium regulation appears to be accomplished by a hypercalcemic pituitary factor (PRL or hypercalcin) and a hypercalcemic factor (hypocalcin) from the corpuscles of Stannius embedded in the kidneys of some bony fishes. Scales are important calcium reservoirs in teleosts. The pituitary factors appear to be the more important regulators in freshwater fishes, and the corpuscles of Stannius appear to regulate calcium in sea water. Mammalian or fish CTs can decrease blood calcium, and the major target appears to be the gills. Salmon CT is a more potent hypocalcemic factor in mammals than is mammalian CT, and this greater potency may be a result of its relative resistance to clearance from the circulation. It is currently used for clinical treatment in some situations of CT insufficiency.

Parathyroid glands are distinct in amphibians and reptiles, and the effects of PTH and parathyroidectomy are similar to those observed in birds and mammals but occur with lessened intensity. The pituitary may have a hypercalcemic role in aquatic amphibians. Amphibians and some reptiles possess endolymphatic sacs, which may be important sites for calcium carbonate storage. CT has been demonstrated in amphibians, and extracts of their ultimobranchial glands have hypocalcemic activity when tested in birds or mammals. Treatment with CT alters calcium and phosphate metabolism as in other vertebrates.

One feature that appears to be dominant throughout the tetrapods examined to date is the strong tendency for innervation of the ultimobranchial glands, although the nature of this innervation (e.g., cholinergic, adrenergic) and biological significance are not clear. Mammalian C cells contain both CT and somatostatin, implying also a paracrine role for these cells.

A second consistent relationship appears in all oviparous species except the cyclostomes and possibly the elasmobranchs. Estrogens increase the liver production of vitellogenic phosphoproteins, which bind Ca^{2+} when they are secreted into the blood. Consequently, there is an elevation in total plasma calcium and phosphate associated with vitellogenesis in females that appears to be a mechanism whereby increased Ca^{2+} becomes available for production of egg shells (birds and reptiles) and/or for incorporation into eggs. Egg stores of Ca^{2+} are used for early developmental processes prior to hatching and feeding by the offspring. A similar mechanism operates in mammals, whereby maternal dietary calcium is directed during development or during lactation to the offspring by actions of PTH with protection of the maternal skeleton afforded through the actions of CT. Estrogens also stimulate production of 1,25-DHC in both birds and mammals.

Suggested Reading

Books

Azria, M. (1989). "The Calcitonins." Karger, Basel.
Bikle, D. D., and Negro-Vilar, A. (1995). "Hormonal Regulation of Bone Mineral Metabolism." The Endocrine Society Press, Bethesda, MD.
Bilezikian, J. P. (1994). "The Parathyroids: Basic and Clinical Concepts." Raven, New York.
Crass, III, M. F., and Avioli, L. V. (1995). "Calcium-Regulating Hormones and Cardiovascular Function." CRC Press, Boca Raton, FL.
Dacke, C. G. (1979). "Calcium Regulation in Sub-Mammalian Vertebrates." Academic Press, New York.
Favus, M. J. (1993). "Primer on the Metabolic Bone Diseases and Disorders of Mineral Metabolism." Lippincott-Raven, New York.
Massry, S. G. and Fujita, T. (1987). "New Actions of Parathyroid Hormone." Plenum, New York.

General Articles

Dempster, D. W., Cosman, F., Parisien, M., Shen, V., and Lindsay, R. (1993). Anabolic actions of parathyroid hormone on bone. *Endocr. Rev.* **14,** 690–709.
Mallette, L. E. (1991). The parathyroid polyhormones: New concepts in the spectrum of peptide hormone action. *Endocr. Rev.* **12,** 110–117.
Orloff, J. J., Reddy, D., de Papp, A. E., Yang, K. H., Soifer, N. E., and Stewart, A. F. (1994). Parathyroid hormone-related protein as a prohormone: Posttranslational processing and receptor interaction. *Endocr. Rev.* **15,** 40–60.
Suda, T., Takahashi, N., and Martin, T. J. (1992). Modulation of osteoclast differentiation. *Endocr. Rev.* **13,** 66–80.
Watson, P. H., and Hanley, D. A. (1993). Parathyroid hormone: Regulation of synthesis and secretion. *Clin. Invest. Med.* **16,** 58–77.

Clinical Articles

Marie, P. J., Hott, M., Launay, J. M., Graulet, A. M., and Gueris, J. (1993). *In vitro* production of cytokines by bone surface-derived osteoblastic cells in normal and osteoporotic postmenopausal women: Relationship with cell proliferation. *J. Cin. Endocrinol. Metab.* **77,** 824–830.

Schneider, D. L., Barrett-Connor, E. L., and Morton, D. J. (1994). Thyroid hormone use and bone mineral density in elderly women. *J. Am. Med. Assoc.* **271,** 1245–1249.

Siris, E. S., and Canfield, R. E. (1991). Paget's disease of bone. *Trends Endocrinol. Metab.* **2,** 207–212.

APPENDIX A

Chordate Evolution

THE PURPOSE of Appendix A is to provide a brief overview of the major evolutionary events in vertebrate phylogeny. A comparative study of vertebrate endocrine systems necessitates knowledge of the major vertebrate groups, their evolutionary history and relationships to one another, as well as the environments in which they arose. Each group of vertebrates—each species, in fact—is a product of individualistic evolutionary change and "progression" as well as a product of adaptations to similar environmental problems faced by unrelated groups. The comparative endocrinologist faces the task of sorting out similarities due to convergent evolution of structures and functions as opposed to similarities due to common ancestry.

Being phylogenetically old does not necessarily mean the organism is primitive as well. An animal species that has existed for millions of years must be adapted superbly to its environment. Even though a species may appear unchanged externally, we must recognize that its physiology and behavior may be very different from those of its ancestors. Therefore, examination of an endocrine mechanism or the structure of gene responsible for a given hormone in a phylogenetically old species may shed no light on its specific ancestory or on how it might be related to other living or extinct species for this "living fossil" may have diverged significantly from its ancestor in some characteristics just as "modern" species have.

The animals we classify as chordates are all members of the phylum Chordata. Earlier classifications were based largely on anatomical and developmental features, but modern schemes rely more heavily on a wide range of shared characteristics of many kinds and attempt to distinguish between parallel, convergent, and divergent evolution.

I. Phylum Chordata

The basic features of the phylum Chordata are **pharyngeal gill slits,** a **dorsal, hollow nerve cord** (that is, the brain and spinal cord), and a supportive endoskeletal element, the **notochord,** which lies beneath the dorsal nerve cord. Each of these features must be present at some stage in the life cycle to qualify for membership in the Chordata. In addition, most chordates possess a **postanal tail.** The aquatic tadpole-like larva is basic to the phylum. In many fishes and amphibians as well as in reptiles, birds, and mammals, however, the larval stage per se no longer exists, having been reduced to a transitory embryonic sequence recapitulating many of the developmental features of the tadpole.

In the more traditional approaches to classification, the phylum Chordata is subdivided into three subphyla: Urochordata (tailed chordates or ascidians or tunicates), Cephalochordata (head chordates), and Vertebrata (vertebrates). Depending on which classification scheme is being considered, subphyla are subdivided further into a variable number of classes, orders, families, genera, and species. However, in the cladistic approach to classification (i.e., phylogenetic systematics), taxonomic groups are designated by sharing derived characteristics, such as the cranium and backbone, that characterize all of the vertebrates. Although the cladistic approach also produces a heirarchy of relationships among groups, it results in different groupings that often transcend the earlier classifications using categories such as classes and orders (see Table A-1 for a comparison of fish classifications). In the following account, only group names are used without designating them as subphyla, classes, orders, etc.

II. The Invertebrate Chordates

A. Urochordates

The all-marine Urochordata is considered to be the most primitive chordate group. A free-swimming aquatic tadpole larva is characteristic. The larva possesses the three basic chordate features, but these are modified when the larva undergoes drastic structural modifications during metamorphosis to become a sessile (that is, attached firmly to some substrate),

Table A-1. Comparison of Living Fish Taxa by Various Traditional Classifiers[a]

Group	Romer (1959)	Orr (1976)	Bond (1980)	Young (1986)
Jawless fishes	Class	Class	Superclass	Superclass
Cyclostomes	Order	Subclass		Order
Hagfishes	Suborder	Order	Infraclass	Suborder
Lampreys	Suborder	Order	Infraclass	Suborder
Cartilaginous fishes	Class	Class		Class
Sharks, rays	Subclass	Order	Class	Subclass
Ratfishes	Subclass	Order	Class	Subclass
Bony Fishes	Class	Class	Class	Class
Lobe-finned fishes	Subclass	Subclass		Subclass
Coelacanth	Order	Order	Subclass	Order
Lungfishes	Order	Order	Subclass	Order
Spiny-rayed fishes	Subclass	Subclass	Subclass	Subclass
Chondrosteans	Superorder	Superorder	Superorder	Infraclass
Holosteans	Superorder	Superorder	Superorder	Infraclass
Teleosts	Superorder	Superorder	Superorder	Infraclass

[a] Blanks indicate no recognition of a comparable grouping. This lack of agreement of level of classification is removed by adoption of the cladistic approach but may split or lump what others have long considered to be recognizable groups.

aquatic adult. The dorsal, hollow nervous system degenerates to a single neural ganglion, the notochord is obliterated, and the animal secretes an exoskeleton or tunic that completely encases the adult.

The adult tunicate (sometimes called a sea squirt because of its habit of ejecting a fine stream of water when disturbed by a curious biologist) retains only one of the unique chordate characteristics: a gill structure called a **branchial basket.** This apparatus is covered with cilia and has a mucus-secreting structure, the **endostyle,** associated with it. Coordinated ciliary movements cause a current of water to flow into the branchial basket (via the mouth of the larva or the incurrent siphon of the adult) and out via the gill slits (the excurrent siphon of the adult). Mucus secreted by the endostyle traps minute food particles, and the mucus plus trapped food is moved by ciliary action into the gut, where both mucus and trapped food particles are digested. This method for obtaining food is common among the invertebrate groups believed to have given rise to the chordates, and organisms possessing such a mechanism are called ciliary-mucus or pharyngeal-filtration feeders.

B. Cephalochordates

The sessile urochordate or tunicate is considered to be an evolutionary dead end, but some ancestral form similar to the larval tunicate may have given rise to the marine Cephalochordata. Certain living tunicates (for example, *Oikopleura*) never undergo metamorphosis to a sessile adult but remain free living and attain sexual maturity while retaining the larval body form. Prolongation of larval life or retention of larval characteristics in sexually mature animals often is termed **paedomorphosis.** If paedomorphosis is brought about by delayed development of nonreproductive (somatic) tissues, it is called **neoteny.** If it is a case of precocious gonadal development, it is called **progenesis.**

The cephalochordates (such as the living amphioxus) have a body plan very similar to that of larval urochordates, including a branchial basket with mucus-secreting endostyle and a persistent notochord throughout their life. Furthermore, cephalochordates anatomically resemble the larvae of cyclostomes, the most primitive members of the Vertebrata (Fig. A-1). Cephalochordate larvae and adults are ciliary-mucus feeders like the urochordates. Similarities among cephalochordates, urochordates, and vertebrates support a common evolutionary origin for all three subphyla, but it is not certain how the groups are related to one another. The ancestral vertebrate may not have been a member of either invertebrate chordate group. Vertebrates have many features not found in either urochordates or cephalochordates, for example, specializations of the head, anterior nervous system, and pharyngeal breathing apparatus that are responsible for the active predaceous life of vertebrates.

III. The Vertebrate Chordates

A number of major groups (previously called classes) comprise the vertebrates (Fig. A-2). Only one major grouping, the placoderm fishes, is entirely extinct; all the others have living members. In addition to possession of the three chordate characteristics, the vertebrates all have **vertebrae,** special cartilaginous or bony structures that surround and protect the spinal cord. Furthermore, there is a special protective case, the **cranium,** that surrounds the enlarged anterior portion of the nervous system, the brain. This latter feature is the basis for another name sometimes applied to vertebrates, the Craniata.

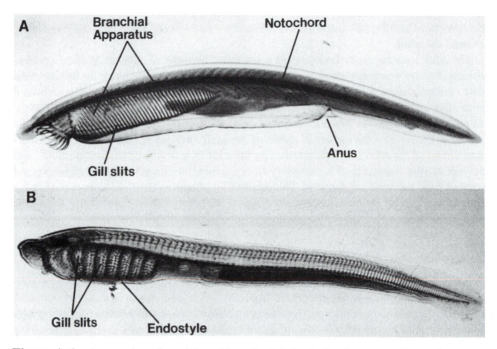

Figure A-1. A comparison of an adult amphioxus (cephalochordate) and ammocetes larvae of a lamprey (cyclostome vertebrate). (A) Amphioxus. Note the prominant notochord, branchial apparatus with ventral endostyle, and post-anal tail. (B) Ammocetes larva. The branchial apparatus resembles that of amphioxus, but has fewer gill slits. The mucus-secreting ventral endostyle is present, but at metamorphosis develops into thyroid follicles.

A. Agnathan Fishes

The agnathans consists of primitive, jawless fishes (Gr., *a,* without; *gnathos,* jaws) believed to have evolved directly from cephalochordates or some cephalochordate-like ancestor. They are divided into two subgroups: the extinct ostracoderms and the extant cyclostomes (Gr. *kyklos,* round; *stoma,* mouth) consisting of the lampreys and hagfishes.

The ostracoderms were all small fishes covered with bony plates (armor). They were limited to the ocean bottom, where they existed primarily as ciliary-mucus feeders. Although the ostracoderms were not sessile like the urochordates, they possibly lived a "sit-and-sift" existence close to the bottom sediments of the oceans. However, recent interpretations of some ostracoderm fossils suggest at least some of them may have been active swimmers and possibly predators.

There are two groups of living cyclostomes, the marine hagfishes (Myxinoidea) and the essentially freshwater lampreys (Petromyzontoidea). Many adult lampreys are parasitic on other vertebrate fishes, but the larvae are ciliary-mucus feeders. The **ammocetes larva** of the lamprey has a branchial basket with an endostyle, and in general, the body form looks much like the cephalochordate, amphioxus (Fig. A-1). Although some biologists suggest these structural similarities imply a close evolutionary relationship, others would argue against such an interpretation. When the ammocetes larva metamorphoses to the adult lamprey, the endostyle differentiates into the thyroid gland of the adult (see Chapter 8). The

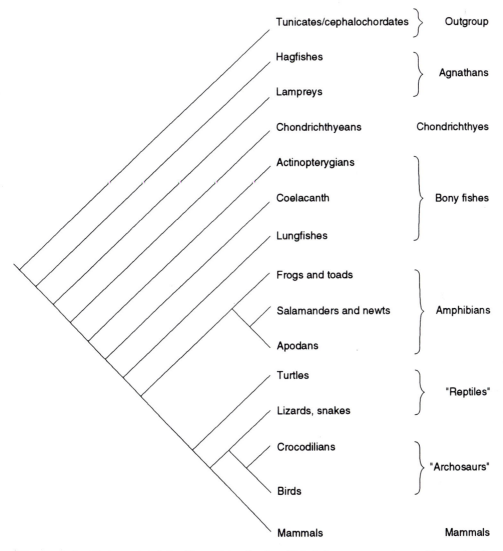

Figure A-2. Phylogenetic relationships of living chordates. This cladogram represents one of the most probable phylogenetic patterns for the living vertebrates with comparison to the urochordates and cephalochordates as outgroups. Each branching point is determined by a number of characters. This classification of the fishes is similar to that provided by traditional classifications that recognize six classes of extant fishes rather than the three classes (Agnatha, Chondrichthyes, Osteichthyes) of most traditional schemes. It correctly aligns the crocodilians and birds into a common group, sometimes referred to as the archosaurs.

Myxinoidea do not have a larval form and have many more primitive features than lampreys. However, modern hagfishes may not be much like the first vertebrate, and their apparent simplicity may be a degenerate condition. Most cladistic analyses place the lampreys closer to the jawed vertebrates than to the hagfishes, indicating that hagfishes and lampreys diverged much earlier from a common ancestor. Consequently, lampreys are probably a better living example of a primitive vertebrate somewhere between the ostracoderms and the first jawed vertebrates.

B. The Placoderm Fishes

The ostracoderm fishes were ancestral to the first jawed vertebrates, the class Placodermi, a heterogeneous collection of extinct, heavily armored fishes. Hinged jaws developed in the placoderm fishes from modifications of the first gill arch, and this same event can be observed early in embryonic development of all jawed vertebrates. Many zoologists consider development of jaws to be the most significant single event in vertebrate evolution. Certainly it was a significant advancement enabling the placoderm fishes to abandon the bottom-dwelling existence of their ancestors for a pelagic, predatory life style. The fossil record suggests that some placoderms retained the ancestral filter-feeding habit. The placoderms became abundant, attained great size, and sported heavy armor. They were the dominant vertebrates of the Devonian period, but they suddenly declined and disappeared, leaving only a strong fossil record and their apparent descendants. During the Devonian, the placoderms gave rise to two important piscine taxa: the bony fishes, Osteichthyes, and the cartilaginous fishes, Chondrichthyes. These predatory descendants retained the jaws of their ancestors but reduced the bony armor to scales, emphasizing speed and agility. The demise of the placoderms was probably due in no small part to the success of these more mobile predators.

C. Chondrichthyean Fishes

The chondrichthyeans (Gr. *chondros,* cartilage; *ichthy,* fish) have skeletons primarily composed of cartilage or calcified cartilage. Of course, the agnathans also had a cartilaginous skeleton but they lacked jaws. True bone is not present in this group. Since cartilage forms prior to bone in the normal developmental sequence, some zoologists suggest that this group arose from the placoderms via neoteny. Included in the Chondrichthyes are the sharks, rays, and skates (Selachii or Elasmobranchii) and the ratfishes or chimaeras (Holocephali). The cartilaginous fishes flourished for a time but then declined. Although in recent geological time they are increasing in abundance, they are believed to represent an evolutionary dead end in not having given rise to any other vertebrate group. The cartilaginous fishes have not been as successful as the higher bony fishes (teleosts) in exploiting the aquatic environment (especially fresh water), and they represent a secondary fish fauna today (Fig. A-3).

D. Osteichthyes: The Bony Fishes

Osteichthyean fishes (Gr. *osteon,* bone) have excelled in exploitation of freshwater and marine habitats. These bony fishes may have had their origin in fresh water and secondarily invaded the marine habitat. Regardless of their origin, it appears that freshwater bony fishes gave rise to the first terrestrial vertebrates, the amphibians.

The bony fishes can be readily separated into two subgroupings: the Actinopterygii, or rayfinned fishes, and the Sarcopterygii, or lobe-finned fishes. The Actinopterygii (spiny wings or fins) have distinct rays that support the fins, whereas the Sarcopterygii (fleshy

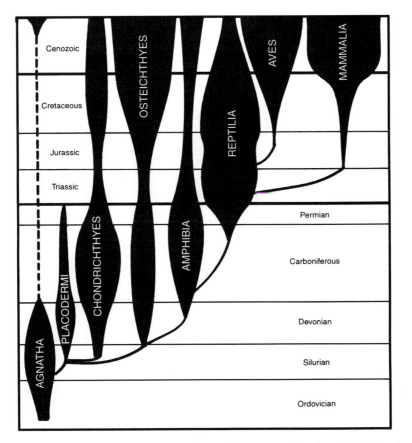

Figure A-3. Relative abundance of some major vertebrate groups through geologic time. The width of the line is roughly proportional to the number of known species within that group. Note how the increased radiation of one group into a new species seems to be correlated with a decline in another group.

fins) have lobed fins with internal skeletal supports (bones) that are homologous to the limb bones of tetrapods.

Early in the evolution of the bony fishes (or possibly in the placoderm group that gave rise to the bony fishes), a pouch developed ventrally off the gut anterior to the stomach, remaining connected to the gut by a duct, and evolved into the air bladder. In the actinopterygian fishes, this air bladder was used as a flotation device or swim bladder. Among the sarcopterygians, the air bladder became modified as an accessory breathing device homologous to the lungs of tetrapods.

1. The Actinopterygii

There are four, distinctive groups of actinopterygians that include most of the living bony fishes: Polypteri, Chondrostei, Holostei, and Teleostei.

The most primitive ray-finned fishes are found in the Polypteri. Two living freshwater genera (*Polypterus* and *Calamoichthyes*) are found in Africa. The Chondrostei includes sturgeons (for example, *Acipenser* spp.) and the spoonbills (*Polydon*) of China and North America. Some zoologists include the Polypteri as the most primitive members of this group. The chondrostean fishes are all freshwater fishes. The Holostei is a group restricted to North America, occurring today only in the Mississippi drainage. This group consists of the bowfin, *Amia calva,* and several species of gar (*Lepisosteus*).

The teleostean fishes (Teleostei) are the most advanced actinopterygians. The teleosts have produced a tremendous adaptive radiation in fresh water and have secondarily invaded the marine habitat, where they are the most abundant and successful vertebrate group. The majority of extant fish species are teleosts, with estimates of 20,000 to 40,000 species. Sometimes the teleosts are divided into the so-called lower teleosts, exemplified by the salmonid fishes, such as trout and salmon, and the higher teleosts, such as the centrarchids (large-mouth bass and bluegill sunfish, etc.). The higher teleosts can be recognized by the strong tendency for the pelvic (ventral) paired fins to occur anteriorly to the vicinity of the pectoral (shoulder) fins.

2. Sarcopterygii

The ancestors of the first four-footed or tetrapod vertebrates were the sarcopterygian fishes. Two groups of sarcopterygian fishes have living representatives: one species of the Crossopterygii (fringe fins) and three genera of the Dipnoi (Gr. *dipnoos,* double breathing). Both of these groups represent side ventures off the main line of evolution within the Sarcopterygii, which gave rise to the first semiterrestrial vertebrates, the Amphibia.

Crossopterygian fishes were known only from their excellent fossil record until 1938 when a living crossoptergyian, *Latimeria chalumnae,* caught by some fishermen off the coast of Madagascar, attracted the attention of some scientists. Since that time a number of these bizarre, bluish giants have been captured and their anatomy, physiology, behavior, and ecology closely scrutinized by comparative zoologists. *Latimeria,* like other crossopterygian fishes, has internal nares and a lunglike air bladder. It may reach 5 to 6 feet in length and is viviparous (live-bearing).

The order Dipnoi consists of three genera of lungfishes restricted to the tropical regions of three continents: *Protopterus* in Africa, *Lepidosiren* in South America, and *Neoceratodus* in Australia. These fishes are gill breathers that use their lungs as accessory breathing structures. Only *Protopterus* survives by breathing air alone. *Protopterus* can secrete a mucus-lined cocoon in which it resides and breathes air during periods of intense drought when its aquatic habitat may disappear altogether. The unusual distribution of these genera of lungfishes relates to the theory of formation of the present continents following the breakup of a "super continent" and a movement or drifting apart of the fragments (that is, continental drift).

E. Amphibia

There are three living groups of amphibians (Gr. *amphi,* both; *bios,* life): the Caudata (Urodela), which includes salamanders and newts; the Apoda (without feet), the limbless caecilians, a tropical group, about which little is known; and the Anura (without tail), the tailless frogs and toads.

The primitive, large-tailed amphibians (labyrinthodonts) most probably had their origin from the crossopterygian fishes that had developed internal nares and lungs for air breath-

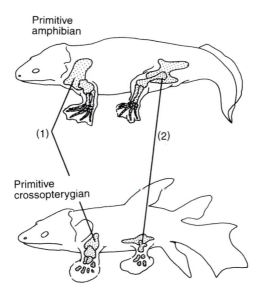

Primitive
amphibian

(1)

(2)

Primitive
crossopterygian

Figure A-4. Comparison of limb structures in a primitive crossopterygian fish and the bones occurring in paired ventral fins in a primitive amphibian. Only the homologous scapulae (1) and femur (2) are indicated, but the other homologies are obvious. Hence the pectoral and pelvic fins of these fishes are homologous to the pectoral and pelvic girdles and limbs of tetrapod vertebrates.

ing and had supportive skeletal elements in their fleshy lobed fins (Fig. A-4). The labyrinthodonts gave rise to the first reptiles as well as to the modern amphibians, which arose via two or three separate lines of evolution.

As the name of this class implies, many amphibians lead double lives: one as aquatic larvae and the second as terrestrial or semiterrestrial adults. This "typical" life history involves laying eggs in fresh water, where they develop into tadpole-like larvae. The larvae remain in fresh water as did their ancestors for a period of growth followed by a remarkable metamorphosis involving drastic structural and biochemical alterations to attain the different adult body form and the physiology to survive in a desiccating environment. The process of metamorphosis is discussed in more detail in Chapter 8, and amphibian life history patterns are described in Chapter 12.

F. The Reptiles

In a sense, it was a mistake for the amphibians to give rise to the reptiles, for the reptiles quickly replaced them as the dominant terrestrial organisms. Reptiles owed their success in exploiting the terrestrial environment in part to the evolution of a unique "land egg," which enclosed the aquatic environment for embryonic development within a membrane (the amnion). This amniote egg could be laid on land, where it was safe from aquatic predators. The reptilian egg is much more resistant to desiccation than the "terrestrial" eggs of amphibians. Furthermore, the reptiles were no longer tied to water for reproduction, which placed fewer restrictions on their movements. The large eggs allow young to hatch at a size considerably greater than is possible from eggs of either oviparous fishes or amphibians.

Birds and mammals have retained many of the features of the reptilian egg, including the amnion and other membranes (the chorion, allantois, and, in some cases, the yolk sac). Bird eggs are not much different from reptilian eggs, and embryonic development is very similar. Mammals have greatly modified the use of these membranes, especially among the placental mammals. Reptiles, birds, and mammals are often referred to collectively as the amniote vertebrates or **amniotes** (that is, they all possess an amnion), whereas fishes and amphibians are termed **anamniotes** (without an amnion).

Primitive amphibians gave rise to the cotylosaurs or stem reptiles, which early in reptilian evolution diverged into several separate pathways. Only four of these pathways have living representatives.

One pathway (the Anapsida) separated early and gave rise to the heavily armored Chelonia—the turtles, tortoises, and terrapins. The chelonians are anatomically a conservative group, having changed little in appearance over several million years. However, it might be a mistake to assume that their physiology has remained equally conservative.

A second pathway (Lepidosaura) produced two important groups, the Squamata (snakes and lizards) and the more ancient Rhynchocephalia, which contains only the New Zealand tuatara, *Sphenodon*. The tuatara is of special interest because it is the only living representative of a very old reptilian group.

The Archosaura represents a third evolutionary line, of which only the Crocodilia (crocodiles and alligators) has living representatives. The extinct dinosaurs were part of this evolutionary line. In addition, the archosaurs (thecodonts) gave rise to bird (Aves). In fact, some zoologists now consider the birds to be archosaurs and more closely related to the crocodilians than to any other vertebrate group. This relationship is supported by molecular, physiological, and anatomical studies.

The final reptilian group that gave rise to extant organisms was the Synapsidia, from which evolved the mammals. This group separated early from the main line of reptilian evolution.

None of the reptilian ancestors of the mammals remains today. Apparently the ability to regulate a relatively constant body temperature developed in the synapsids (specifically in the therapsid reptiles) independently of its development in the thecodont reptiles, which gave rise to the birds. Physiological temperature regulation may have evolved among the dinosaurs, too.

G. Birds

Birds are characterized by having feathers, no teeth, a relatively constant body temperature (ranging from 37–43°C in different species), a four-chambered heart analogous to that of mammals, and numerous structural modifications for flight. Nevertheless, birds are not much more than warm, feathered reptiles that lack teeth and probably should be classified with the crocodilians, as indicated in Fig. A-2. Although viviparity has developed in all other tetrapod classes (as well as some fishes), all birds lay eggs. However, birds have been successful at exploiting the terrestrial habitat in such a way as to avoid undue competition with mammals and hence exhibit an impressive adaptive radiation. There are many diverse groups within the Aves, but a more detailed account is beyond the scope of this presentation.

H. Mammals

The most distinctive and uniform features of mammals are the possession of hair and the mammary glands, for which the group is named. Mammals may be separated into three, distinct subgroups: Prototheria (Gr. *protos,* first; *thereon,* animal), Metatheria (Gr. *meta,* middle), and Eutheria (Gr. *eu,* good).

The most primitive mammals are the monotremes (Monotremata) of the Prototheria. This group includes the duckbill platypus (*Ornithorhynchus anatinus*) and the spiny ant-eaters, *Tachyglossus* spp. All members of this group lay eggs, and they are found only in Australia.

The Metatheria consists of the Marsupialia, the pouched mammals or marsupials such as the kangaroos and wallabys of Australia and the opossum of North America. The fossil record indicates that the marsupials were just beginning their adaptive radiation when continental drift began to separate the continents. This explains in part their present skewed distribution, with the vast majority of extant species being found in Australia.

The Eutheria consist of the placental mammals. The most primitive group of eutherian mammals is the insectivores, from which 13 other groups have evolved. The eutherian mammals evolved in the Old World, entered North America from Asia, and eventually entered South America as well. Although marsupials once were common elements of the New World, they were replaced almost completely by the eutherians. The survival of marsupials in Australia today probably is a consequence of Australia separating from the supercontinent via continental drift before the evolution of eutherians. The Primates are considered (by humans, of course) to be the most highly evolved group of mammals, that group showing the most advanced evolutionary adaptations and the highest ecological success.

Suggested Reading

Books

Carroll, R. L. (1988). "Vertebrate Paleontology and Evolution." W. H. Freeman, New York.
Musick, J. A., Bruton, M. N., and Balon, E. K. (1991). "The Biology of *Latimeria chalumnae* and Evolution of Coelacanths." Kluwer, Dordrecht.
Pough, F. H., Heiser, J. B., and McFarland, W. N. (1996). "Vertebrate Life," 4th Ed. Macmillan, New York.

Article

Meyer, A. (1995). Molecular evidence on the origin of tetrapods and the relationships of the coelacanth. *Trends Ecol. Evol.* **10,** 111–116.

APPENDIX B

Vertebrate Tissue Types

I. The Origin of Vertebrate Tissues

DURING EMBRYONIC development, the process of gastrulation defines the three **primary germ layers:** ectoderm, mesoderm, and endoderm. The **ectoderm** gives rise to the nervous system, including the neural crest and its derivatives, the epidermis, the lining of the oral cavity, and parts of certain sense organs. The **endoderm** gives rise to the mucosal lining of the gut and a number of derivatives of the gut, including the lungs, thyroid gland, liver, and pancreas. **Mesoderm** is the source of muscle, dermis, linings of the coelomic cavity (peritoneum, pleura, and pericardium) and blood vessels (endothelium), and special organs such as the kidneys, adrenal cortex, and the gonads.

The primary germ layers give rise to four primary tissues: epithelium, connective tissues, muscle, and nervous tissue. These tissue types are defined below. Ectoderm gives rise to nervous tissue and certain epithelia. Endoderm gives rise to both covering epithelia and glandular epithelia. Mesoderm gives rise to special epithelia (mesothelia and endothelia), in addition to glandular epithelia, the elements of the various connective tissues, and all muscle. The origin of primordial germ cells that eventually reside in the gonads in at least some vertebrates has been traced to endoderm, although it is possible that mesoderm may be involved in some species.

II. Epithelium

An epithelium consists of closely associated cells organized into sheets that develop into coverings of either outer and inner surfaces or is modified into a glandular structure. Little intracellular material is found between the cells of epithelia but they are typically associated with a distinct basement membrane associated with the basal portion of the cells. Epithelia also may occur as tubes (**ducts, cords**) or spheres (**follicles, acini**). Individual cells may

584

differ markedly in shape, and some cells may exhibit specialized adornments such as cilia or microvilli (brush border). Cilia may be responsible for producing currents to aid the flow of materials through a duct or other passageway as a result of coordinated rhythmic beating. Sometimes cilia are sensory structures, too. Microvilli greatly increase the total surface area of cells. The presence of microvilli is a clue to an epithelium's involvement in transport of molecules between the cells and extracellular fluids. Epithelia may consist of a single layer or sheet of cells (**simple**) or several layers of one (**stratified**) or several (**compound**) types of simple epithelia appearing in layers. Some examples of epithelia are as follows:

Simple squamous epithelium: Thin, flat cells organized in sheets such as the peritoneum

Simple cuboidal epithelium: Cube-shaped cells comprising the lining of certain ducts such as portions of the nephron

Simple columnar epithelium: Tall, rectangular cells that may also be found lining certain ducts

Stratified squamous epithelium: Epidermis of skin

Glandular epithelium: Cords of cells such as in the adrenal cortex, acini (solid balls of cells with a central duct) such as found in the exocrine pancreas, follicles or hollow balls of endocrine cells such as those comprising the thyroid gland

III. Connective Tissues

The cells of connective tissue are generally separated from one another by extensive intracellular material (matrix) that they have secreted. Mesenchyme derived from mesoderm gives rise to four basic kinds of connective tissue: blood and lymph-forming tissues, connective tissue proper, cartilage, and bone.

A. Blood-Forming Tissues

The blood-forming elements give rise to circulating **erythrocytes** or red blood cells (RBCs) and **leukocytes** or white blood cells (WBCs) in adult vertebrates. The RBCs of mammals are unique among the vertebrates in that the mature circulating cell lacks a nucleus. The WBCs are further subdivided into cells with granular cytoplasm, the **granulocytes** (eosinophils, basophils, and neutrophils) and **agranulocytes (lymphocytes, monocytes,** and **plasma cells).** Circulating eosinophils are identical to those associated with many organs such as the uterus and the lung. Basophils may be identical to **tissue mast cells** that secrete histamine.

B. Connective Tissue Proper

A large number of tissue types are lumped under the title of **connective tissue proper,** including **loose fibrous connective tissue, dense fibrous connective tissue** (e.g., tendons), **elastic and reticular connective tissue,** and **adipose tissue** or fat. The fibrous components are proteins (collagen, elastin, and reticular fibers) and are found in the different connective tissue types to various extents.

Adipose tissue consists of cells that contain large amounts of fat restricted to a central vacuole, which confines the cytoplasm to a thin rim adjacent to the plasmalemma. In rodents there are two readily distinguishable types of adipose tissue, termed white and brown.

White adipose tissue shows regular variations in the amount of stored fat with nutritional state. **Brown fat** is found in particular locations and does not vary with nutritional state. It has been correlated with hibernating behavior. In most mammals it is not possible to differentiate brown and white types of adipose tissue.

C. Cartilage

Cartilage cells or **chondrocytes** secrete a matrix consisting of a glycoprotein, chondromucoid, that contains the sulfonated polysaccharide chondroitin sulfate. The extensive matrix between the chondrocytes may have a few fibrous components (**hyaline cartilage**) or may contain collagen (**fibrocartilage**) or elastin fibers (**elastic cartilage**). Cartilage may also be strengthened by the addition of calcium salts.

D. Bone

Bone is the strongest connective tissue and the most dense. The extensive matrix of bone is composed of crystalline calcium salts, primarily calcium phosphate, and very little water. Bone occurs in a uniformly dense, compact form (**compact bone**) and in a less dense, more easily modified form (**cancellous** or spongy **bone**). The bone-forming cells are known as **osteoblasts** and are rich in phosphatase. Osteoblasts that have become embedded within the bone matrix are termed **osteocytes.** Giant multinucleated cells, the **osteoclasts,** produce hydrolytic enzymes and are responsible for bone destruction (resorption). The osteoblasts are responsible for bone forming, resorbing, and reforming in accordance with physical stresses placed on bone and with physiological demands and sources of calcium and phosphate.

IV. Muscle

Mesoderm gives rise to three basic muscle types: smooth, striated, and cardiac. **Smooth muscle** frequently is termed involuntary muscle since it is under control of the autonomic nervous system. Primarily it is associated with internal organs and is found in such places as the gut wall, blood vessels, various ducts, and the wall of the uterus. Individual smooth muscle cells are smaller than striated muscle cells, and the contractile elements (myofibrils) are not highly organized within the cell. Smooth muscle is characterized by slow, rhythmic contractions.

    Striated or **skeletal muscle** is the so-called voluntary muscle tissue that is under conscious control, although it also may be influenced by the autonomic system. Striated muscle cells are large cells with myofibrils so highly organized as to produce regular bands or striations when the cells are viewed with the aid of a microscope. Movements of the skeleton are controlled by striated muscles that are attached to the bone by dense fibrous connective tissue. In addition, a few sphincter muscles are also of this type and hence are under conscious control (e.g., the external urinary bladder sphincter).

    Cardiac muscle possesses properties of both skeletal muscle (striations due to highly organized myofibrils) and smooth muscle (rhythmic contractions that are innate properties of cardiac muscle cells). Instead of inserting on bones, the cardiac cells connect directly to one another through specialized tendinous attachments known as *intercalated disks.*

V. Nervous Tissue

Nervous tissue is specialized for integrative functions and conduction of information throughout the body. **Neurons** are specialized cells for conducting electrochemical neural impulses to coordinate body processes. Most neurons release chemical neurotransmitters to control the activity of other neurons, muscle cells, and glands. **Neurosecretory neurons** produce neurohormones that are secreted into the blood vascular system and constitute a second type of control mechanism. The central nervous system also contains several types of **glial cells** (neuroglia), which perform many supportive functions for the neurons. One important role of glial cells is the secretion of myelin that forms the white matter of the central nervous system. The ependymal cells that line the brain ventricles are another form. The **Schwann cell,** a type of glial cell in the peripheral nervous system, secretes the myelin sheath characteristic of many peripheral neurons.

VI. General Tissue Responses

In response to various stimuli a given tissue may exhibit no morphologically observable response, degeneration (atrophy, resorption), or growth. The last response may be due simply to an increase in cell size (**hypertrophy**) or to increased cellular divisions with an actual increase in cell numbers (**hyperplasia**) or both. Tumors or neoplasms are abnormal proliferations of tissues (**neoplasia**) having no normal physiological function. Such growths may be classified as either **benign** (harmless) or **malignant** (very harmful or likely to cause death; i.e., a cancer). The terms **adenoma** refers to a benign tumor of glandular origin that may or may not synthesize and release abnormal amounts of hormones. Connective tissue tumors are called **sarcomas,** whereas a lymphatic tumor is a **lymphoma.** Malignant growths of any epithelial tissue, including glandular epithelia, are termed **carcinomas.** Some adenomas or carcinomas also produce excessive quantities of hormones or hormone-like substances, as in the production of excessive amounts of growth hormone by pituitary adenomas or secretion of gastrin (a hormone that stimulates hydrochloric acid secretion in the stomach) by a pancreatic carcinoma.

APPENDIX C

Amino Acids and
Their Symbols

Amino acid	Old system	New system	Basic AA	Acidic AA
Alanine	Ala	A		
Arginine	Arg	R	+	
Asparagine	Asn	N	+	
Aspartate	Asp	D		+
Cysteine	Cys	C		
Glutamate	Glu	E		+
Glutamine	Gln	Q	+	
Glycine	Gly	G		
Histidine	His	H		
Isoleucine	Ile	I		
Leucine	Leu	L		
Lysine	Lys	K	+	
Methionine	Met	M		
Phenylalanine	Phe	F		
Proline	Pro	P		
Serine	Ser	S		
Threonine	The	T		
Tryptophan	Trp	W		
Tyrosine	Tyr	Y		
Valine	Val	V		

APPENDIX D

Abbreviations of Endocrine Terms

Alphabetical by Abbreviation

1,25-DHC	1,25-Dihydroxycholecalciferol
3β-HSD	3β-hydroxy-Δ⁵-steroid dehydrogenase
5-HIAA	5-Hydroxyindole acetic acid
5-HT	Serotonin
5-HTP	5-Hydroxytryptophol
5-MP	5-Methoxytryptophol
AAAD	Adrenal ascorbic acid depletion
ABP	Androgen-binding protein
ACE	Angiotensin-converting enzyme
ACELA	Angiotensin-converting enzyme-like activity
ACh	Acetylcholine
ACTH	Corticotropin
ADD	Attention-deficit hyperactivity disorder
ADH	Antidiuretic hormone
AMH	Antimüllerian hormone (same as MIS)
α-MSH	α-Melanotropin
ANG-I	Angiotensin I
ANG-II	Angiotensin II
ANG-III	Angiotensin III
ANP	Atrial natriuretic peptide
APUD	Amine precursor uptake and decarboxylation
AVP	Arginine vasopressin
AVT	Arginine vasotocin
B	Corticosterone
βARK	β-Adrenergic receptor kinase
β-LPH	β-Lipotropin
BNP	Brain natriuretic peptide
BSD	Behavioral sex determination
CAH	Congenital adrenal hyperplasia
cAMP	Cyclic adenosine monophosphate
CAT	Choline acetyltransferase
CBG	Corticosteroid-binding globulin
CC	Chorionic corticotropin
CCK	Cholecystokinin
CG	Chorionic gonadotropin
cGMP	Cyclic guanosine monophosphate
cGnRH-I	Chicken-I GnRH
cGnRH-II	Chicken-II GnRH
CGRP	Calcitonin gene-related peptide
CLIP	Corticotropin-like peptide
CO	Carbon monoxide

COH	Compensatory ovarian hypertrophy	GRB2	Growth factor receptor-binding protein
COMT	Catechol *O*-methyltransferase	GRP	Gastrin-releasing peptide
CREB	cAMP regulatory element-binding protein	GSD	Gonotypic sex determination
		GTH	Gonadotropin
CRH	Corticotropin-releasing hormone	GTH-I	Gonadotropin I
		GTH-II	Gonadotropin II
CS	Chorionic somatomammotropin	GTP	Guanosine triphosphate
		HDL	High-density lipoprotein
CT	Calcitonin, chorionic thyrotropin	HIOMT	Hydroxyindole-*O*-methyltransferase
DA	Dopamine	HPLC	High-performance liquid chromatography
DAG	Diacylglycerol		
DHEA	Dehydroepiandrosterone	ICSH	Interstitial cell-stimulating hormone (same as LH)
DHEAS	Dehydroepiandrosterone sulfate		
DHT	5α-Dihydrotestosterone	IDDM	Insulin-dependent diabetes mellitus
DIT	Diiodotyrosine		
DNA	Deoxyribonucleic acid	IDL	Intermediate-density lipoprotein
DSIP	Delta sleep-inducing protein		
DOC	Deoxycorticosterone	IGF-I	Insulin-like growth factor I
DOPA	Dihydroxyphenylalanine	IGF-II	Insulin-like growth factor II
EDRF	Endothelium-derived relaxing factor	IL-1	Interleukin 1
		IL-2	Interleukin 2
EGF	Epidermal growth factor	IL-6	Interleukin 6
ELISA	Enzyme-linked immunoabsorbent assay	IP_3	Inositol trisphosphate
		IRMA	Immunoradiometric assay
EOP	Endogenous opioid peptide	IST	Isotocin
EPTH	Extrapancreatic tumor hypoglycemia	LAF	Luteinization of atretic follicles
		LATS	Long-acting thyroid stimulator
F	Cortisol	LDL	Low-density lipoprotein
FSH	Follicle-stimulating hormone	LH	Luteinizing hormone
γ-LPH	γ-Lipotropin	LHRH	Luteinizing hormone-releasing hormone (same as GnRH)
GABA	γ-Aminobutyric acid		
GAP	Gonadotropin-releasing hormone-associated peptide	LPH	Lipotropin
		LVP	Lysine vasopressin
GH	Growth hormone	MAO	Monoamine oxidase
GHRH	Growth hormone-releasing hormone, somatocrinin	MAPK	Mitogen-activated protein kinase
GHRIH	Growth hormone release-inhibiting hormone, somatostatin	MCH	Melanophore concentrating hormone
		MIS	Müllerian-inhibiting substance (same as AMH)
GIP	Glucose-dependent insulinotropic peptide, gastric inhibitory peptide	MIT	Monoiodotyrosine
		MMC	Migrating motor complex
GLP-I, -II	Glucagon-like peptides	MRH	Melanotropin-releasing hormone
Gn	Gonadotropin		
GnRH	Gonadotropin-releasing hormone	MRIH	Melanotropin release-inhibiting hormone

mRNA	Messenger RNA		SCO	Subcommissural organ
MSH	Melanotropin		SIAD	Syndrome of inappropriate
MST	Mesotocin			diuresis
NAT	*N*-Acetyltransferase		SL	Somatolactin
NE	Norepinephrine		SON	Supraoptic nucleus
NEFA	Nonesterified fatty acids		SOS	Son of sevenless protein
NGF	Nerve growth factor		SS	Somatostatin; also SS_{14}, SS_{28},
NID	Dorsal infundibular nucleus			SS_{34}
NIDDM	Non-insulin-dependent diabe-		StAR	Steroidogenic acute regulatory
	tes mellitus			protein
NIV	Ventral infundibular nucleus		STP	Steroidogenic stimulating
NLT	Nucleus lateralis tuberis			protein
NO	Nitric oxide		SU4885	Metyrapone
NOS	Nitric oxide synthetase		T_3	Triiodothyronine
NPY	Neuropeptide Y		T_4	Thyroxine (tetraiodothyronine)
NT	Neurotensin		TBA	Thyroid-binding albumin
OAAD	Ovarian ascorbic acid depletion		TBG	Thyroid-binding globulin
$P\text{-}450_{c18}$	Aldosterone synthase		TBPA	Thyroid-binding prealbumin
PAG	Pineal antigonadotropic		TETRAC	Tetraiodothyroacetic acid
	peptide		TGF-α	Transforming growth factor α
PEPCK	Phosphoenolpyruvate		tGLP-I	Truncated GLP-I
	carboxykinase		TPO	Thyroid peroxidase
PHI	Peptide histidine isoleucine		TR	Thyroid receptor
PIP_2	Phosphoinositol diphosphate		TRAP	Thyroid receptor auxiliary
PIPAS cell	Calcium-sensitive cell			protein
PKC	Protein kinase C		TRE	Thyroid response element
PMSG	Pregnant mare serum		TRH	Thyrotropin-releasing hormone
	gonadotropin		TRIAC	Triiodothyroacetic acid
PNMT	Phenylethanolamine-*N*-		TSD	Temperature-dependent sex
	methyltransferase			determination
POA	Preoptic area		TSH	Thyrotropin
POMC	Proopiomelanocortin		TU	Thiourea
PON	Preoptic nucleus		VIP	Vasoactive intestinal peptide
POS	Polycystic ovarian syndrome		VLDL	Very low-density lipoprotein
PP	Pancreatic polypeptide		VSCC	Voltage-sensitive calcium
PRL	Prolactin			channels
Pst	Pancreostatin			
PTH	Parathyroid hormone			
PTHrP	Parathyroid hormone-related			
	protein			

Alphabetical by Term

PTU	Propylthiouracil		1,25-Dihydroxycholecalciferol	1,25-DHC
PVP	Phenypressin		5-Hydroxyindole acetic acid	5-HIAA
PYY	Peptide YY		5-Hydroxytryptophol	5-HTP
RIA	Radioimmunoassay		5-Methoxytryptophol	5-MP
RNA	Ribonucleic acid		5α-Dihydrotestosterone	DHT
rT_3	Reverse T_3		α-Melanotropin	α-MSH
SCG	Superior cervical ganglion		Acetylcholine	ACh
SCN	Suprachiasmatic nucleus		Adrenal ascorbic acid depletion	AAAD
			Aldosterone synthase	$P\text{-}450_{c18}$

Amine precursor uptake and decarboxylation	APUD
Androgen-binding protein	ABP
Angiotensin I	ANG-I
Angiotensin II	ANG-II
Angiotensin III	ANG-III
Angiotensin-converting enzyme	ACE
Angiotensin-converting enzyme-like action	ACELA
Antidiuretic hormone	ADH
Antimüllerian hormone (same as MIS)	AMH
Arginine vasopressin	AVP
Arginine vasotocin	AVT
Atrial natriuretic peptide	ANP
Attention-deficit hyperactivity disorder	ADD
β-Adrenergic receptor kinase	βARK
β-Lipotropin	β-LPH
3β-Hydroxy-Δ^5-steroid dehydrogenase	3β-HSD
Behavioral sex determination	BSD
Brain natriuretic peptide	BNP
Calcitonin	CT
Calcitonin gene-related peptide	CGRP
Calcium-sensitive cell	PIPAS cell
cAMP regulatory element binding protein	CREB
Carbon monoxide	CO
Catechol O-methyltransferase	COMT
Chicken-I GnRH	cGnRH-I
Chicken-II GnRH	cGnRH-II
Cholecystokinin	CCK
Choline acetyltransferase	CAT
Chorionic corticotropin	CC
Chorionic gonadotropin	CG
Chorionic somato-mammotropin	CS
Chorionic thyrotropin	CT
Compensatory ovarian hypertrophy	COH
Congenital adrenal hyperplasia	CAH
Corticosteroid-binding globulin	CBG
Corticosterone	B

Corticotropin	ACTH
Corticotropin-like peptide	CLIP
Corticotropin-releasing hormone	CRH
Cortisol	F
Cyclic adenosine monophosphate	cAMP
Cyclic guanosine monophosphate	cGMP
Dehydroepiandrosterone	DHEA
Dehydroepiandrosterone sulfate	DHEAS
Delta sleep-inducing peptide	DSIP
Deoxycorticosterone	DOC
Deoxyribonucleic acid	DNA
Diacylglycerol	DAG
Dihydroxyphenylalanine	DOPA
Diiodotyrosine	DIT
Dopamine	DA
Dorsal infundibular nucleus	NID
Endogenous opioid peptide	EOP
Endothelium-derived relaxing factor	EDRF
Enzyme-linked immunoabsorbent assay	ELISA
Epidermal growth factor	EGF
Extrapancreatic tumor hypoglycemia	EPTH
Follicle-stimulating hormone	FSH
γ-Lipotropin	γ-LPH
γ-Aminobutyric acid	GABA
Gastric inhibitory peptide	GIP
Gastrin-releasing peptide	GRP
Genotypic sex determination	GSD
Glucagon-like peptides	GLP-I, -II
Glucose-dependent insulinotropic peptide	GIP
Gonadotropin	Gn
Gonadotropin	GTH
Gonadotropin I	GTH-I
Gonadotropin II	GTH-II
Gonadotropin-releasing hormone	GnRH
Gonadotropin-releasing hormone-associated peptide	GAP
Growth factor receptor-binding protein	GRB2

Growth hormone	GH	Mitogen-activated protein	
Growth hormone release-		kinase	MAPK
inhibiting hormone,		Monoamine oxidase	MAO
somatostatin	GHRIH	Monoiodotyrosine	MIT
Growth hormone-releasing		Müllerian-inhibiting substance	MIS
hormone, somatocrinin	GHRH		(same as
Guanosine triphosphate	GTP		AMH)
High-density lipoprotein	HDL	N-Acetyltransferase	NAT
High-performance liquid		Nerve growth factor	NGF
chromatography	HPLC	Neuropeptide Y	NPY
Hydroxyindole-O-		Neurotensin	NT
methyltransferase	HIOMT	Nitric oxide	NO
Immunoradiometric assay	IRMA	Nitric oxide synthetase	NOS
Inositol trisphosphate	IP_3	Nonesterified fatty acids	NEFA
Insulin-dependent diabetes		Non-insulin-dependent diabe-	
mellitus	IDDM	tes mellitus	NIDDM
Insulin-like growth factor I	IGF-I	Norepinephrine	NE
Insulin-like growth factor II	IGF-II	Nucleus lateralis tuberis	NLT
Interleukin 1	IL-1	Ovarian ascorbic acid depletion	OAAD
Interleukin 2	IL-2	Pancreatic polypeptide	PP
Interleukin 6	IL-6	Pancreostatin	Pst
Intermediate-density		Parathyroid hormone	PTH
lipoprotein	IDL	Parathyroid hormone-related	
Interstitial cell-stimulating		protein	PTHrP
hormone	ICSH	Peptide histidine isoleucine	PHI
Isotocin	IST	Peptide YY	PYY
Lipotropin	LPH	Phenylethanolamine-N-	
Long-acting thyroid stimulator	LATS	methyltransferase	PNMT
Low-density lipoprotein	LDL	Phenypressin	PVP
Luteinization of atretic follicles	LAF	Phosphoenolpyruvate	
Luteinizing hormone	LH	carboxykinase	PEPCK
Luteinizing hormone-releasing		Phosphoinositol diphosphate	PIP_2
hormone	LHRH	Pineal antigonadotropic	
	(same as	peptide	PAG
	GnRH)	Polycystic ovarian syndrome	POS
Lysine vasopressin	LVP	Pregnant mare serum	
Melanophore-concentrating		gonadotropin	PMSG
hormone	MCH	Preoptic area	POA
Melanotropin	MSH	Preoptic nucleus	PON
Melanotropin release-inhibiting		Prolactin	PRL
hormone	MRIH	Proopiomelanocortin	POMC
Melanotropin-releasing		Propylthiouracil	PTU
hormone	MRH	Protein kinase C	PKC
Mesotocin	MST	Radioimmunoassay	RIA
Messenger RNA	mRNA	Reverse T_3	rT_3
Metyrapone	SU4885	Ribonucleic acid	RNA
Migrating motor complex	MMC	Serotonin	5-HT

Somatolactin	SL	Thyroid-binding prealbumin	TBPA
Somatostatin; also S_{14}, SS_{28}, S_{34}	SS	Thyroid peroxidase	TPO
Son of sevenless protein	SOS	Thyroid receptor	TR
Steroidogenesis-stimulating protein	STP	Thyroid receptor auxiliary protein	TRAP
Steroidogenic acute regulatory protein	StAR	Thyroid response element	TRE
		Thyrotropin	TSH
Subcommissural organ	SCO	Thyrotropin-releasing hormone	TRH
Superior cervical ganglion	SCG		
Suprachiasmatic nucleus	SCN	Thyroxine (tetraiodothyronine)	T_4
Supraoptic nucleus	SON	Transforming growth factor α	TGF-α
Syndrome of inappropriate diuresis	SIAD	Triiodothyroacetic acid	TRIAC
		Triiodothyronine	T_3
Temperature-dependent sex determination	TSD	Truncated GLP-I	tGLP-I
		Vasoactive intestinal peptide	VIP
Tetraiodothyroacetic acid	TETRAC	Ventral infundibular nucleus	NIV
Thiourea	TU	Very low-density lipoprotein	VLDL
Thyroid-binding albumin	TBA	Voltage-sensitive calcium channels	VSCC
Thyroid-binding globulin	TBG		

INDEX

Entries followed by f or t denote figures and tables, respectively.